CTS-D®
Certified Technology Specialist-Design

EXAM GUIDE

Second Edition

ABOUT THE AUTHOR

Andy Ciddor is a technical writer, consultant, educator, and systems technician. His unhealthy interest in how things work began when he terminally dismembered an army field telephone at the age of eight. Ciddor's interests eventually expanded into wireless, audio, lighting, electronics, AV, IT, software, control systems, communications, technical production, pyrotechnics, cosmology, rocket science, etc., as the opportunities arose. He has contributed on a wide range of subjects to publications across the English-speaking world and was the founding editor of *AV Technology* magazine. For AVIXA and McGraw Hill he has edited *CTS® Certified Technology Specialist Exam Guide, Third Edition* (2020) and *CTS-I® Certified Technology Specialist-Installation Exam Guide, Second Edition* (2021). His favorite place is at the bottom of a steep learning curve.

About the Technical Editor

Greg Bronson, CTS-D, is Director of Standards for AVIXA. He has worked as a technician, service manager, and system designer/project manager over a 30+ year career in AV. Bronson is a longtime volunteer and subject matter expert for AVIXA, including leadership roles within its standards, education, committee, and council programs.

CTS-D®
Certified Technology Specialist-Design

EXAM GUIDE

Second Edition

Andy Ciddor

New York   Chicago   San Francisco
Athens   London   Madrid   Mexico City
Milan   New Delhi   Singapore   Sydney   Toronto

McGraw Hill is an independent entity from AVIXA and is not affiliated with AVIXA in any fashion. This publication and online content may be used in assisting students to prepare for the CTS-D Exam. Neither AVIXA nor McGraw Hill warrants that use of this publication or online content will ensure passing any exam. Practice exam questions were not subject to the same item development process as questions on the actual exam and may or may not reflect the actual exam experience.

McGraw Hill books are available at special quantity discounts to use as premiums and sales promotions, or for use in corporate training programs. To contact a representative, please visit the Contact Us pages at www.mhprofessional.com.

CTS-D® Certified Technology Specialist-Design Exam Guide, Second Edition

Copyright © 2023 by McGraw Hill and AVIXA. All rights reserved. Printed in the United States of America. Except as permitted under the Copyright Act of 1976, no part of this publication may be reproduced or distributed in any form or by any means, or stored in a database or retrieval system, without the prior written permission of publisher, with the exception that the program listings may be entered, stored, and executed in a computer system, but they may not be reproduced for publication.

All trademarks or copyrights mentioned herein are the possession of their respective owners and McGraw Hill makes no claim of ownership by the mention of products that contain these marks. McGraw Hill is not associated with any product or vendor mentioned in this book. AVIXA®, CTS®, and related marks are registered trademarks of AVIXA in the United States and/or other countries.

1 2 3 4 5 6 7 8 9   LCR   27 26 25 24 23

Library of Congress Control Number: 2022949886

ISBN   978-1-260-13612-8
MHID   1-260-13612-4

Sponsoring Editor Tim Green	**Technical Editor** Greg Bronson	**Composition** KnowledgeWorks Global Ltd.
Editorial Supervisor Patty Mon	**Copy Editor** Lisa McCoy	**Illustration** KnowledgeWorks Global Ltd.
Project Manager Nitesh Sharma, KnowledgeWorks Global Ltd.	**Proofreader** Paul Tyler	**Art Director, Cover** Jeff Weeks
Acquisitions Coordinators Caitlin Cromley-Linn, Emily Walters	**Indexer** Ted Laux	**Creative Director, Cover** Lillie Fujinaga-Obioha, AVIXA
	Production Supervisor Thomas Somers	**Design, Cover** William Murillo, AVIXA

Information has been obtained by McGraw Hill from sources believed to be reliable. However, because of the possibility of human or mechanical error by our sources, McGraw Hill, or others, McGraw Hill does not guarantee the accuracy, adequacy, or completeness of any information and is not responsible for any errors or omissions or the results obtained from the use of such information.

Image credits:
Page 111, photograph: SOPA Images / Getting Images
Page 156, online amplifier power requirement calculator: Courtesy of Biamp
Page 162, ceiling-mounted multimicrophone: Courtesy of Audio Technica
Page 221, programmable control panel: Courtesy of Philips Dynalite
Page 226, illuminated emergency exit sign: Courtesy of Clevertronics
Page 288, slotted cable tray and cable ladder: vav63 / Getty Images
Page 348, Thunderbolt socket: Getty Images
Page 372, VU meter: Courtesy of cirilla / Getty Images
Page 372, digital and analog versions of a PPM: (left) Courtesy of spencerdare / Getting Images; (right) Courtesy of Tondose / Wikimedia (no changes) under the terms of: https://creativecommons.org/license/by-sa/4.0/legalcode and https://www.gnu.org/licenses/fdl-1.3.en.html
Page 396, CPU master unit: Courtesy of Crestron
Page 397, touch-screen control panel: Courtesy of Crestron
Page 442, wireless access point: thongseedary / Getty Images
Page 521, Panasonic AV camera: Courtesy of Panasonic Corporation

CONTENTS AT A GLANCE

Part I — The Certified Technology Specialist-Design
- **Chapter 1** What Is a Certified Technology Specialist-Design? 3
- **Chapter 2** The CTS-D Exam 9

Part II — Environment
- **Chapter 3** Communicating Design Intent 31
- **Chapter 4** Ergonomics in AV Design 79
- **Chapter 5** Visual Principles of Design 95
- **Chapter 6** Audio Principles of Design 135

Part III — Infrastructure
- **Chapter 7** Communicating with Allied Trades 187
- **Chapter 8** Lighting Specifications 199
- **Chapter 9** Structural and Mechanical Considerations 229
- **Chapter 10** Specifying Electrical Infrastructure 255
- **Chapter 11** Elements of Acoustics 295

Part IV — Applied Design
- **Chapter 12** Digital Signals 323
- **Chapter 13** Digital Video Design 335
- **Chapter 14** Audio Design 367
- **Chapter 15** Control Requirements 393
- **Chapter 16** Networking for AV 409
- **Chapter 17** Streaming Design 465
- **Chapter 18** Security for Networked AV Applications 491
- **Chapter 19** Conducting Project Implementation Activities 505

Part V Appendixes and Glossary

Appendix A	Math Formulas Used in AV Design	545
Appendix B	AVIXA Standards	559
Appendix C	AVIXA AV Standards Clearinghouse	563
Appendix D	Video References	565
Appendix E	About the Online Content	567
	Glossary	571
	Index	619

CONTENTS

Foreword xxiii
Acknowledgments xxv

Part I The Certified Technology Specialist-Design

Chapter 1 What Is a Certified Technology Specialist-Design? 3
Introducing AVIXA 3
Why Earn Your CTS-D Credential? 4
What Does a CTS-D Do? 6
Are You Eligible for the CTS-D Exam? 7
Chapter Review 7

Chapter 2 The CTS-D Exam 9
The Scope of the CTS-D Exam 9
Exam Preparation Strategies 11
Mathematical Strategies 13
    Order of Operations 14
Electrical Calculations 16
    Electrical Basics 16
The CTS-D Exam Application Process 20
    Getting to the Testing Center 20
    Identification Requirements 21
    Items Restricted from the Exam Room 21
    About the Exam 21
    During the Exam 22
    Dismissal or Removal from the Exam 23
    Hazardous Weather or Local Emergencies 23
    Special Accommodations for Exams 24
    Exam Scoring 24
    Retesting 24
CTS-D Exam Practice Questions 25
    Answers to CTS-D Practice Questions 26
Chapter Review 27

Part II Environment

Chapter 3 Communicating Design Intent 31
Creating an Exceptional AV Experience 32
  Types of AV Experience 32
  Components of an AV Experience 33
The Phases of an AV Design Project 33
  Concept Design/Briefing/Program Phase 35
  Design Phase 36
  Construction Phase 37
  Verification Phase 37
Reading Construction Drawings 38
  Scaled Drawings 39
  Drawing Types 42
  Architectural Drawing Symbols 46
  Common Architectural Drawing Abbreviations 50
  AV Drawing Symbols 51
The AV Design Package 54
  Front-End Documentation 54
  Architectural and Infrastructure Drawings 55
  AV System Drawings: Facility Drawings 58
  AV System Drawings: System Diagrams 59
  AV System Specifications 59
The Basics of AV-Enabled Spaces 60
  Audience and Presenter Areas 60
  Control and Projection Areas 62
Programming 63
The Needs Analysis 63
  Who Are the End Users? 63
  Clients on a Project 64
Developing the Concept Design/Program Report 66
  Step 1: Ask Questions 66
  Step 2: Review Existing Documentation 67
  Step 3: Evaluate the Site Environment/Benchmarking . 68
  Step 4: Conduct Program Meetings 71
  Step 5: Create the Concept Design/Program Report ... 71
A Closer Look at the Concept Design/Program Report 72
  Executive Summary 73
  Space Planning 73
  System Descriptions 73
  Infrastructure Considerations 74
  Budget Recommendations 74
  Distribution and Approval 75
Chapter Review 76
  Review Questions 76
  Answers 78

Chapter 4	Ergonomics in AV Design	79
	Human Dimensions and Visual Field	80
	The Horizontal Visual Field	81
	The Vertical Visual Field	82
	Head Rotation	83
	Sightlines	84
	Human Sightlines	85
	Eye Height	85
	Seating Layouts	86
	Floor Layouts	87
	Furniture	88
	Tables and Chairs	89
	Lecterns	90
	Other Furniture	91
	Chapter Review	92
	Review Questions	92
	Answers	93
Chapter 5	**Visual Principles of Design**	**95**
	Determining Image Specifications	96
	Viewing Requirements	96
	Determining Image Size	97
	Determining Text Size	99
	Visual Acuity and the Snellen Eye Chart	99
	Element Height	99
	Viewing Angles	101
	Viewing Distance	101
	Display Device Selection	103
	Video Resolution	104
	Aspect Ratio	105
	Calculating Aspect Ratio	105
	Calculating Screen Diagonal	106
	Display Types	106
	Direct-View Displays	107
	Front Projection	108
	Rear Projection	111
	Videowalls	115
	Display Environment	118
	Measuring Light	118
	System Black	120
	Contrast Ratio	120
	ANSI/AVIXA V201.01:2021	121
	Five Viewing Positions	121
	ISCR Conformance	121

Projector Positioning	122
Projector Light Path	123
Projection Throw	123
Predicting Projector Brightness	124
Screen Gain	125
Light Source Life	125
Projector Lens	126
Ambient Light Levels	127
Calculating Required Projector Brightness	127
Measuring OFE Projector Brightness	128
Task-Light Levels	129
Calculating Task-Light Levels	130
Chapter Review	131
Review Questions	131
Answers	133

Chapter 6 Audio Principles of Design   135

Introduction to the Decibel	136
Calculating Decibel Changes	137
Decibel Reference Levels	137
Perceived Sound Pressure Level	139
SPL Meter Weighting: Spectrum Analysis	141
Loudspeaker Systems	142
Point-Source Systems	142
Distributed Systems	142
Designing Loudspeaker Coverage	142
Calculating Loudspeaker Coverage	143
Distributed Layout Options	145
Edge-to-Edge Coverage	145
Partial Overlap Coverage	146
Edge-to-Center Coverage	146
Ohm's Law Revisited	147
Loudspeaker Impedance	147
Wiring Loudspeakers	148
Loudspeakers Wired in Series	148
Loudspeakers Wired in Parallel, Same Impedance	149
Loudspeakers Wired in Parallel, Different Impedances	151
Loudspeakers Wired in a Series and Parallel Combination	151
Measuring Impedance	152
Transformers	153
Loudspeaker Taps	154
Specifying a Power Amplifier	155
Headroom Requirements	156
Loudspeaker Sensitivity	156

Power Amplifiers 157
    Specifying a Power Amplifier for Directly
        Connected Systems 157
    Specifying a Power Amplifier for Distributed Audio 157
Microphones 158
Microphone Polar Response 162
Polar Plot 163
Microphone Frequency Response 165
Microphone Signal Levels 165
Microphone Sensitivity 166
Microphone Pre-Amp Gain 167
Microphone Mixing and Routing 168
    Automatic Microphone Mixers 168
Microphone Placement: A Conference Table 169
Microphone Placement: The 3:1 Rule 170
Reinforcing a Presenter 171
    Microphones and Clothing 172
    Polar Plots for Reinforcing a Presenter 173
Audio System Quality 174
    It Must Be Loud Enough 174
    It Must Be Intelligible 175
    It Must Remain Stable 176
PAG/NAG 177
More Variables: NOM and FSM 179
PAG/NAG in Action 180
Chapter Review 180
    Review Questions 181
    Answers 183

Part III Infrastructure

Chapter 7 Communicating with Allied Trades 187

Communicating with Stakeholders 188
Tracking the Project 189
    Work Breakdown Structure 190
    Gantt Chart for Project Schedules 191
    Logic Network Diagram 192
Industry Standards as Common Language 193
Hierarchy of Design Consultation 195
Showing Workmanship 196
Chapter Review 196
    Review Questions 197
    Answers 198

Chapter 8	Lighting Specifications	199
	Basics of Lighting	200
	Brightness	200
	Color Temperature	201
	Color Rendering Index	203
	Energy Consumption	205
	Lighting the Space	206
	Task Lighting	206
	Shades and Blackout Drapes	206
	Choosing Light Sources	207
	Choosing Luminaires	209
	Luminaire Specifications	209
	Lighting Design	210
	Creating a Zoning Plan	213
	Documenting Luminaires	216
	Lighting Control	218
	On/Off vs. Dimmable	218
	Lighting Scenes	220
	Lighting a Videoconference	221
	Glare	222
	Light Balance	223
	Color Balance	224
	Wall and Table Finishes	224
	Illuminated Exit Signs	225
	Chapter Review	226
	Review Questions	226
	Answers	228
Chapter 9	Structural and Mechanical Considerations	229
	Codes and Regulations	230
	Designing for Equal Access	231
	Electric and Building Codes	232
	Mounting Considerations	233
	Mounting Options	233
	Load Limit	236
	Mounting Hardware	237
	Designing the Rack	239
	Rack Sizes	241
	Ergonomics	242
	Weight Distribution	243
	Signal Separation Within a Rack	244
	Block Diagrams	245
	Heat Load	245
	Calculating Heat Load from Power Amplifiers	247
	Cooling a Rack	247

	HVAC Considerations	249
	HVAC Issues That Impact Design	249
	Fire and Life Safety Protection	250
	Fire Isolation	251
	Energy Management	252
	Chapter Review	252
	Review Questions	253
	Answers	254

Chapter 10 Specifying Electrical Infrastructure 255

Circuit Theory		256
Circuits: Impedance and Resistance		258
Inductors, Capacitors, and Resistors in a Series Circuit		261
Inductors, Capacitors, and Resistors in a Parallel Circuit		262
Specifying Electrical Power		263
Established Terms		264
Codes and Regulations		264
Electrical Distribution Systems		265
Power Distribution Basics		266
Protective Earth/Safety Ground/EGC		266
Ground/Earth Faults		269
The Dangers of Three-to-Two-Pin Adapters		271
Grounds for Confusion		271
Power Distribution Systems		272
The Neutral/Return in Three-Phase Distribution		273
Power Distribution Systems in North America		275
Specifying AV Circuits		276
Branch Circuit Loads		277
Calculating the Number of Circuits: An Example		278
Power Strips and Leads		280
Isolated Ground		282
Interference Prevention and Noise Defense		282
Magnetic-Field Coupling		283
Electric-Field Coupling		284
Electromagnetic Shielding		285
Ground Loops		286
Balanced and Unbalanced Circuits		286
Transformer-Balanced Circuits		288
Cable Support Systems		288
Cable Tray		288
Conduit		289
Cable Duct		289
Hook Suspension Systems		289

	Conduit Capacity	290
	Jam Ratio	291
	Chapter Review	292
	Review Questions	292
	Answers	293

Chapter 11 Elements of Acoustics ... 295

Acoustic Engineering	296
Sound Production	296
Sound Propagation	297
Sound Intensity	298
Particle Displacement	299
Sound Interaction	299
Reflection	300
Absorption	306
Transmission	309
Sound Reception	313
The Integration Process	313
Background Noise	316
Chapter Review	318
Review Questions	318
Answers	319

Part IV Applied Design

Chapter 12 Digital Signals ... 323

Digital Signals	324
Digital Audio Bandwidth	324
Digital Video Bandwidth	325
Bit Depth	326
YUV Subsampling	326
Bandwidth: Determining Total Program Size	327
Content Compression and Encoding	328
Digital Media Formats	329
Codecs	329
Digital Audio Compression: MP3	330
Digital Video Compression	330
Chapter Review	332
Review Questions	332
Answers	333

Chapter 13 Digital Video Design ... 335

Digital Video Basics	336
High-Definition and Ultra High-Definition Video	337
The Cliff Effect	339

	Video Signal Types	340
	Serial Digital Interface	341
	Transition-Minimized Differential Signaling	341
	DVI	342
	HDMI	342
	HDBaseT	344
	DisplayPort	344
	USB4, USB3.2, and USB Type-C Connectors	345
	Thunderbolt	346
	Video over IP Networks	348
	Introduction to EDID	348
	EDID Packets	349
	How EDID Works	349
	EDID Table	350
	Developing an EDID Strategy	352
	EDID Data Tables	353
	Resolving EDID Issues	354
	Managing EDID Solutions	356
	EDID and Displays	356
	Digital Rights Management	357
	High-Bandwidth Digital Content Protection	357
	HDCP Interfaces	357
	How HDCP Works	358
	HDCP Device Authentication and Key Exchange	358
	HDCP Locality Check	359
	HDCP Session Key Exchange	359
	HDCP and Switchers	359
	HDCP Authentication with Repeaters	360
	HDCP Device Limits	361
	HDCP Troubleshooting	361
	Chapter Review	362
	Review Questions	363
	Answers	364
Chapter 14	**Audio Design**	**367**
	Analog vs. Digital Audio	368
	DSP Architectures	369
	Signal Monitoring	371
	Analog vs. Digital Signal Monitoring	372
	Setting Up the System	373
	Where to Set Gain	374
	Common DSP Settings	377
	Introduction to Equalization	383
	Parametric Equalizers	385
	Pass Filters	386

	Crossover Filters	386
	Feedback-Suppression Filters	387
	Noise-Suppression Filters	388
	Graphic Equalizers	388
	Chapter Review	389
	Review Questions	389
	Answers	391

Chapter 15 Control Requirements ... 393

	Types of Control Systems	394
	Control System Components	395
	Central Processing Unit	395
	Control Interfaces	396
	Control Points	399
	Control System Design	400
	Best Practice for User Experience Design	401
	Needs Analysis	401
	CPU Configurations	402
	Programming for Control	403
	Establishing Control Points	405
	Verifying System Performance	405
	Chapter Review	405
	Review Questions	405
	Answers	406

Chapter 16 Networking for AV ... 409

	What Is a Network?	410
	Network Components	410
	Clients and Servers	412
	Network Interface Cards	413
	Network Devices	414
	Links	415
	Fiber-Optic Cable	417
	Wireless Connections	422
	The OSI Model	425
	Layers of the OSI Model	426
	The Layers	426
	Data Transmission and OSI	427
	Ethernet	428
	Local Area Networks	428
	Topology	428
	What Is Ethernet?	431
	Isolating LAN Devices	432
	Wide Area Networks	434
	WAN Topologies	435
	Private and Public WANs	436

Network-Layer Protocols	437
Internet Protocol	437
IP Addressing	437
IPv4 Addressing	438
Types of IP Addresses	439
IPv6 Addresses	442
Address Assignment	443
Static and Dynamic IP Addresses	443
Dynamic Host Configuration Protocol	444
Reserve DHCP	445
Automatic Private IP Addressing	445
Domain Name System	446
Dynamic DNS	447
Internal Organizational DNS	447
Transport Protocols	448
TCP Transport	448
UDP Transport	449
TCP vs. UDP	449
Ports	449
The Host Layers	451
The Session Layer	451
The Presentation Layer	451
The Application Layer	452
Bandwidth	452
Quality of Service	452
Security Technologies	454
Network Access Control	454
Firewall	455
AV over Networks	457
AV over Ethernet	457
AV over IP Networks	458
AES67 Interoperability	458
Dante	458
SMPTE ST2110	459
RAVENNA	459
Crestron DM NVX	460
AMX SVSI	460
Extron NAV	460
BlueRiver and SDVoE	460
NewTek NDI	461
Q-SYS	461
Chapter Review	461
Review Questions	462
Answers	463

Chapter 17	Streaming Design	465
	Streaming Needs Analysis	466
	Streaming Tasks	466
	Audience	467
	End Points	468
	Content Sources	468
	Using Copyrighted Content	469
	Streaming Needs Analysis Questions	470
	Streaming Design and the Network Environment	471
	Topology	471
	Bandwidth: Matching Content to the Network	472
	Image Quality vs. Available Bandwidth	472
	Streaming and Quality of Service	473
	Latency	474
	Network Policies and Restrictions	475
	Cheat Sheet: Streaming Network Analysis Questions	475
	Designing the Streaming System	476
	Real-Time Transport Protocol	477
	Other Streaming Protocols	477
	High-Quality Streaming Video	478
	Unicast and Multicast	480
	Unicast	480
	Multicast	480
	Unicast vs. Multicast	482
	Implementing Multicast	483
	IPv4 Multicast Addressing	485
	IPv6 Multicast Addressing	486
	Streaming Reflectors	486
	Chapter Review	487
	Review Questions	488
	Answers	489
Chapter 18	Security for Networked AV Applications	491
	Security Objectives	492
	Identifying Security Requirements	493
	Determining a Security Posture	494
	Stakeholder Input	496
	Assessing Risk	497
	Risk Registers	497
	Mitigation Planning	498
	Change Default Passwords	499
	Use Two-Factor Authentication	499
	Create Multiple User Roles	500
	Accounts for Every User	500

	Disable Unnecessary Services	501
	Keeping Systems Updated	501
	Air-Gapping	501
	Enable Encryption and Auditing	501
	AVIXA RP-C303.01:2018 Recommended Practices for Security in Networked AV Systems	502
	Chapter Review	502
	Review Questions	503
	Answers	504
Chapter 19	**Conducting Project Implementation Activities**	**505**
	Performance Verification Standard	506
	System Verification Process	508
	Regional Regulations	509
	Resources for Regional Codes	510
	Verification Tools	510
	Audio System Verification	511
	Audio-Testing Tools	511
	Video System Verification	517
	Verifying the Video Signal Path	517
	Signal Extenders	517
	Verifying Video Sources	519
	Camera Adjustments	519
	Display Setup	524
	Audio/Video Sync	530
	Verifying Audio/Video Sync	531
	Correcting Audio/Video Sync Errors	531
	Conducting System Closeout	532
	Closeout Documentation	532
	Troubleshooting	536
	Customer Training	537
	Client Sign-Off	539
	Chapter Review	539
	Review Questions	539
	Answers	541
Part V	**Appendixes and Glossary**	
Appendix A	**Math Formulas Used in AV Design**	**545**
	Using the Proper Order of Operations	545
	Steps to Solving Word Problems	545
	Step 1: Understand the Problem	546
	Step 2: Create a Plan	546
	Step 3: Execute Your Plan	547
	Step 4: Check Your Answer	548

	Rounding	548
	AV Math Formulas	549
	Estimated Projector Throw	549
	Projector Lumens Output	549
	ANSI Brightness of a Projector	550
	Luminance from Illuminance	550
	Decibel Formula for Distance	550
	Decibel Formula for Voltage	551
	Decibel Formula for Power	551
	Current Formula (Ohm's Law)	551
	Power Formula	552
	Series Circuit Impedance Formula	552
	Parallel Circuit Impedance Formula: Loudspeakers with the Same Impedance	552
	Parallel Circuit Impedance Formula: Loudspeakers with Different Impedances	553
	Series/Parallel Circuit Impedance Formulas	553
	Needed Acoustic Gain	553
	Potential Acoustic Gain	554
	Audio System Stability (PAG/NAG)	554
	Conduit Capacity	554
	Jam Ratio	555
	Heat Load Formula (Btu)	555
	Heat Load Formula (kJ)	555
	Power Amplifier Heat Load (kJ)	556
	Power Amplifier Heat Load (Btu)	556
	Required Amplifier Power (Constant Voltage Loudspeaker Systems)	556
	Wattage at the Loudspeaker	557
	Simplified Room Mode Calculation	557
	Loudspeaker Coverage Pattern (Ceiling Mounted)	557
	Loudspeaker Spacing (Ceiling Mounted)	558
	Digital Video Bandwidth	558
Appendix B	AVIXA Standards	559
	Published Standards, Recommended Practices, and Technical Reports	559
	Standards in Development	561
Appendix C	AVIXA AV Standards Clearinghouse	563
Appendix D	Video References	565

Appendix E	About the Online Content	567
	System Requirements	567
	Your Total Seminars Training Hub Account	567
	Privacy Notice	567
	Single User License Terms and Conditions	567
	TotalTester Online	569
	Technical Support	569
	Glossary	571
	Index	619

FOREWORD

What does it take to create an exceptional audiovisual experience? What creates a spectacle that wows people or the type of integrated collaboration space that makes companies more productive? It takes the right combination of content, space, and technology, but it also takes you. It takes trained professionals—integrators, designers, manufacturers, distributors, and technology managers.

AVIXA®, the Audiovisual and Integrated Experience Association, believes in the value of creating experiences through technology—experiences that lead to better outcomes through enhanced communication, entertainment, and education. To deliver these experiences and outcomes, our industry must continually master its craft. That means staying well-trained, using and understanding standards, and committing to professional certifications. Building integrated audiovisual experiences is guided by the science of sight and sound, activated by technology, and realized by dedicated specialists.

For more than 30 years, AVIXA has administered the Certified Technology Specialist™ (CTS®) program, which is recognized as the leading AV professional credential. There are three CTS credentials: general (CTS), design (CTS-D), and installation (CTS-I). There are currently more than 13,500 CTS holders in more than 100 countries. The CTS program is accredited by the American National Standards Institute (ANSI) to meet the International Organization of Standardization (ISO) and International Electrotechnical Commission (IEC) ISO/IEC 17024:2012 certifications of personnel standards.

As you will learn in this guide, a CTS-D works with clients to understand their needs, designs AV systems that meet those needs, prepares the necessary design documents, coordinates with other professionals to create AV systems, and ultimately ensures that the final product meets the clients' requirements. Those who hold the CTS-D credential, like its counterpart for the AV installer community, the Certified Technology Specialist–Installer (CTS-I), are members of a special group of AV professionals who have gone beyond foundational experience and dedicated themselves to quality work, focused expertise, and the confidence of the people they work with.

So, congratulations on your decision to pursue your CTS-D certification. We hope this guide helps you reach your goals and continue to advance this exciting industry.

What's inside these pages? For starters, everything you might expect. *CTS-D® Certified Technology Specialist-Design Exam Guide, Second Edition* includes the latest information that AV designers need to create solutions in a converged AV/IT world. Yes, all the bread-and-butter skills are here, too, such as how to perform a thorough needs assessment and document an AV design. But so are tips and background information for configuring and distributing AV, such as AV over IP (AVoIP) and HDBaseT—two important attributes of today's high-definition video systems. Plus, you'll find important information about current multimedia transmission technologies, such as High-Definition Multimedia Interface (HDMI), and USB through version USB4. And with all the talk about ultra-high-definition "4K," "8K," and even "16K" video, AV designers need to understand when it's really needed, how it's achieved, and how it impacts an overall system design. This book has you covered.

Finally, CTS-D holders—and AV designers in general—must know how to integrate and control AV gear on a network. They must understand Internet Protocol, network security, and other IT-related issues to help ensure that networked AV systems operate as promised without impacting other IT services. This book includes several chapters that highlight important AV/IT skills.

All told, the document in your hands encapsulates the AVIXA knowledge base for aspiring CTS-D holders. Even if you have never taken the CTS-D exam, this guide represents a handy reference for those in the field. It reflects the necessities of being a modern, expert AV designer; as such, it also reflects the CTS-D exam as it has changed over time. For more on the exam itself, see Chapter 2.

The *CTS-D® Certified Technology Specialist-Design Exam Guide, Second Edition* will prepare you for the CTS-D exam, but it is not required reading, and for good reason.

The CTS-D credential is accredited by ANSI under the ISO/IEC 17024 General Requirements for Bodies Operating Certification Schemes of Persons program. For many, an ANSI/ISO/IEC certification is an additional mark of distinction. There are roughly 1 million ANSI-certified professionals across different industries. In accordance with globally recognized principles, no single publication or class will necessarily prepare you for the CTS-D exam, nor are you obligated to enroll in AVIXA courses to take the exam. If, however, you are interested in other ways that AVIXA can help establish your professional qualifications, visit us at www.avixa.org/ctsd.

This is an exciting time to be part of the AV industry. By certifying your skills, you've shown already that you are committed to your own success and to the success of AV professionals everywhere. Now you're ready to go a step further and be a leader in this industry. We thank you for your commitment and wish you luck in your certification journey.

David Labuskes, CTS, CAE, RCDD
Chief Executive Officer
AVIXA

ACKNOWLEDGMENTS

Thank you to the countless audiovisual professionals who have contributed to the knowledge base that resides at AVIXA. These professionals have also recognized the need for advanced study and experience in the field of AV systems integration, resulting in the Certified Technology Specialist-Design (CTS-D) exam and certification.

The repository of knowledge at AVIXA continues to grow as inventors, manufacturers, engineers, system designers, system integrators, subject matter experts, AV technicians, and instructors share their expertise and develop new solutions. This edition of the *CTS-D® Certified Technology Specialist-Design Exam Guide* draws heavily on that accumulated knowledge and expertise, together with help from the volunteer subject matter experts whose efforts assisted in steering this book down the ever-changing technology path that is twenty-first-century AV.

I would like to thank AVIXA for giving me the opportunity to work on this exciting and challenging project, with special thanks to the project supervisor and technical editor, Greg Bronson, CTS-D, and Charles Heureaux and Will Murillo, who stepped out of their normal roles to prepare the artwork for this edition. At McGraw Hill, my profound thanks go to Tim Green, Caitlin Cromley-Linn, Emily Walters, and Patty Mon for once more patiently guiding me through the process of creation. My thanks to copy editor Lisa McCoy, proofreader Paul Tyler, and to Nitesh Sharma and his team at KnowledgeWorks Global Ltd. for wrangling the manuscript through the production process.

Most of all, I need to offer my heartfelt thanks to the countless legions of technicians, directors, designers, engineers, trainers, educators, clients, and innocent bystanders who have patiently (and sometimes not so patiently) answered my questions during a lifetime of inquiring as to what it is, how it works, and why they did it that way.

Like every twenty-first-century technical writer, I must acknowledge the contributions of Vannevar Bush and Ted Nelson, whose prescient concept of hypertext, and its child, the World Wide Web, made researching this work possible. I must also tip my hat to Jimmy Wales and his imperfectly amazing Wikipedia project, which serves as such a valuable launching point for journeys into the unknown.

This guide happened only because of my partner Val's unfailing care, encouragement, support, and tolerance through the serious health issues I have faced while getting this project to the finishing line. Thanks also to my adult offspring, Rivka, Lachlan, and Rhian, for their relentless moral support and jeering from the peanut gallery.

Congratulations for extending your skills to the CTS-D level, and best wishes for success in the exam.

—Andy Ciddor, 2022

Special thanks go to the AVIXA volunteer technical review task group members Jesse Anderson, CTS-D, Tufts University; Bob Higginbotham, CTS-D, CTS-I AV SME Retired; and Mike Tomei, CTS-D, CTS-I, Tomei AV Consulting, for sharing both time and expertise throughout the technical edit process.

2022 AVIXA Board of Directors

Leadership Search Committee Chair: Jon Sidwick, President, Collabtech Group
Chairman: Samantha Phenix, CEO, Magwire, Founder, Phenix Consulting
Vice Chair: Martin Saul, CEO, ICAP Global
Secretary-Treasurer: Cathryn Lai, Chief Commercial Officer OpenBet
Directors:
- Michelle Grabel-Komar, Vice President of Sales, Full Compass Systems
- Tze Tze Lam, Executive Director, Electro-Acoustics Systems Pte Ltd
- Tobias Lang, CEO, Lang AG
- Dena Lowery, COO, Opus Agency
- Alexandra Rosen, Senior Director, Venture Forward, GoDaddy
- Jatan Shah, Chief Operating and Technology Officer, QSC
- Mradul Sharma, Managing Director, 3CDN Workplace Tech Pvt Ltd
- Brad Sousa, Chief Technology Officer, AVI Systems

AVIXA Staff

David Labuskes, CTS, CAE, RCDD, CEO

Dan Goldstein, Chief Marketing Officer

Pamela M. Taggart, CTS, Vice President, Content Creation

Joé Lloyd, Senior Director, Communications

Jodi Hughes, Director of Content Delivery

Nicole R. Verardi, Senior Director Marketing

Lillie Obioha, Creative Director

Kelly Smith, Manager Marketing

Zachary Fisher, CTS, Instructional Design Manager

Chuck Espinoza, CTS-D, CTS-I, Instructor Lead

Leslie Rivera, Developer On-Demand Content

Charles Heureaux, Instructional Designer

William Murillo, Senior Designer

John Pfleiderer, CTS-D, CTS-I Standards Developer

Loanna Overcash, Standards Manager

PART I

The Certified Technology Specialist-Design

- **Chapter 1** What Is a Certified Technology Specialist-Design?
- **Chapter 2** The CTS-D Exam

CHAPTER 1

What Is a Certified Technology Specialist-Design?

In this chapter, you will learn about
- AVIXA certifications and audiovisual (AV) industry standards
- The benefits of earning a Certified Technology Specialist-Design (CTS-D) credential
- What types of work an AV designer does
- Eligibility criteria for taking the CTS-D exam

You might be holding this book for many reasons. You could be a professional audiovisual systems designer in search of a handy reference. You could be a user or operator of AV systems at a company, school, house of worship, or other organization and need in-depth information to make sure those systems work properly and deliver the best possible experience. Or perhaps you are a Certified Technology Specialist (CTS). You have already proven yourself to be an expert AV professional, committed to a higher standard of workmanship and to keeping abreast of the fast-moving technology and best practices that characterize this highly dynamic industry. But now you're ready for more.

As a current CTS holder, you've decided to take your skills to the next level and become a Certified Technology Specialist-Design. CTS-D is a specialized industry credential, recognized by employers, customers, and international standards bodies. A CTS-D demonstrates the broad expertise of a CTS but also the skills and knowledge required to design AV systems. You demonstrate the skills and knowledge by excelling on the CTS-D exam. Throughout this book, you will learn what you need to know to be a CTS-D holder, but first let's align your ambition with what's expected of a CTS-D.

Introducing AVIXA

AVIXA created and administers the Certified Technology Specialist program. Founded in 1939, AVIXA is the leading nonprofit association serving the professional audiovisual communications industry worldwide. Through activities that include tradeshows, education, certification, standards development, government relations, public outreach, and information services, AVIXA promotes the industry and enhances members' ability to conduct business successfully and competently.

AVIXA has offered certification programs for nearly 35 years, as well as industry-specific and general business training and education for people seeking careers in professional AV. Every year, AVIXA certifies more qualified AV professionals than anyone else in the industry.

AVIXA is also an American National Standards Institute (ANSI)–accredited standards developer, creating voluntary performance standards for the AV industry. AVIXA develops both independent and ANSI-approved standards, as well as joint standards with other professional associations. It is important for certified professionals to recognize, understand, and apply relevant standards when designing AV systems. Although implementing AVIXA standards is not a requirement for being a CTS-D, the standards themselves are available to CTS-D exam item writers and could be referenced on the exam. The following are the current approved standards:

- Audio Coverage Uniformity in Enclosed Listener Areas
- Recommended Practices for Security in Networked Audiovisual Systems
- Audiovisual Systems Energy Management
- Audiovisual System Performance Verification
- Cable Labeling for Audiovisual Systems
- Standard Guide for Audiovisual Systems Design and Coordination Processes
- Display Image Size for 2D Content in Audiovisual Systems
- Image System Contrast Ratio
- Rack Building for Audiovisual Systems
- Rack Design for Audiovisual Systems
- Sound System Spectral Balance

The following were developed in cooperation with other professional organizations:

- Lighting Performance for Small-to-Medium-Sized Videoconferencing Rooms
- Audio, Video, and Control Architectural Drawing Symbols

Many more standards are in development. Visit AVIXA's standards website at www.avixa.org/standards to learn more.

Why Earn Your CTS-D Credential?

Certification shows your commitment to being among the best in a professional field. This benefits you, your company, and your clients.

In the field of AV and information communications, the CTS credential is recognized worldwide as the AV leading credential. Being a CTS holder shows your professionalism and technical proficiency. It increases your credibility and boosts customers' confidence in your work.

There are currently three available CTS certifications:

- Certified Technology Specialist (CTS)
- Certified Technology Specialist-Design (CTS-D)
- Certified Technology Specialist-Installation (CTS-I)

All three of AVIXA's certifications have achieved accreditation through the International Organization of Standardization (ISO) and the International Electrotechnical Commission (IEC) as administered by ANSI in the United States. They have been accredited by ANSI to the ISO/IEC 17024:2012 personnel standard—the AV industry's only third-party accredited personnel certification program. These are the only certifications in the AV industry to achieve ANSI accreditation.

The certification programs are administered independently by AVIXA's certification committee. You can learn more about how the exams are developed and administered, as well as how to maintain your certification, by visiting the certification website at www.avixa.org/certification.

Although certification is not a guarantee of performance by certified individuals, CTS holders at all levels of certification have demonstrated AV knowledge and skills. They adhere to the CTS Code of Ethics and Conduct and maintain their status through continued education. Certification demonstrates commitment to professional growth in the audiovisual industry and is strongly supported by AVIXA.

Why take the next step toward specialized certification? Simply put, you're ready for more. You have a deep understanding of the many aspects of a successful AV system, from the AV components themselves to the other building systems and networks with which they integrate. You know how to translate what clients say they need into a technology solution that helps them achieve their goals. And you know how to build a team of professionals from inside and outside the AV industry that can execute on time and on budget. We've just described a CTS-D.

A career in AV design is a commitment. You're dedicating your professional life to a higher level of excellence that can be achieved only through education and expertise in the AV field. The continuing education that accompanies CTS-D certification will help keep you up to date on changing technologies and position you as a major player on project teams. In short, pursuing advanced certification is an excellent decision for your career and your company.

What Does a CTS-D Do?

A CTS-D is a leader in the AV industry. As a designer of AV systems, from conference and display spaces to learning spaces and performance venues, a CTS-D often takes the reins early in a project and performs specific tasks to assess a client's needs, design appropriate AV systems, prepare supporting documents, and coordinate and collaborate with other professionals to create systems that satisfy the client's requirements. A CTS-D is more than a technologist and recognizes that the goal of an AV system is to create an experience for the client that combines content, space, and technology so that the client and the client's clients can communicate better, work more efficiently, or be entertained.

To identify specifically what a CTS-D does, AVIXA developed a job task analysis (JTA). The JTA is a comprehensive list of the key responsibilities (referred to as *duties*) and tasks in which an AV designer should demonstrate proficiency. You will learn more about the JTA in Chapter 2, but in general, the many tasks that a CTS-D must perform fall into four general categories:

- Conducting a needs assessment and identifying the scope of work
- Collaborating with other professionals, including architects, engineers, electricians, interior designers, and more
- Developing AV designs, drawings, and documentation to describe the required audio, video, and network systems
- Conducting project implementation activities, from verifying system performance to troubleshooting

Based on the JTA, AVIXA's independent certification committee created a CTS-D exam content outline. Both the JTA and outline are available at the organization's website and are included in the free *CTS-D Candidate Handbook,* which is available online at www.avixa.org/ctsd.

It is important to note that the content and practice exercises in this book do not follow the CTS-D exam content outline perfectly. Nor do they follow the order in which actual CTS-D exam questions may be presented. Instead, the book follows the real-world course of an AV design, from conducting the needs analysis to commissioning and supporting AV systems. It is organized into three parts:

- Environment, which covers information for laying the groundwork of an AV design
- Infrastructure, which covers acoustic, lighting, mechanical, and other considerations that affect an AV design
- Applied design, which details specific aspects of an AV design, including audio, video, communications, networking, security, and other specifications

Upon completing this book, you will have been exposed to the knowledge and skills identified by the JTA and included in the exam outline as it was defined at the time of publication. When the exam JTA and exam outline are updated, the new information will be available from the AVIXA CTS-D web page at www.avixa.org/ctsd.

NOTE A JTA is a study conducted to identify the knowledge, skills, and abilities necessary for professional competence in a particular field. Such an analysis is often conducted to determine the content and competencies that should be included in a certification or exam. AVIXA's independent certification committee conducts periodic JTAs to make sure the various CTS exams and certification processes align with the real-world skills required of AV professionals.

Are You Eligible for the CTS-D Exam?

To be considered eligible to take the CTS-D exam, you must meet the following prerequisites:

- Hold a current CTS certification
- Be in good standing with the certification committee (in other words, have no ethics cases or sanctions)
- Have two years of audiovisual industry experience in design

There are several other prerequisites, such as an application form, proof of identity, and application fee. You can find information about all requirements and more in the *CTS-D Candidate Handbook*.

VIDEO Throughout this book, you will see references to online videos that reinforce what you read and offer additional insight into the CTS-D exam. To get you started, watch a video on demystifying AVIXA certification and beginning preparation for the CTS-D exam. Check Appendix D for a link to the AVIXA video library.

Chapter Review

A CTS-D must have both broad and deep knowledge and understanding about pretty much every aspect of an AV project. Proficient in audiovisual and other technology principles, a CTS-D also understands how to coordinate the efforts of many different trades and translate the stated needs of users into a solution that meets their goals.

Achieving CTS-D certification demonstrates an AV designer's comprehensive range of knowledge and skills encompassing everything from conducting a needs analysis, to coordinating audiovisual and network technologies, to training users on a system once it's been installed. The CTS-D exam measures an aspiring professional's skills and knowledge in a series of tasks identified by experienced industry peers. Good luck in your pursuit of CTS-D accreditation and in your professional AV career.

The CTS-D Exam

In this chapter, you will learn about
- The scope of the CTS-D exam
- The skills and knowledge that the exam covers
- Exam preparation and math strategies
- The process required to apply for, schedule, and complete the CTS-D exam
- The types of questions you might encounter on the exam

Now you're ready. You understand the role of an AV designer and appreciate the professional commitment it takes to earn your CTS-D. It's time to start preparing.

This chapter takes you inside the CTS-D exam, from what it covers in general terms to specific examples of questions you might see on the test. Along the way, we'll describe helpful preparation strategies, point you to additional resources, and detail exactly how to apply for the exam (and plan for the big day). We will even take you through a quick-and-dirty review of some basic mathematical concepts you'll absolutely need to know to succeed on the exam and in your chosen field—AV design.

The Scope of the CTS-D Exam

As noted earlier, the CTS-D exam tests the knowledge and skills required by an AV professional to earn CTS-D certification. To create the CTS-D exam, a group of volunteer audiovisual subject-matter experts (SMEs), guided by professional test development experts, participated in a job task analysis (JTA) focused on AV design. The results of this study form the basis of a valid, reliable, fair, and realistic assessment of the skills, knowledge, and abilities required for certified AV design professionals.

In creating the JTA, the group of volunteer SMEs identified major categories, or *duties*, as well as tasks within each duty, based on the tasks that a certified individual might perform on an AV design job. The exam development team examined the importance, criticality, and frequency of AV design tasks and used the data to determine the number of CTS-D exam questions related to each duty and task.

Based on the JTA, the CTS-D exam content outline divides design tasks into four duties, each of which will be addressed on the exam. The *CTS-D Certified Technology Specialist-Design Exam Guide* was written to cover the exam content in a manner that spends the most time on areas that make up the largest portion of the exam. At the same time, topics are covered in a logical manner, meaning the book doesn't go in exactly the same order as the content outline. As you proceed through the book, you will see Duty Check information at the beginning of each chapter, which describes how the material in the chapter relates to the duties and tasks in the CTS-D exam content outline.

The complete CTS-D exam content outline is shown in Table 2-1.

CTS-D Duties and Tasks	% of Exam	# of Items
Duty A: Conducting a Needs Assessment	**16.8%**	**21**
Task 1: Identify Decision-makers and Stakeholders	1.6%	2
Task 2: Identify Skill Level of End Users	2.4%	3
Task 3: Educate AV Clients	1.6%	2
Task 4: Review Client Technology Master Plan	3.2%	4
Task 5: Identify Client Expectations	4.0%	5
Task 6: Identify Scope of Work	4.0%	5
Duty B: Coordinating with Other Professionals to Develop Project Documentation	**23.2%**	**29**
Task 1: Review A/E (Architectural and Engineering) Drawings	4.0%	5
Task 2: Coordinate with Architectural/Interior Design Professionals	4.0%	5
Task 3: Coordinate with Mechanical Professionals	1.6%	2
Task 4: Coordinate with Structural Professionals	1.6%	2
Task 5: Coordinate with Electrical Professionals	3.2%	4
Task 6: Coordinate with Lighting Professionals	2.4%	3
Task 7: Coordinate with IT and Network Security Professionals	4.0%	5
Task 8: Coordinate with Acoustical Professionals	1.6%	2
Task 9: Coordinate with Life Safety and Security Professionals	0.8%	1
Duty C: Developing AV Designs	**48.0%**	**60**
Task 1: Create Draft AV Design	10.4%	13
Task 2: Confirm Site Conditions	8.0%	10
Task 3: Produce AV Infrastructure Drawings	12.0%	15
Task 4: Produce AV System Drawings	12.0%	15
Task 5: Finalize Project Documentation	5.6%	7
Duty D: Conducting Project Implementation Activities	**12.0%**	**15**
Task 1: Participate in Project Implementation Communication	3.2%	4
Task 2: Conduct System Performance Verifications	5.6%	7
Task 3: Conduct Project Closeout Activities	3.2%	4
Total	**100.0%**	**125**

Table 2-1 CTS-D Exam Duties and Tasks

Exam Preparation Strategies

You can prepare for the CTS-D exam in many ways, including studying this book. But studying this book alone may not be enough to help you pass the exam. You are required to have two years of design experience and a valid CTS, so real-world training and experience are key to your success. That said, another way to start your CTS-D studies is by performing a self-assessment of your existing AV design knowledge to identify your strengths and weaknesses. There are free online practice questions that are similar to the questions presented on the CTS-D exam. You will also find helpful questions in the "Chapter Review" sections throughout this book.

Keep in mind that because the CTS-D exam is designed to comply with American National Standards Institute (ANSI) standards, the CTS-D practice exam cannot include actual exam questions. In fact, the practice exam questions may not be informed by the exam itself. Any practice question you find here or elsewhere is written to be *similar* to an actual CTS-D exam question.

AVIXA provides other resources to help you prepare for the CTS-D exam. The glossary toward the end of this book covers most of the acronyms, technical terms, and other language you will need to be versed in to navigate the CTS-D exam and life as a certified designer. Also, refer to Appendix A, Math Formulas Used in AV Installations, for a handy reference to important math formulas used in AV design. AVIXA also offers online courses, such as *AV Math Online, AV Math for Design Online,* and *AV Design Levels 1–3 Online* for a fee. (Read more about math strategies later in this chapter.)

As you prepare for the CTS-D exam, keep the exam content outline and JTA handy. Figure out not only where you're strongest but also where the JTA has placed the most emphasis.

Not surprisingly, the part of the exam that covers developing AV designs (Duty C) counts the most (48 percent). So, spend extra time on sections that inform the five tasks in Duty C, which have a lot to do with drawings and documentation—critical components of a good AV design. You will find information about drawings and documentation, for example, in Part II, Environment, where we discuss the needs analysis, program reports, AV documentation, and more. You will find more information about drawings and documentation near the end of the book, in Part IV, Applied Design, where we discuss closing out a project and handing over documentation to the client. Documenting an AV design is an important part of a CTS-D's job and occurs throughout a project.

In addition, as you prepare to take the CTS-D exam, refer frequently to the *CTS-D Candidate Handbook,* which you can download from www.avixa.org/ctsd. In addition to information about the exam, the JTA, and the exam content outline, the *CTS-D Candidate Handbook* lays out the important knowledge and professional attributes required of CTS-D candidates. From there, download and refer to the entire JTA, also available at www.avixa.org/ctsd, for a complete breakdown of knowledge, attributes, and skills required across all duties and tasks.

 NOTE Make sure to download the most up-to-date free edition of the *CTS-D Candidate Handbook* for important policy and procedure updates by going to the AVIXA website at www.avixa.org/ctsd.

For example, under Duty A: Conducting a Needs Assessment, Task 4: Review client technology master plan, the detailed JTA indicates CTS-D candidates should have good written and verbal skills, plus knowledge of the following:

- Basic fiscal planning terminology (return on investment and so on)
- Client's structured cabling system
- Equipment life cycles
- Sustainability and other green issues
- Basic networking equipment, connectivity, and terminology

As another example, under Duty B: Coordinating with Other Professionals to Develop Project Documentation, Task 2: Coordinate with architectural/interior design professionals, the JTA indicates CTS-D candidates should demonstrate the following attributes:

- The ability to calculate area
- The ability to identify three-dimensional interference issues from two-dimensional plans
- The ability to visualize spatial relationships from drawings
- An understanding of AV maintenance requirements
- An understanding of AV systems operational requirements
- An understanding of equipment space and access requirements
- An understanding of ergonomic best practices
- An understanding of lighting levels
- An understanding of signage requirements
- The ability to utilize reference materials

This type of information is detailed in the JTA and can help guide your thinking as you work through the *CTS-D Certified Technology Specialist-Design Exam Guide*. The content and practice questions in this book are based on the same JTA and exam content outline as the CTS-D exam. However, because the exam questions are confidential, there can be no guarantee that this book will cover every question on the exam or that the exam will address every topic in this book. This exam guide prepares you for a career as a certified AV designer—not just for the credentialing exam.

And as you might expect, technology and best practices change over time. Therefore, the CTS-D exam must change. When AVIXA revises the CTS-D exam, it will publish such revisions on its website, www.avixa.org/ctsd.

 NOTE No book, course, or other study material is required to take the CTS-D exam. By the same token, no book, course, or other study material can guarantee you will pass the exam.

Mathematical Strategies

Some of the questions on the CTS-D exam can be answered only by solving math equations. When you earned your CTS certification, chances are you had to brush up on a pair of important mathematical concepts, which we will review briefly: the order of operations and Ohm's law. Both are important math principles for AV professionals and will serve you long after you've taken the CTS-D exam.

There are a lot of mathematical formulas associated with AV design, but you don't have to memorize them all. On test day, CTS-D candidates have access to relevant math formulas on the computer screen while they test. The formula sheet does not cover every possible formula that you may encounter on the exam, but it does contain many common (and complicated) formulas. You can download the list ahead of time and use it to practice AV math functions before test day. Find it at www.avixa.org/ctsd.

You are not allowed to bring a calculator into the CTS-D exam room. You will, however, have access to a virtual, computer-based calculator that simulates a Texas Instruments TI-30XS MultiView calculator (see Figure 2-1). The TI-30XS is a scientific calculator, designed to perform complex operations that are not common in everyday mathematics (for example, exponent and square root). Many AV design tasks call for complex calculations, so you should learn how to use a scientific calculator.

Figure 2-1
The TI-30XS MultiView calculator

As you work your way through this exam guide, it is a good idea to use a physical TI-30XS MultiView calculator or an on-screen emulator to familiarize yourself with its functions. A search of the Texas Instruments Education website will locate a free, downloadable TI-30XS emulator for either Windows or Macintosh computers.

As you familiarize yourself with your TI-30XS MultiView calculator, make sure you can find the following buttons:

- Log
- Antilog
- Reciprocal
- Exponent
- Square Root

The more you use the TI-30XS before you take the exam, the faster you will be able to locate the buttons during the exam.

Order of Operations

Many AV math formulas use addition, subtraction, multiplication, division, exponents, and logarithms. These formulas require a solid foundation in the order of operations. The order of operations helps you correctly calculate the desired result by prioritizing which part of the formula to use first. It is a way to rank the order in which you work your way through a formula.

This is the order of operations:

1. Any numbers within parentheses or brackets
2. Any exponents, indices, or orders
3. Any multiplication or division
4. Any addition or subtraction

If there are multiple operations with the same priority, then proceed from left to right: parentheses, exponents, multiplication, division, addition, and subtraction. You can remember the order of operations by using the acronyms BODMAS (brackets, orders, division, multiplication, addition, subtraction) or PEMDAS (parentheses, exponents, multiplication, division, addition, subtraction).

Practice Exercise 1

Solve the following equation using the order of operations:

$$2 + (5 \div 8^2 \times 9)$$

Step 1 Anything inside parentheses is processed first. Inside the parentheses, calculate the exponent first.

$$2 + (5 \div 8^2 \times 9)$$
$$2 + (5 \div 64 \times 9)$$

Step 2 Inside the parentheses are now two operations that have the same priority: multiply and divide.

Because they are the same priority, begin solving them from left to right.

$2 + (5 \div 64 \times 9)$
$2 + (0.078125 \times 9)$
$2 + (0.703125)$

This step is typically where a mistake might be made. If the formula is at a stage where the operations are of the same priority, continue solving from *left to right*. If this step is not completed, you will arrive at the wrong answer.

Examine the following *incorrect* processing shown here:

$2 + (5 \div 64 \times 9)$
$2 + (5 \div 576)$
$2 + (0.009)$

In this example of incorrect processing, multiplication was performed first. The incorrect processing of right to left within the parentheses means that the final answer will be wrong.

Step 3 Now the only remaining operation is addition.

$2 + (0.703125) = 2.703125$

Answer Rounded to one decimal place, the result is 2.7.

Use the next practice exercise to reinforce your calculation skills.

Practice Exercise 2

Solve the following equation using the order of operations:

$6^2 + 4 \div (3 \times 8) - 8$

Step 1 Anything inside parentheses is processed first. Inside the parentheses, multiply first.

$6^2 + 4 \div (24) - 8$

Step 2 There are no more operations in parentheses. Calculate the exponent next.

$36 + 4 \div 24 - 8$

Step 3 The next operator is divide. Calculate the division.

$36 + 0.16667 - 8$

Step 4 The only two operations remaining are addition and subtraction. Both addition and subtraction are the same priority. Remember to process *from left to right*.

$36.16667 - 8 = 28.16667$

Answer Rounded to one decimal place, the result is 28.2.

Electrical Calculations

Now that you have practiced basic mathematical principles, you can apply them to a fundamental concept in AV installation: Ohm's law, which you will have studied to attain your CTS certification.

Electrical Basics

From your prior electrical knowledge, you should be able to recall concepts and terms such as Ohm's law, voltage, current, resistance, impedance, and power, which are summarized in Table 2-2.

If you are not comfortable with these terms, please consider revisiting your electrical fundamentals before proceeding with the CTS-D exam.

Ohm's Law

Voltage, current, and resistance/impedance are all related. The current in an electrical circuit is proportional to the applied voltage. An increase in voltage means an increase in current, provided resistance/impedance stays the same. The relationship between current and resistance/impedance is inversely proportional, meaning that as one increases, the other decreases. An increase in resistance/impedance produces a decrease in current, provided voltage stays the same.

This relationship between voltage, current, and resistance is defined in mathematical form by Ohm's law.

$$I = V \div R$$
Current = Voltage ÷ Resistance

So, 1 amp is equal to the steady current produced by 1 volt applied across a resistance of 1 ohm.

The relationship between power, current, and voltage is defined in the power equation:

$$P = I \times V$$
Power = Current × Voltage

Parameter	Symbol	Definition	Unit of Measurement	Abbreviation
Voltage or Electromotive Force	V or E	Potential difference (electrical pressure)	Volt	V
Current	I	Electron flow	Ampere (Amp)	A
Resistance	R	Opposition to electron flow due to the materials in the circuit	Ohm	Ω
Impedance	Z	Total opposition to electron flow in a circuit	Ohm	Ω
Power	P	Rate at which work is done	Watt	W

Table 2-2 Characteristics of Voltage, Current, Resistance, Impedance, and Power

So, 1 watt is the power produced by a current of 1 amp driven by a potential difference of 1 volt.

Figure 2-2 shows Ohm's law and the power equation and gives the formulas for calculating any one parameter if the other two are known. If you know the values of two of these variables, you can easily calculate the third.

- If the two known variables are on top of each other, divide the top variable by the bottom.
- If the two known variables are next to each other, multiply them.

Combining Ohm's law and the power equation allows us to make other calculations such as these:

$$P = I^2 \times R$$
$$R = P/I^2$$
$$P = V^2/R$$
$$I = \sqrt{(P/R)}$$

You will not be provided with any kind of formula sheet to reference during the CTS-D exam. You should use these simple diagrams during the exam because they are much easier to memorize.

For example, if you want to determine the power (P) resulting when there is 3 amps of current (I) at 12 volts (V), simply cover the P in the power formula diagram, which leaves V and I next to each other in a multiply relationship. Multiply the values for V and I (12 × 3), and the result (36) is the number of watts.

Apply the math operation shown here:

$$P = I \times V$$
$$P = 12 \times 3$$
$$P = 36$$

There are 36 watts of power.

 VIDEO Watch AV Math Online: *Ohm's Law and the Power Formula* is a short video tutorial that explains how to use the Ohm's law and power law relationships. Check Appendix D for the link to the AVIXA video library.

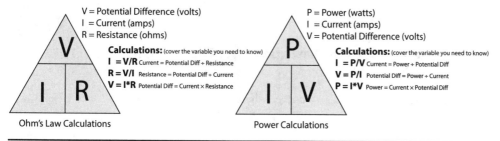

Figure 2-2 Ohm's law and power law formulas

The best way to master these calculations is to keep practicing. If you find the following practice exercises simple to do, look for more challenging equations to solve at your place of work or in an online course.

Practice Exercise 1: Ohm's Law Calculation

Perform the following calculation using Ohm's law:

Calculate the current in a circuit where the voltage is 2 volts and the resistance is 8 ohms.

Step 1 In this example, voltage and resistance are known, and you are solving for current. The correct formula to use is $I = V \div R$.

Step 2 Divide voltage by resistance.

$I = 2 \div 8$
$I = 0.25$

Answer

$I = 0.25A$ (250mA)

Practice Exercise 2: Ohm's Law Calculation

Perform the following calculation using Ohm's law:

Calculate the voltage in a circuit where the current is 4 amps and the resistance is 25 ohms.

Step 1 In this example, current and resistance are known, and you are solving for voltage. The correct formula to use is $V = I \times R$.
Multiply current times resistance.

$V = 4 \times 25$

Step 2 Calculate.

$V = 100$

Answer

100V

Practice Exercise 3: Ohm's Law and Power Formula Calculation

Perform the following calculation using Ohm's law and the power formula:

Calculate the resistance in a circuit where the voltage is 4 volts and power is 2 watts.

Step 1 In this case, you have only one known Ohm's law variable: voltage. However, you have two known power equation variables: voltage and power. You can use the power formula to derive another Ohm's law variable, current, and then solve for resistance.

First, use the power equation to solve for current.

$I = P \div V$
$I = 2 \div 4$
$I = 0.5$

The current is 0.5A (500mA).
Now that you know the current, use Ohm's law to solve for resistance.

$R = V \div I$
$R = 4 \div 0.5$
$R = 8$

Answer The resistance is 8 ohms.

In fact, using the two relationship diagrams, you can easily derive any two variables from any other two, unless your only two given variables are power and resistance. Just in case, it might be a good idea to memorize $V = \sqrt{(P \times R)}$ for the CTS-D exam.

Practice Exercise 4: Ohm's Law and Power Formula Calculation

Perform the following calculation using Ohm's law and the power formula:

Calculate the voltage in a circuit where the resistance is 16 ohms and power is 4 watts.

Step 1 In this example, resistance and power are known, and you are solving for voltage. Because you already have values for power and resistance, the correct formula to use is the following:

$V = \sqrt{(P \times R)}$
Enter the values from the question.
$V = \sqrt{(16 \times 4)}$

Step 2 Calculate the value in parentheses first.

$V = \sqrt{64}$

Step 3 Calculate the square root.

$V = 8$

Answer The voltage is 8 volts.

NOTE If you encounter a question or topic in this book that is unfamiliar to you, write it down for further study. That way, you know to focus on these topics when you develop your personal study plan.

The CTS-D Exam Application Process

When you're ready to take the CTS-D exam, review the application process. All the information you need is posted on AVIXA's website at www.avixa.org/ctsd and in the *CTS-D Candidate Handbook*. Bear in mind the following:

- You must meet all eligibility requirements as of the date of the application.
- You may apply for the exam using the application in the *CTS-D Candidate Handbook* or by downloading the most recent application. You may apply online at www.avixa.org/ctsd.
- Applications will not be processed unless all required information on the application is completed and the application fee is received.
- You must provide phone and e-mail contact information to facilitate e-mail confirmation of receipt of the application and any necessary phone contact prior to or following the exam.
- AVIXA will review and respond to your application within approximately 10 business days. For applications that are incomplete or lack documentation and/or payment, AVIXA will contact the applicant regarding the missing requirements. Once you have been approved for eligibility, AVIXA will notify you within one day of notifying Pearson VUE of your eligibility. You may then contact Pearson VUE after a 24-hour period to make your testing appointment. You can find the list of available testing locations at home.pearsonvue.com/avixa.
- Your application is approved for a period of 120 days from the date of the eligibility approval notice, and you must arrange for and be tested during that 120-day period. The exam application fee must be paid by using a major credit card or by check at the time the application is submitted.

Getting to the Testing Center

On the scheduled day of the CTS-D exam, you should report to the exam center as instructed in your appointment confirmation letter. Plan to arrive at least 30 minutes prior to your scheduled start time. Allow extra travel time for unforeseen events, such as traffic delays. If you arrive after your assigned exam time, you will be considered a "no-show" and will not be admitted. To take the exam, you will need to reapply by contacting AVIXA and paying a reinstatement fee.

It is not necessary (although it is preferred) to bring your e-mail or letter of confirmation with you. However, you must have proper identification (as described shortly). The name on the ID must match the information on file with AVIXA and the vendor responsible for presenting the exam.

TIP If you live more than an hour from the exam center, consider staying at a nearby hotel the night before so you can get a good night's rest and make sure you arrive on time. It may also be a good idea to visit the testing center prior to the exam to ensure you know exactly where to go and how to get there.

Identification Requirements

Candidates must check in at the testing center with two forms of valid ID, one of which must be a government-issued photo ID with signature (driver's license, government ID, or passport). (See the "On the Day of the Exam" section in the *CTS-D Candidate Handbook*.) For testing center identification purposes, you must bring both a valid government-issued ID and a secondary ID that has a matching signature to the name on the government ID. *The first and last names on the ID must match exactly the name submitted on the application or you will be denied admission.*

Candidates can make changes to their names by contacting AVIXA (certification@avixa.org) prior to scheduling their exam appointment. Candidates will also be required to provide a digital signature and have a digital photo taken when checking in. This information is retained in a secure database for no more than five years from the last exam date. The candidate's electronic signature is not linked to the candidate's personal identification information, such as address or credit card information.

For certain countries, including China, Hong Kong, and Taiwan, identification requirements may differ. Please see the *CTS-D Candidate Handbook* for more details.

Items Restricted from the Exam Room

You are not allowed to bring anything into the exam room. Secure lockers are provided to store personal items while taking the exam. The following are examples of items that are *not* permitted in the exam room or testing center:

- Slide rules, papers, dictionaries, or other reference materials
- Phones and signaling devices such as pagers
- Alarms
- Recording/playback devices of any kind
- Calculators (a calculator will be displayed on the test computer screen)
- Photographic or image-copying devices
- Electronic devices of any kind
- Jewelry or watches (time will be displayed on the computer screen and wall clocks in each testing center)
- Caps or hats (except for religious reasons)

This list is not exhaustive. To be safe, you shouldn't expect to be allowed to bring *any* item into the testing room.

About the Exam

You will take the exam on a computer. The exam includes multiple-choice questions. For each question, the computer will display four possible answers (A, B, C, and D). One of the answers represents the single correct response, and credit is granted only if you select that response.

Candidates get 180 minutes to answer 135 questions, 10 of which don't count because they're pilot questions used to study future additions to the CTS-D exam. They won't be scored, but you will not know which are the pilot questions.

There is a brief on-screen computer-based tutorial just prior to starting the exam and a brief online survey at the end of the exam. The time necessary to complete the tutorial and survey does not count against the 180 minutes you're given to finish the exam.

To get familiar with the testing interface, use the online tutorial practice exam from Pearson VUE's AVIXA Testing page at home.pearsonvue.com/avixa. The questions you find there are not AV-related; the tutorial is simply intended to familiarize you with the testing interface. You can use this free resource any time before you take the CTS-D exam.

During the Exam

What is it like inside the exam room? Knowing ahead of time should help you plan for any eventuality. Bear in mind the following:

- Candidates should listen carefully to the instructions given by the exam proctor and read all directions thoroughly.
- Questions concerning the content of the exam will not be answered during the exam.
- The exam center proctor will keep the official time and ensure that the proper amount of time is provided for the exam.
- Restroom breaks are permitted, but the clock will not stop during the 180 minutes allotted for the actual exam.
- During the exam, candidates will be reminded when logging in to the testing center computer screen and prior to being allowed to take the exam that they have agreed to follow the CTS Code of Ethics and Conduct and nondisclosure agreements presented earlier in the application process.
- Candidates will have access to a computer-based calculator and a wipe-off note board provided by the testing center.
- Candidates will have the ability to provide comments for any question, as well as mark questions and return to them for review.
- There will be an on-screen reminder when only five minutes remain to complete the exam.
- No exam materials, notes, documents, or memoranda of any kind can be taken from the exam room.

For best results, pace yourself and periodically check your progress. This will allow you to adjust the speed at which you answer questions, if necessary. Remember that the more questions you answer, the better your chance of achieving a passing score. In other words, don't leave any question unanswered. If you are unsure of a response, eliminate as many options as possible and choose from the answers that remain. You will also be allowed to mark questions for review prior to the end of the exam.

TIP Be sure to record an answer for every question, even if you're not sure which answer is correct. You can note which questions you want to review and return to them later. There is no penalty for wrong answers, so marking an answer to all questions will maximize your chances of passing.

Dismissal or Removal from the Exam

During the exam, the exam proctor may dismiss a candidate from the exam for any of the following reasons:

- The candidate's admission to the exam is unauthorized
- A candidate creates a disturbance or gives or receives help
- A candidate attempts to remove exam materials or notes from the testing room
- A candidate possesses items that are not permitted in the exam room
- A candidate exhibits behavior consistent with attempting to memorize or copy exam items

Any individual who removes or attempts to remove exam materials, or is observed cheating in any manner while taking the exam, will be subject to disciplinary or legal action. Sanctions could result in removing the credential or denying the candidate's application for any AVIXA credential.

Any unauthorized individual found in possession of exam materials will be subject to disciplinary procedures in addition to possible legal action. Candidates in violation of AVIXA testing policies are subject to forfeiture of the exam fee.

Hazardous Weather or Local Emergencies

Hey, you never know. In the event of hazardous weather conditions or any other unforeseen local emergencies occurring on the day of the exam, the exam presentation vendor will determine whether circumstances require cancellation. Every attempt will be made to administer all exams as scheduled.

When an exam center must be closed, the vendor will contact all affected candidates to reschedule the exam date and time. Under those circumstances, candidates will be contacted through every means available: e-mail and all phone numbers on record. This is an important reason for candidates to provide and maintain up-to-date contact information with AVIXA and the exam vendor.

Special Accommodations for Exams

AVIXA complies with the Americans with Disabilities Act (or your country's equivalent) and is interested in ensuring that no individual is deprived of the opportunity to take the exam solely by reason of a disability. Two forms must be submitted to receive special accommodations:

- Request for AVIXA (CTS, CTS-D, CTS-I) Exam Special Accommodations
- AVIXA (CTS, CTS-D, CTS-I) Exam, Healthcare Documentation of Disability Related Needs

Applicants must complete both forms and submit them with their application information to the AVIXA certification office no later than 45 days prior to the desired exam date.

Requests for special testing accommodations require documentation of a formally diagnosed or qualified disability by a qualified professional who has provided evaluation or treatment for the candidate.

You can find these forms, along with more information about the process, at the AVIXA CTS-D website and in the *CTS-D Candidate Handbook*.

Exam Scoring

Candidates receive their results immediately upon exam completion. The final passing score for each examination is established by a panel of SMEs using a criterion-referenced process. This process defines the minimally acceptable level of competence and takes into consideration the difficulty of the questions used on each examination.

Candidates who do not pass the exam receive their score and the percentages of questions they answered correctly in each duty area. AVIXA provides these percentages to help candidates identify their strengths and weaknesses, which may assist them in studying for a retest. It is not possible to arrive at your total exam score by averaging these percentages because there are different numbers of exam items in each duty area on the exam.

Retesting

If you do not pass the CTS-D exam, you may take it again two more times following your original exam date. There is a minimum 30-day waiting period between each retest. After your retest application has been approved, you have 120 days from the date of the reissued eligibility notice to retake the exam.

If, after two retests, you still have not passed the exam, you must wait 90 days before restarting the application process. This period gives you time to adequately prepare and prevents overexposure to the exam. Candidates must meet all eligibility requirements in effect at the time of any subsequent application. You can find the CTS-D Exam Retest Application form and current retest fees at the AVIXA website.

Currently certified CTS-D individuals may not retake the CTS-D exam, except as specified by AVIXA's CTS-D renewal policy.

CTS-D Exam Practice Questions

To reiterate, the CTS-D exam consists of 125 multiple-choice questions that address each of the duties and tasks listed in Table 2-1. The questions focus primarily on issues that an AV professional may encounter when working on a specific job or task, rather than on general AV technology knowledge. The exam also includes an additional 10 unscored pilot items, for a total of 135 questions. The pilot items may be from any duty and task area, and the test taker will not know which items are scored and which are pilot.

Let's take a look at examples of CTS-D exam questions. For each question, the duty and task from which the question is drawn are identified first. Not to sound like a broken record, but remember that these are sample questions, not actual CTS-D exam questions. They may be similar to exam questions, but because of the way the CTS-D exam is designed to meet ANSI standards, there can be no guarantee that these practice questions reflect the actual exam. That said, both the practice questions and the exam questions are guided by the same JTA.

1. [Conducting a Needs Assessment/Review Client Technology Master Plan]

    How do you determine the client's long-term plans and needs in terms of equipment and support maintenance?

    A. Ask advice from other AV industry experts and vendors
    B. Contract with a consulting company to determine client needs
    C. Consult the owner, examine standards and design manuals, and create the project report
    D. Design a generic system that fits any organization of that size

2. [Conducting a Needs Assessment/Identify Scope of Work]

    In a proposed presentation room with a flat floor and theater-style seating, which has the greatest influence on the visibility of a projection screen?

    A. Ceiling height
    B. Projection room location
    C. Location of windows
    D. Width of the room

3. [Coordinating with Other Professionals to Develop Project Documentation/Coordinate with Architectural/Interior Design Professionals]

    For a flat-floor auditorium with rows of staggered seats evenly spaced for proper horizontal viewing, which row or rows will determine the preferred minimum height of the bottom of the image on a display at the front of the room?

    A. First three front rows of the room
    B. Ends of the back row of the room
    C. Rear three rows of the room
    D. Second row from the back of the room

4. [Developing AV Designs/Create Draft AV Design]

   Which of the following signals can safely coexist in the same conduit without interference?

   A. Serial control and video

   B. Speaker level audio and serial control

   C. Video and microphone-level audio

   D. Data network and line-level audio

5. [Conducting Project Implementation Activities/Participate in Project Implementation Communication]

   In an AV project involving a courtroom, sightlines from the jury box are obscured by an architectural feature. How should the issue be addressed?

   A. Communicate with the project manager

   B. Contact the architect with recommended changes

   C. Let the judge manage the situation

   D. No consult necessary; move forward with your recommendations

NOTE If you encounter a question or topic in this book that is unfamiliar to you, write it down for further study. That way, you know to focus on these topics when you develop your personal study plan.

Answers to CTS-D Practice Questions

Did you peek or read ahead? Here are the answers to the preceding practice questions:

1. **C.** When determining a client's long-term plans and needs in terms of equipment and support maintenance, you should consult the owner, examine standards and design manuals, and create the project report.

2. **A.** Ceiling height has the greatest influence on the visibility of a projection screen in a presentation room with a flat floor and theater-style seating.

3. **C.** For a flat-floor auditorium with rows of staggered seats evenly spaced for proper horizontal viewing, the rear three rows of the room will determine the preferred minimum height of the bottom of the image on a display at the front of the room.

4. **A.** Serial control and video signals can coexist in one conduit without interference.

5. **A.** If you have sightline issues from the jury box of a courtroom, communicate with the project manager.

Chapter Review

Upon completion of this chapter, you should have a clear understanding of the following:

- How the CTS-D exam is designed
- The skills, knowledge, and AV design-related tasks that the exam covers and why
- How and why this guide covers the exam material differently than the exam content outline
- How to study and prepare for the CTS-D exam
- How to solve AV-related mathematical equations using the order of operations
- How to solve calculations using Ohm's law
- How to apply for the exam
- What to expect on the day of the exam
- What the CTS-D exam questions might look like (but not exactly)

Now you're ready to delve into the specific knowledge, skills, and responsibilities you will need to succeed in your career as a CTS-D–certified AV designer.

PART II

Environment

- **Chapter 3** — Communicating Design Intent
- **Chapter 4** — Ergonomics in AV Design
- **Chapter 5** — Visual Principles of Design
- **Chapter 6** — Audio Principles of Design

CHAPTER 3

Communicating Design Intent

In this chapter, you will learn about
- What makes exceptional AV experiences
- The phases of an AV design project
- The documentation necessary for communicating a design to the necessary stakeholders
- Creating accurate AV design drawings
- Reading architectural drawings and identifying the elements that affect your AV design
- Identifying end users' requirements through a needs analysis process
- Creating a concept design/program report for an AV system installation

AV systems are complex. They're integrated with other building systems, such as network; electrical; fire detection; security; mechanical; heating, ventilation, and air conditioning (HVAC); and building automation/energy conservation systems. In many cases, AV systems provide operational functionality to an owner, requiring a thoughtful and well-organized approach to commonly accepted planning, design, and integration procedures. As you proceed through a design, you must constantly review and adjust assumptions to ensure the needs of the customer are met given the available environment.

Any seasoned AV designer will tell you that a design project is not linear. That is, you don't complete tasks 1, 2, and 3 in order. Rather, the process is more like, "Do task 1 first, then task 2, then go back to task 1, before jumping to task 3 and then task 2 again. Oh, and simultaneously finish task 4." What's more, the AV design and systems integration processes may span and parallel a lengthy construction cycle, resulting in input and review from key personnel in many different trades and disciplines.

Throughout everything, it is critical that an AV designer be able to communicate—in words and drawings—the intent of a design. Such design intent comprises everything from a client's expectations to the details of a system that meets those expectations, from how a design works within the architecture of a space to how it fits within a client's budget. Successfully communicating design intent requires a phased approach.

> **Duty Check**
>
> This chapter relates directly to the following tasks on the CTS-D Exam Content Outline:
>
> - Duty A, Task 1: Identify Decision-makers and Stakeholders
> - Duty A, Task 2: Identify Skill Level of End Users
> - Duty A, Task 3: Educate AV Clients
> - Duty A, Task 4: Review Client Technology Master Plan
> - Duty A, Task 5: Identify Client Expectations
> - Duty A, Task 6: Identify Scope of Work
> - Duty B, Task 1: Review A/E (Architectural and Engineering) Drawings
> - Duty B, Task 2: Coordinate with Architectural/Interior Design Professionals
> - Duty B, Task 3: Coordinate with Mechanical Professionals
> - Duty B, Task 4: Coordinate with Structural Professionals
>
> In addition to skills related to these tasks, this chapter may also relate to other tasks.

Creating an Exceptional AV Experience

AV design is all about creating exceptional AV experiences. The goal of an AV experience usually lies somewhere along the continuum between transferring information to the participant and delighting, amusing, amazing, and entertaining them. In some AV experiences the participants are actively engaged in shaping or creating the content of the experience, while in others the participants sit back and observe content.

Types of AV Experience

AV experiences can be broadly categorized into four types:

- **Collaboration**   Participants work together to achieve a common goal.
- **Instruction**   Knowledge is transferred to the participant.
- **Exploration**   Participants discover and interact with real or virtual content.
- **Sensation**   Participants are delighted, amazed, and amused.

Collaborative and instructional experiences are both focused on informing the audience, but a collaborative experience strives for the audience to interact with material, while instructional experiences want the audience to focus on absorbing content.

Explorative and sensational experiences are both focused on entertaining the audience, but an explorative experience seeks for the audience to interact with material, while sensational experiences want the audience to absorb content.

Determining which type best suits your AV experience will help to clarify the client's end goal, which will help you identify what's needed to deliver an exceptional experience to the end users.

When you achieve excellence in an AV experience, participants are inspired to shape the outcome with others in a collaborative experience, absorb or apply what they've learned in an instructional experience, make their own path through the content in an explorative experience, and remember the experience and tell others about it in a sensational experience.

Components of an AV Experience

The three components of an AV experience are content, space, and tech:

- **Content**   This is the material being shaped or conveyed and can include presentations, exhibits, narratives, performances, or music.
- **Space**   This is the area in which the experience occurs and can be a board room, a concert hall, a laboratory, a museum, a sports field, a shopping complex, or a digital device.
- **Tech**   This is the means by which content is delivered and can include cameras, microphones, replay systems, signal processors, projected and direct-view video, loudspeakers, atmospheric effects, pyrotechnics, staging effects, and lighting.

As an AV designer you may only have limited control over the content and some input into the design and construction of the space, but you will usually have significant input and control of the means of delivering the content. As a designer of exceptional AV experiences, your mission is to pull these three components together to design a memorable, engaging, and fit-for-purpose AV system.

The Phases of an AV Design Project

Even though design is iterative and there is no precise step-by-step process, there are still logical phases to the project. ANSI/AVIXA D401.01:202X, *Standard Guide for Audiovisual Systems Design and Coordination Processes*, lays out the steps of an AV design. If you follow them, you will find yourself laying the five important building blocks of a successful design, as shown in Figure 3-1.

The ANSI/AVIXA D401.01:202X standard was created to describe the methods, procedures, tasks, and deliverables typically recommended by AV systems designers and systems integrators. By using the standard, clients, designers, and construction team members can be confident that everyone on the project team delivers on expectations.

 VIDEO   Watch a video from AVIXA's online video library about planning an AV design. Check Appendix D for the link to the AVIXA video library.

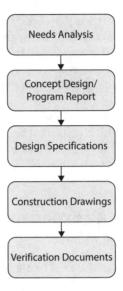

Figure 3-1 The building blocks of an AV design

As a designer, your job is to analyze and define performance needs, functional requirements, and system budgets prior to ordering and installing equipment. You need to record clearly the client's needs and expectations through written documents and formal architectural drawings. To help you accomplish this goal, consider using the phases described in the ANSI/AVIXA D401.01:202X standard as your guide. Let's take a quick look at design phases before going into more detail later in the chapter.

A Word About Meetings

All parties involved in an AV project want the finished product to meet the design intent—on time and within budget. But before you can design anything, you need to hold planning and coordination meetings, sometimes called *pre-construction meetings*, each with a potentially different focus. These will be the first of many meetings throughout the project.

You should adjust the type, timing, and quantity of meetings to reflect project requirements, and there should be a contract that specifies the number and location of meetings to be held throughout the design process. It is important to come out of every meeting with some form of written documentation. Meeting minutes, notes, and contracts help reinforce discussions and record any decisions that are made.

Meetings will serve a variety of purposes during a project and should include the following:

- Technology programming meetings, to understand the client's goals and level of technical sophistication
- Design coordination and review meetings, held at major milestones to review with the client the progress of the design and to coordinate with the rest of the team
- Pre-bid and pre-construction meetings, held with the client and potential contractors to review the scope of the project
- Construction meetings, to observe the progress of the work and address issues onsite
- Post-construction meeting, held with the client to review the project

NOTE Here is an example of the nonlinear nature of a design project: Early in a project you determine that a large video monitor would be best for your customer's needs. However, later in the project, you discover that the large screen does not fit in the building's elevators and cannot be delivered to the space. In such a case, you would need to review the design selection to ensure it is still the best choice given the limitation. Moreover, changing screen specifications may impact other parts of your design.

Concept Design/Briefing/Program Phase

Once you have met with your customers and determined the shape and structure of your project, you can begin the *briefing, concept design,* or *program phase* of your design. The purpose of this phase is to discuss, clarify, and document the client's needs, concerns, expectations, and constraints. These will guide the design team in its approach to the functionality and cost of the AV systems.

There are two key components of this phase: the needs analysis and concept design/program report. Keep in mind, this phase of the project is not for discussing particular equipment; it's to define the needs, contexts, methods, and wants of the client.

For architects, *programming* is the process by which the overall requirements of building are defined. Architects document the needs of owners and occupants as a preliminary step to putting lines on paper in what will become a set of drawings or other engineering designs.

The architectural program document describes the facility in terms of space sizes, space configuration, and overall building quality. For AV professionals, *programming* may also refer to the separate process of software coding of programmable devices, such as control systems and digital signal processing (DSP) equipment.

Design Phase

Designers use the results of the concept design/briefing/program phase, including the concept design/program report, to create two more sets of documents in what's known as the *design phase*. These documents provide the backbone for communication and coordination among the design team, the contractors, and the clients. The design phase and its associated documents are meant to identify the functional, physical, and system design requirements that meet the clients'—or building owners'—needs.

The first set of documents includes specification and AV system documents. They describe functional, operational, and technical performance specifications (e.g., frequency response, speech intelligibility goals, and image contrast ratio). The specifications may also provide information about equipment manufacturers, models, and quantities required to implement the design, as well as installation, testing, and verification procedures (as found in ANSI/AVIXA 10:2013), warranty information, and other details.

The second set of documents is a collection of drawings. These drawings comprise a graphical representation of the facility (architecture and infrastructure) and indicate how the AV equipment will integrate with the built environment—for instance, the locations of such visible items as loudspeakers, displays, control panels, projectors, cameras, interface panels, equipment racks, and so on. The drawings also include system diagrams that show what equipment will be provided and how it will connect.

VIDEO Watch a video about the design phase of an AV project in the AVIXA online video library. Check Appendix D for the link to the AVIXA video library.

It's worth noting that architects and some other trades break down design into the subphases discussed in the following sections.

Conceptual Design Phase

Following architectural programming, the architect sometimes creates a conceptual design—a diagram that graphically portrays the program information for space shapes, adjacencies, and sizes.

Schematic Design Phase

The conceptual design is developed to a more detailed level, beginning to show more detail such as double lines for walls, door locations, occupancies, and space orientations. In addition, the architect defines the overall "massing," or general shape and size of the buildings, and a schematic narrative generally describes the major systems to be included in the building.

Design Development Phase

The goal of design development is to move beyond major coordination issues to the basic floor plans. During this phase, all major design decisions are made and finalized with the owner so the building floor plan is set, engineering systems are selected, and detailing can commence. This is an intense period of design consulting and decision making for the design team and the owner. The end result is the final architectural and engineering design.

Within the architectural design process, the design development phase is a go/no-go decision point. Usually, enough design information has been gathered by the end of this phase to know with a fair degree of accuracy how much the facility construction will cost. For some projects this is the stage at which the owner may decide that a project should be abandoned because of budget or other issues before proceeding into the construction phase.

Construction Phase

Let us return to the AV design phases. In the *construction phase*, you will turn your design documents into construction drawings. These drawings, provided in locked Portable Document Format (PDF) or other unalterable document format, should include sufficient detail to convey to the AV installation team the physical configuration of the AV systems. These drawings are also known as *submittal drawings, workshop drawings,* or *shop drawings*. As the AV designer, you must make sure these documents are handed off to the team responsible for implementing your design.

Throughout the construction phase, accurate drawings that correctly depict the system and its components are critical. Inaccurate documentation leads to change orders, which can delay a project and frustrate everyone involved. Changes *will* occur, but minimizing them is a major goal of accurate construction documentation.

The construction phase involves three processes (or *efforts* in project management terminology) that are inextricably interconnected: coordination, procurement, and installation. These processes feed off documents created during the design phase, namely, drawings, specifications, and equipment schedules. These documents will be used to create more detailed drawings, called *shop, workshop,* or *installation drawings,* which will be used for fabrication at the contractor's workshop and onsite installation. They will also be used to procure equipment and materials, and they will lay the groundwork for system testing, verification, adjustment, and training when the project nears completion.

Verification Phase

After you have handed off the design documents to the installation team, your job is not over. You are still responsible for ensuring that the client's new AV system is installed according to your specifications. This means having a process in place to check that every component was installed properly and performs as expected.

When the project installation is coming to a close, using standards for the verification and optimization of AV systems demonstrates that the installation meets the design intent. Years after the project is completed, the next AV provider can refer to your design and verification documentation and know the initial delivery was professionally installed.

ANSI/AVIXA 10:2013, *AV Systems Performance Verification,* offers a framework and set of processes for determining which elements of an AV system need to be verified. It also details the timing of systems verification within the project delivery cycle, a process for determining verification metrics, and reporting procedures.

The Design Process: A Summary

As mentioned earlier, AV design is an iterative process, which means you often find yourself repeating steps to get the interrelated parts of a design seamlessly integrated. That said, if it helps you remember what you're trying to accomplish during an AV design, think in terms of these four actions:

1. **Think it**   Take time to speak with clients. Find out what they need so you can address their problems with your design.

2. **Draw it**   Draw or write down what clients tell you and then show them what you're thinking. This way, everyone can verify that they're on the same page.

3. **Calculate it**   Once a client has agreed with your idea, make sure your design works. Don't waste time designing systems for ideas that won't work in a space or with other equipment. Make calculations, determine specifications, and take measurements. After you've determined that your plans will or will not work, communicate the results to the client.

4. **Detail it**   Document all your ideas in a formal design that AV installers will use to build the systems. This step includes determining specific equipment, creating detail drawings, and coordinating your AV system with allied trades.

This process is not part of the ANSI/AVIXA D401.01:202X standard, and it is not meant to limit you to an inflexible structure. But by thinking in these terms, you may find yourself moving fluidly between the different phases and seeing the "big picture" of your design. How will the system address your client's needs? How will the gear work together to create a seamless, productive AV experience?

Reading Construction Drawings

Throughout an AV project, designers need to be comfortable reading construction drawings. This means being familiar with the following:

- Drawing scales
- Drawing types

- Drawing symbols
- Drawing abbreviations

As you develop a plan and communicate with everyone involved in the project, you will inevitably have to review construction drawings. To understand these drawings, however, you need to be familiar with drawing scales. Accurate conversions between scaled drawings and the real object are necessary for implementing a design according to the documentation.

Scaled Drawings

Construction drawings are usually created in a computer-assisted design (CAD) system in actual size (1:1), but output at a reduced, more manageable size. The image output on a screen or on paper has a relationship, or proportion, between a length on the drawing and the actual length of an object. This relationship is the *scale* of the output image. As an example, for a drawing output at 1:100 scale, every 1 millimeter on the printed drawing represents an actual length of 100 millimeters.

TIP When reading an architect's or engineer's drawings, make sure to note the scale of the drawing. In addition, check that the physical size of a printed drawing is accurate. Besides checking the scale noted on the drawing, it is always a good idea to confirm the scale using the dimensions of common building elements of known size, such as doors or ceiling grid tiles. Drawings may have been unintentionally rescaled during printing, copying, or PDF output.

Scale Rulers

A scale ruler provides a quick method for measuring an object drawn to scale on paper and interpreting its true size in the actual space. The numbers along the scale indicate the actual length of an object when measured against its scaled representation on the drawing.

Most metric scale rulers look similar to normal rulers, except that the marking has different numbers, with a different scale, along each edge. A ratio at the left end indicates the scale measured using that side.

U.S. customary scale rulers are often prism-shaped tools. A whole or fractional number to the left or right edge of the measurement tool indicates the scale those numbers represent.

The selected scale of a drawing is usually written in the title block in the lower-right corner of the drawing but may be located anywhere on the sheet. Converting the scale length to actual length is required to determine height, length, and width, as well as cable run estimates.

VIDEO Watch a video from the AVIXA online video library about how to use a scale ruler. Check Appendix D for the link to the AVIXA video library.

Converting Dimensions on Scaled Drawings

Being able to accurately measure using a scale ruler is an important skill when working with scaled drawings. Scale drawings are used to communicate the dimensions of a full-size project on a paper or in an electronic document. A set of drawings may include more than one scale, so you must check each drawing page for its scale. Some drawings may even have multiple scales on the same sheet. Occasionally, several details may need to be called out, but they may not be large enough to require their own sheet. Check the identifying information adjacent to the item to confirm its scale. Converting the scale length to actual length is required to determine actual height, length, width, and cable run estimates.

Two different scale systems can be found on drawings: metric and US customary (imperial) measuring systems.

Drawings using metric measurements usually state the scale as a single ratio, such as 1:50 (see Figure 3-2). Millimeters and meters are the standard units used in architectural drawings and construction language for commercial projects, and 1:50 and 1:100 are the most common scales. This means that for a drawing using 1:100, 1 unit of length on a drawing equals 100 units of the same length in reality.

1:20, 1:10, and other larger scales are generally used only for details. 1:200, 1:500, and 1:1000 are usually used for small-scale drawings with little detail, such as the whole floor of a large building or a site plan. 1:25 is widely used for venue and production drawings for event, live production, and broadcast projects. The metric or International System of Units (SI) is used throughout the world with the exception of the United States, Myanmar, and Liberia.

Drawings using U.S. customary measurements usually state a scale using a particular fraction of an inch on the drawing to represent a foot in reality. For example, you might find a scale on a drawing that shows 1/4-inch equaling 1 foot (a ratio of 1:48). This means that a 1/4-inch length on the drawing is equal to a 1-foot length in reality. The "1/4in. = 1ft" scale is the most common used in the United States and is also referred to as 1/48 size. The "1/2in. = 1ft" (1:24) scale is widely used for venue and production drawings for event, live production, and broadcast projects.

Figure 3-3 shows two different U.S. customary scales. The 1/4-inch scale on the left indicates that for every 1/4 inch (") measured there is 1 foot (') of real distance.

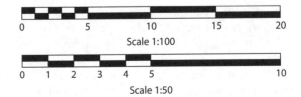

Figure 3-2 Examples of metric scale keys that you may find on a drawing set

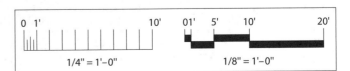

Figure 3-3 Examples of scale keys in U.S. customary units that you may find on a drawing set

The following are typical U.S. unit scales:

3/32 inch = 1 foot (1:128)	¼ inch = 1 foot (1:48)	¾ inch = 1 foot (1:16)
3/16 inch = 1 foot (1:64)	3/8 inch = 1 foot (1:32)	1 inch = 1 foot (1:12)
1/8 inch = 1 foot (1:96)	½ inch = 1 foot (1:24)	1½ inches = 1 foot (1:8)

The following are some common metric (SI) scales:

1:1/1:10	1:2/1:20	1:5/1:50
1:25/1:250	1:100/1:1000	1:500/1:5000

Always note the scale on the drawing and check that the printed physical scale of the drawing is correct.

Consider what would happen if you took a drawing that is the correct scale on a piece of U.S. letter-sized paper that is 8.5in. × 11in. (216mm × 279mm) and tried to print it on metric A4 paper that is 210mm × 297mm (8.27in. × 11.7in.). The printer driver may rescale the image to fit the different sized paper, which will result in an inaccurate printout. All your measurements taken from these printed copies would be incorrect.

Before taking off measurements or making conversions you should always verify that the document was printed accurately. Make sure all construction drawings are printed so that 1 unit of the printed document equals 1 unit of the scaled-output drawing.

Drawings may be inaccurate because of any form of copying or processing. While most original plots from a CAD file should be accurate (though human mistakes can and do happen), anything that has been plotted to "fit" the sheet of paper, rather than to a specific scale, is likely to be scaled incorrectly.

In addition to providing detailed information that is essential to ensuring the AV system design is accurately depicted, construction drawings also tell you where in a space to install an AV system.

CAD Systems and Drawing Units

In CAD systems, the two-dimensional drawings and three-dimensional models are normally input and stored at full size, and they are output at a scaled size only when required for printing hard-copy drawings or creating digital drawing files.

As CAD object dimensions are stored in "drawing units," it is up to the drafting team what each drawing unit represents in the full-size world. In the United States, it is common practice for the drawing unit to be either the U.S. customary unit of the inch or the U.S. customary unit of the foot. In the majority of disciplines in every other country, the common practice is to use the millimeter as the drawing unit. None of this matters when drawings and files are being output to document projects, but it becomes critical when drawings or objects are being exported from, or imported into, existing CAD systems. If your drawing is in millimeter units but the manufacturer's CAD drawing of the projector lift or winch motor is in inch units (or vice versa), it is necessary to rescale/resize these elements as they are imported into the CAD system's object store.

Drawing Types

Designers create drawings to show where each component will be located and how each component should be physically installed. The design documentation package typically contains functional diagrams, connection details, plate and panel details, patch panel details, equipment diagrams, rack elevation diagrams, and control panel layouts.

The following are a few of the different types and subsets of drawings you will typically encounter:

- Plan view drawings
- Reflected ceiling plans
- Elevation drawings
- Section drawings
- Detail drawings

Plan View Drawings

Plan views provide an orientation to the space. This is a "top view" taken from directly above, such as a floor plan or site plan. A floor plan is typically represented as a slice through the building at 1m above floor level (4ft in the United States).

As shown in Figure 3-4, the plan view identifies the room locations and layout dimensions, including locations of walls, doors, and windows. A floor plan view typically contains indicators to other detailed views of the site. For example, the arrows in the plan drawing indicate the direction of view to other detailed views of the site.

Reflected Ceiling Plans

A reflected ceiling plan is used to illustrate elements in the ceiling with respect to the floor. It should be interpreted as though the floor is a mirrored surface, reflecting the features within the ceiling.

As shown in Figure 3-5, the reflected ceiling plan view shows the locations of non-AV elements such as ventilation diffusers and light fixtures, as well as the position of AV components such as ceiling-mounted loudspeakers, display screens, and projectors. This view is of particular interest to the AV system designer because it helps ensure that the AV systems will meet performance standards. It also indicates which other trades the AV designer may need to coordinate with on specific rooms or areas during the pre-installation phase of the project.

Elevation Drawings

An elevation drawing (see Figure 3-6) is a drawing that looks at the environment from a front, side, or back view. Elevations provide a true picture of what the interior wall will look like. Elevations show anything that might be on the walls, such as display panels, cameras, electrical outlets, windows, doors, AV faceplates, and chair rails.

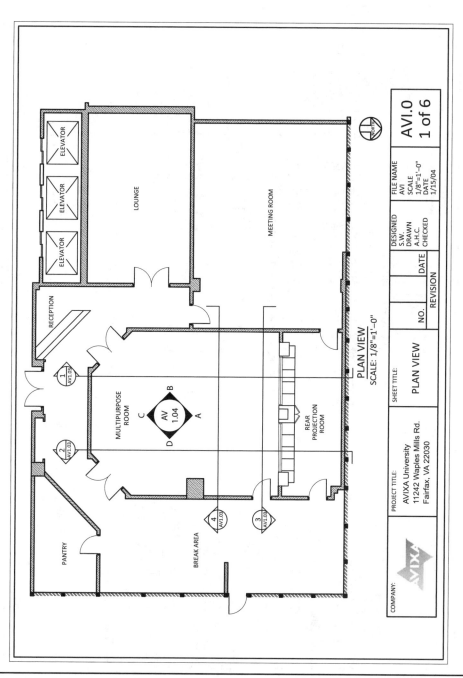

Figure 3-4 A plan view drawing

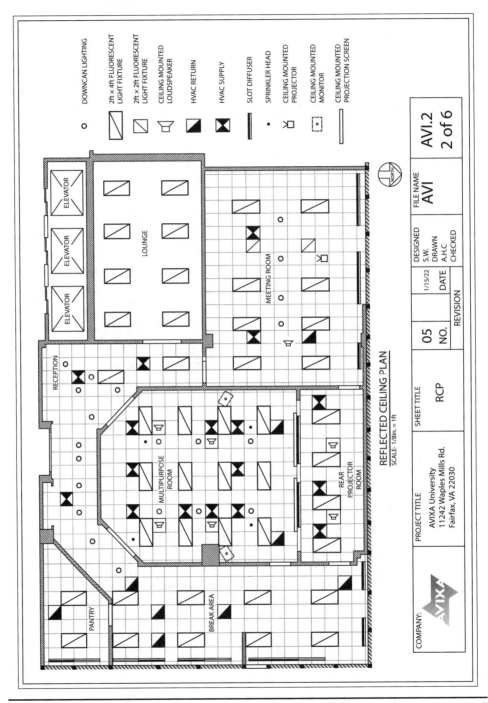

Figure 3-5 Example of a reflected ceiling plan (shows the position of loudspeakers, monitors, and rear-projection system screens). Note the symbol key on the right.

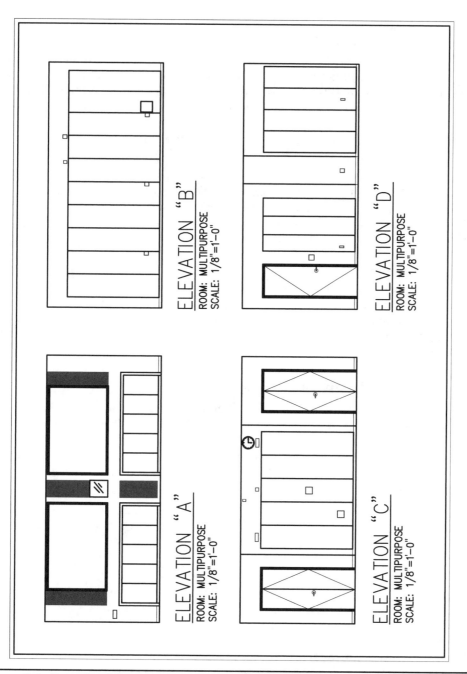

Figure 3-6 Examples of elevation views of the plan in Figure 3-4

Section Drawings

A view of the interior of a building in the vertical plane is called a *section drawing* (see Figure 3-7). A section drawing shows the space as if it was cut apart, and the direction you are looking is indicated by an arrow in the plan drawing. Section drawings show walls bisected, which allows you to view what is behind the wall and the internal height of the infrastructure. A section drawing can be rendered at an angle, so study it carefully.

AV professionals use section drawings to plan for installation needs, such as mounting locations and cable runs.

NOTE It may help to understand the difference between elevation drawings and section drawings by thinking of elevations as pictures and sections as cutouts.

Detail Drawings

Detail drawings depict items too small to see at the project's typical drawing scale. Details may show how small items are put together or illustrate mounting requirements for a specific hardware item, such as a ceiling projector mount. For example, Figure 3-8 indicates where a projection screen housing should be installed and fastened to the grid.

Architectural Drawing Symbols

Each project drawing has different symbols and icons that are used to depict specific elements of the project design or the relationship between multiple drawings, such as where a space depiction in one drawing is continued on another drawing.

Architectural symbols are standardized by industry, country, or region, and normally a drawing set will include a legend. As long as they are defined in the legend, the actual symbol object does not matter.

Typical symbols you will see include the following:

- Column lines
- Match lines
- Elevation flags
- Section cut flags
- Detail flags

Column System Symbols

A grid system is used to indicate the locations of columns, load-bearing walls, and other structural elements within the building layout, prior to the room locations being defined. Grid lines are used to reference the schedule and for dimensioning. Vertical grid lines should have designators at the top and be numbered from left to right.

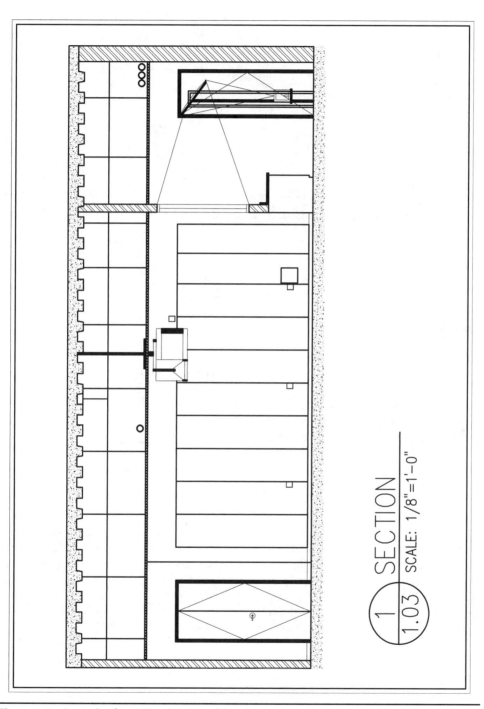

Figure 3-7 Example of a section drawing of the plan in Figure 3-4

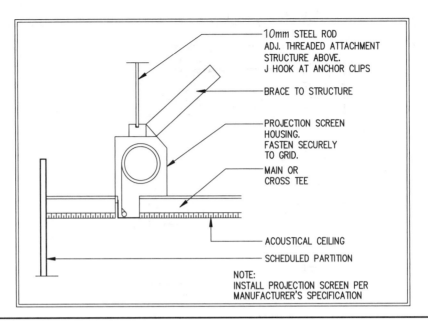

Figure 3-8 Example of a detail drawing

Horizontal grid lines should have designators at the right and be alphabetized from bottom to top. Figure 3-9 shows an example of this system.

The grid system can be used to find your way around a work site in the early construction phase. Contractors will refer to a point in the space with respect to where it exists on the drawing, as in "4 meters west of B6." Since the space may not yet be divided into rooms, identifying a point by the grid system ensures that everyone at the site understands the location.

Figure 3-9 Example of grid system symbols

Match Lines

Match lines are used to line up, or rebuild, a drawing that has been cut into separate drawings because it does not fit onto a single page. For example, it may take multiple pages to depict one area of a building. To assemble the separate drawings so you can see the whole picture, the individual pieces are aligned using the match lines as a guide.

There may also be a single drawing, at a much larger scale, that indicates the relative positions of the pages in such a drawing set. In Figure 3-10 the drawing number reference is the side that is considered.

Figure 3-10 Example of a match line

Elevation Flags

As shown in Figure 3-11 elevation flags are used on plan drawings to indicate related elevation drawings. The center text indicates the identification number of the elevation drawing and the sheet where that elevation drawing appears. The text at the apex of the triangles identifies the elevation drawing on that page. The direction of the apex of the triangle gives an approximate orientation of the elevation from the viewpoint of the symbol on the plan.

Figure 3-11 Example of an elevation flag

Each elevation view for a room is drawn in order and presented in a clockwise manner. This way, you can "look around the room" as if you were standing on the elevation flag and viewing the walls.

Section Cut Flags

Section cut flags indicate which section drawing depicts a section of a master drawing in more detail, as shown in Figure 3-12.

The bottom number (A3.0) indicates the page number where the section drawing can be found. The top number (D1) is the section drawing identification, because multiple section drawings may be on the same page. A line extends from the center of the symbol indicating the path of the section cut. The right angle of the triangle is an arrow indicating the view direction of the section drawing.

Figure 3-12 Example of a section cut flag

Section views are similar to elevation views. They depict specific wall sections, based on the section cut line indicators on the master floor plan drawing.

Detail Flags

Detail flags indicate the small items that need to be magnified to show how they need to be installed. These items are too small to draw or see at the project's typical drawing scale.

In a large system, there may be hundreds of areas requiring more detail to read accurately. Each instance is numbered. The bottom number indicates the page number where the detail drawing can be found. The top number is the detail drawing identification, because multiple detail drawings may be on the same page. In Figure 3-13, the symbol refers to detail number 1, where you can find more information about drawing AV1.06.

Figure 3-13 Example of a detail flag

Drawing Dates

Drawing dates are critical to ensuring that you are working with the most current information. It is important to recognize that dates may come in one of several formats that can be ambiguous:

- **MM/DD/(YY)YY**   Month/day/year is the only format used in the United States. It is also a minor alternative format used in Canada, some parts of Southeast Asia, and some parts of Africa.

- **DD/MM/(YY)YY**   Day/month/year is the only format used in South America, most of Europe, North Africa, South Asia, Oceania, and most of Southeast Asia. It is also the major alternative format used in Canada, the remainder of Europe, the remainder of Southeast Asia, and some parts of Africa.
- **(YY)YY/MM/DD**   Year/month/day is the only format used in North Asia and some parts of Eastern Europe. It is the minor alternative format in some other countries. This format is less likely to be confused with the alternatives.

Dates may easily be confused during some parts of the year, particularly if only two digits are being used for the year number. Care should be taken to verify which format is being used. As the United States is one of the few countries using nonmetric measurements and nonmetric drawing scales, other elements of a drawing's title block give a good indication of the likely date format should the drawing date be ambiguous.

Common Architectural Drawing Abbreviations

See Table 3-1 for common architectural drawing abbreviations.

Abbreviation	Definition
AFC	Above finished ceiling
AFF	Above finished floor
AS	Above slab
AVC	Audiovisual contractor
CL	Center line
CM	Construction manager
DIA	Diameter
(E) or EXG	Existing
E.	East
E.C.	Empty conduit
EC	Electrical contractor
EL	Elevation
ELEC	Electrical
EQ	Equal
FB	Floor box
FUT	Future

Table 3-1 Common Architectural Drawing Abbreviations (*continued*)

Abbreviation	Definition
GC	General contractor
ID	Inside diameter
IR	Infrared
LV	Low voltage
LVC	Low voltage control
MISC	Miscellaneous
NIC	Not in contract
NTS	Not to scale
OC	On center
OD	Outer diameter
OFCI	Owner furnished, contractor installed
OFE	Owner-furnished equipment
OFOI	Owner furnished, owner installed
PM	Project manager
RCP	Reflected ceiling plan
SECT	Section
TB	Table box
TYP	Typical
VIF	Verify in field

Table 3-1 Common Architectural Drawing Abbreviations

AV Drawing Symbols

There is an established AV drawing symbol standard: ANSI-J-STD-710, published jointly by AVIXA, the Consumer Electronics Association (CEA), and the Custom Electronic Design and Installation Association (CEDIA) and accredited by the American National Standards Institute (ANSI). Symbols do not necessarily follow a single standard, so the meanings of specific symbols are defined in the drawing legend or symbol key. An example of a drawing legend can be seen on the right of the reflected ceiling plan in Figure 3-5. Symbol keys or drawing legends are often located in a separate labeled box toward an edge of the drawing.

The ANSI-J-STD-710 defines the use of standardized and easily interpreted symbols for the complete documentation of an AV system design, including AV and information technology (IT) equipment, services, cabling points, and many electrical devices, including some U.S. power distribution equipment. Some of the ANSI-J-STD-710 standard's recommended symbols for AV objects in a drawing, together with some common symbol attribute abbreviations and their meanings, can be seen in Figure 3-14. Figure 3-15 is an example of a floor plan drafted to this symbol standard.

1	Audio - Video Systems				
Category-Symbol #	Name (Abbreviation)	Symbol	Technology Attribute Abbreviation Examples *For more abbreviations see Annex C*	Stretchable	Source
1-1	Loudspeaker (SPKR)		HF – HIGH FREQUENCY M – MONITOR LAR – LINE ARRAY P – POWERED SPEAKER S – SUBWOOFER LCR – LCR BAR ST – STEREO LR – LR BAR	Yes	CEA
1-2	Display Monitor (VMON) (TV)	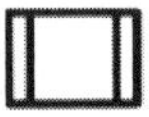	M – MIRROR TV TV – TELEVISION VM – VIDEO MONITOR WP – WEATHERPROOF TV	Yes	J-STD 710
1-3	Video Projector (PROJ)		LCD – LIQUID CRYSTAL DISPLAY DLP – DIGITAL LIGHT PROCESSING LED – LIGHT EMITTING DIODE LCOS – LIQUID CRYSTAL ON SILICON	No	CEA
1-4	Projection Screen (SCRN)		F – FIXED M – MOTORIZED R – REAR PD – PULL-DOWN PU – PULL-UP MLB – MOBILE	Yes	CEA
1-5	Video Camera (CAM)		D – DOCUMENT IP – IP CAM PT – PAN/TILT PTZ – PAN/TILT/ZOOM	No	CEA
1-6	Remote AV Source (R-AV)		A – AUDIO SOURCE AV – AUDIO VIDEO SOURCE V – VIDEO SOURCE	No	J-STD 710
1-7	White Board (WB)		A – ACTIVE I – INTERACTIVE OVLY – OVERLAY PAS – PASSIVE	Yes	J-STD 710
1-8	Microphone (MIC)		B – BOUNDARY CLP – CLIP GNK – GOOSENECK HH – HANDHELD SGN – SHOTGUN ST – STEREO WLS – WIRELESS	No	J-STD 710
1-9	Junction Box (JBOX)		AV – AUDIO VIDEO D – DATA J – JUNCTION BOX WP – WEATHER PROOF	No	NCS

Figure 3-14 Some AV object symbols from the ANSI-J-STD-710 standard

Chapter 3: Communicating Design Intent

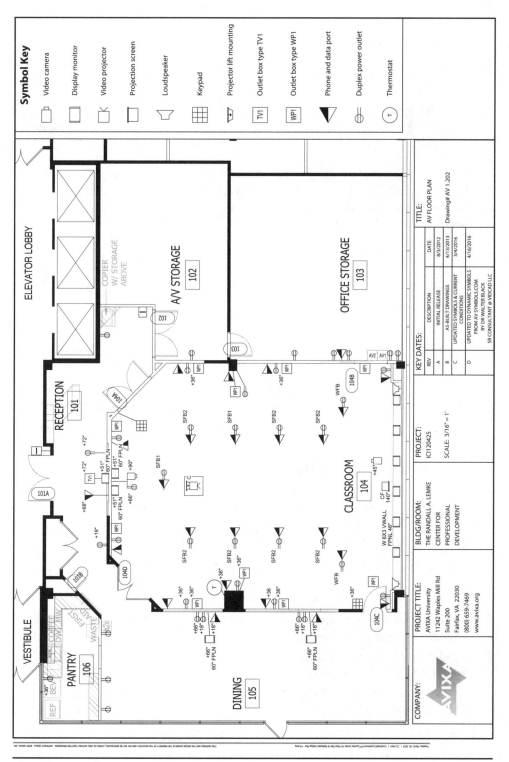

Figure 3-15 A floor plan drafted to the ANSI-J-STD-710 symbol standard

The AV Design Package

You can't design or install AV systems without considering other system designs. Because in the course of your work you will be communicating with members of the architecture, engineering, IT, electrical, and consultant teams, refer to the standards of all applicable organizations for guidelines on effective communication.

The *AV design package* represents a written method of communicating design intent to allied trades. It also conveys architectural details that may impact the AV-enabled environment.

Design documents must include enough information to convey the intent of the AV design. In addition to system design information, certain administrative contract language will form the basis of the actual installation contract. The design package is the basis of most—if not all—of the contract that an AV integrator signs when the project is awarded.

Major components of the design package include the following:

- Administrative front-end specifications
- Architectural and infrastructure drawings
- AV system drawings
- AV system specifications

 VIDEO There is a video about the components of an AV design package in the AVIXA online video library. Check Appendix D for the link to the AVIXA video library.

Front-End Documentation

A construction contract typically has an administrative front-end section. This section explains how the contractor and owner will interact by specifying what documentation, meetings, reports, and record drawings are required.

Front-end documentation usually includes the following:

- **Invitation to bid**   This is a cover letter that formally invites potential bidders to respond to the bid package. It includes contract information, dates, times, and schedules associated with the bid process.
- **Request for qualifications**   This is a statement that invites potential bidders to submit a "resumé" of their qualifications for the job.
- **Bid response form**   This organizes responses from potential bidders and indicates what information is required.
- **General conditions**   The architect for the base building contract provides the general conditions. Typical sections of the general conditions include the following:
    - Project work conditions and terms
    - Definitions of terminology

- Safety and accident prevention
- Permits, regulations, and taxes
- Environmental responsibilities
- Insurance
- Overall quality control
- Submittals
- Substitutions schedule
- Changes in the scope of work
- Initiating and processing change orders
- Warranty
- Nondiscrimination and affirmative action
- Contract termination options
- Arbitration
- Invoicing and payments
- **Money matters in the contract**   Within the general conditions—and sometimes within a consultant's specifications section—are the terms related to invoicing and payment.

Architectural and Infrastructure Drawings

Architectural drawings indicate where the systems will be installed, what infrastructure exists, and what may be required for AV systems. These indicate available power, signal raceways and conduits, data outlets, and structural accommodations for AV equipment.

Architectural drawings provide a technical illustration of all construction details, including the following:

- Site work
- Foundation
- Structure
- Power and lighting systems
- Plumbing
- Fire detection and sprinklers
- Telecommunications and data networks
- Audiovisual equipment
- Electrical
- Mechanical
- Finishes
- Details

3D Modeling and CAD Drawings

When working on an AV design, you will need to express your needs to various allied trades. For ease of communication, you should present your design in a format that will be compatible with documentation from those allied trades.

Design firms and allied trades use 3D or solid modeling in CAD systems to record, visualize, display, and document their designs. CAD systems allow multiple trades to communicate their needs and flag possible conflicts. For example, if a light fixture was placed in the lighting plan in the same place where a loudspeaker was placed in an AV plan, a quick comparison of each trade's CAD drawings would be the easiest way to spot the conflict.

3D modeling helps establish efficient communication. Changes to 3D-modeled documents are reflected quickly and accurately. In addition, 3D modeling allows engineers to explore a greater number of design iterations during product development. When utilized in conjunction with, or incorporated into, building information modeling (BIM) systems, 3D modeling allows for concurrent engineering, whereby engineering and manufacturing processes are enabled simultaneously through shared data.

3D modeling also leads to lower unit costs because of reduced development and prototype expenses. All of these advantages lead to a quality project.

Building Information Modeling

The term BIM describes a data repository for building design, construction, and maintenance information used by multiple trades on a single project. It includes CAD drawings and models, as well as information such as bid and contract documents, bills of materials (BOMs), timelines, specifications, price lists, installation and maintenance guides, cable lists, and cable label guidelines.

The ISO 19650:2019 standard defines BIM as the "[u]se of a shared digital representation of a built asset to facilitate design, construction, and operation processes to form a reliable basis for decisions." A BIM is seen as a shared knowledge resource for information about a facility, forming a reliable basis for decisions during its life cycle, defined as existing from earliest conception to demolition.

Many buildings use a wide variety of real-time data feeds into a live BIM system to monitor, maintain, operate, repair, secure, and control the operations of a building throughout its life. For example, energy consumption is monitored in real time throughout the day, with the BIM communicating with HVAC, lighting, AV, security, and IT systems to optimize energy use as areas change in occupancy and application.

As an AV designer, BIM may impact your design in several ways. Because many of the allied trades you will be working with—especially architects—will probably use BIM, you need to ensure that your documentation is compliant and compatible with theirs, making it important that you have a firm grasp of the fundamentals of BIM systems and principles. Early in the design, get in touch with your project architects and other allied trades. This will keep your design compatible and compliant from the beginning.

The systems you design may be required to communicate with—and possibly be controlled by—other building systems that form part of the BIM. You need to discuss the cost and technology implications of BIM integration on the AV systems and obtain full documentation on the communications protocols and interface technologies involved.

Detail Drawings: Custom or Integrated Carpentry Work

When designing a room that contains audiovisual equipment, be aware of the architectural elements that are commonly used to house such equipment. Two types of carpentry that fit that classification would be ready-made profiled and patterned materials such as moldings, architraves, trims, doors, etc., sometimes referred to as *millwork,* and built-in cabinetry and furniture such as cabinets, cupboards, credenzas, benches, and lecterns, sometimes called *casework.*

AV designers often design the carpentry detail for structures that hold a projection screen in place, for example, as shown in Figure 3-16. As you determine the locations of a screen or loudspeaker, you need to ensure that the architectural elements will allow them to be placed in exactly the right position.

There is a range of integration questions associated with the casework/cabinetry that an AV designer will need to help answer, including the following:

- What equipment is expected to be installed in the cabinetry?
- How large should the cabinetry components be?
- What connections are required to interface with the AV system?
- Is the cabinetry adequately sized for the equipment, including mounting, cable management, and serviceability?
- What are the electrical and signal connection requirements for the equipment to be installed in the cabinetry?
- What are the cooling and air flow requirements, and how will they be implemented?
- Does the cabinetry meet the necessary user-access requirements?

AV designers often partner with interior designers to create detailed drawings for the carpentry and cabinetry required to accommodate AV equipment. These drawings communicate to allied trade the exact dimensions, style, shape, color, finishes, ventilation, and so on that the AV systems require. Therefore, you should understand how to

Figure 3-16
Detail drawing showing cabinetry detail for a projection screen

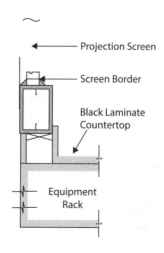

communicate these requirements to the architect. The level of detail required will vary depending on the project.

> ### Color Schemes
>
> AV designers should work with the interior designer and architect to help ensure that a room's visual design is compatible with the AV presentation needs. The interior designer will be responsible for the color and texture of the finishes. Your job is to remind them that contrast around an image is important. For example, visual design is important when considered from a video camera's point of view. This means verifying that a room is designed with complementary colors, minimal patterns, and no strong lines.
>
> Interior design coordination includes the following:
>
> - Locations for key AV components
> - Connection plate finishes
> - Connection plate mounting heights
> - The level of exposure or hidden integration of AV equipment
> - Shapes, finishes, and placement of furniture
>
> Coordinate color, finishes, materials, and patterns with the architect and interior designer to ensure compatibility with the overall interior design.

AV System Drawings: Facility Drawings

AV system drawings depict the AV system itself. These should reflect equipment configurations, interconnections, details, plate layouts, and other graphic depictions of the system installation. Typical components of the design drawings package include the following:

- Title and index
- Power, earthing/grounding, and signal wiring details
- Floor and reflected ceiling plans showing device locations
- System functional diagrams
- Rack and equipment cabinet elevations
- Custom plate and panel details
- Speaker-aiming information
- Lighting focus information
- Large-scale plans, such as equipment or control room plans

- Architectural elevations showing AV devices, their locations, and their relationships to other items on the walls
- Custom-enclosure or mounting details for devices, including, screens, displays, projectors, microphones, loudspeakers, and media players
- Furniture integration details
- Any special circumstances or details that may be required for the installers to understand the design intent

Architectural and facilities drawings are used to perform sightline studies and determine viewing areas, audience listening areas, camera angles of view, and so on.

AV designers use these drawings to coordinate and locate AV elements, including displays; screens; projectors; equipment racks; loudspeakers; specialized lighting; microphones; lecterns; equipment storage; raceways, ducts, and conduits for AV cabling; wall-, floor-, and table-boxes; connection plates; and electrical and data receptacles needed for AV equipment.

AV System Drawings: System Diagrams

For the AV system itself, AV designers create system diagrams. These system diagrams could be as simple as a one-line conceptual drawing that shows the system's intent but not the specific connections. More detailed system drawings show all specific connection points and other important system details. These drawings include the following:

- Video system flows
- Audio system flows
- Data network flows
- Control system flows
- Wireless coverage plans
- Rack elevations
- Connection plate details
- Digital signal processor settings

AV System Specifications

Specifications are like the project's manual in that they explain how AV systems should be installed and tested. They describe the system, components, codes, references, and other requirements. They also contain information concerning submittal requirements, shop drawing requirements, components, and system testing requirements.

VIDEO Now that you have studied all the components of design documentation, take a video tour of the AV design package in AVIXA's online video library. Check Appendix D for the link to the AVIXA video library.

The Basics of AV-Enabled Spaces

AV designers must take a big-picture view of a space to ensure that they understand exactly how it will be used. They also must be able to read plans and drawings to identify architectural details that may affect AV and communicate those issues to other professionals. For example, the size of the space and how people move around in it will affect design. Keep the following in mind:

- The form of the space should follow its function.
- The number of people in the space will help define the size.
- Circulation and work patterns within the space will help define the form.

All AV-enabled spaces have as their primary function the communication of ideas and information. This concept of a space's function should be at the center of all design decisions, and all other design characteristics, such as HVAC, lighting, and windows, should be integrated with that function.

Many AV-enabled spaces include audience, presenter, control, and display areas. You should also look at other rooms and areas (such as break areas, reception, etc.) to determine whether those spaces might affect the overall performance of the space you're designing. In short, work with the architect to ensure that the layout of the space and associated spaces is appropriate for the tasks that will take place there.

Audience and Presenter Areas

When planning the layout of a space, you need to consider both the audience and presenter. While there are some considerations unique to each area, many are common to both. The following are some of the commonalities:

- **Equipment and materials**   Is there enough room for presentation equipment, people's personal devices, books, and writing materials?
- **Connection points**   Where are the AC power, network connections (wired or wireless), and display connections required?
- **Comfort**   What furniture will the space need? What level of comfort is required to remain seated for the duration of the typical session? Consider ease of movement for both access and egress.
- **Sightlines**   Will everyone be able to adequately see the presenter and/or materials being presented?
- **Audio coverage**   Will everyone be able to clearly hear the content of the session?

When it comes specifically to the presenter area, there are several design considerations you need to address. The presenter area refers to the designated location in a space where presenters (teachers, performers, speakers, etc.) will operate. The area typically includes a lectern, microphones, and a control system interface. The following are some of the design issues to consider in the presentation area:

- **Presenter workstation** Is the lectern/console/control interface well laid out for the range of presentation needs identified by the client?
- **Presenter equipment** Is there enough room for the presenter and the presenter's equipment, such as notebook computers, tablets, autocues, or presentation notes? Also keep in mind that a presenter may want to move around during a presentation and not be anchored to the lectern.
- **Power, voice, and data** Are the power, data, communications, video, and computer connections in front of the space adequate to meet the needs of presenters?
- **Sightlines** Specific to the presenter area, can the audience see both the presenter and the displayed images? For example, if a presenter is too close to the screen, it may be necessary to control the amount of the lighting on the presenter because ambient light may interfere with the image. And if that's the case, cutting back on light may make it difficult to see the presenter.
- **Versatility** Have you met *all* the needs of the client? For example, the client may not want a lectern permanently attached to a specific location. Will other formats of presentation or performance be conducted in the space? They may intend to move the presenter area to different locations when changing the space orientation to accommodate different audiences. Can they?

Looking out from the presenter area is the audience area—where people sit to take in a presentation. The following are some of the design issues to consider in the audience area:

- **Visuals** Can everyone in the audience see the presenter and/or the presentation? Consider screen size, image resolution, screen-to-presenter contrast ratios, task requirements, sightlines, and more.
- **Sound** Can everyone in the audience hear the presentation with sufficient intelligibility? Consider the acoustics of the space and determine the appropriate configuration of microphones, loudspeakers, and hearing assistance facilities.
- **Ease of movement** Can everyone in the audience enter and exit with ease? Is there sufficient space to meet emergency egress requirements? A properly designed seating layout is essential.
- **Audience comfort** Will people be comfortable in their seats? Ensure that there is sufficient room between the seats and in the aisles for the required number of persons, given the desired audience seating layout. And be sure to consider the needs of people with disabilities, in compliance with the local authority having jurisdiction.

Control and Projection Areas

Depending on the size and scope of a space, the AV system may require a dedicated control or projection room. Think of the control/projection room at the back of an auditorium, for example. The design issues associated with a control or projection area are usually technical in nature and include the following:

- **Equipment space**   In some cases, AV and other technical operators may be required to support a presentation. Is there enough room for both operators and all the required equipment? If operators are required to remain within the control area for long periods, then the room should be designed to provide an adequate level of comfort.

- **Heating and cooling**   Can the heat generated by the equipment and the operators be removed from the space? Coordinate the design requirements with the architect and HVAC designer to ensure adequate ventilation.

- **Power requirements**   Is there sufficient power for the required equipment? Allowance should be made for ancillary equipment such as media servers, cameras, effects equipment, and mixing consoles that may be brought in to support presentations.

- **Voice and data**   Does the area include appropriate communications and data capacity? Control rooms may require network feeds for Internet and AV streaming and wired communications with the presentation area and other technical areas. Consider the possibility that wireless communications (including cellular/mobile telephone networks, Wi-Fi, Bluetooth, DECT cordless phone systems, and handheld transceivers) may not be effective due to the hostile radio frequency environment in an equipment area.

- **Task lighting**   There needs to be enough lighting for operators to see system controls without adversely affecting the quality of the presentation.

- **Sound isolation**   Noise from the area, such as cooling fans, monitoring, and technician communications, must be controlled and minimized so that it does not distract the audience. If the audio is being mixed in the control room, the audio operator usually needs to be able to hear the same audio as the audience. In spaces that require several technical operators, it may be necessary for the control area to have both an acoustically open area for the audio operator and an acoustically isolated area for other technical operators, such as lighting, video, effects, staging, cameras, and production coordination.

- **Monitoring**   Operators must be able to clearly see and hear what's going on in the main space to support the presentation or other activity. They shouldn't be "tucked away in a closet," as sometimes happens. It may be necessary to provide feeds of closed-circuit audio and video of the space to enable remote monitoring for all involved in technical, audience management, security, and production operations.

NOTE There are other spaces that support an AV-enabled space, its occupants, and the activities in a building. These include such spaces as lounges, breakout rooms, reception areas, bars, cloak rooms, break rooms, self-catering areas, and restrooms/toilets. Often, the design of these associated spaces can impact the success of the spaces an AV designer creates. For example, a reception desk adjacent to the presenter area of a space can be a source of noise that may disrupt the presentations.

Programming

In AV and IT fields, the term *programming* often refers to software coding of programmable devices, such as computers, network devices, control systems, and signal processing equipment. But for architects programming is the process by which the overall requirements for the building are defined. Architects document the needs of the owner and end user as a preliminary step to putting lines on paper or CAD that represent the space plan. The architectural program document discusses the facility in terms of the size of areas, space configuration, and overall quality of the building.

Programming an AV design begins with careful planning and time spent identifying and discussing the end user's needs. Therefore, let's take an in-depth look at this crucial step of the AV design process.

The Needs Analysis

A needs analysis requires identifying the activities that end users need to perform using AV and developing the functional descriptions of the systems that support those needs. Conducting a needs analysis is the most critical stage of the design process. It determines the nature of the systems, their infrastructure, and the project's budget.

When you ask a client and the client's end users why they need an AV system, they may not always know. However, they do know what tasks they need to accomplish to do their jobs. This is where an AV design educates clients and helps fill in the missing information, making connections between jobs, tasks, and the capabilities of an AV system.

Figure 3-17 shows the needs analysis pyramid, a visual representation of how needs relate to an AV system's functionality, through the applications they enable and the tasks they help accomplish.

But first, it's important to understand who we're talking about when we talk about your client's end users and why you need their input on a project.

Who Are the End Users?

The client is the person paying for the AV system. It might be a business owner, a chief technical officer, or a facilities manager.

End users are people who operate a system day to day. They can be technical end users who know a lot about the system, such as a technology manager, AV technical operator, or production supervisor. Or they can be nontechnical people who need the system to

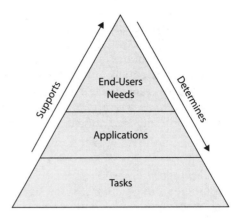

Figure 3-17 The needs analysis pyramid

work at the push of a button, such as a teacher, museum manager, or conference system user. You need to understand the needs of both the person holding the purse strings (the client) and the person using the system (the end user).

End-user input is essential when designing an AV system. The fact is, other client representatives may not really understand how a space and its AV system will be used. Ignore end users at your peril—you're certain to hear from them if they're dissatisfied with the final product.

Asking end users probing questions helps fill in the blanks when determining what they really need. Ask not only what they plan on doing in a space but also how they plan to do it. And depending on who you're talking to, work out who can get you what you need. For example, ask for scale drawings of the space and its surrounding infrastructure. Get a feel for estimated or fixed budgets. And establish whether there will be any constraints placed on the AV team by the client, such as work hours during the project (i.e., do you have to install systems on weekends?), acceptable ambient noise levels, or what else will be going on in the building at the same time.

Clients on a Project

The actual who's who of a project team will depend on the size and nature of the project. An AV designer has other project clients, too, including the project manager (if not the architect or a member of the organization for whom the project is being built) and other design specialists.

It is important to identify any differences in approach and opinion among end users, decision makers, funding groups, support groups, and even external contractors, such as project and design team members. The difference between a good and unsatisfactory project outcome can be in the AV designer's ability to recognize and manage varying expectations.

Some end users may be low in an organization's hierarchy but have the best knowledge of what an AV system should achieve. Others, with more authority but less understanding, may have different expectations or place timing and budget constraints on the project. The AV designer must find ways to communicate with, educate, and understand the perspectives of each stakeholder.

For example, at times, end users will have unrealistically high expectations and ask for functionality (equipment and features) that is not justified by their needs or that are not within the project budget. The AV designer has to know when to support end users and argue for better funding and when to recognize that a budget is realistic and help end users appreciate that some of the things they are asking for are "wish list" items.

The following sections discuss some of the key players you may encounter as you work on your AV design.

The Owner Team

Depending on the type and size of the owner organization, participants may consist of a small number of people or a large group. An owner can be a corporation, a government body, or an individual. In some cases, the owner of a building is fitting out a space to be leased to another company that is the actual purchaser of the AV systems. Some owner configurations are more complicated, such as investor/builder consortiums, conference centers, and some university systems, and some are simple, as in the case of a small private company, such as a law firm or a small manufacturer.

The owner typically has its own representatives in the building design process. Representatives range from nontechnical administrative employees to chief executive officers (CEOs) and from facilities managers to in-house architects if the client owns or leases a lot of properties. Sometimes a team or committee represents the owner. The owner team can include one or many of the following types of people.

End Users

As noted earlier, end users are the people who will ultimately use the AV systems and the spaces where they will be installed. These are the producers, presenters, and meeting participants who use AV systems for their intended purpose: information communication. These may be teachers, trainers, students, producers, performers, salespeople, CEOs, or anyone who participates in an event that utilizes an AV system. It is important to note that in many cases end users may not be employed by the owner's organization.

Facility Managers

Larger organizations have a facility manager heading up a department that covers everything from janitorial services to renovation work and from planning to operations. Facility managers are concerned with physical facility standards (standard finishes, cabling schemes, and electrical/mechanical services) and construction schedules from the owner's operational perspective. Their areas of AV concern may include space allocation, cabling standards, and installation schedules because large AV installations often overlap with occupancy periods, after a building contractor has finished its own work.

AV Manager or Technology Manager

Many organizations have someone on staff who manages the current AV systems, schedules usage, and maintains the AV spaces. Most are familiar with the IT network and how the AV systems relate to it; some are also responsible for the IT systems. The AV manager usually coordinates training for end users once the system has been installed. They may also act as the owner's representative or point of contact throughout the AV project.

Smaller organizations may not have a dedicated AV position. In that case, responsibility often rests with someone who has technical experience or expertise, often in the IT department.

Buyer, Purchasing Agent, or Contract Representative

Medium to large organizations often have a department or individual whose role is to manage contracts, select vendors, and establish or negotiate contract pricing and terms. These personnel may be involved in the construction document preparation (particularly the administrative portions); the bidding and contract negotiation processes for designers and installers; and the administration of billing, paying, and contract close-out tasks.

Building by Committee

Sometimes, a project involves several divisions or departments of an organization, including a large number of administrative and technical stakeholders. In this case, the owner may create an internal committee to steer the process of building a new facility. This type of committee should have representation from the technical staff, which can provide valuable input to the design process.

Developing the Concept Design/Program Report

Now you know who is involved in the needs analysis. Next you'll talk to stakeholders, review existing documentation, evaluate the site environment, conduct program meetings, and create the concept design/program report. That's a lot, but each step is important. We'll break them down.

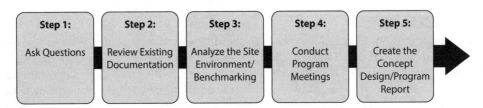

Step 1: Ask Questions

First, you'll talk to as many stakeholders as feasible. Knowing that there are many ways clients can use AV, how can you find out what they need so you recommend just the right solution? Just ask. But ask in the right way.

When you create a good dialogue with a client and a client's end users, everyone will better understand the project's goals. Good dialogue is also friendlier and less threatening.

Questions foster dialogue. Questions also help gather important information, get the client involved, and persuade more effectively (and softly) than statements. Use them often and intelligently. When asking questions, employ the following:

- Open-ended questions
- Closed-ended questions
- Directive questions

When you want the client to give a full, informative response—or just start talking—ask open-ended questions that are hard to answer in a word or two. Here are a few examples:

- How do people in your organization like to present content in meetings?
- What kinds of meetings are held in this facility?
- How would you describe the current effectiveness of your training?
- What would you like to see be done differently or better?

Answers to questions like these tell you how your client thinks, how much they know, and what they consider important. If you need factual or specific information, ask closed-ended questions, such as the following:

- How many people will the space need to seat?
- Are there any windows or columns in that space?
- When do you expect construction to start?

When it's time to bring a client to your point of view or persuade them that one of your ideas will work well, ask directive questions. These are questions that suggest their own answer—sometimes the one you want to hear—such as the following:

- Can you see the added value of a maintenance agreement to protect your equipment investment?
- Wouldn't it be great if everyone knew how to take full advantage of this space?
- Would you get more use out of a flat-panel display or a projector and screen?

Step 2: Review Existing Documentation

Once you've talked to key stakeholders, your next step is to gather and review the available documentation. This documentation is usually available with information about the existing physical, organizational, and technical aspects of the project.

Ask the client for the existing documentation. It may include items such as scale and engineering drawings (hardcopy or CAD drawings), architectural program documents, organizational project directories, design manuals, standards, best practices, and other owner and end-user information. Ensure that you have a full set of drawings for the spaces your AV design will address. Be sure to examine all elements that might potentially affect the layout, mounting, installation, and operation of AV system components. If no documentation is available, you may need to create new material based on existing conditions.

At this point, it's also important to ask the client about budget. Providing a fixed budget and setting expectations will go a long way toward addressing concerns the client may have. Be sure to ask which part of the budget is for AV and which part is for other contractors' work. The customer may not be aware that other trades share in the project.

Step 3: Evaluate the Site Environment/Benchmarking

If the AV systems are to be installed in an existing facility, it is important to tour those areas during the needs analysis process to gather information about physical characteristics and how the spaces are currently being used. It is important to learn about any issues that may impact your ability to work at the client site once the design and installation tasks begin, such as building security and access.

In many cases, the design and installation teams can identify a method to work around constraints, such as working during certain hours or in areas where they are less likely to disrupt operations.

The following sample checklist illustrates the type of general site information you should collect.

Onsite Survey Checklist

- ☐ Site contact name _____
- ☐ Site contact phone number _____
- ☐ Site contact mobile number _____
- ☐ Exact address of job site _____
- ☐ Best route to job site _____
- ☐ Travel time from your base to the site, taking into account normal traffic conditions for your scheduled trip _____
- ☐ Type of loading/unloading/parking access (hour or time restrictions) _____
- ☐ Location of loading/unloading access _____
- ☐ Access route from delivery dock to storage area _____
- ☐ Elevator dimensions _____
- ☐ Security concerns (such as whether the area can be locked and who has a key) _____
- ☐ The primary function of the work site space(s) _____

- ☐ The space's proximity to other functions in the same building and area
  _____
  _____

- ☐ Name and contact number of facility owner's representative, site's maintenance chief, or AV technician (who will know where power connections and ducts are)
  _____
  _____

- ☐ Potential for electrical and radio frequency interference with other equipment
  _____
  _____

- ☐ Potential for any other problems regarding space location (for example, ambient noise from outside the space) _____
  _____
  _____

- ☐ Potential for any problems regarding traffic patterns during installation
  _____
  _____

- ☐ Dimensions of the space(s): ceiling height, space length, and width
  _____

- ☐ Ceiling type (drywall/plaster sheet, suspended ceiling, location of joists)
  _____

- ☐ Wall material (drywall/plaster sheet, masonry, block, etc.)_____
- ☐ Ambient noise measurement (using a meter)_____

- ☐ Acoustical properties of the space (echoes, loud mechanical noise, outside noise, voices or sounds from adjacent space, etc.)
  _____
  _____

- ☐ Existing sound system (if there is one and, if so, what kind is it?)
  _____

- ☐ Ambient light measurement (using a meter)
  _____

- ☐ Natural light from windows (can it be masked if necessary?)
  _____

- ☐ Existing lighting _____
- ☐ Existing security lights that might make lighting difficult
  _____

- ☐ Electrical capacity of the space _____
- ☐ Location(s) of electrical distribution panel(s) serving the space _____
- ☐ Existing AV features or equipment _____

(continued)

- ☐ Existing IT network and location of network ports in the space

- ☐ Location of network data center or server/equipment space

- ☐ Possible obstructions to audience view (such as chandeliers, sliding walls that are not completely retractable, seating pitch, and pillars)

- ☐ Suitability of space(s) to accommodate the audience size and the type of equipment being considered

- ☐ Seating capacity of space(s) according to requested setup

- ☐ Space shape(s) and orientation of the requested setup

It is a good idea to take extensive pictures of the site (including panoramic views) for future reference in case questions come up later in the project. Digital "film" is virtually free, so don't be afraid to waste it. You can always delete the images you don't need. If you don't have the plans for the space you're surveying, a few minutes spent with a laser measuring device can quickly give you some dimensions to work with until the actual plans are available.

You should also consider visiting other similar facilities for review and comparison. This is known as *benchmarking* and gives the owner and the design team a common (and sometimes expanded) vision of what the user wants and needs. Seeing a number of locations of similar size, type, and usage establishes a benchmark or guide on which to base the new facility design.

Benchmarking offers the following benefits:

- It provides an opportunity to see varying approaches to design versus budget.
- It may inspire new design ideas.
- The team can identify successful (and unsuccessful) designs and installations similar to the project at hand.
- It can help determine which functions and designs are most applicable to the current project.
- It allows stakeholders to establish a communication path with other building managers and end users about what they learned in the design and construction process and to discuss what they would do the same or differently if they needed to do it over again.

Benchmarking can't occur on all projects. Sometimes you can save the time and money required to visit similar sites by letting the client look through a portfolio of your firm's work on similar projects or by asking them to provide you with pictures of projects they've seen and liked.

Step 4: Conduct Program Meetings

Once you've collected feedback from stakeholders, reviewed existing documentation, and evaluated the environment, you can schedule program meetings.

Program meetings should include representatives of both the design team and the owner. During these meetings, the architect, AV professionals, and other design team members discover more about the end users' needs by examining required applications and the tasks and functions that support those applications. In other words, you've learned what they want; now you're discussing in-depth how they'll get it. If you don't know who you should be meeting with, go back to the "Clients on a Project" section.

As you already understand, one or more people will use the AV system you design. Others may manage or maintain it, and still others may pay for it. So, it follows that some will care about functionality, others about upkeep, and others about cost. A safe way to find out how decisions will be made during program meetings is to ask your contact, "Who, in addition to you, will be involved in this decision?" Words are important: You don't want to suggest that your contact has no authority.

A good follow-up question might be, "What is their primary interest or concern?" This will help you target your message to each stakeholder based on what they care about most. Yes, they all care about the whole project, but they also have their own areas of focus: operations, security, maintenance, financing, and so on.

Here's another good question for a program meeting: "How far along are you in your budgeting process?" This asks in a nonthreatening way whether the client has a realistic budget in place or if they're able to make the investment a certain design decision may require. It will also help you learn more about the client's planning and decision-making processes.

AV solutions are often—although not always—scalable to a client's needs and budget. But it's unsafe to guess about expectations because you will usually guess wrong. Not only does this waste your time and theirs, it can create hard feelings and damage relationships. The basic rule is never make a guess about what you can find out for certain with careful questioning.

TIP In each meeting with your client, make it a point to learn who else has an interest in AV and look for opportunities to meet them. Never go around your contacts or over their heads. Offer to arrange meetings and introductions between your colleagues and theirs. Naturally, each meeting needs a purpose; it can't only be something nice to do.

Step 5: Create the Concept Design/Program Report

At the conclusion of the program meetings, information is captured in a written report of the findings, including an interpretation of the users' needs with respect to the AV systems. The report should include a conceptual system description, along with necessary information about its effect on spaces that have already been programmed, designed, or built.

The concept design/program report is a functional description of the system that your design team intends to deliver. It functions as your project's scope of work, in that it contains the milestones, reports, deliverables, and end products that you agree to deliver to the project customer. It should also lay out change order, escalation, sign-off, and payment procedures. Many designers will also include a list of equipment in the concept design/program report because it helps with the cost estimation, though a concept design/program report should contain much more information than a simple bill of materials.

The objectives of the concept design/program report are

- Communicate to decision makers about the overall systems and budget
- Communicate to users the system configurations that would serve their needs as identified during the program meetings
- Communicate to the design team a general description of the AV systems and what impact they may have on other trades
- Communicate the scope and functionality of the AV systems to be designed and installed

NOTE A scope of work can describe many elements of an AV project, including aspects of the design, such as the concept design/program report. The scope of work is often the most comprehensive description of the project and may include any or all of the following: information about all contractors, a description of the solution (with drawings and products), key assumptions, pricing, project milestones, responsibilities, project management procedures, warranty information, and terms and conditions.

A Closer Look at the Concept Design/Program Report

This nontechnical document describes the client's needs, the AV system's purpose and functionality, and your best estimate of probable cost, and it ultimately documents the client's approval.

NOTE The concept design/program report may also be referred to as the AV narrative, the discovery phase report, or the return brief.

The contents of a concept design/program report generally include the items discussed in the following sections.

Executive Summary

This section provides an overview of the project, the programming process, systems, special issues, and overall budget. Keep in mind that the report may be read by a wide range of stakeholders: the chief executive officer (CEO), chief information officer (CIO), chief technology officer (CTO), architect, AV manager, IT manager, end users, electrical engineer, mechanical engineer, facilities department manager, and so on. It should briefly highlight the findings from the various interviews you conducted, as well as the systems you will design.

The executive summary should include

- An overview of the project (what the facility is; how it serves the owner; and why it is being built, upgraded, or renovated)
- An overview of the programming process (who was interviewed at the departmental or organizational level and other individuals who were important during the information-gathering step)
- An overview of the systems (what and where the systems are, their type and quality, and which users they serve)
- An overview of any special issues
- A reference to the overall budget required for the systems

The client should be made aware of any special issues related to the proposed AV system design, installation, or operation. Examples include major project obstacles or limitations, project schedule issues, options for specific spaces, or overall system configuration options.

Space Planning

This section provides advice to the design team, where necessary, about the space requirements of AV systems. This may include equipment closets, projection rooms, observation rooms, dressing rooms, internal room layouts, room adjacencies, and so on.

System Descriptions

This section is a nontechnical description of the client's desired functionality for each system. It may include AV sketches, drawings, diagrams, photos, product data, and other graphics, such as touch-panel interfaces, to illustrate the capabilities of the proposed systems.

Depending on the project size, the number of systems, and the different types of users, systems descriptions may be organized by space, user, system type, or a combination of the three. The nature of the project will determine how the system descriptions should best be presented. For example, some systems may be specific to a single space, or there may be systems that are in several similar spaces that span a number of departments. There may also be facility-wide or BIM systems that connect some or all of the AV systems together or that provide overall control, monitoring, interconnection, or help-desk functions to a building or campus.

Infrastructure Considerations

This section describes the electrical, voice, communications, data, mechanical, lighting, acoustic, structural, and architectural infrastructure needed to support the AV systems. The documentation identifies site environment issues that will impact the system design and installation and any recommended site modifications that may be necessary to facilitate optimum system operation.

AV systems often have a significant impact on architectural, mechanical, and electrical systems, so awareness of where the impact may occur is critical to a complete understanding of the project. The following areas should be addressed:

- Lighting
- Electrical
- Mechanical (both noise and heat)
- Acoustical
- Data/telecommunications
- Security
- Networking
- Structural
- Architectural (space plans, adjacencies, allocations, and other architectural issues)
- Interiors (finish requirements)
- Coordination of other trades with AV installation needs
- The budget implications of these issues

Note that a general discussion of infrastructure budgets may be included in the infrastructure section. For example, if AV applications and systems will require bandwidth on the IT network, it may be necessary to add bandwidth capacity. But there may be more space- or trade-specific issues that require additional discussion or illumination. If the section on infrastructure considerations is running long, you can address them in a section about special issues.

Budget Recommendations

This section outlines the probable costs to procure, install, and commission the proposed AV systems, as well as any additional costs, such as tax, "builders work in connection" (BWIC), markups, service, support, and contingencies. The concept design/program report should include an opinion or estimate of how much the AV system will cost. As you form your opinion, pay attention to the following budget terms. Using them incorrectly may lead to confusion and financial consequences later in the project.

Opinion of Probable Cost

This term describes an early attempt to determine the cost of a system before there is enough detailed design to produce a line-item estimate. The opinion of probable cost is an "educated guess" based on experience and some line-item costs for large equipment, such as video walls, projectors, or large-matrix switchers. Final costs cannot be applied until the system is designed and the actual equipment selected.

Estimate

An estimate implies that there is a more objective basis for the cost provided. It is an "approximate calculation" that includes a line-item analysis of equipment and labor (perhaps including taxes and other ancillary costs); it would be more accurate than an opinion of probable cost.

Quote

A quote is a detailed and enforceable estimate. It should be provided and identified as such only if it's based on a concept design/program report and an AV system design. If it isn't, the client, the end users, and the AV provider are at risk of a painful mismatch of needs, capabilities, and cost.

Budget

Although the term *budget* is often used in the context of terms described earlier, by strict definition it applies only to what the owner or project team has allocated for a particular system, trade, or facility. In the correct relationship of these terms, the budget should be established based on an opinion of probable cost or an estimate. A quote is then submitted by a provider based on a request for proposal (RFP). The quote is subsequently compared to the budget before acceptance by the owner.

NOTE Your concept design/program report will vary from project to project. There is no single correct way to order the information, but using the list of items outlined in the ANSI/AVIXA D401.01:202X standard as a table of contents is a reasonable starting point.

Distribution and Approval

A concept design/program report may contain sensitive information. It should be distributed on a need-to-know basis and never to anyone outside the project team without permission from its writer and the client (or authorized representative). Distribution to unauthorized parties could reveal confidential information about the client, project, or other details. In particular, it should not be distributed to individuals and organizations with connections to the AV industry. Doing so could undermine the competitive bid process or a project where an integrator is not already engaged. Apart from leaked cost information, which could potentially affect the owner, unauthorized distribution of the report opens a door to suppliers and others inappropriately lobbying project team members and/or manipulating prices.

VIDEO You can view a video about writing a concept design/program in the AVIXA online video library. Check Appendix D for the link to the AVIXA video library.

Assuming you distribute the concept design/program report to the appropriate parties, you need to establish a set period of time for comments and responses. At its conclusion, any necessary revisions should be made and the report either redistributed or submitted for approval if only minor changes were required. Depending on the nature of the feedback, more program meetings may be required to address any major issues or deficiencies.

Once the concept design/program report is signed and approved, it can be used as the basis for the design of the AV systems and their supporting infrastructure. In this formalized, approved state, it protects the owner, consultant, and designer by ensuring everyone defines the project in the same, unambiguous way.

Chapter Review

In this chapter, you learned about the phases of an AV design project, including the program, design, construction, and verification phases. You also learned about what goes into an AV design package, including drawings and diagrams. AV designers must be able to read and interpret such drawings and diagrams to collaborate with other professionals on an AV project.

Perhaps the most important part of the AV design is the needs analysis, which occupies a significant chunk of the program phases. Being able to discern the various stakeholders on a project and elicit information about their needs and wants is critical to designing the best possible AV system to meet those needs and ensuring it delivers as promised. All the documentation that comes out of the program phase will serve to guide the project to its successful completion.

Review Questions

The following questions are based on the content covered in this chapter and are intended to help reinforce the knowledge you have assimilated. These questions are similar to the questions presented on the CTS-D exam. See Appendix E for more information on how to access the free online sample questions.

1. What is the primary goal of a needs assessment?
    A. Identifying the intended use of a space
    B. Selecting the right equipment
    C. Establishing the "first use" date
    D. Meeting the budget requirements

2. The total length of a measurement from a metric (SI) scaled drawing of 1:50 is 100mm. What is the actual length?
    A. 5 meters
    B. 50 meters
    C. 0.5 meters
    D. 150 meters

3. Which of the following architectural drawing symbols is a detail flag?

A.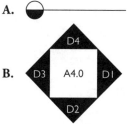

B.

C.

D.

4. The total length of a measurement from a U.S. customary scaled drawing of 1/8 inch is 3.5 inches. What is the length of the real object?
   A. 18 feet
   B. 25 feet
   C. 28 feet
   D. 35 feet

5. What type of drawing would you use to find out exactly how a projector should be mounted to the ceiling?
   A. Reflected ceiling plan
   B. Mechanical drawing
   C. Detail drawing
   D. Section drawing

6. Which abbreviation on a drawing tells you that you will not need to install a specific display screen?
   A. OFOI
   B. OC
   C. OD
   D. NTS

7. Technical end users are sometimes referred to as:
   A. Teachers
   B. Technology managers
   C. Clients
   D. Architects

8. Which of the following is *not* a step in conducting a needs analysis?

    A. Review existing documentation

    B. Evaluate the site environment

    C. Conduct program meetings

    D. Educate allied trades

9. Benchmarking is a process by which AV designers and clients:

    A. Evaluate the performance of AV systems

    B. List specifications required on AV equipment

    C. Examine the AV designs of other facilities

    D. Compare expected results to actual results

10. Which of the following would not be considered part of the system drawings for an AV design?

    A. DSP settings

    B. Rack elevations

    C. Control system flows

    D. Color schemes

Answers

1. **A.** The primary goal of a needs assessment is to identify the intended use of a space.
2. **A.** When the total length of a measurement from a metric (SI) scaled drawing of 1:50 is 100mm, the actual length is 5 meters.
3. **C.** Figure C shows a detail flag.
4. **C.** When the total length of a measurement from a U.S. customary scaled drawing of 1/8 inch is 3.5 inches, the length of the real object is 28 feet (336 inches).
5. **C.** You would use a detail drawing to find out exactly how a projector should be mounted to the ceiling.
6. **A.** OFOI is an abbreviation for "owner furnished, owner installed." If found on a drawing it would indicate that you do not need to supply or install that specific item of equipment.
7. **B.** Technical end users are sometimes referred to as technology managers.
8. **D.** Educating allied trades is not part of the needs analysis.
9. **C.** Benchmarking is a process whereby AV designers and clients examine the AV designs of other facilities.
10. **D.** Color schemes would not be considered part of the system drawings for an AV design.

Ergonomics in AV Design

CHAPTER 4

In this chapter, you will learn about
- What ergonomics means in the context of AV design
- Identifying the limitations of a viewer's visual field
- Creating optimal sightlines based on the seating layout of a space
- Selecting the furniture for an AV design project

AV systems are tools for improving human communication; therefore, the human element—and specifically, human comfort—is an important consideration in AV system design.

Ergonomics, also known as *human-factors engineering,* is the scientific study of the way people interact with a system. The purpose of ergonomics is to limit injury, fatigue, and discomfort in a work environment. Factored properly into an AV design, ergonomics can help the users of AV systems enjoy greater effectiveness, efficiency, productivity, comfort, and safety, while at the same time reducing the errors, fatigue, frustration, and stress that could otherwise come from interacting with technology.

> **Duty Check**
>
> This chapter relates directly to the following tasks on the CTS-D Exam Content Outline:
>
> - Duty B, Task 2: Coordinate with Architectural/Interior Design Professionals
> - Duty B, Task 4: Coordinate with Structural Professionals
> - Duty C, Task 1: Create Draft AV Design
>
> In addition to skills related to these tasks, this chapter may also relate to other tasks.

AV designers should strive to ensure that a system follows design and ergonomic principles. Most jurisdictions also have accessibility requirements that will need to be addressed. For example, consider eye and head levels when conducting sightline studies, and factor in reach distances when deciding where to place a control or input panel. The goal is to create an excellent experience that takes into account users, technology, content, and the space. Achieving that end requires the proper design and layout of physical equipment, displays, and system interfaces.

Human Dimensions and Visual Field

Because so much of the audiovisual experience today is about what people see, it's important that the video portion of an AV design create the proper experience. (You will learn about the human perception of sound and subjects such as loudspeaker coverage in Chapter 6.)

When you design a visual system, the screens or displays need to be placed in locations that are comfortable for viewers. This means accounting for visual ergonomics and ensuring that the displayed images sit within what is called the *visual field*.

The visual field is the volume of space that can be seen when a person's head and eyes are absolutely still. It is specified as an angle, usually in degrees. The visual field of a single eye is called *monocular vision,* and the combined visual field encompassing the overlapping images from both eyes is called *binocular vision*.

Each human eye, when looking straight ahead, can see comfortably about 15 degrees side to side without any head movement, as shown in Figure 4-1. This range is the optimum field of vision, meaning that when you design your visual system, the image should fall comfortably within this range so that audience members can easily experience it.

Of course, many variables go into designing a visual system for optimal viewing. Chief among them is the content that the system will display. Let's drill down further into the visual field so you understand more about how viewers experience your systems.

 VIDEO To give you a better understanding of ergonomics for AV design, watch Ergonomics Parts 1 and 2 in AVIXA's online video library. Check Appendix D for the link to the AVIXA video library.

Figure 4-1
Optimum eye rotation

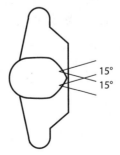

The Horizontal Visual Field

The human visual field spans both the horizontal and vertical planes. Let's first explore the visual field along the horizontal plane (side to side).

The horizontal (zero-degree) line is the standard line of sight—what is directly in front of a person. This is the area of sharpest focus and where depth perception occurs. A person's total field of vision—without rotating their head—typically extends 60 degrees from the line of sight in either direction. That doesn't mean you have 60 degrees of visual field to work with. Depending on the content that the visual system will display, there are limits to what viewers can recognize or comprehend within the field of vision (see Figure 4-2). The following list is a simple set of guidelines:

- People can recognize words, fine details, and color information up to about 10 to 20 degrees from the standard line of sight. This is the limit of word recognition. If the main content will include words, the visual information should be placed within this range.
- Depending on size, shape, and detail, people can identify symbols between 5 and 30 degrees from the standard line of sight. This is the limit of symbol, or picture (image), recognition. For graphical displays of information, you have a slightly wider visual field to work with.
- The limits of color discrimination are between 30 and 60 degrees from the standard line of site, depending on the color. While a viewer may be able to see color information in this area, they are not able to discern symbols or words across this entire field. There may be applications, such as digital signage, where wide-angled colorful images can be used to attract the attention of viewers who are then able to see the important words or symbol information in the central area of the display.

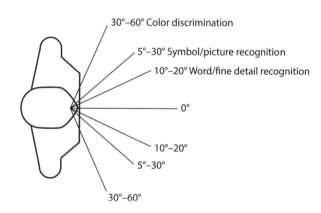

Figure 4-2
Limits to the visual field in the horizontal plane

The Vertical Visual Field

From a vertical perspective (up and down), the standard line of sight is straight ahead (zero vertical degrees). However, the *natural* line of human sight is not perfectly perpendicular to the body, but actually angles downward by approximately 10 degrees below the vertical when standing and about 15 degrees when sitting. When a viewer is relaxed or lounging, the natural line of sight may drift even farther downward. It's just more comfortable for people to rotate their eyes downward than upward.

- Optimal eye rotation in the vertical plane is 30 degrees below the standard line of sight. The maximum eye rotation above the standard line of sight is 25 degrees. It's more comfortable to cast a gaze downward.
- The limit of color discrimination is between 40 degrees below the standard line of sight and 30 degrees above the standard line of sight.
- The limit of the total visual field in the vertical plane is 70 degrees below the standard line of sight and 50 degrees above it.

The important thing to remember is that people view objects most comfortably in the lower visual field. Therefore, placing a screen more than 25 degrees above the standard line of sight would be tiring and uncomfortable for viewers (see Figure 4-3).

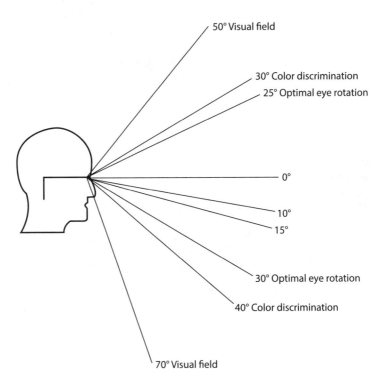

Figure 4-3
Limits to the visual field in the vertical plane

Head Rotation

As you've no doubt noticed, the earlier descriptions of visual field assumed, quite erroneously, that a viewer keeps their head perfectly level and still. As human necks have a range of motion, the real-world viewable area is affected by the range of eye *and* head movement. For the most comfortable and ergonomically designed visual system, you should not assume a wider visual field based on neck movement. Think of the front rows of a traditional movie theater. Yes, people could tilt their heads upward for the entire length of the feature, but is it comfortable?

That said, it's important to know how much rotation may be reasonable. In the horizontal plane the average human head can move 55 degrees from side to side, but *comfortable* head movement is more like 45 degrees in either direction (see Figure 4-4).

Along the vertical plane, the average human head can move within the 50- to 40-degree area in Figure 4-5. Comfortable head movement, however, falls between the 30-degree lines (remember the movie theater).

Figure 4-4
Head rotation in the horizontal plane

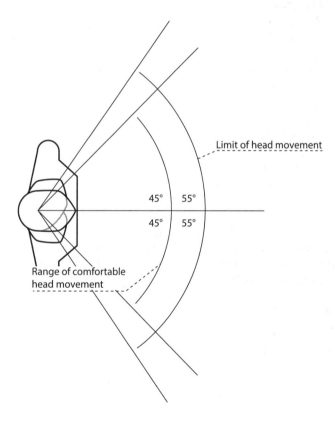

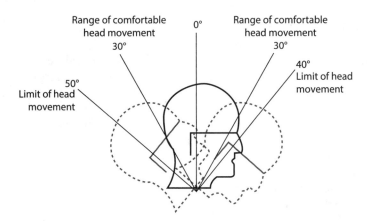

Figure 4-5
Head rotation in the vertical plane

Sightlines

Your understanding of the visual field informs one of the most important ergonomic considerations of an AV design: the sightline. A *sightline* is the unobstructed view between a person and the object or content that he or she needs to see. A *sightline study* helps determine the most appropriate seating layout, for example, to ensure a clear field of view. Such a study identifies such AV design criteria as the lowest visible point on a display, the nearest viewers' line of sight, the farthest viewers' line of sight, the possible distortion of an image from off-axis viewing locations, and other ergonomic factors that affect preferred field of vision and viewing comfort tolerances. Some of these design issues are covered by the ANSI/AVIXA standard *V202.01:2016 Display Image Size for 2D Content in Audiovisual Systems.*

In a conference room, viewing theater, classroom, or other AV space, the designer is responsible for designing a seating layout that creates the best viewing environment for the audience. Sightline studies are used to verify that everyone in the audience will have a clear view of the presentation area (see Chapter 3). These studies determine the lowest visible point on the front wall of the presentation area, for example. This information helps the designer determine how low a screen can be mounted.

Sightline studies are based on three factors:

- Seating types
- Floor types
- The limits of comfortable viewing (ergonomics)

Sightlines should not be confused with viewing angles, which are used to verify that a given display in the presenter area can be viewed on an axis as much as possible from the audience. In other words, it is possible to have great sightlines for viewing a presenter at a lectern but poor sightlines for a screen on the wall (screen too small, too high, too low, off-center).

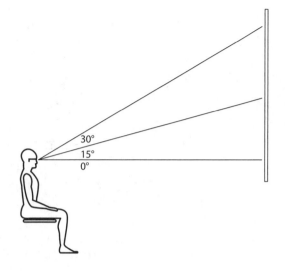

Figure 4-6 Sightlines from the top of a projected image to the eye of the closest viewer

Human Sightlines

When doing a sightline study, you can determine the minimum distance between the nearest viewer and a display. As the limit of comfortable vertical head movement is 30 degrees, this can be determined by drawing a 30-degree sightline from the top of the projected image to the eye of the closest viewer (see Figure 4-6).

Conversely, if you have already placed your audience members, you can use sightlines to determine the maximum image height. Simply draw a sightline from the eye of the closest viewer at a 30-degree angle. The point at which that line intersects the wall will be your maximum image height.

The more perpendicular a sightline is to the display plane, the greater the viewing comfort will be. The center of the screen should therefore fall at a 15-degree angle above the standard line of sight.

Eye Height

The eye height of viewers is a critical variable in a sightline study. Typical seated eye level varies between populations, with studies finding the median (50th percentile) of eye height above seat level in Hong Kong being approximately 780mm (30.7in.), while in Japan it's 785mm (30.9in.), in the UK it's 790mm (31.1in.), and in the United States it's 800mm (31.5in.). The age of the viewers is also a major consideration, with seating for kindergarten, primary, and secondary school children presenting very different eye heights. Check the *anthropometrics* of the population you will be designing your system for, taking careful notice of the range of variation within that population.

 TIP If terms such as "normal distribution," "percentile," "median," and "standard deviation" are unfamiliar to you, it would be a good idea to find basic tutorials on statistics and the normal distribution curve before diving too deeply into ergonomics and anthropometrics.

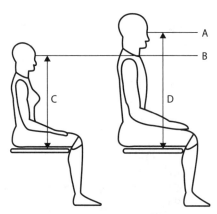

A. Sightline: Tall Viewer
B. Sightline: Short Viewer
C. Seated Eye Height: Small Viewer
D. Seated Eye Height: Tall Viewer

Figure 4-7 Seated sightlines of typical tall and short North American viewers. The seated eye height "C" is 714mm (28.1in.) above the seat for a woman in the 5th percentile, and the seated eye height "D" is 861mm (33.9in.) above the seat for a man in the 95th percentile.

In Figure 4-7, the person on the left represents the stature of a short viewer, while the person on the right represents the stature of a tall viewer. If the large viewer were to sit in front of the short viewer, the short viewer would not be able to see over the large viewer's head and shoulders. In such a case you can stagger your seating layouts or even adjust the floor of a space into a tiered or sloped layout, as discussed in the next section.

Seating Layouts

When designing a visual system for a larger space, you have two primary options for arranging viewers: aligned or staggered.

In an *aligned seating arrangement*, viewers are placed directly behind the viewer in front of them (see Figure 4-8). On a flat floor, the view of people in the back rows will obviously be obstructed. As you move farther back through an aligned seating arrangement, you will need to raise sightlines higher to clear the bodies of viewers seated closer to the screen or presenter.

Compare this with a *staggered seating arrangement* (see Figure 4-9). In this layout, people are placed so they can view the image between the shoulders of the viewers in front of them. This lowers the sightlines significantly in comparison to an aligned seating layout.

NOTE If your needs analysis determines that the people using the AV room will be seated at least part of the time, you must consider seating ergonomics. You need to ensure that your end users will be able to see any displays or presenters comfortably from their seated positions.

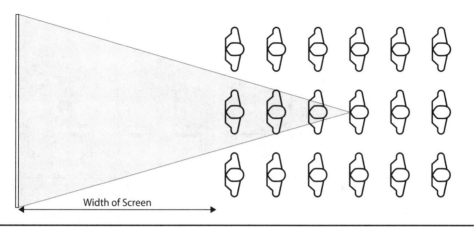

Figure 4-8 Aligned seating

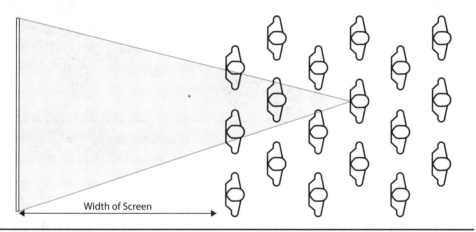

Figure 4-9 Staggered seating

Floor Layouts

Ultimately, the type of seating you choose for your AV design will depend on the space's infrastructure. A key element of that infrastructure is the floor.

Flat floors are the most common. However, as you can see in Figure 4-10, flat floors make it difficult for viewers in the back to see what is being displayed or communicated. Sightlines for the farthest viewers need to clear obstructions, such as nearer viewers' heads, to see displayed images. Poor seating alignment, such as aligned seating on a flat floor, will force those sightlines higher and higher.

Sloped and tiered floors, which are typically found in theaters, auditoriums, arenas, training facilities, amphitheaters, and large-capacity meeting spaces, represent an ideal design solution to challenging sightlines. Figure 4-11 shows how tiered seating can

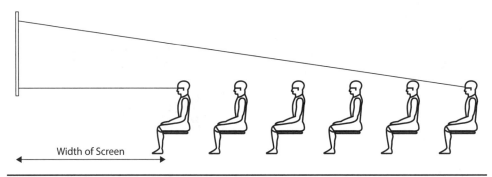

Figure 4-10 Flat floor sightlines

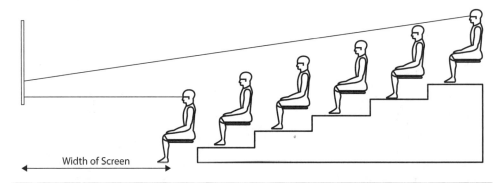

Figure 4-11 Tiered floor sightlines

dramatically lower sightlines for the farthest viewers, even in cases of aligned seating. If you combine a tiered or sloped floor with a staggered-seating alignment, you could lower sightlines even more because viewers can look between the shoulders of the people in front of them.

Furniture

It may not seem like advanced technology, but furniture is an integral part of an AV design, and you need to apply ergonomic principles when selecting furniture. For example, the layout of furniture within a space will depend on the space's functions and design considerations, such as your sightline study.

When communicating furniture needs to an architect or interior designer, be sure to include dimensions and coordinate color schemes and finishes. If furniture is custom designed, make sure to include detailed drawings as part of the audiovisual drawing package, as shown in Figure 4-12. That way, other contractors can build exactly what you need to complement the rest of your AV design.

Let's briefly consider some of the furniture that factors into an AV design.

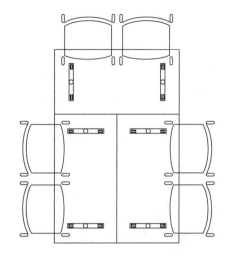

Figure 4-12
Furniture drawing for the design package

Tables and Chairs

Tables and chairs are some of the most basic elements of an AV design project. For example, certain types of tables may be required by the architect or interior designer to support specific presentation needs. The AV designer may be required to provide tables with AV system input/output connections. In a room that will host videoconferencing or any other on-camera applications, table tops should be light in color to reflect light up onto participants' faces. The size and location of tables will depend on activities that users intend to conduct in the room. These are the types of considerations an AV designer is trusted to understand.

Chairs for both presenters and the audience come in a wide range of styles and sizes. The architect or interior designer will likely pick out the chairs for a room, but they should be comfortable, plain, medium in color, and with no specular features that might be distracting in video applications. Swiveling chairs can be a source of distraction in video applications, as participants may unconsciously fidget by twisting in the chair and cause problems with auto-focus or microphone noise.

Stackable, multipurpose chairs are common, as are ergonomic chairs. Keep in mind that a room layout may need to accommodate the use of storage dollies to deliver and remove chairs and tables. Depending on the AV requirements for the space, this could affect equipment placement or other aspects.

In addition, when collaborating with the architect or interior designer, consider the amount of time people will be sitting in the room and what tasks they will be engaged in. Hour-long meetings, full-length performances, feature-length video viewing, and multiday training sessions usually have different seating requirements. During your needs analysis you will have determined how much of each the client intends to use the room for. Communicate this to other project team members.

Many types of furniture have standard dimensions, designed to match the size of the average adult human body:

- Tables are typically 735mm (29in.) high but can range from 635 to 815mm (25 to 32in.).
- Chair widths are typically 460 to 560mm (18 to 22in.) for most fixed-seating situations and 560 to 660mm (22 to 26in.) for most movable chairs.
- Counters are typically 915mm (36in.) high but can range from 815 to 1070mm (32 to 42in.).

When creating a layout for a space to be used by adults, the AV designer can usually assume that furniture will comply with these standard dimensions.

As you determine where furniture will go in a space, consider the total volume of space each person will occupy. Specifically, table and chair configurations should be designed so each occupant feels comfortable in the space provided.

Seated people need to be able to sit down and stand up easily. Experts recommend a minimum distance of 1220mm (48in.) between the edge of a table and the nearest physical obstruction, such as a wall. This amount of space accounts for chair clearance and circulation (moving around). If the design plans for less space than that, you run the risk of seated users having to sidestep or interfere with each other when moving around.

Standing people also need circulation space. For passageway clearance, such as in an aisle or walkway, specify about 750 to 900mm (30 to 36in.) of circulation space.

NOTE Circulation space is the amount of area a person needs to turn around without disturbing other people. According to best practice, this area is 1 to 1.2 square meters (10 to 13 square feet) per person.

In addition to circulation space, seated people should have enough room for materials such as microphones, data and power connection points, meeting notes, reference books, electronic devices, and possibly also refreshments. The width of a seated person's area (a desk or shared table) should also accommodate human dimensions and body position (to account for elbows, for example).

The optimal range of table space width is 600 to 750mm (24 to 30in.) per seated person. The optimal range of table space depth is 450 to 600mm (18 to 24in.).

Lecterns

Despite a growing range of open-stage presentation formats, lecterns remain important to many AV situations, including boardrooms, classrooms, media conferences, awards presentations, political meetings, shareholder presentations, and hotel conference spaces.

The location and design of a lectern are important. AV designers should ensure that they follow ergonomic principles when integrating a system with a lectern. The goal is to create an optimal environment for users, not only in the way they present their material but also in the way they interact with the audiovisual equipment.

The architect or interior designer may want to craft the lectern or at least select it for use within the room. Lecterns can be premanufactured or custom-designed to meet specific design requirements. No matter how nice the lectern looks in the room, AV designers need to coordinate with architects and interior designers on where to put it within a space. Nothing will ruin a new room faster than a beautiful, custom-built lectern that blocks the audience's view of part of the presenter area, is in a place that doesn't fit with the presentation, or can't accommodate the AV controls the user wants.

Remember that some presenters may have limited mobility and ability to access the facilities on the lectern. Taking mobility and access issues into account may change lectern design specifications. Most jurisdictions have accessibility requirements that will need to be addressed.

When presenters are at a lectern, they may want access to system controls either built into or adjacent to the lectern. This requirement should have surfaced during the needs analysis. Keep in mind that presenters shouldn't have to search for common controls or when trying to use technology.

In addition, consider the placement of microphones, video displays, user control interfaces, signal connections, digital signage, equipment ventilation, all required power, and space for presentation media. Lecterns offer a good location for presentation source equipment, such as document cameras, personal computers, and other digital devices—just make sure your design includes provisions for wiring and equipment integration when using a lectern as an AV source.

Other Furniture

Other types of AV-related furniture may factor into your design:

- **Carts**   These can be used to hold projectors and monitors. A monitor should be secured to a cart to minimize the potential for damage should the monitor slip from the cart.
- **Whiteboards**   These are required in many installations. They can take up an entire wall or just part of a presentation area.
- **Electronic whiteboards/interactive smart boards**   These items provide a method for interactively sharing or transferring information from the writing surface to a computer, display, or other end point. Interactive smart boards can be wall-mounted or portable; just remember that they need a main power source and data connections.
- **Flip charts**   Flip charts are simple tools still used by a wide range of clients and found in many meeting rooms. The AV designer may have nothing to do with a flip chart, but if there is going to be one, its location and the space required to utilize it could affect other parts of the design such as lighting and audio.

NOTE   Whiteboards generally have a glossy finish and thus make poor projection surfaces. They should not be considered a substitute for a purpose-made front-projection screen.

Chapter Review

In this chapter, you learned about ergonomics, human dimensions, sightlines, and the role of furniture decisions in an AV design. You learned that the foundation of an AV design begins with planning for human comfort. Having completed this chapter, you should be able to begin preparing a sightline study, which is critical to planning a visual system. You will build on this knowledge in Chapter 5.

Review Questions

The following questions are based on the content covered in this chapter and are intended to help reinforce the knowledge you have assimilated. These questions are similar to the questions presented on the CTS-D exam. See Appendix E for more information on how to access the free online sample questions.

1. When considering viewers' visual field in the horizontal plane, people can recognize words, fine details, and color at what angle from the standard line of sight?

    A. 0 to 10 degrees

    B. 10 to 20 degrees

    C. 25 to 30 degrees

    D. 40 to 50 degrees

2. Which of the following methods can be used to improve sightlines in an auditorium?

    A. Lowering the height of the display

    B. Using an aligned seating layout

    C. Providing a sloped or tiered-seating layout

    D. Installing seats with taller dimensions

3. In a sightline study, you can determine the minimum distance between the nearest viewer and a display by drawing a sightline from the top of the projected image to the eye of the closest viewer at what angle?

    A. 10 degrees

    B. 20 degrees

    C. 30 degrees

    D. 40 degrees

4. People view objects more comfortably when the objects are located _____.

    A. To the left of their standard line of sight

    B. To the right of their standard line of sight

    C. Higher than their standard line of sight

    D. Lower than their standard line of sight

5. Circulation space is _____.
   A. The area a person needs to turn around without disturbing other people
   B. The area around AV equipment that allows it to ventilate
   C. The area below a lectern where AV sources are housed
   D. The space behind a rear-projection screen

Answers

1. **B.** When considering viewers' visual field in the horizontal plane, people can recognize words, fine details, and color 10 to 20 degrees from the standard line of sight.
2. **C.** Providing a sloped or tiered-seating layout can improve sightlines in an auditorium.
3. **C.** You can determine the minimum distance between the nearest viewer and a display by drawing a sightline from the top of the projected image to the eye of the closest viewer at a 30-degree angle.
4. **D.** People view objects more comfortably when the objects are located lower than their standard line of sight.
5. **A.** Circulation space is the area a person needs to turn around in without disturbing other people.

CHAPTER 5

Visual Principles of Design

In this chapter, you will learn about
- Determining image specifications for a video system, including character height, farthest and nearest viewing locations, and good and acceptable viewing areas
- Calculating the aspect ratio of a display, given its height, width, or diagonal
- The advantages and disadvantages of front-screen and rear-screen projection
- The basics of videowalls
- Measuring reflected light bouncing off a screen and into a viewer's eyes
- Calculating lumens, projector brightness, and task-light levels for a given projection system
- Using industry standards to calculate contrast ratio correctly

The visual display components of an AV system drive space design. Designers must configure the space to ensure that an audience has a high-quality viewing experience. They must then design the other elements of the room to complement that layout.

The display system (also referred to as the *video system*) consists of the display devices themselves (flat-panel display, projector and screen, etc.) and the components that support those devices (cabling, sources, processors, switchers, etc.). Every aspect of the video system affects what the audience sees; therefore, it all requires careful consideration by the AV designer.

> **Duty Check**
> This chapter relates directly to the following tasks on the CTS-D Exam Content Outline:
>
> - Duty A, Task 2: Identify Skill Level of End Users
> - Duty A, Task 3: Educate AV Clients
> - Duty A, Task 6: Identify Scope of Work
>
> *(continued)*

> - Duty B, Task 1: Review A/E (Architectural and Engineering) Drawings
> - Duty B, Task 2: Coordinate with Architectural/Interior Design Professionals
> - Duty B, Task 3: Coordinate with Mechanical Professionals
> - Duty B, Task 4: Coordinate with Structural Professionals
> - Duty C, Task 1: Create Draft AV Design
> - Duty C, Task 2: Confirm Site Conditions
>
> In addition to skills related to these tasks, this chapter may also relate to other tasks.

Determining Image Specifications

When designing a video system, your first task is to determine image specifications. After all, you can't select effective video equipment if you don't know what needs to be displayed and who's going to be displaying/looking at it. For example, where will the viewers be located? Will they be sitting? Where should the screen(s) go so the most people can see them? How tall should any displayed text or symbols be so viewers can read them? What type of information will viewers be seeking to acquire from the system?

If you determine this information first, you can use it later when you select the display equipment required to produce the right image to communicate the intended message.

An AV designer must consider a range of parameters before selecting an actual display: text size, farthest viewing distance, image complexity, image brightness, minimum contrast ratio, nearest viewing distance, and good or acceptable viewing areas.

Viewing Requirements

In creating the standards for specifying image properties and specifications, the AVIXA standards, developers have established four categories of viewing requirements:

- *Passive viewing* is where the content does not require assimilation and retention of detail, but the general intent is to be understood (e.g., noncritical or informal viewing of video and data).
- *Basic decision-making* (BDM) requires that a viewer can make decisions from the displayed image, but comprehending the informational content is not dependent upon being able to resolve every element detail (e.g., information displays, presentations containing detailed images, classrooms, boardrooms, multipurpose rooms, product illustrations). Decisions are made by comprehending the informational content itself and are not dependent on the resolution of every element of detail,

- *Analytical decision-making* (ADM) is where the viewer is fully analytically engaged with making decisions based on the details of the content right down to the pixel level (e.g., medical imaging, architectural/engineering drawings, fine arts, forensic evidence, electronic schematics, photographic image inspection, and failure analysis).
- *Full-motion video* is where the viewer is able to discern key elements present in the content, including detail provided by the cinematographer or videographer necessary to support the story line and intent (e.g., home theater, business screening room, live event production, broadcast postproduction).

These categories are used in AVIXA standards, including V202.01:2016 *Display Image Size for 2D Content in Audiovisual Systems* (DISCAS) and V201.01:2021 *Image System Contrast Ratio* (ISCR), which provide recommendations for a range of image parameters, including image element height, screen dimensions, screen placement, viewer placement, and image contrast ratios.

Determining Image Size

Visual acuity is the ability of your eye to see details. There is a certain distance at which a given height of text can no longer be read clearly, which is a point beyond your eye's acuity. In AV design, you want the text size to be comfortable to read; the audience members should not strain their eyes. Logically, as the image is set farther away from the viewer, the screen size and content must be larger to accommodate comfortable viewing parameters. The size of the display needs to be sufficient for the farthest viewer in the room. Either the screen size or the farthest viewer distance can be the starting point for calculations when laying out a room.

The maximum viewing distance is dependent on the amount of detail in the image. AVIXA has developed a set of online calculation tools based on the DISCAS standard, which provide a calculation and assessment tool for determining proper display image size and optimal viewing distances. These tools cater to viewer requirements in BDM and ADM situations.

Calculations based on the DISCAS standard can be used to do the following:

- Plan and design new displayed image systems
- Determine image size relative to space and viewing requirements
- Determine closest and farthest viewer positions
- Determine horizontal angles of view
- Provide metrics for content design

The online tools available on the AVIXA website perform the calculations contained in the standard. Figures 5-1 and 5-2 show the calculation tools available on the AVIXA website for performing the DISCAS calculations. If you would like more information on the underlying assumptions and formulae for these calculations, the complete DISCAS V202.01:2016 standard is available from the Standards section of the AVIXA website at www.avixa.org.

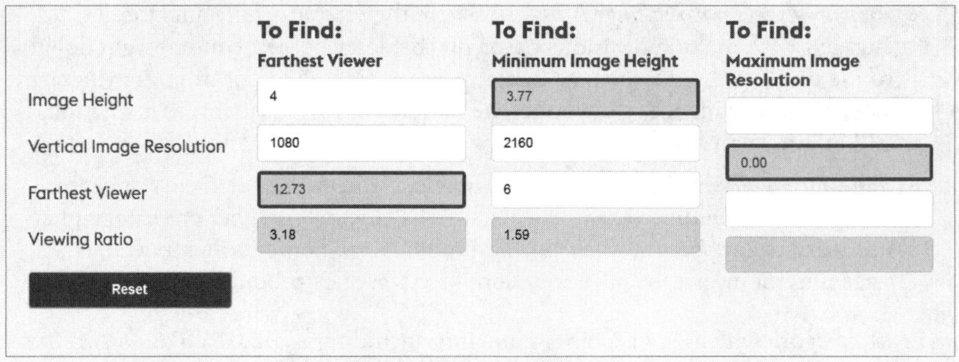

Figure 5-1 The DISCAS analytical decision-making (ADM) calculator on the AVIXA website

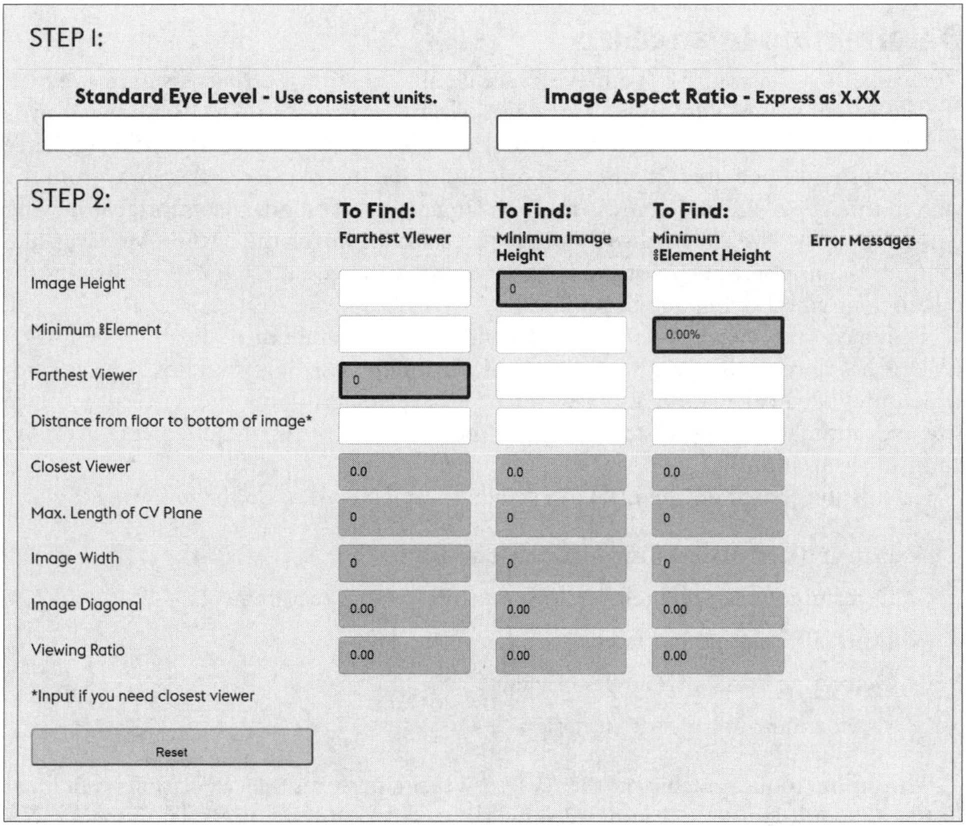

Figure 5-2 The DISCAS basic decision-making (BDM) calculator on the AVIXA website

Determining Text Size

When designing a video system, you should first consider the size of any text or symbols the user wants to display. After all, people should be able to comprehend the information on the display from a specified distance. And if customers can't read digital signs, what good are they? Once you determine the required text size, you can explore the presentation medium.

Displays are characterized by resolution; the human eye is characterized by acuity. Resolution describes a video system's ability to reproduce highly detailed information. Visual acuity is an eye's ability to discern fine details. Put another way, visual acuity is what an individual set of eyes can see, but screen resolution is what an electronic display can output.

There are several different kinds of acuity, including *resolution acuity,* which is the ability to detect that there are two stimuli, rather than one, in a visual field. Resolution acuity is measured in terms of the smallest angular separation between two stimuli that can still be seen as separate.

Recognition acuity is the ability to correctly identify a visual target. In terms of text on screen, it's the ability to differentiate between a *G* and a *C* or an *E* and an *F*, for example. Usually—but not always—recognition acuity is measured in terms of the angular dimension of the smallest target that can be discriminated. A clinical eye chart, where visual-acuity testing is performed using letters, is a form of recognition-acuity testing.

VIDEO Watch a video from AVIXA's online video library about determining image specifications. Check Appendix D for the link to the AVIXA video library.

Visual Acuity and the Snellen Eye Chart

You may be familiar with the eye charts used for testing visual acuity. Figure 5-3 shows a Snellen chart. This chart displays several characters, called *optotypes,* which look like letters. An optotype is a character used to assess a person's visual acuity. It looks like a block letter and is drawn with specific and rigid geometric rules. Only ten optotypes are used on the traditional Snellen eye chart: C, D, E, F, L, N, O, P, T, and Z. In this chart, the thickness of the lines equals the thickness of the white spaces between lines and the thickness of the gap in the letter *C*. In addition, the height and width of the letter are five times the thickness of the line. This is derived from the minimum vertical resolution possible for the letter E, which requires three black and two white lines to display it.

Element Height

When determining the size of text or symbol elements, your design goal is to create elements of a height that is easy to read and won't cause eye strain or fatigue. Imagine an audience looking at PowerPoint presentations for three days.

The minimum acuity required for human vision is the ability to resolve the five-by-five matrix of an optotype. At a distance of 6 meters (20 feet), a person with normal vision can just resolve a vertical angle of 2 minutes of arc to differentiate between the five horizontal lines in the letter E. (There are 60 minutes in a degree of arc.)

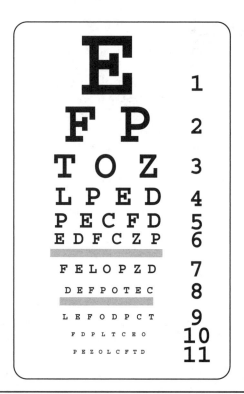

Figure 5-3 A Snellen eye chart

The element height calculations in the DISCAS standard for BDM refers to the height of an element as a percentage of the overall image height. The percentage element height (%EH) factor represents the ratio of element height to screen height expressed as a percentage (e.g., a %EH of 1 represents an element height of 1 unit relative to 100 units of screen height). The standard provides a range of percentages; users of the standard should vary this percentage according to the content.

The element used for DISCAS calculations for ADM is the height of the smallest elements that a human eye can resolve on a display: a line pair (two adjacent rows of pixels).

 NOTE Normal visual acuity is often described as being 6/6 (20/20) or 1.0. This means that a person with normal vision will resolve objects spaced 1.75mm (0.069in.) apart at a viewing distance of 6 meters (20 feet). A higher visual acuity of 6/3 (20/10) or 2.0 means that the person has the acuity to resolve at 6 meters (20 feet) what a person with normal acuity could resolve at 3 meters (10 feet). A lower visual acuity of 6/10 (20/30) or 0.6 means that the person has the acuity to resolve at a distance of 6 meters (20 feet) what a person with normal acuity could resolve at 10 meters (30 feet).

Image Height

The minimum image height for a screen is based on the element height:

- **BDM**   *Image height = Number of vertical pixels × Pixel height*
- **ADM**   *Image height = 1 ÷ (Element height × %Element height)*

Viewing Angles

The DISCAS standard recommends the following viewing angles:

- **BDM**   In a basic decision-making situation it is recommended that the top of the image should be no more than 30 degrees above the eye level of the closest viewer, as shown in Figure 5-4. Horizontally, it is recommended that viewers are located within a 60-degree arc of the far edge of the screen being viewed, as shown in Figure 5-6.
- **ADM**   In an analytical decision-making situation, as a viewer may need to closely approach the image to resolve pixel-level detail, there are no recommended vertical viewing angles. In the horizontal plane, it is recommended that viewers are located within a 60-degree arc of the far edge of the screen being viewed, as shown in Figure 5-5.

Viewing Distance

The recommended viewing distances calculated using the DISCAS standard are based on the level of detail in the image.

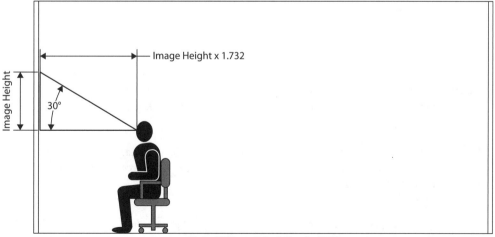

Closest Viewer Calculation for BDM
No Scale
Image Height Based on 16:9 (1.78:1) Aspect Ratio

Figure 5-4   Vertical viewing angle for BDM

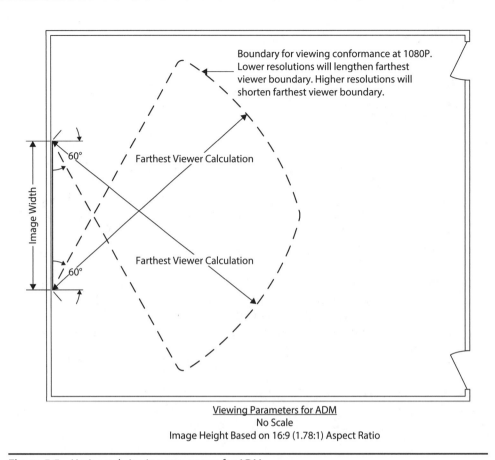

Figure 5-5 Horizontal viewing parameters for ADM

Viewing distances in DISCAS are quoted in terms of the closest and farthest viewers. The *closest viewer* is a viewing position that defines how close one can comfortably sit in relation to a display and still assimilate content. The *farthest viewer* is a viewing position that defines how far away one can sit from a display and still discern the requisite level of detail.

- **BDM** In a basic decision-making situation, viewing distance calculations are based on percent element height (%EH), which is the percentage of the display that an element fills.

  %EH = Element height / Total Image height

  If a text character is 38 pixels high on an HD display (1920×1080), it has a %EH of 3.5. A symbol with a %EH of 5 is 54 pixels high. For any text font, element heights are taken from the lowercase characters, without including descenders. Figure 5-6 shows farthest viewer boundaries for a range of percentage element heights.

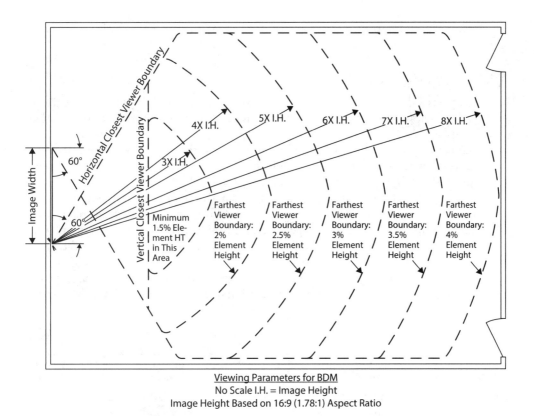

Figure 5-6 Horizontal viewing boundaries for BDM

- **ADM** In an analytical decision-making situation, viewing distance calculations in DISCAS are based on a single-pixel element height, as the viewer must be able to resolve every pixel in the image.

Display Device Selection

Once you've determined the specifications needed for your images, you are ready to determine which equipment you need for your system, starting with display sources. In digital video systems, the term *format* refers to the source of the image. Digital sources can include Blu-ray players, DVD players, streaming media players, personal computers, cloud-based services, and personal devices, such as tablets and smartphones.

Knowing your customer's source media, resolution, refresh rate, bit depth, codec, color space, audio format, and more will help you identify the proper player for that media. Remember, as you specify display devices, it is important to employ the highest-quality signal available while still maintaining system functionality.

To that end, let's start breaking down some of the important characteristics of display equipment.

Video Resolution

Graphic processors output many different display resolutions. It's up to the AV professional to select a resolution output that meets the requirements of the content being displayed.

A display device has a "native" resolution, which is the resolution that is matched, pixel for pixel, with the display. For example, if a display has 3840 pixels horizontally and 2160 pixels vertically, it has a native resolution of 3840×2160.

If an image is larger or smaller than the display's native resolution, it might not display well. If a source image is different in height or width from the display, the display's scaler may down-sample or stretch the image, or create unsightly borders to directly map the pixels. Whenever possible, you should match the resolution of the source to the native resolution of the display.

A higher-resolution display means more pixels, greater detail, and better image quality. It also means the system requires more information to build an image. Table 5-1 shows a range of display resolutions.

Name	Description	Width	Height	Aspect Ratio
CGA	Color Graphics Adapter	320	200	16:10
SDTV	Standard-Definition Television	720	480	3:2
XGA	eXtended Graphics Array	1024	768	4:3
WXGA	Widescreen eXtended Graphics Array	1152	864	4:3
WXGA	Widescreen eXtended Graphics Array	1280	800	16:10
SXGA	Super eXtended Graphics Array	1280	960	4:3
SXGA	Super eXtended Graphics Array	1280	1024	5:4
WXGA	Widescreen eXtended Graphics Array	1366	768	16:9
WSXGA	Widescreen Super eXtended Graphics Array	1440	900	16:10
UXGA	Ultra eXtended Graphics Array	1600	1200	4:3
WSXGA	Widescreen Super eXtended Graphics Array Plus	1680	1050	16:10
HDTV	High-Definition Television	1920	1080	16:9
WUXGA	Widescreen Ultra eXtended Graphics Array	1920	1200	16:10
DCI 2K	Digital Cinema – 2K Uncropped	2048	1080	1.9:1
UHDTV1	4K Ultra-High-Definition Television	3840	2160	16:9
DCI Scope	Digital Cinema – Cinemascope (4K)	4096	1716	2.39:1
UHDTV2	8K Ultra-High-Definition Television	7680	4320	16:9

Table 5-1 Media Formats, Sizes, and Aspect Ratios

Aspect Ratio

Human vision is oriented horizontally. Your eyes' angle of view is wider than it is high—landscape rather than portrait. Therefore, people naturally frame a scene horizontally. This perceptual quality is translated into a mathematical equivalent known as *aspect ratio*, which is the ratio of image width to image height.

Aspect ratio is expressed as two whole numbers, separated by a colon (for example, 16:9), or by the decimal equivalent (for example, 1.78:1). Common video aspect ratios include the following:

- 4:3 (1.33:1) for standard-definition (SD) video
- 16:9 (1.78:1) for high-definition (HD) video

TIP As both of these ratios are infinitely repeating decimals (16 ÷ 9 = 1.777777777777... and 4 ÷ 3 = 1.333333333333...), they are usually written rounded to two decimal places. However, for accurate results from calculations involving large screen dimensions (e.g., 2,560mm), it is best to use at least four decimal places for the ratio (1.3333 and 1.7778).

Some computer and video production displays operate at a 16:10 (1.6:1) aspect ratio, while the aspect ratio for cinemascope cine images is 2.39:1. Your ability to determine the correct aspect ratio for any image, screen, or display will help you specify the right product for your design, match a projector's image to the correct screen size, or select the correct image size for a given environment. Where multi-image blending or arbitrary-sized customized displays are required, the aspect ratio may be unique and specific to the situation. Without a solid understanding of aspect ratio, you could make a costly error.

The worst-case scenario is a mismatch between a screen and projector *and* a mismatch from image to projector. In such a scenario, the size of the image will be reduced. When an image gets smaller, the audience will have to move closer to get a good view. Moving closer will, in turn, change viewing angles and the closest/farthest viewer calculations.

Calculating Aspect Ratio

An HDTV display with a 1920×1080 resolution has an aspect ratio of 1920:1080 or 16:9 or 1.78:1. Using this aspect ratio, if you know one of the image dimensions, you can easily find the other.

Let's say you need a projected HD image to fill a custom space above the chair rail in a room. You measure and find the height to be 1,600mm (63 in.). By multiplying 1,600 (63) by 1.7778, you will get the width for a 16:9 HD image, which is 2,844mm (112in.). Conversely, if you know the width and are looking for the height, you simply divide the width by 1.7778 to find the height required to show a 16:9 image. If you have a width of 2,600mm (102in.) and divide that by 1.7778, you will get a height of 1,462 mm (57.6in.).

So, let's try it. Find the width of an image 1,500 millimeters high with a 16:9 aspect ratio.

width ÷ height = 16 ÷ 9:

$X \div 1{,}500 = 16 \div 9$

Cross-multiply:

$9 \times X = 1{,}500 \times 16$
$9 \times X = 24{,}000$

Divide both sides by 9:

$(9 \times X) \div 9 = 24{,}000 \div 9$
$X = 2{,}666.6666...$

Rounded to the nearest millimeter, the image width is 2,667 millimeters.

Calculating Screen Diagonal

The width and height of an image can also be used to determine a screen's diagonal measurement using the Pythagorean theorem: $H^2 + W^2 = D^2$, where H is the height of a screen, W is the width of a screen, and D is the diagonal length of a screen. Doing the calculation, a 16:9 image has a diagonal of 18.3575, while a 4:3 image has a diagonal of exactly 5.0000, and a 16:10 image has a diagonal of 18.8680.

If you know the aspect ratio and any dimension—width, height, or diagonal—you can calculate the other dimensions using the "cross-multiply and divide" method. Let's give it a shot. A screen has an aspect ratio of 16:9. The diagonal of the screen is 2,500 millimeters. What is the width?

Cross-multiply:

$2{,}500 \times 16 = X \times 18.3575$
$40{,}000 = X \times 18.3575$

Divide both sides by 18.3575:

$40{,}000 \div 18.3575 = (X \div 18.3575) \div 18.3575$
$X = 2{,}178.9459$

Rounded to the nearest millimeter, the width is 2,179 millimeters.

Display Types

There are many display options on the market. The type an AV designer chooses will depend on a wide range of factors. In general, display options include projected, direct-view, and interactive.

Direct-view display systems, such as LCD or LED panels and LED modules, can provide bright, high-quality images but may be constrained in size by weight, rigging, power, and signal distribution requirements. Still, direct-view display systems offer excellent contrast ratios, and there is no projected light path for viewers or presenters to block.

Given the right lighting conditions, projection systems can be inexpensive and create large images. Increasingly, AV designers are mapping projected images onto complex surfaces with little resemblance to a conventional projection screen.

Interactive displays allow users to make annotations directly on an image or interact with a graphical user interface. An interactive display can be based on a projection system, or it may be based around such technologies as a screen overlay, look-through cameras, or an external bezel on a direct-view display. Some interactive displays are large touch screens.

Direct-View Displays

Monitors, flat-panel displays, flat screens, videowalls—they are all terms for a common type of display technology. The prominent flat-panel technologies include LCD, OLED, LED, and microLED. The brightness of direct-view displays is typically measured in candelas per square meter (cd/m^2), or nits, with brightness levels ranging from around 350 candelas per square meter for hospitality or meeting room use up to 6,500 candelas per square meter for displays intended for outdoor signage or other applications with a lot of ambient light.

LCD

Liquid-crystal displays (LCDs) use a grid of light-controlling cells to create images on a screen. Each pixel on the screen is filled with a liquid crystal compound that changes its optical properties in the presence of an electric field.

To create an LCD pixel, light must first pass through a polarizer. The polarized light then travels through a sandwich of transparent switching transistors and liquid crystals. The liquid crystals within each pixel act like light shutters, passing or blocking varying amounts of light. As polarizing filters cannot block 100 percent of the incoming light, the "black" from LCD panels is actually dark gray. In some special high-contrast or high-dynamic-range applications two light control matrix layers may be vertically stacked.

A color LCD has three filters in each imaging pixel array, one for each primary color: red, green, and blue. The backlight may be LEDs, mini-LEDs, micro-LEDs, quantum dots, or compact fluorescent tubes. The backlight LEDs can be clustered in groups with red, green, and blue elements or only white LEDs. LEDs have lower power consumption than compact fluorescent tubes, higher brightness, wider color range, better white balance, and longer life, and they can include local area dimming for increased contrast. LED backlights are also more compact than compact fluorescent tubes, allowing for much slimmer panel profiles.

LED

Light-emitting diode (LED) displays are a matrix of red, green, and blue LEDs, grouped to form the pixels of the display. LEDs are emissive devices, meaning that they create their own light, as opposed to LCDs, which require a separate light source: the backlight.

As a result, LED displays have much higher contrast ratios, faster pixel switching times, operational lives in the tens of thousands of hours, and they are capable of higher brightness and a wider color gamut than LCDs.

Individual LED chips or full-pixel clusters of RGB LED chips are mounted into modules that carry physical framing and linkages in addition to power and signal distribution systems. Multiple modules can be connected to form screens of arbitrarily large sizes and shapes. Modular LED screens are often seen in large spaces, in concert productions, and at public events.

MicroLEDs are very small (1 to 10 µm) crystalline LEDs with the same characteristics as their larger brethren. As microLED pixels are quite physically small, the spaces between them are relatively large. This enables a black background between pixels to absorb ambient light and produces high-contrast images and much deeper blacks than LCD panels.

OLED

OLED technology is based on organic, carbon-based, chemical compounds that emit light when a current flows through the device. There are separate organic compounds for red, green, and blue.

OLED devices use less power than LCDs, have much higher contrast ratios, and may be capable of higher brightness and a wider color range than LCDs.

OLEDs are imprinted on a very thin substrate. The active-matrix silicon-integrated circuits are imprinted directly under the display, controlling the power to each organic point of light diode (pixel) and performing certain image control functions at a very high speed. OLED's capability to refresh in microseconds, rather than the milliseconds taken by LCDs, allows for highly dynamic motion video. As with microLEDs, the small physical size of OLED pixels facilitates high-contrast images with rich blacks.

Compared to crystal LEDs, OLEDs have a relatively short life span, exhibiting progressive degradation of output, particularly in the shorter wavelengths—seen as blue fading.

Front Projection

Although flat-panel, direct-view displays have become the primary screens used for small to medium-scale video systems, projection is still popular for certain types of installations, and about the only possible choice for mapped-projection applications. As such, AV designers need to know how to integrate these two-element projection systems into spaces.

Front-screen projection generally requires less space than rear projection. It offers simplified equipment placement. In front projection, the projector is in the same space as the viewer. This can result in problematic ambient noise from cooling fans in the projector. Another downside is that people, especially the presenter or instructor, can block all or part of the image by walking in front of the projected light. Ultra-short-throw projectors have helped to address this problem by allowing substantial screen coverage from throws less than 500mm (18in.).

Front-Projection Screens

A screen's surface coating greatly influences the quality and brightness of a video projector's image. A well-designed projection system incorporates a screen type that reflects projected light over a wide angle to the audience, while minimizing the reflection of stray light. Stray light causes a loss of contrast and detail, most noticeably in the dark areas of the image.

Projection screens are passive devices and cannot amplify or create light rays. They can reflect light rays back at wide or narrow angles, thereby providing gain, or brighter images.

Screen gain is the ability of a screen to redirect light rays into a narrower viewing area, making projected images appear brighter to viewers sitting on-axis to the screen. The higher the gain number of a screen, the narrower the viewing angle over which it provides optimal brightness.

A matte white screen is usually made of magnesium carbonate or a similar substance that provides a perfect diffuser for the redistribution of light. In other words, the light energy striking the screen surface is scattered identically in all directions. Matte white is a reference surface, with a screen gain of 1.0 (also known as *unity gain*).

Ambient light rejection generally increases as the gain of the screen increases. This is because as the screen gain increases, the angle at which light hits the screen becomes more important. Ambient light that is not on-axis with the projector and viewer is reflected away from the viewer. This makes the screen appear darker, increasing the contrast.

There are also angularly reflective screen surfaces, which reflect light at various angles, rather than a uniform 180 degrees. Such surfaces perform like a mirror and have a gain greater than 5. The light is reflected at the same angle—it strikes the screen but on the other side of the screen's axis. If a projector is mounted at a height equal to the top center of the screen, the light would be directed back at the audience in a downward direction. This screen type works well for motion video.

Specifying Front-Screen Projection

A designer must consider a range of factors when specifying front-projection screens. These include the following:

- Will the screen be fixed, manual, or electric? Electric screens have motors that are offset or inserted into the roll. Offset motors require a larger cutout than the screen to accommodate the motor.

- How will the screen be mounted? For example, will it be mounted to the wall or ceiling? Will it be recessed into the ceiling? Will it be foldable? Roller screens are the most common mounted screens. Tension screens are made of a thinner material and have "arms" that gently stretch the material flat. Fixed-mount screens are stretched and mounted to the wall.

- Where in a room will the screen be mounted? If it's mounted near air diffusers, will the screen vibrate as air blows on the screen? Will thermal gradients cause air turbulence in the projection or viewing path? The image can also be shaken by structural vibrations. You need to pay attention to the location of other wall-mounted devices, such as alarm indicators, thermostats, clocks, light switches, power and data outlets, exit signs, and whiteboards.

- How far does the screen need to lower to get into the proper position? Will it require extra fabric, called *extra drop*, at the top or bottom? Designers should specify whether extra drop is required to compensate for ceiling or projection height. This drop should be communicated to the screen manufacturer and written into the specs.
- Do you need to specify reverse wrap? Usually, screens wrap (roll up) toward the back and around. This allows the screen to be placed close to the wall. With a reverse wrap, the screen wraps toward the front and around, which puts it a few centimeters out from the wall. Such placement, with reverse wrap, allows the screen to be located where it avoids other mounted items, such as chalk trays or whiteboards.
- What size screen do you need, and can it be easily delivered to the required location through doors and stairways or up lifts/elevators? Will multiple people be required to deliver it?
- Do you need to specify a masking system? A masking system (active or passive) allows you to change the aspect ratio of the visible area of a screen.
- Will the screen be retracting into a ceiling space that is considered part of an air circulation system? This may require the screen components or their enclosures to be made from approved plenum-rated materials.

Mapped Projection

Mapped projection, also known as *projection mapping,* is an application of front projection where the projection surface is an irregularly shaped, three-dimensional object such as a scenic element, an architectural structure, or a part of the landscape.

In some mapped-projection projects it is possible to have a degree of control over the reflective properties of the mapped objects through the choice of surface finishes, such as paint, or arranging for the cleaning of the surfaces of the objects, if practicable. Where buildings are the projection surfaces, it is worth considering temporarily inserting reflective materials of similar reflectivity to the rest of the structure behind uncovered blank windows.

The precision of the 3D-mapped images projected requires very high accuracy of projector placement and robust and rigid equipment mountings. As mapped projections are frequently on the exteriors of structures and landmarks, weather protection and physical security are major considerations for the design of projector mounting, maintenance access, and support systems.

Most mapped-projection projects require multiple projectors to provide adequate coverage of the geometry of the projection surfaces. This necessitates the provision of power and data to multiple projection points, for systems control, security, air filtration, communications, data networks, safety lighting, monitoring, and HVAC, in addition to the program audio and video feeds.

Some mapped-projection displays or performances may include extensive effects lighting, laser shows, pyrotechnic displays, atmospheric effects, and house lighting.

These are usually co-located with projection points and will require the provision of additional mountings, access capabilities, power, and data feeds at some projection locations.

Rear Projection

Rear-screen projection describes a presentation system in which the image is projected from behind and through a translucent screen, toward the audience on the opposite side. It is a transmissive system, in which light passes through the screen toward the viewer.

Rear projection requires more space for installation because it needs a projection room or cabinet behind the screen to accommodate the projector. Mirrors are commonly used to save space in rear-projection applications by folding the optical path. Generally, in spaces that suffer from high ambient light, a rear-projection system can deliver better contrast and color saturation than a front-projection system. Many of the advantages of rear-projection systems are shared with direct-view flat-panel displays, which have largely replaced rear projection for general-purpose applications.

Rear-projection systems have a number of characteristics that make them especially suited to specific AV installations, including the following:

- **High tolerance of ambient light** The images on a rear-projection screen can usually be seen clearly without requiring the room to be as dark as it might be for a front-projection screen.
- **Unobtrusive equipment placement** Because the projector is out of sight in a rear-projection area, it will not interfere with the room layout or sightlines.

- **System black level** Rear-projection screens can create a system black level (the level of brightness at the darkest—blackest—part of an image) that is superior to front-projection screens. This can translate into brighter images and greater contrast.
- **Contrast** Images produced by a rear-projection system usually benefit from a better contrast ratio than those produced by a front-projection system. Sometimes, rear-projection systems can also focus projected light on viewers, further reducing ambient light and improving contrast.
- **Low noise** Because the projector and associated equipment go into a rear-projection room, the noise they produce is isolated.
- **User mobility** Presenters can walk in front of a rear-projection screen without disturbing the projected image, making such screens more appropriate for certain types of rooms, such as command-and-control centers or visualization spaces.

All that said, rear-projection systems have their downsides too, including the following:

- **High cost** On average, the mirrors, mounts, and screen material required for a rear-projection system add up to cost more than a front-projection or some flat-panel systems would cost. This is why the needs analysis is so important when deciding between a rear-projection system and a front-projection system.
- **Required floor space** Because rear-projection systems need space to house the projector behind the screen, they require more floor space than front-projection systems do. Moreover, the projector room needs to be kept clear of objects that could block the light path or push the projector out of alignment. Users can't think of their rear-projector room as extra storage space!

Rear-Projection Screens

Diffusion screen material for rear projection is usually rigid plastic, glass, or vinyl fabric. These materials provide a diffused, coated, or frosted surface on which to focus an image. The ambient light rejection of these materials is considered moderate, based on the viewer-side material's reflectivity or sheen. Assuming the projection room is finished in flat black, most ambient light behind the screen will be absorbed while still in the room.

Rear-projection screen materials offer a wide viewing angle—horizontally and vertically—but with little or no gain. The light from the projector is transmitted through the screen with relatively little bending. You may see some hot-spotting with rear projection, where an image is unevenly illuminated because of the transparency of the screen fabric and the vertical placement of the projector in relation to the audience.

Optical-pattern rear-projection screens capture and concentrate light to maximize the amount passing through to the audience. An image can appear less bright if viewers or the projector are moved off-axis. Therefore, the projector must be at the focal point indicated by the manufacturer.

Figure 5-7
Types of rear-projection screen material

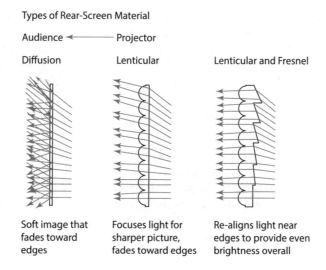

For permanent rear-projection applications, you'll want to use optical screen material. This type of screen system comprises a series of lenses formed into the screen material. The most common is a two-lens system, in which the lens that faces the projector is a *Fresnel lens*. It is usually a flat glass or polymer lens in which the curvature of a normal lens surface has been collapsed, creating concentric circles on the lens surface. This lens gathers the light from the projector. The second lens is a *vertical lenticular lens*. It must be flat to avoid hot spots. The lenticular lens faces the viewing audience and spreads the light horizontally, providing a relatively large viewing area. See Figure 5-7.

When creating specifications for a rear-projection screen, AV designers should consider the following:

- **The required size of the screen**   The raw screen should be of sufficient size to account for mounting and framing. This means the screen will need to be larger than the desired viewing area. Other size issues to consider include the type of framing to specify and whether the screen will fit through the doors, passageways, and lifts/elevators of the building (because a rigid screen cannot be folded or broken down into separate pieces). The size will also factor into the size of the wall opening required for installation.

- **Delivery**   Some screens will be larger than the doors to the space they're intended for and should be delivered prior to wall completion. Rear-projection screens are delicate and must be protected if stored onsite while a space is being completed. If the screen is installed early in the construction process, it must be well protected from accidental damage by other trades.

- **The side that will face the audience**   There are two sides of a projection screen— the projector side and the audience side. Typically, the projector side is smooth and clear, while the audience side is rougher. Always refer to the manufacturer of the screen to identify which side is designed to face the audience. Failure to do so will reduce the quality of the image.

- **Other attributes**   Rear-screen material can be lighter or darker in color. The darker the screen, the darker the blacks in an image will be. Whites will be less bright, but overall contrast should improve. In addition, optical specifications of screens need to be matched to the lens of the projector. There are screens designed for short-throw projectors and other special applications.

NOTE   When specifying the size of a rear-projection screen, anticipate mounting requirements. Most screen manufacturers offer mounting and framing options. Remember, the frame will reduce the actual viewing area slightly. You will need to provide the builder/building contractor/general contractor with the "rough opening" size for your screen and frame assembly. Given the room dimensions and the preferred projector and lens, most screen manufacturers can help with screen recommendations, mirror assembly drawings, framing options, and more, to make the design and installation go smoothly.

Rear-Projection Design Considerations

A designer must consider a number of factors when designing a rear-projection system, including the following:

- **Screen materials**   Diffusion rear-projection screens need to be specified with the required density, tinting, and protective coatings.
- **Screen frames**   Frames of the appropriate type, finish, and size create borders around a rear-projection screen.
- **Space requirements**   Because of the need for a dedicated and secure rear-projection room, designers must determine whether the advantages are worth the amount of floor space required. This is where direct-view flat-panel displays offer a substantial advantage.
- **Mirrors**   Mirror systems can help reduce the size of a rear-projection room by folding the optical path, but they require more time and skill to design and install. The first-surface mirrors required for folding rear-projection optical paths are expensive; readily damaged; and require special handling, protection, and cleaning—all factors that need to be accounted for during the project design.
- **Lens focal length**   Short focal-length lenses can also help reduce the size of a rear-projection room. But they can also create distortion and reduce screen brightness compared to longer focal-length lenses, which have a light path more closely aligned with the viewers.
- **Sources of light reflections**   Shiny objects in a rear-projection room can reflect light onto the screen and reduce viewability for the audience. All surfaces in the projection room need to be specified matte black, including doors, floors, and hardware. Indicator lamps and status screens on devices located inside the projection room may require masking out.

Videowalls

A videowall features multiple adjacent monitors, video screens, display cubes, video projectors, LED/microLED modules, or televisions that are configured to form a single, large, contiguous display.

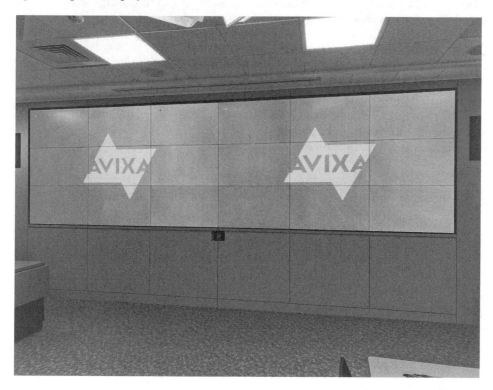

Depending on the component technology used, a videowall can give clients a large display surface with uniform brightness and clarity that can be harder to accomplish with traditional, single-projector systems.

Because videowalls often comprise multiple smaller displays, they can be designed in many sizes and shapes depending on the display environment. Not only will the size of the room dictate the size and shape of a videowall (a high or low ceiling, for example, is often a determining factor), but so will the available wall space. Available space will also determine the display technology used. One of the advantages of a videowall designed to incorporate flat display panels, for example, is that it requires less physical space than a videowall that incorporates projection technology or video cubes, which are often deeper than flat panels.

When designing a videowall, it is important to specify displays that can be color-matched in applications where one or a few images will span multiple screens. Remember that the sum of the parts will need to appear as one display when all is said and done, so the reds on one screen need to match the reds on another.

In a simple videowall system, a videowall processor, either built in or external, will take a single image and divide it among many screens. Alternatively, an external processor can route several different sources to different parts of the videowall. The processor maps the appropriate image (or portion of an image) to each display and synchronizes them. More advanced videowall processors can adjust the output image size for display on videowall screens that aren't necessarily identical in shape, size, or pixel density.

Common Videowall Applications

The choice to design a videowall is driven by the application. Videowalls can be more expensive and more intricate than more traditional video systems, so there should be a compelling reason to include one. That said, video technology has become better and less costly, so more and more clients see videowalls as a way not only to accomplish their business goals but also to create visually engaging experiences—the "wow" factor. Here are some typical videowall applications:

- Digital signage
- Corporate lobbies
- Sporting arenas and stadiums
- Boardrooms
- Command-and-control centers
- Emergency operations centers
- Television studios
- Museums
- Scientific visualization
- Public spaces
- Simulation and training
- Live productions
- Sports bars
- Lecture and presentation spaces

In general, videowalls are fertile ground for AV designers to be creative and tailor video solutions to their clients' needs. Designers can mix and match technologies, screen orientations, source content, and more in unique solutions.

Videowall Design

Because videowalls can vary widely in their composition, it is especially important to understand what the client intends to use a videowall for. A videowall can be a significant investment, so a thorough needs analysis is required to determine whether it's the best solution for the video application. Many of the questions you will ask are similar to those for other video systems:

- How much room is there for a videowall?
- Is there an equipment closet or other space for processors, switchers, and so on, as necessary?
- What type of information will be displayed? Is any of it protected content?
- How many people will view it, and what will the layout of the space be?
- What and how many video sources will feed images to the videowall?

Most of the principles that dictate other video systems design also apply to videowall design. You will need to determine image specifications, viewing distances, and viewing angles. You will also need to take into account ergonomics and human factors (see Chapter 4). But videowalls also require some unique considerations, including but not limited to the following:

- **Bezels** Because some systems consist of individual screens, videowalls can include gaps or borders (mullions) where the screens come together. This is most prominent when flat-panel displays are used; projection displays, LED/microLED modules, and display cubes don't typically show noticeable gaps. It is important to determine from the client how much of a gap or border around screens will be tolerated. The more the client is trying to create a single, uninterrupted image, the more necessary it may be to specify displays with thin to no bezels—the frames that hold the screen in place. Videowall processors or technology built into many videowall screens can help compensate for gaps, which can make moving or still imagery look poor or unnatural when it spans multiple screens across bezels.

- **Hours of operation** Some videowalls are intended to be on 24/7—whether for digital signage or a command-and-control room. Knowing this will help guide your design. A more mission-critical videowall will require sturdier, more-reliable components. In addition, should there be a technical problem with one or more screens in a videowall, the client's reliability requirements may guide you to a solution that is easier to maintain. For example, some videowall displays can be serviced from the front of the wall.

- **Source inputs and processing** Depending on the application, videowall sources—and their location—can vary greatly. In a more centralized design, the sources are local to the videowall and feed directly either into the processor (or, in the case of a simple videowall, directly into the displays) or into a switcher, or both. But videowall sources can also be remote, as in the case of an emergency operations center that needs to be able to patch in remote feeds to view situations in the field. In this case, the format of video sources can vary significantly, and they likely will be delivered over a network. Such a scenario may require a more distributed processing model.

- **Projector blending** Videowalls can include video from multiple projectors, combined to create one seamless image. When this is the case, the projected images overlap at the edges. A video design that includes projectors with blending capabilities or processing to blend the overlapping edges is necessary.

Videowalls are a rapidly emerging specialty of AV design. If, during the needs analysis, you and the client determine a videowall is a proper solution, you will determine other factors beyond the scope of this guide, such as screen orientation, wall composition, source resolutions and formats, pixel density, user control, and more.

Display Environment

Once you've performed image specification and display selection, the next step is to consider how the display environment will affect a video system. Mostly, this comes down to making calculated predictions and measurements about how light will affect your design. For example, you need to know how much reflected light might bounce off a screen and into viewers' eyes. You should be able to calculate precisely projector brightness given ambient light, screen area and gain, contrast ratio, and other factors. Much of how an AV design accounts for the display environment comes down to the designer's understanding of light.

Measuring Light

Light affects every aspect of the user's visual experience, which makes it important to be able to accurately measure and quantify all aspects of light.

The intensity of light is measured using two types of meters:

- *Incident meters* measure the light coming directly from a source such as a lamp, projector, lighting fixture, digital sign, videowall, or monitor.
- *Reflected meters,* or *spot meters,* measure the light that bounces off an object like a projection screen, a wall, a participant in a video conference, or a work surface.

The units of measurement for light vary by geographic region. Every country except for the United States, Myanmar, and Liberia uses the SI international system of units. You need to be able to recognize U.S. and international units for both direct light and reflected light.

Fundamental Units

Two fundamental units are used in light measurement:

- **Candela** The *candela* is the base unit of luminous intensity. It is the amount of light emitted by a light source that just happens (after a bit of manipulation of the specification) to have a similar output to a historic standard wax candle. That's why it's called a candela, which is the Latin word for candle.

- **Lumen** The lumen is the unit of *luminous flux*—the quantity of photons emitted from a 1 candela light source through an area of 1 square meter on a sphere 1 meter in radius surrounding that light source (1 steradian). Lumens are the most common unit of measurement of light output from a projector or a light source. However, different organizations measure lumens by different methods. Although more accurate and more complex methods for calculating brightness have been developed, and despite being retired in 2003, the most commonly quoted standard for the lumen measurement worldwide is still the method specified by the American National Standards Institute (ANSI) in ANSI/NAPM IT7.228-1997 and ANSI/PIMA IT7.227-1998.

Direct Light Measurement

Using an incident light meter, you can measure the brightness, or *illuminance,* of an emitting light source. Two units of measure are commonly used for illuminance in the AV industry:

- **Lux** Lux is the international unit of illuminance, the measure of the luminous flux (lumens) falling on a specified area. One lux is equal to 1 lumen/square meter. Generally, illuminance is measured at a task area such as a video screen, a note-taking location, or a work area.
- **Foot-candle** The foot-candle (fc) is the U.S. customary measurement of illuminance. One foot-candle is equal to 1 lumen/square foot. One foot-candle equals approximately 10.76lux. One lux equals approximately 0.093fc.

Figure 5-8 shows the U.S. and SI units for measuring illuminance.

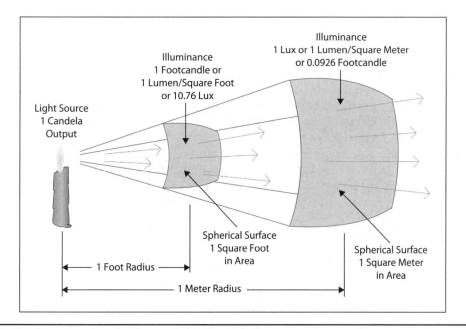

Figure 5-8 U.S. and SI units for measuring illuminance

Reflected Light Measurements

Using a reflected light or spot meter, you can measure the *luminance* of light emitted from a surface area. Two units are commonly used for luminance in the AV industry:

- **Candela per square meter or nit**   Sometimes referred to as the nit (nt), the candela per square meter is the international unit of luminance. A nit is equal to a luminance of 1 candela per square meter.
- **Foot-lambert**   The foot-lambert (fl) is the U.S. customary unit of measurement for luminance. It is equal to $1/\pi$ candela per square foot. A foot-lambert is approximately 3.43 candelas per square meter.

System Black

Projectors can't project black light. Instead, a bright projected image tricks the brain into perceiving a black area by making whites brighter.

System black is the lowest level of luminance a system is capable of producing for a task under defined operating conditions. System black is a function of three parameters:

- The material of the display screen
- The ambient light level
- The light from the display with a solid black image input

Contrast Ratio

In the AV industry, the level of contrast between black and white in a displayed image is critical. A person who views a display needs to be able to distinguish clearly the edges of objects and the subtle changes of color that result from proper contrast. If the difference between black and white is not great enough, the resulting images are difficult to see and text is difficult to read.

Contrast ratio describes the dynamic video range of a display device as a numeric relationship between the brightest color (typically white) and the darkest color (typically black) that a system is capable of producing.

Ensuring a high-contrast ratio will make projected images appear crisper and seem to have more depth. High-contrast ratios will make reading text easier because viewers' eyes will be able to find the edges of the characters.

A 16-zone checkerboard pattern projected onto a screen shows eight areas of white and eight areas of black (see Figure 5-9). Using a reflected light meter, AV professionals take measurements of the whites and blacks. When both sets of measurements are averaged and expressed as a ratio (white:black), you arrive at the contrast ratio.

According to the V201.01:2021 standard (often referred to as ISCR), the formula for calculating contrast ratio is as follows:

$$\text{Contrast ratio} = \text{Luminance}_{\text{average max}} / \text{Luminance}_{\text{average min}}$$

where:

- $\text{Luminance}_{\text{average max}}$ = the average level of luminance of all eight white squares
- $\text{Luminance}_{\text{average min}}$ = the average level of luminance of all eight black squares

Figure 5-9 Checkerboard pattern for measuring contrast ratio in accordance with ANSI/AVIXA V201.01:2021 Image System Contrast Ratio (ISCR) standard

ANSI/AVIXA V201.01:2021

Because images are often the centerpiece of AV systems, the image contrast ratio is one of the most important criteria. The ANSI/AVIXA V201.01:2021, *Image System Contrast Ratio (ISCR),* standard was developed to ensure high-quality images.

This standard applies to image display systems, including both projection and direct-view displays. It defines acceptable contrast ratios for the four application types previously described, based on criteria within a space, including ambient room light. The following are the minimum contrast ratios:

- 7:1 for passive viewing
- 15:1 for basic decision-making
- 50:1 for analytical decision-making
- 80:1 for full-motion video (home theater)

The standard includes a set of simple measurements to verify that a system conforms to the desired contrast ratio of the viewing task. We will explore the ISCR standard in more detail in the following sections.

Five Viewing Positions

The ANSI/AVIXA V201.01:2021 ISCR standard calls for taking measurements in five viewing locations: closest left and right, farthest left and right, and center. If there is an obstruction in the center, such as a table, designers should use the first available position behind the obstruction.

Designers—or during actual installation, AV technicians—should measure contrast ratio using a 16-position checkerboard test pattern and calculate the contrast ratio at each viewing location using a spot meter. See Figure 5-10.

ISCR Conformance

Standards conformance means you meet the conditions laid out in a standard. Conforming to the ANSI/AVIXA V201.01:2021 ISCR standard means the contrast ratio of your display design matches the standard specifications from all viewing locations.

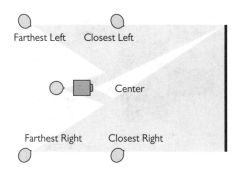

Figure 5-10 Five viewing positions for measuring contrast ratio, according to the *Image System Contrast Ratio (ISCR)* standard

Based on the contrast ratio determined by the measurement results, the image system either

- **Conforms** The contrast ratios at all five measurement (viewing) locations meet or exceed the contrast ratios required by the identified viewing category.
- **Fails to conform** The contrast ratio at any one of the measured locations falls below the identified viewing category. Should the space fail to conform, probable causes should be noted on the measurement form.

Projector Positioning

Calculating the correct projector position can be complex. The formula for projector position depends on the specific equipment. Manufacturer specifications include instructions for positioning the gear.

Projection positioning calculations are usually made in three dimensions, as shown in Figure 5-11. The dimensions represent the following:

- **H** The horizontal offset of the projector to the screen
- **V** The vertical offset of the projector to the screen
- **T** Throw distance

If you don't have access to a manufacturer-specific formula (web search engines are your friends), estimating the third variable, throw distance, can be useful all on its own. Estimating throw distance reveals how far away the projector needs to be from the screen. Your goal is to have an image fill the entire screen surface. Projectors with optical zoom and/or lens shift capabilities allow more flexible choices for positioning. Throw distances for ultra-short-throw projectors have very little margin for error due to the short distances and the extremely wide-angle lenses.

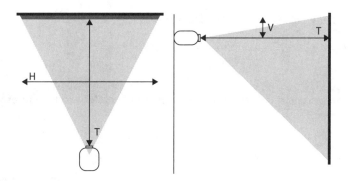

Figure 5-11 Three dimensions are relevant to projector positioning: horizontal offset (H), vertical offset (V), and throw distance (T).

Projector Light Path

When designing a front-projection system, you need to make sure that the projector sightlines are unobstructed. Obstructions can be found in the ceiling, such as lighting fixtures, monitors, truss systems, fire sprinklers, microphones, soffits, uneven ceiling, or loudspeakers. They're also on the floor, in the form of standing or sitting people, lecterns, tables, autocue/teleprompters, and microphone stands.

If you know that a room will be used for presentations, place presenters and their teleprompters and lecterns out of the projector's light path. Viewers may become frustrated if the image is frequently obscured by presenters. From the presenter's point of view, staring directly into a bright projector for an extended period of time will be uncomfortable and will limit their ability to read their notes and see the audience.

Projection Throw

Throw distance refers to the distance that a specific combination of projector and lens needs to be from the projection screen to project a specific image size.

The *throw ratio* is the ratio of the throw distance to the image size for a specific projector and lens combination. Throw ratios vary widely depending on the model, manufacturer, and lens type of a projector.

Manufacturers tend to have specific formulas that provide the best results for their projector and lens combination. Refer to the owner's manual of your specific projector and lens combination to find the most accurate throw distance and projector position relative to the screen. The projector manual will contain the correct lens information and reference point of measurement (e.g., screen to front of lens). Remember that all lenses are not the same. Also, the distance from the projection screen may vary because screen size and lens zoom ratio will affect projector location. In addition, the image processors in a projector can create images of different sizes.

The formula for estimating throw distance is as follows:

Throw distance = Screen width × Throw ratio

where:

- *Throw distance* is the distance from the front of the lens to the closest point on the screen.
- *Screen width* is the width of the projected image.
- *Throw ratio* is the ratio of throw distance to image width.

You may need to refer to the owner's manual of your chosen projector and lens combination to find an accurate formula for your specific projector.

Let's try the throw distance formula. Say a projector has an installed lens with a throw ratio of 2:1 (you'll get that information from the manufacturer documentation). The width of the screen in the room is 2 meters. What's the throw distance?

Distance = Screen width × Throw ratio
Distance = 2m (78.7in.) × 2
Distance = 4m (157.4in. or 13.1ft)

And let's try it again. A projector has an installed lens with a throw ratio of 2.5:1. The width of the screen in the room is 2 meters. What's the throw distance?

Distance = Screen width × Throw ratio
Distance = 2m (78.7in.) × 2.5
Distance = 5m (196.8in. or 16.4ft)

Compare those two calculations. Notice that the screen width stays the same, but the throw ratio determines how far away the projector is from the screen.

 TIP When using a lens that has an optical zoom capability, use two throw ratio calculations that provide the range from closest distance (minimum) to farthest distance (maximum) for a given combination of projector and zoom lens.

Predicting Projector Brightness

Image brightness is affected by a range of factors, including

- Brightness of screen
- Screen gain
- Lumen output of the projector
- Age of projector light source
- The maintenance status of the projector
- The lens aperture and f-stop settings

Designers will need to perform more calculations to determine a suitable brightness level for a viewing environment. Let's examine several of the factors that will go into brightness calculations.

Screen Gain

A projection screen is a passive device and does not have the capacity to amplify or create brightness. But the surface of a screen can focus reflected light to help increase the apparent projector brightness.

Screen gain is the ability of a screen to redirect projected light to make the image appear brighter within the viewer area. As a result of this ability, off-axis viewing is diminished. Screen gain refers to the amount of light reflected by the screen materials, not the amplification of light.

The higher the gain number of a screen, the brighter the picture viewed on-axis. When someone is seated within the good viewing area, a screen should provide uniform brightness over the entire image area, with no dim areas or hot spots.

Ambient light rejection generally increases as screen gain increases. This is because the light becomes more directional as screen gain increases and reflected ambient light appears outside the viewing cone.

NOTE Ambient light will have its own viewing cone, which should be outside the projected image's viewing cone. If this is not the case, you'll need to adjust the lights.

A gain of 1.0 refers to a matte white screen surface that does not absorb or channel light, is uniformly reflective, and can be seen from all directions. A glass-beaded screen that has a gain of 1.3 or 1.5 is highly reflective, is less uniformly reflective, and has more hot spotting.

Hot Spotting

As gain increases, so does hot spotting; it manifests as a brighter area on the screen. For example, when shining a flashlight on a wall, the reflected spot is not of uniform brightness. The brightest area in that spot is the hot spot.

Hot spotting is one of the trade-offs for screen gain in front-projection screens and some rear-projection screens and is most noticeable when viewing a projected image off-axis. Screen manufacturers work to develop surfaces that yield high gain with minimal hot spotting.

Light Source Life

A projector's brightness will be greatest on the first day of installation and decline over the life of the projector's light source (see Figure 5-12). When calculating light levels, you will need to account for derating the light output (lumens) of the projector. This will allow you to factor in the average amount of decay over the life of a projector lamp.

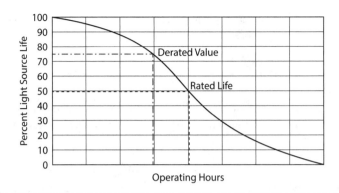

Figure 5-12 Light source life derating over time. Operating hour figures are not displayed because each light source behaves differently.

Light source life is defined as the time it takes for the light source to reach 50 percent of its initial light output. With some light sources, as they pass the 50 percent point, there is also a noticeable shift in the color balance of the output.

For the purposes of derating a projector's specified output, we select the halfway point between 100 percent and 50 percent brightness: 75 percent. This means the projector should be derated by a factor of 0.75 of its specified brightness. Some solid-state light sources have an operational life approaching the total operational life of the projector and thus may be derated by less than 0.25 of the initial output. The derating factor will be important to calculating a projector's required brightness, which we will discuss in the section "Calculating Required Projector Brightness."

Projector Lens

When a projector manufacturer writes its specifications, it usually describes the lens and electronics as a complete package. The projector's lens has a specific f-stop, or *focal ratio*, which is the ratio of the lens's focal length to the diameter of its aperture (opening). The f-stop value indicates the amount of light that is lost as it travels through the lens (see Table 5-2).

If you select a lens that's not the lens normally supplied with the projector, it is important to compare the new lens f-stop with that of the old lens. The difference in f-stops can mean the projector's light output goes up or down. A full f-stop—going down from f/2 to f/2.8, for instance—will cut the specified light output in half. Similarly, going up from f/4 to f/2.8 will double your light output. Either way, if you don't plan around the f-stop of a projector lens, it can lead to major issues with the system's performance.

Decreasing aperture size and transmission →								
f/1.4	f/2	f/2.8	f/4	f/5.6	f/8	f/11	f/16	f/22
← Increasing depth of focus								

Table 5-2 Some Common F-Stop Values Where Each Increasing Step Transmits Half as Much Light as Its Predecessor

Ambient Light Levels

Let's consider a simple front-projection environment. Light arrives at the screen from multiple sources:

- The projector, which is focused directly on the screen to display an image
- Overspill and reflections from the task light required for the audience to see around the room and take notes
- Overspill and reflections from the lights required for the audience to see a presenter
- All the ambient light from other sources, including that from lighting fixtures, the sun, the presenter's lectern lights, video monitors, equipment indicator lights and control panels, autocue/teleprompters, and light from the projected image that's reflected around the room

When designing a display environment, there is natural competition between ambient light levels and the projector light levels as they affect the screen. Reducing ambient light at the screen location is the most important element of creating a good contrast ratio for the projected image. It can also be the most difficult thing to do.

In new construction, communicate often with architects, interior designers, lighting designers, and electrical contractors. These allied trades have more flexibility to control ambient lighting. The lighting designer will be able to give you the planned light levels so you can factor them into your projector brightness specifications. This is also an opportunity for you, as the AV designer, to discuss issues that are important to the AV design, such as screen placement, lighting zones, and lighting control systems.

As a designer, it's important that you develop an awareness of lighting levels by using a light meter on all projects. Taking accurate measurements and making specific recommendations will help you communicate your needs to allied trades.

Calculating Required Projector Brightness

When you select a projector, you need to make sure the one you choose is bright enough for your viewing environment. That means taking a number of variables into account.

Some projection technologies, such as those based on LCD panels, do not completely block the light for black pixels and thus project a dark gray, not an absolute black. A white screen can only get brighter when light is applied, but the screen can never become darker. So, with black from the projector, what you have is a screen that is as black as it will ever be. To calculate what the white level would be with a contrast ratio of 15:1, for example, you simply multiply the black level by 15.

Divide the previous answer by the gain of the screen. You divide because the light will need to be less if the gain is higher; conversely, if the gain is low, you will need a brighter projector.

Dividing the final result by the derating factor (0.75) will increase the projector brightness to account for the decrease in light source performance over its life.

The formula for required projector brightness is as follows:

$$\text{Lumens} = [(L_A \times C \times A) \div S_G] \div D_R$$

where:

- *Lumens* = The required brightness of the projector in lumens.
- L_A = The ambient light level.
- A = The screen area.
- C = The required contrast ratio.
- S_G = The gain of the screen. Assume a screen gain of 1 unless otherwise noted.
- D_R = The projector derating value. Assume a derating value of 0.75 unless otherwise noted.

NOTE The units for ambient light (L_A) and screen area (A) measurements must match. If your ambient light measurement is in lux (lm/sqm), the screen area must be in square meters, but if your ambient light measurement is in footcandles (lm/sqft), the screen dimensions must be in square feet.

Measuring OFE Projector Brightness

As you see, it's possible to calculate brightness requirements for a given application to specify a projector. However, how can you be sure that a projector you already have on hand (owner-furnished equipment, or OFE) is actually producing the amount of light needed? When using a client's existing equipment in a design, you should test it to ensure it still meets its original specifications. Older devices may meet specifications but may not be suitable for reuse due to current environmental policies and regulations covering such matters as the energy efficiency or the chemical content of the light source. Even the performance of new equipment should be verified.

Calculating the output of a projector using the ANSI method requires an incident light meter, which is used to measure the light reaching the screen in each of the nine zones of a projected standard ANSI lumens graphic, as shown in Figure 5-13. The average of these measurements is then used to calculate the total amount of light, in lumens, being delivered to the screen. This is one of the primary issues when considering the amount of light that should be produced by the projector for the viewing environment.

The formula for calculating the ANSI brightness of a projector is as follows:

$$\text{ANSI lumens} = [(Z_1 + Z_2 ... Z_9) \div 9] \times (S_H \times S_W)$$

where:

- Z_1 to Z_9 = The incident light measured on the nine zones of a screen
- S_H = Height of the image
- S_W = Width of the image

Figure 5-13
ANSI nine-zone measurement grid

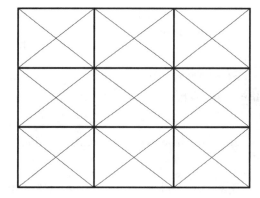

  NOTE When calculating projector brightness, the units used for incident light (Z_n) and area measurements (S_H and S_W) must match. If the incident light measurement is in lux (lm/sqm), the image dimensions must be in meters, but if the incident light measurement is in footcandles (lm/sqft), the image dimensions must be in feet.

Task-Light Levels

The difference between the white levels of an image and a viewer's work surface, or task area, should be within a range of three times greater to three times less than the white in the image. The purpose of keeping this level within this range is to keep eye strain at a minimum. Greater differences between the light levels (screen and task) require a viewer to adjust to differing light levels too frequently, resulting in eye strain and visual discomfort.

In many cases, task lighting can raise the ambient light in the room. This is especially true if the task lighting is broadly directed and the room treatments (wall coverings, table surfaces, ceiling, and floor) reflect a significant proportion of the incident light. In a properly designed room, the task-light measurements should be independent of the ambient-light measurements at the screen. Consider moving or directing light fixtures closer to the task area and away from the screen. Ideally, the lighting system would combine directed fixtures and dimming control.

Luminance differences are normally necessary for vision. Print, for instance, can be seen and interpreted because of the difference in luminance between the white page and the black print. In a display space, luminance variations are a function of both the reflectance of the surfaces and the distribution of light on those surfaces.

Large variations in luminance can be problematic. Office interiors should be lit to provide good visibility without distracting glare from either direct or reflected sources. It is, however, important to provide enough variation in luminance (or color) to contribute to a stimulating, attractive environment. Where there are no prolonged visual tasks, as in lobbies, corridors, reception areas, lounges, and auditoriums before and after presentations, greater variations in luminance are a good idea, using attractive colors and appropriate focal points of high contrast to direct the eye.

In a working office environment, luminance near each task area and other parts of the office within the field of view should be balanced with task luminance. Two separate phenomena are influenced by luminance ratios within the field of view:

- **Transient adaptation**   The requirement to adapt the eye to differing light levels.
- **Disability glare**   The loss of visual contrast due to a light source directly striking the eye.

To limit the effects of these phenomena, the luminance ratios generally should not exceed the following:

- **Between paper task and adjacent screen**   Between 3:1 and 1:3
- **Between task and adjacent dark surroundings**   Between 3:1 and 1:3
- **Between task and remote (nonadjacent) surfaces**   Between 10:1 and 1:10

It is neither practical nor aesthetically desirable to maintain these ratios throughout the environment. For visual interest and distant focus (for periodic eye muscle relaxation throughout the day), small visual areas that exceed the luminance-ratio recommendations are desirable. This would include artwork; accent finishes on walls, ceilings, or floors; small window areas; accent finishes on chairs and accessories; decorative lighting effects; and accent focal lighting.

Calculating Task-Light Levels

Once you know the projector brightness, you can solve for task lighting. The objective is to have a task-light level that is in a range of plus or minus three times the white level. Once you have selected a projector, you can find its lumen output within the projector's specifications. From this, you can calculate how bright the white level should be on the screen.

The following are the formulas for task-light levels:

$$\text{Task Light}_{upper} = [(\text{Projector Lumens} \div A) \times S_G \times D_R] \times 3$$

and

$$\text{Task Light}_{lower} = [(\text{Projector Lumens} \div A) \times S_G \times D_R] \div 3$$

where:

- *Task Light$_{upper}$ /Light$_{lower}$* = The upper and lower limits of the acceptable range
- *Projector Lumens* = Specified projector brightness
- *A* = Area of the screen
- S_G = Screen gain (usually 1)
- D_R = Projector derating factor (usually 0.75)

NOTE The units for Task Light$_x$ and screen area (A) measurements must match. If your task light measurement is in lux (lm/sqm), the screen area must be in square meters, but if your task light measurement is in footcandles (lm/sqft), the screen area must be in square feet.

Chapter Review

In this chapter, you studied the fundamental principles needed to determine image specifications, including text size, farthest and nearest viewers, and viewing area. All are critical components to consider early in your design.

With an understanding of image specifications, you're ready for display selection. Available options include front-projection, rear-projection, direct-view displays, and videowall systems. It is always important to remember how aspect ratios play a role in ensuring a quality image in your design.

Finally, you learned how environmental factors affect image quality and how to apply measurements to perfect your design. With the information you gained from this chapter, you can appropriately position and select equipment for your display design.

Review Questions

The following questions are based on the content covered in this chapter and are intended to help reinforce the knowledge you have assimilated. These questions are similar to the questions presented on the CTS-D exam. See Appendix E for information on how to access the free online sample questions.

1. A conference room requires videoconferencing and AV presentation capabilities. Which of the following design elements are defined by the architectural dimensions and orientation of the room?

   A. Display size

   B. Display type

   C. Display aspect ratio

   D. Display brightness

2. What device is required to measure the light reflected from a test slide on a projection screen to determine the contrast ratio of a projector?

   A. Spot meter

   B. Incident light meter

   C. Time domain reflectometer

   D. Illuminance meter

3. If the height of a symbol on a 1920×1080 screen used for BDM applications is 43 pixels, what is the percentage element height (%EH) of the symbol for use in the DISCAS calculator?

   A. 2 percent
   B. 5 percent
   C. 4 percent
   D. 10 percent

4. A matte white screen surface has a gain of _____.

   A. 0.5
   B. 0.75
   C. 1.0
   D. 1.25

5. A 16:9 screen will be installed in a lecture hall. The screen's diagonal is 1,830mm (72in.). What is its width?

   A. 1,270mm (50in.)
   B. 1,570mm (62in.)
   C. 1,780 mm (70in.)
   D. 1,830 mm (72in.)

6. If you have a projector that is 2,500 ANSI lumens on a 5.5m² standard matte white screen, what should be the limits of the task lighting? Note that the projector specification has a 0.75 derating factor.

   A. 114 to 1,023 lux
   B. 120 to 175 lux
   C. 515 to 1,001 lux
   D. 527 to 632 lux

7. Lamp life is defined as the time it takes for a lamp to reach _____ percent of initial output.

   A. 10
   B. 25
   C. 50
   D. 75

8. You need to project an image that has a 4:3 aspect ratio. You measure the screen height and determine that it is 2,286mm high and 3,048mm wide. Can you use this screen for the 4:3 image?

   A. Yes
   B. No

C. Not enough information to determine

D. All of the above

9. You require a 16:9 screen with a height of 1,520mm (60in.). What should the screen's diagonal be?

   A. 2,290mm (90in.)

   B. 1,520mm (60in.)

   C. 3,100mm (122in.)

   D. 2,540mm (100in.)

10. In videowall design, _____ is the process of adjusting for the edges of multiple projected images.

    A. Blending

    B. Pixel mapping

    C. Processing

    D. Orienting

Answers

1. **A.** For visual systems, the architectural dimensions and orientation of the room will determine display size.

2. **A.** A spot meter (reflected light meter) is required to measure the light reflected from a test slide on a projection screen to determine the contrast ratio of a projector.

3. **C.** The percentage element height (%EH) of a 43-pixel symbol on a 1920×1080 screen is 100 ÷ (1080 ÷ 43) = 3.9814, which rounds to 4 percent.

4. **C.** A matte white screen surface has a gain of 1.0.

5. **B.** Width/diagonal = Width ratio/diagonal ratio, and width/1830 = 16/18.3575. Therefore, width = 1,570mm (62in.).

6. **A.** Task light = [(2,500/5.5) × 1 × 0.75]. For the upper limit, multiply by 3; for the lower limit, divide by 3. Your task-light range, to the nearest lux, is 114 to 1,023 lux.

7. **C.** Lamp life is defined as the time it takes for a lamp to reach 50 percent of initial output.

8. **A.** Yes, because 3,048:2,286 is in the ratio of 4:3, or 3,048/2,286 = 1.3333 (4/3).

9. **C.** At 16:9, width = 2702mm. Diagonal^2 = 1,520^2 + 2,702^2. Diagonal = 3,100mm (122in.).

10. **A.** In videowall design, blending is the process of adjusting for the edges of multiple projected images.

CHAPTER 6

Audio Principles of Design

In this chapter, you will learn about
- Calculating the change in a sound and signal level using decibel equations, as well as power and distance using logarithmic functions
- Plotting loudspeaker coverage in a room and wiring loudspeakers
- Transformers and amplifiers
- Different types of microphones and polar-response patterns
- The benefits of using an automatic microphone mixer
- The qualities of an effective sound-reinforcement system
- Determining whether an audio system will be stable by comparing potential acoustic gain (PAG) and needed acoustic gain (NAG) calculations

Much of what you need to know as an audiovisual professional revolves around sound: how sounds are generated, how sound moves through a medium (such as air), how you might control its propagation, how sound interacts with the environment, and how you receive and process sound. As an AV designer, you need to apply math concepts and accurate measurements to support the human perception of sound as intended by your design. With an understanding of sound and sound systems, you will be better positioned to meet the needs of your clients.

For example, if you determine a loudspeaker system is necessary, your design will need to specify where the loudspeakers will be located. Your goal is to provide adequate audio coverage for listeners. To accomplish this, you will need to evaluate coverage patterns, loudspeaker locations, and power requirements.

Microphones play a vital role in ensuring your audio design supports your client's message, giving the audience an exceptional listening environment. Special calculations, such as PAG and NAG, reduce the possibility of audio feedback within a system. PAG and NAG determinations are among the many that AV designers make when creating a stable, effective audio system design.

> **Duty Check**
>
> This chapter relates directly to the following tasks on the CTS-D Exam Content Outline:
>
> - Duty A, Task 6: Identify Scope of Work
> - Duty B, Task 5: Coordinate with Electrical Professionals
> - Duty B, Task 8: Coordinate with Acoustical Professionals
> - Duty C, Task 1: Create Draft AV Design
> - Duty C, Task 2: Confirm Site Conditions
>
> In addition to skills related to these tasks, this chapter may also relate to other tasks.

Introduction to the Decibel

When discussing audio, you need a way to talk about how people experience the sound the system produces. The most common unit of measurement for sound is the decibel (dB), which is one-tenth of a bel. The bel (B) is the name given to the base-ten logarithmic ratio between two numbers (e.g., a 10 to 1 ratio equals 1 bel, and 1,000 to 1 ratio equals 3 bel). However, as the bel is a relatively large unit, the decibel has been adopted for most applications.

The advantage of a logarithmic ratio for comparisons is that a comparatively small range of ratios can describe variations over a range of values, varying over many orders of magnitude. For example: a 2 to 1 ratio is 3 decibels, while a 1 million (1,000,000 or 10^6) to 1 ratio is just 60 decibels, and a 1 trillion (1,000,000,000 or 10^9) to 1 ratio is only 90 decibels, thus saving on the writing and misreading of countless strings of zeros.

VIDEO Watch a video from AVIXA's online video library about logarithms. Check Appendix D for the link to the AVIXA video library.

A decibel does not really measure sound pressure levels (SPLs); it measures the difference between two SPLs, where one of those levels is the reference point. Because the human ear's response to sound is logarithmic rather than linear, expressing SPLs in decibels allows for a more linear representation of changes to sound levels.

Here are some accepted generalities in relation to human hearing:

- A "just noticeable" change—either louder or softer—requires a 3dB (2 times) change in SPL. For example, an increase from 85dB SPL to 88dB SPL would just be noticeable to most listeners.

- A 10dB (10 times) change in SPL is required for you to subjectively perceive either a change twice as loud as before or a change one-half as loud as before. For example, an increase from 85dB SPL to 95dB SPL is perceived by most listeners to be twice as loud as before.

Calculating Decibel Changes

Since a decibel is a comparison of two values, you need a starting point and an end point to compare. Those values will be real physical measurements in linear units, such as pascals (pressure), volts (potential difference/EMF), watts (power), or amps (current). The pascal (Pa) is the international unit for pressure, and 1 pascal = 0.00015 pounds per square inch.

As an example, calculating the decibel change from one voltage or power measurement to another will tell you how much louder or softer the output of a sound system will seem after that change.

To calculate the change in decibels between two power measurements, use the 10 log equation. For example, the equation

$$dB = 10 \times \log_{10}(P_1 \div P_2)$$

would give you the difference in decibels if you compared one power in watts (P_1) against another power in watts (P_2).

Because power, and hence SPL, from a source varies as the square of the voltage and current in a circuit ($P = V^2 \div R$ and $P = I^2 \times R$) and because field strength decreases as the inverse square of the distance from a source ($1 \div d^2$) due to the inverse-square law, a different equation is required to calculate ratios with these measurements. This quadratic (square) relationship requires using a 20 log formula for the difference in decibels when comparing two voltages, currents, or distances. For example, the equation

$$dB = 20 \times \log(V_1 \div V_2)$$

would give you the difference in decibels if you compared one voltage (V_1) to another voltage (V_2).

This equation

$$dB = 20 \times \log(D_1 \div D_2)$$

would give you the difference in decibels if you compared the sound pressure level at one distance from the source (D_1) to the sound pressure level at some other distance away from the source (D_2).

Now that you know how to perceive differences in sound levels logarithmically and that you use the decibel for a logarithmic scale, let's explore further how the decibel is used.

Decibel Reference Levels

A decibel can be a comparison between two arbitrary values, or it can be a comparison of a value to a predetermined starting point, known as a *reference level*. This is also sometimes referred to as a *zero reference*. For example, you could compare two voltages using the 20 log formula to calculate the decibel ratio between them. Or you could find

Quantity	Decibel Abbreviation	Reference Level
Sound pressure	dB SPL	20µPa at 1kHz
Voltage (often used in consumer electronics)	dBV	1V
Voltage (mostly used with professional AV equipment)	dBu (dBv)	775mV
Power	dBW	1W
Power	dBm	1mW

Table 6-1 Common Reference Levels Used in Audiovisual Applications

an absolute voltage measurement in decibels by comparing a measured voltage to its reference voltage level.

For sound pressure, the reference level is the threshold of human hearing at 1 kilohertz, which is 20 micropascals (µPa). Human beings perceive that sound pressure level as silence. Any unit you might quantify in decibels has its own reference level.

As shown in Table 6-1, decibels will be abbreviated differently to indicate the reference level used. For example, dB SPL indicates that the reference level is a sound pressure level of 20µPa at 1kHz, whereas dBV indicates that the reference level is a voltage of 1V.

Some units, such as volts and watts, have more than one zero reference, as shown in Table 6-1. Use the one that makes sense for the application. If you are taking power measurements at a radio station, use dBW. If you are measuring the power of wireless microphones, use dBm.

TIP You will encounter specifications expressed as negative decibel values. For instance, you might see a microphone sensitivity specification of −54dBu. That doesn't mean the microphone has negative voltage. Remember, a negative decibel measurement means "less than before" or, in this case, "less than the reference level." The specification of −54dBu just means that the mic signal voltage is 54dB less than 775mV, the reference voltage level for 0dBu.

Pro-audio line level is often expressed as +4dBu, which expresses a voltage above the 775-millivolt reference point, or 1.23 volts. If you have a microphone level of −50dBu, that does not mean the device has "negative voltage." Instead, it is expressing a voltage level less than the 775-millivolt reference point, or 2 millivolts.

0dBV is equivalent to 1 volt.

The consumer line level is often expressed as −10dBV, or 316 millivolts. A good clue that a device's level might be −10dBV is the use of the phono or RCA connector. Portable personal device and computer outputs are usually 3.5mm (1/8in.) jack connectors at a level of −10dBV.

As an AV professional, you should commit certain decibel values to memory:

- Sound pressure level should always fall between 0 and 140dB SPL.
- Microphone level, which is typically measured in dBu, should be −60 to −50dBu, well below the zero reference of 775 millivolts for the dBu.

- Line level, on the other hand, should be between 0 and +4dBu for pro audio.
- The consumer audio level is −10dBV (316mV).

NOTE The dBu uses a lowercase *u*, while the dBV uses an uppercase *V*. This is to avoid confusion between the two.

Perceived Sound Pressure Level

Not all sounds with the same sound pressure level seem equally loud. Very low- and high-frequency sounds are harder to hear than sounds in the middle of the audible frequency range. In addition, the quieter a sound at the fringes of human perception is, the harder it is to hear.

Figure 6-1 shows a graph of the equal loudness curves. The threshold of human hearing is 0dB SPL at 1kHz. This graph shows how loud different frequencies must be for the human ear to perceive them as equally loud as another. For instance, what sound pressure level would be required of a 40Hz tone for it to be to be perceived equally as loud as a 1kHz tone? The answer depends on the tone's actual sound pressure level.

The dotted curve on Figure 6-1 represents the threshold of human hearing, which is what the human ear perceives when listening to a 1kHz reference tone. The x-axis of the graph shows actual frequency, and the y-axis shows actual dB SPL. At the threshold of

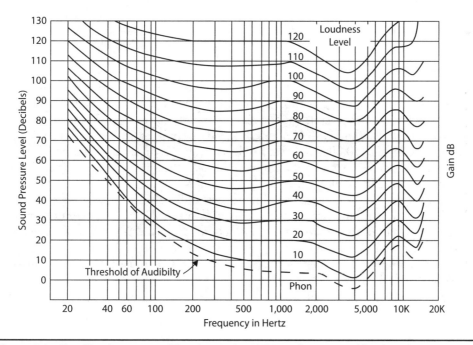

Figure 6-1 Equal loudness curve. (The phon is the ANSI unit for perceived loudness.)

human hearing, the 40Hz tone must be about 50dB SPL louder before the human ear can perceive it equally as loud as a 1kHz tone. A 200Hz tone, however, would have to be only 15dB SPL louder to be perceived equally as loud as a 1kHz tone.

Notice that at overall louder listening levels, the hearing response curve begins to flatten. A 90dB, 40Hz tone seems as loud as a 70dB, 1kHz tone. Human perception of the energy across the audible spectrum is more even at overall louder listening levels. Also note that your ears are more sensitive in the normal speech frequency range, 500Hz to 4kHz. And your ears are most sensitive to higher-pitched sounds, such as a crying baby.

You will use the decibel to measure many aspects of audio system performance: ambient noise, audio signal level, gain structure, loudspeaker performance, and so on. As you build and verify an audio system, it is important to keep in mind that humans and machines respond differently. An audio system may have a "flat" response, reproducing all frequencies equally. The human audio system, our ears, does not.

TIP As part of your needs assessment, you should document your customer's SPL needs. Speech and program audio may have different sound-level requirements. Your rationale for determining the level above ambient noise, for example, should take into account your customer's input. Designers must document such agreed-upon levels so installers can set them.

VIDEO Watch a video from AVIXA's online video library about human perception of equal loudness across frequencies. Check Appendix D for the link to the AVIXA video library.

Measuring Background Noise

Although there are better metrics for quantifying background noise levels and their effect on listeners, a simple method is to measure the SPL of the background noise. To gauge background noise levels, the SPL should be measured using an A-weighted scale.

As an example, the ANSI/ASA S12.60 standard specifies that for learning spaces, 566 cubic meters (20,000 cubic feet) or less in volume, maximum; one-hour background SPL levels, including that from building services (HVAC, etc.), should not exceed 35dB SPL A-weighted.

Given the maximum background noise level of 35dB SPL A-weighted specified in the standard and the fact that a space should acoustically have a minimum 25dB acoustic signal-to-noise ratio (the level of a desired signal compared to the level of background noise), you arrive at a targeted speech level of 60dB SPL A-weighted. However, a sound reinforcement system for speech may more typically operate in the range of 70 to 75dB SPL.

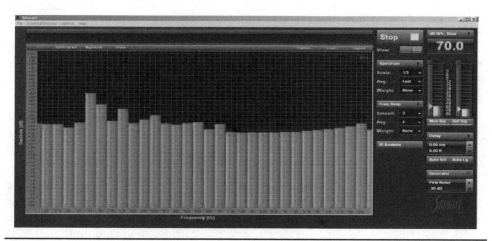

Figure 6-2 Unfiltered audio signal on a spectrum analyzer

SPL Meter Weighting: Spectrum Analysis

Let's take a look at two spectrum-analyzer readings taken from the same room. The first reading, shown in Figure 6-2 and fairly flat, was taken with no weighting applied. Notice that the overall dB SPL reading is 70.

The second reading, shown in Figure 6-3, was taken using an A-weighted filter. The overall level, especially the low-frequency energy, is shown as reduced; now the meter reads 53.6dB SPL. This reading reflects how a listener would actually perceive the noise in the room.

 TIP If you use a filter when taking an SPL reading, you must note which filter you applied. When recording the readings in Figures 6-2 and 6-3, you would write the first reading down as "70dB SPL" because no filter was used. The second reading should be written "53.6dB SPL A-wtd" to indicate that you used an A-weighting filter when you took the reading.

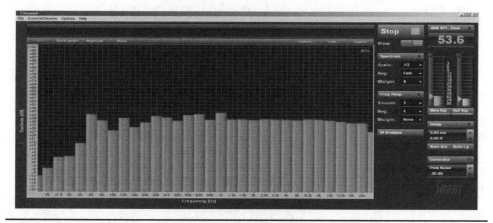

Figure 6-3 Audio signal on a spectrum analyzer with an A-weighted filter

Loudspeaker Systems

All loudspeakers have coverage patterns in which they project sound waves at specific frequencies. Therefore, different loudspeaker types and configurations are needed for various audio applications. There are two categories of loudspeaker systems: point source and distributed.

Point-Source Systems

Point-source systems are used where the directionality of the sound source is important. If the sound should appear to originate from a particular place in space, one or more loudspeakers, or clusters of loudspeakers, are positioned to give the illusion of sound coming from that place. Sometimes the system consists of a single cluster of loudspeakers directly above or behind the source point, but more frequently the system consists of multiple loudspeaker clusters with their signals phase-, time-, and amplitude-aligned to give the illusion of point-source directionality.

The basic two-source stereo speaker arrangement for the audio component of a video replay system may extend out to become dozens of loudspeaker clusters, distributed throughout a space, for the replay of highly spatialized, surround-sound content for cinematic screenings and visualization spaces.

For many productions in large spaces, phase-aligned arrays of loudspeakers are placed above, below, and to the sides of the stage to generate a coherent wave-front for the entire audience. These are often supplemented with additional time-aligned speaker arrays located deeper into the audience space to augment the sound levels for more distant listeners.

Distributed Systems

A distributed system delivers even sound coverage of the same program material throughout a space. They are used where even coverage is more important than drawing people's attention to the directional source of the sound or where physical limitations prevent the use of point-source systems. A distributed loudspeaker system employs multiple loudspeakers that are separated from each other by some distance. This is most usually accomplished by installing the speakers in the ceiling above an audience area.

Designing Loudspeaker Coverage

To design a distributed layout, you first must know how much area the sound from each of your selected loudspeakers will cover. The polar-pattern directivity information provided by the loudspeaker manufacturer enables you to create a view of the loudspeaker pattern. This will allow you to identify a circular area that each unit will cover.

Loudspeaker coverage patterns are frequency dependent. This means that the exact pattern of coverage will depend on frequency. In other words, a loudspeaker rated at 90 × 40 degrees only covers the quoted 90 × 40 degrees at a certain frequency, not across the entire audible frequency range.

Balloon plots are typically part of a loudspeaker's specifications and are often available in computer files for use in various modeling programs, such as Enhanced Acoustic

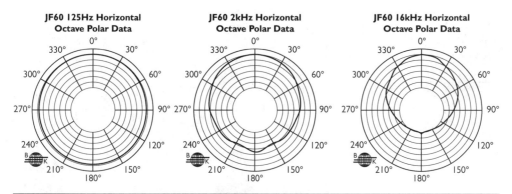

Figure 6-4 Polar plots of different frequency bands for a loudspeaker

Simulator for Engineers (EASE). Computer modeling is commonplace (and necessary) in all but the simplest installations.

Figure 6-4 shows some of the plots from the EAW JF60 loudspeaker. Each concentric circle on these charts represents a change of 5dB SPL. A polar pattern shows how far off-axis a loudspeaker's coverage pattern extends at a given frequency. You can see that at high frequencies this loudspeaker's energy drops off steeply as you move off-axis. As you move off-axis, look for the point at which the range of frequencies you're using drops off by 6dB SPL. The "6dB down" point is typical for defining a loudspeaker's coverage pattern.

The specification sheet for this loudspeaker says its "nominal beam width" (coverage pattern) is 100 degrees horizontal by 100 degrees vertical. The axial grids are 5dB divisions. Notice that the coverage pattern (dispersion) at 2kHz is about 100 degrees wide. In other words, when comparing the level on-axis with the level 50 degrees off-axis, you should find a reduction in level of about 6dB at 2kHz.

The pattern shown at 125Hz disperses in an omnidirectional pattern. Does this mean that low frequencies are omnidirectional? Not at all. It has to do with wavelength. It takes a larger device to control the dispersion pattern associated with the longer wavelengths of lower frequencies.

Realistically, the pattern shown for 2kHz is the only one of these three that comes close to being 100 degrees by 100 degrees.

 TIP Coverage is typically stated at the 6dB down points. This means the level at the edge of the stated coverage pattern would be found to be 6dB less than the energy measured on-axis.

Calculating Loudspeaker Coverage

To create a loudspeaker layout, you must first determine two things: loudspeaker coverage angle and listener ear level.

By referencing the polar pattern information, you find the angle at which your highest target frequency is 6dB below the on-axis level. For example, this might be at 40 degrees off-axis, which would provide a full 80 degrees of coverage for that target frequency.

Figure 6-5
Calculating loudspeaker coverage

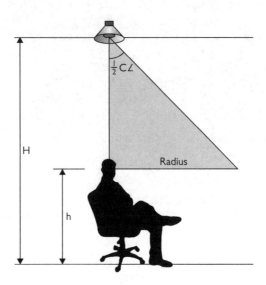

Next, you must determine the listening ear height—the highest level being of most interest. For example, if you are designing a multipurpose room where the audience may be standing for some presentations and seated for others, design for the standing audience. The size of a loudspeaker coverage circle can be dramatically different when you factor in a low ceiling and a standing audience versus a higher ceiling and a seated audience, even with the same loudspeaker. See Figure 6-5 to help visualize your loudspeaker coverage calculation.

The formula for calculating the diameter (twice the radius) of the circle that represents the coverage area of a loudspeaker is as follows:

$$D = 2 \times (H - h) \times \tan(C\angle \div 2)$$

where:

- D is the diameter of the coverage area.
- H is the ceiling height.
- h is the height of the listeners' ears.
- $C\angle$ is the loudspeaker's angle of coverage in degrees.

For example, if your loudspeaker provides 80-degree coverage, your ceiling height is 4 meters, and the audience is seated with an ear level of 1.2m, the following would give you your coverage area:

$$D = 2 \times (H - h) \times \tan(C\angle \div 2)$$
$$D = 2 \times (4 - 1.2) \times \tan(80 \div 2)$$
$$D = 2 \times (2.8) \times \tan 40$$
$$D = 5.6 \times 0.839$$
$$D = 4.70 \text{ meters, or a radius of } 2.35 \text{ meters}$$

Distributed Layout Options

When distributing loudspeakers, the goal is to place them in a strategic pattern to create a uniform sound source. You might design a distributed loudspeaker system when it's not possible to implement a point-source system. For example, a certain ceiling height may be inadequate for a centralized, point-source speaker system because all listeners don't have a good line of hearing to a central speaker or cluster.

You can use a number of patterns for a distributed loudspeaker system. As the amount of background noise in a space increases or the reverberation becomes high, loudspeaker pattern control becomes very important. Ultimately, the pattern you choose may be a compromise between the ideal and the attainable. Often, the available space for loudspeakers—or the budget—is less than desirable.

Uniformity of coverage is the greatest difference between distributed loudspeaker patterns. Generally, the denser the pattern, the more uniform the coverage will be, but increasing density will increase interaction between loudspeakers.

Designers should strive to minimize coverage where it's not needed or where it can create problems. For example, you shouldn't place loudspeakers where the audio will strike a wall before reaching listeners' ears. This can cause uneven frequency response and phase cancellation at the listener position. Too much sound at the head of a conference table can cause feedback, so designers should make sure to implement a way to reduce levels or turn off offending loudspeakers in that area—or don't put them there in the first place.

Figure 6-6 shows six common arrangements for distributed ceiling systems.

Edge-to-Edge Coverage

Edge-to-edge coverage places the loudspeakers in such a way that the farthest extent of their acoustic energy comes together at listeners' ear level. There is no overlap with edge-to-edge patterns and therefore significant gaps. One loudspeaker's coverage area is simply adjacent to the next; therefore, the distance between loudspeakers in this layout is two times the radius of the loudspeaker's coverage pattern: (*Distance = 2 × r*).

Figure 6-6 Six common coverage configurations for distributed loudspeakers

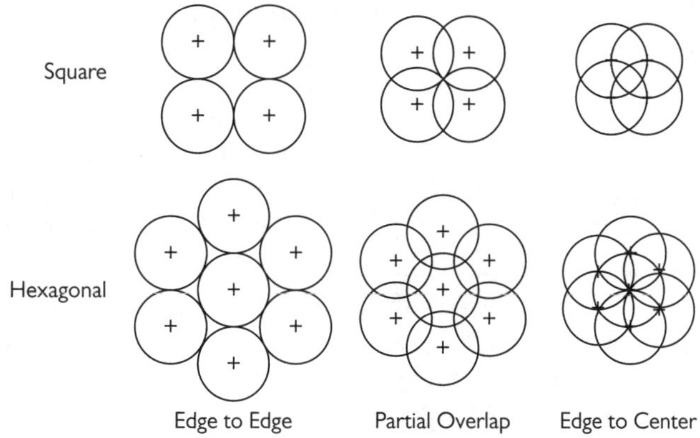

This approach is inexpensive and results in minimum interaction between loudspeakers within a room. But an edge-to-edge configuration may result in an uneven SPL with low spots in the corners of the coverage area. This is least favorable in a business communication setting but may be appropriate for general (noncritical) paging or simple background music.

Maximum to minimum coverage variations for this configuration are around 4.35dB. When deployed in a hexagon edge-to-edge pattern (the lower edge-to-edge pattern), coverage variations are around 5.4dB.

Partial Overlap Coverage

Coverage patterns with minimal or *partial overlap* are among the most common methods of laying out distributed loudspeakers. In partial overlap systems, each loudspeaker's coverage pattern overlaps about 20 percent of the adjoining speaker's coverage pattern. Specifically, the minimum overlap in a square layout should be the radius of the loudspeaker polar pattern, multiplied by the square root of 2: (*Distance = r × 1.41*). In a hexagonal layout, that distance is the radius of the loudspeaker polar pattern, multiplied by the square root of 3: (*Distance = r × 1.73*).

Partial overlap provides good coverage at most frequencies, with 3dB of variation, and ensures few or no "low" spots. However, this approach may not produce a perfectly even frequency response because dispersion patterns vary according to the frequency of the sound. In addition, partial overlap may create some acoustic interference between adjacent loudspeakers.

Edge-to-Center Coverage

An *edge-to-center* layout, or 50 percent overlap, will provide excellent coverage at most frequencies (with 1.4dB variation). However, this is a costly approach because it requires many loudspeakers and will likely provide more coverage than required. Additional loudspeakers will likely also require more power amplification and additional installation labor.

Such a large amount of pattern overlap may also result in some acoustic interference with the sound from nearby loudspeakers, producing an uneven frequency response. This is because a dense overlap pattern may create too much pattern overlap within the space. Adding more acoustic energy into a space can also reduce intelligibility, which will make it difficult for listeners to understand presenters.

In 50 percent overlap systems, each loudspeaker's coverage pattern overlaps half of the adjoining coverage pattern, which can provide even SPL coverage. Whether using a square or hexagonal layout, the spacing distance between loudspeakers in a 50 percent overlap system equals the radius of the coverage circle: (*Distance = r*).

TIP To assist in verifying the uniformity of coverage of a loudspeaker system, the AVIXA standard A102.01:2017 *Audio Coverage Uniformity in Listener Areas* provides a procedure to measure and classify the uniformity of early arriving sound from a sound system across a listener area.

Ohm's Law Revisited

Ohm's law and the power equation are used to calculate and predict four properties of an electrical circuit: voltage, current, resistance, and power. They can help calculate the amount of current required to power the AV equipment within a rack or determine signal level at the end of a long cable run.

Although Ohm's law defines the electrical relationships in direct current (DC) circuits, it also applies to alternating current (AC) circuits if the total value of all opposition to current flow (*impedance*) is used in place of resistance. Before we go further in our discussion of loudspeakers, you may want to review Ohm's law in Chapter 2 and review your notes from preparing for your CTS exam.

Loudspeaker Impedance

Impedance (Z), measured in ohms (Ω), is the total opposition to current flow in a circuit. It includes the resistance (R) of the materials in the circuit plus their *reactance* (X). Reactance is the opposition to current flow in the circuit caused by changes in the voltage and current and is directly related to the frequency of the current. Reactance has two components: *capacitive reactance* (X_C) due to the capacitance in the circuit and *inductive reactance* (X_L) due to the inductance in the circuit. As audio signals are AC waveforms, the impedance in a circuit varies with the frequency of the audio signal.

On the back of almost any loudspeaker, you will see a nominal impedance rating for it, in ohms. The impedance of a loudspeaker is related to many factors, including the construction of the speaker, the circuitry included in the speaker enclosure, the signal frequency range, and the wiring configuration of the total speaker system.

Most common loudspeakers will have a nominal impedance rating of 4, 8, or even 16 ohms. You may even see some that have an impedance of 6 ohms. You will need to determine the resulting impedance of the entire loudspeaker line when you connect and wire together a loudspeaker system.

Impedance and Amplifiers

How you wire loudspeakers together determines the circuit's impedance. You need to be certain that the power amplifier you choose to drive those speakers is rated for that load.

If you're wiring only a couple of loudspeakers, determining the circuit's impedance and checking the power amplifier's rating are simple processes.

Amplifiers have specified impedance that is expected to be connected to their output terminals. Matching this specified impedance with the loudspeaker load maximizes the energy transfer from amplifier to loudspeaker to acoustic energy and reduces the possibility of the amplifier being overdriven.

Wiring Loudspeakers

A number of different schemes are used to connect and drive loudspeakers in audiovisual installations:

- **Networked**   Networked loudspeakers are generally fed with a power supply and digital audio via a looped (or occasionally star topology) data network connection. Some lower-powered networked loudspeakers are powered via their digital audio feed on an overlaid power-over-Ethernet (PoE) system, which is similar in concept to audio phantom power.
- **Active (powered) analog**   These loudspeakers are usually fed power from a distributed mains power supply and looped (parallel) line-level audio signals from the outputs of such signal processing devices as mixers, divider networks, distribution amplifiers, digital signal processors (DSPs), or loudspeaker controllers. They may also require data connections for monitoring and control.
- **Passive**   Passive loudspeakers are connected to power amplifiers (either directly or via a transformer) in some combination of series and/or parallel circuits. Some passive loudspeaker units may be equipped with Internet of Things (IoT) monitoring capabilities, which will require additional power and/or network data connections.

In designing and specifying loudspeaker systems, one of your first tasks is to verify the expected impedance of any passive, directly connected loudspeaker circuits. If you don't, you risk damaging equipment, personal injury, or simply poor system performance.

Ideally, you can complete the impedance calculations for the passive loudspeaker networks before the speakers are installed and commissioned, but sometimes, circumstances require you to make calculations in the field. The formula you use depends on how the loudspeakers will be wired: in series, in parallel, or a combination of the two.

TIP   Parallel loudspeaker circuits are easier to install and troubleshoot than series circuits.

Loudspeakers Wired in Series

In a series loudspeaker circuit, each loudspeaker's coil is connected to the next loudspeaker in the series, in sequence, as shown in Figure 6-7. The power amplifier's positive output terminal connects to the positive terminal of the first loudspeaker. The first loudspeaker's negative terminal connects to the second loudspeaker's positive terminal.

Figure 6-7
A series loudspeaker coil is connected to the next loudspeaker's coil in sequence.

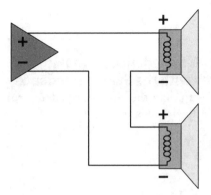

The second loudspeaker's negative terminal would connect to a third loudspeaker's positive terminal, and so on. The last loudspeaker's negative terminal completes the circuit by connecting to the amplifier's negative terminal.

> ### Series Circuit Impedance Formula
> The formula for calculating the total impedance of a series loudspeaker circuit is as follows:
>
> $$Z_T = Z_1 + Z_2 + Z_3 + \ldots Z_N$$
>
> where:
>
> - Z_T is the total impedance of the loudspeaker circuit.
> - Z_N is the impedance of each loudspeaker.

Figure 6-7 shows two loudspeakers wired in series—plus to minus to plus to minus. If you want to calculate the total impedance of this loudspeaker circuit, you simply add the impedances of the individual loudspeakers. If each of these loudspeakers has a nominal impedance of 8 ohms, the total impedance is 16 ohms.

Loudspeakers Wired in Parallel, Same Impedance

Another method of connecting loudspeakers in a circuit is to wire them in parallel. This means that the positive output of the amplifier connects to every loudspeaker's positive terminal, and each loudspeaker's negative terminal connects to the amplifier's negative terminal.

Loudspeakers wired in parallel can have the same or different impedances.

> ### Parallel Circuit Impedance Formula:
> ### Loudspeakers with the Same Impedance
> The formula for finding the circuit impedance for loudspeakers wired in parallel with the same impedance is as follows:
>
> $$Z_T = Z_1 \div N$$
>
> where:
>
> - Z_T is the total impedance of the loudspeaker system.
> - Z_1 is the impedance of one loudspeaker.
> - N is the number of loudspeakers in the circuit.

Figure 6-8 shows two examples of loudspeakers wired in parallel. If all the loudspeakers have the same impedance, the impedance of the loudspeaker divided by the number of loudspeakers wired in parallel equals the impedance of the circuit. If you have two loudspeakers wired in parallel, each rated at 8 ohms, the circuit's impedance is 4 ohms (8 ÷ 2).

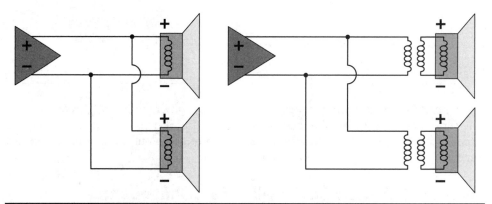

Figure 6-8 In a parallel loudspeaker circuit, the positive output of the amplifier connects to every loudspeaker's positive terminal.

Loudspeakers Wired in Parallel, Different Impedances

If loudspeakers wired in parallel have different impedances, you must use a different formula for finding the total impedance of the circuit.

> ### Parallel Circuit Impedance Formula: Loudspeakers with Different Impedances
> The formula for finding the circuit impedance for loudspeakers wired in parallel with differing impedance is as follows:
>
> $$Z_T = \frac{1}{\frac{1}{Z_1} + \frac{1}{Z_2} + \frac{1}{Z_3} \cdots \frac{1}{Z_N}}$$
>
> where:
>
> - Z_T is the total impedance of the loudspeaker circuit.
> - $Z_{1 \ldots N}$ is the impedance of each individual loudspeaker.

If you have three loudspeakers wired in parallel, with the first rated at 4 ohms, the second rated at 8 ohms, and the third rated at 16 ohms, the circuit's impedance is 2.29 ohms, or $Z_T = 1 \div [(1/4 \text{ ohms}) + (1/8 \text{ ohms}) + (1/16 \text{ ohms})]$.

Loudspeakers Wired in a Series and Parallel Combination

In a series/parallel loudspeaker circuit, groups of loudspeakers, called *branches,* are wired together in series. Typically, loudspeakers in the same branch have the same impedance. Each branch is connected to the positive and negative lines of the amplifier in parallel, as shown in Figure 6-9.

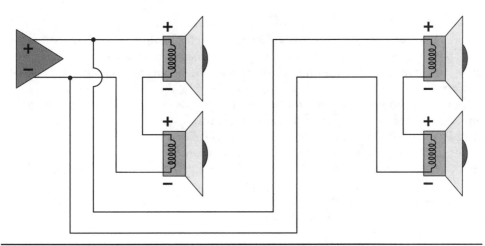

Figure 6-9 Loudspeakers wired in a combination circuit

To calculate the total impedance of a series/parallel circuit, you must do the following:

- Calculate the total impedance of each branch using the series circuit impedance formula, as described earlier.
- Calculate the total circuit impedance of the circuit using the parallel circuit impedance formula, as described earlier.

> **Series/Parallel Circuit Impedance Formula**
>
> The formula to calculate the expected total impedance of a parallel circuit may also be used to calculate the total impedance of a series/parallel circuit:
>
> $$Z_T = \frac{1}{\frac{1}{Z_1} + \frac{1}{Z_2} + \frac{1}{Z_3} \ldots \frac{1}{Z_N}}$$
>
> where
>
> - $Z_1 \ldots Z_N$ = The total impedance of each circuit branch
> - Z_T = The total impedance of the loudspeaker circuit
>
> This formula is used *after* the series portions of the circuit have been calculated.

NOTE It is rare to encounter loudspeakers wired in a series/parallel combination in the field. Although the idea of implementing a series/parallel combination would be to present a proper load to the output of the power amplifier, such systems are difficult to troubleshoot.

Measuring Impedance

Before electrically testing the loudspeaker circuit, refer to your impedance calculations and compare them to readings from an impedance meter. An impedance meter is a piece of test equipment used to measure the true impedance in an entire loudspeaker or loudspeaker circuit. Most are portable and battery powered, similar to a multimeter.

In addition to measuring the total impedance of the connected loudspeaker system, impedance meter readings can indicate short circuits, open loudspeaker lines, transformers installed backward, and low-impedance loudspeakers on a high-impedance system.

When working with large systems, check the line impedance often. Do not wait until all the loudspeakers are wired to find out that one is bad. As you install a group of loudspeakers, test it to make sure there are no problems before you install the next group. Be sure to write down the final measured impedance. This final measurement will be useful for future system service and maintenance.

It is important to check the system load prior to connecting the loudspeakers to an amplifier. Too little impedance connected to the amplifier will cause the amplifier to drive too much current into the loudspeakers and may cause the amplifier to fail.

To measure impedance, take the following steps:

1. Disconnect the wires from the amplifier.
2. Calibrate the meter by doing the following (analog meters only):
   a. Connect the test leads to the meter.
   b. Select the scale that is appropriate for your expected value for greatest accuracy.
   c. Hold the test leads together so the tips are touching, or pressing.
   d. Adjust the calibration control until the reading indicates zero.
3. Connect one lead to each of the wires on the first loudspeaker in the chain. Polarity is not important.
4. Observe the reading and compare it with the expected reading. It should be within a reasonable tolerance (≈10%); otherwise, there may be problems requiring further investigation.
5. Reconnect the loudspeaker wires and power on the amplifier.

Transformers

The voltage of an alternating current can be manipulated in an electrical circuit through the use of transformers. Transformers are common electrical devices that are used in power supplies, in audio and video circuits, and, particularly for audiovisual use, in loudspeaker systems. Transformers transfer energy from one circuit to another without physical connection, using the principle of magnetic induction between two windings, or coils of wire. One winding is connected to the power source (primary winding), and the other winding is connected to the load (secondary winding). The ratio of the number of coils (or turns) of wire in the primary and secondary sides determines the impedance and/or voltage ratio. Some energy is dissipated in the imperfect conversion of electrical to magnetic energy and back again within the transformer. This is known as *insertion loss*.

Types of Transformers

Transformers have the ability to increase or decrease the voltage in a circuit or keep it the same.

The following types of transformers are available (see Figure 6-10):

- **1:1 transformer**   This type of transformer has an equal number of primary and secondary windings. It is used for electrical (galvanic) circuit isolation to solve problems such as ground loops causing audible hum, buzz, or rolling hum bars on a display.
- **Step-up transformer**   This type has more windings on the secondary side than on the primary side. Voltage and impedance increase, while available current decreases.

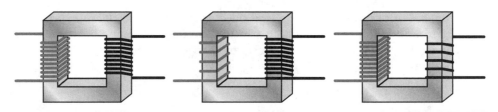

Figure 6-10 L to R: 1:1, step-up, and step-down transformer windings

- **Step-down transformer** This type has fewer windings on the secondary side than on the primary side. Available current increases, while voltage and impedance decrease.

Loudspeaker Taps

Most speaker transformers for use in constant-voltage loudspeaker systems have multiple wires on the primary side that allow you to adjust the voltage level to each loudspeaker. These wires are commonly referred to as *taps*. Taps are intermediate connections to the transformer windings that allow you to select different power levels from the transformer (see Figure 6-11).

Either the transformer manufacturer will identify these wires in some way and provide a chart for their values or you can write the values on the wires themselves. The taps can be selected for the appropriate amount of power in watts or impedance steps that they will deliver to the loudspeaker and must comply with the designer's intentions for the desired performance.

Many multi-tap speaker transformers are packaged with a switching device for the simplified selection of the required tap point.

Many transformer manufacturers pre-strip the tap wires for termination. When terminating, be sure that the wires from the amplifier are connected to the tap points specified on the designer's drawings. The consequences of connecting the wrong wire include exceeding the capability of the amplifier or decreasing the signal to a barely audible level. You will connect only one tap value and the common connection on the transformer to the loudspeaker.

Figure 6-11
Multi-tap
step-down
transformer

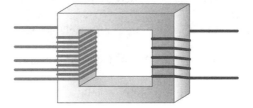

Specifying a Power Amplifier

Once loudspeaker locations have been determined, the next step is to calculate how much power is needed at each loudspeaker to provide adequate SPL at the listener location. This calculation should also include the necessary headroom appropriate to the application. Headroom is the difference in dB SPL between the peak and average-level performance of an audio system.

To determine the power required at the loudspeaker, you need to know the following:

- The sound pressure level required from the sound system at the listener position. For speech applications, this is typically 70dB SPL.
- The headroom required. For speech applications, 10dB is considered adequate. For music applications, 20dB or more may be required.
- Loudspeaker sensitivity. Typically, this will be stated as SPL in decibels expected at a distance of 1 meter from the loudspeaker with 1 watt applied.
- Distance to the farthest listener from the loudspeaker.

Once you have this information, you can calculate the amount of power needed at the loudspeaker, also known as the electrical power required (EPR), or wattage at the loudspeaker. You calculate EPR using the formula:

$$EPR = 10 \left(\frac{[L_p + H - L_s + (20 \times \text{Log}(D_2 \div D_r))]}{10} \right) \times W_{ref}$$

where:

- *EPR* is the electrical power required at the loudspeaker.
- L_p is the SPL required at distance D_2.
- *H* is the headroom required.
- L_s is the loudspeaker sensitivity reference, usually 1 watt at 1 meter.
- D_2 is the distance from the loudspeaker to the farthest listener.
- D_r is the distance reference value, usually 1 meter.
- W_{ref} is the power reference value; assume a W_{ref} of 1 watt, unless otherwise noted.

Many amplifier power requirement calculators are available online to perform these calculations for you—an example is shown in Figure 6-12. Web search engines are your friends (search term: amplifier power required calculator).

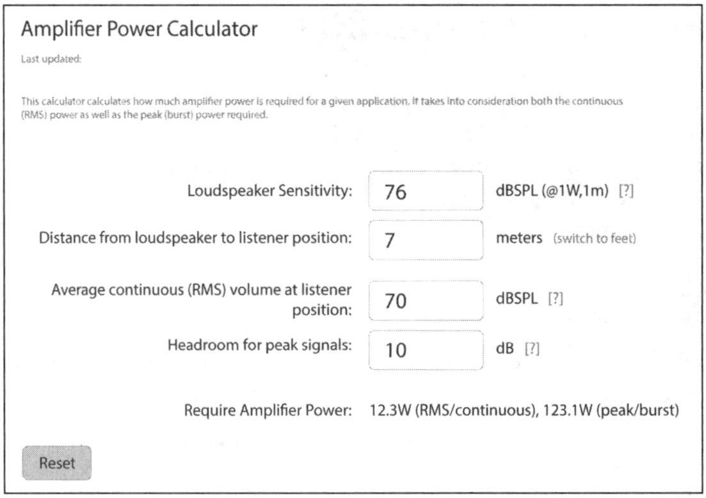

Figure 6-12 An online amplifier power requirement calculator (Courtesy of Biamp)

Headroom Requirements

A sound reinforcement system needs to be loud enough for the listeners to hear it. When choosing a loudspeaker, you must verify that the loudspeaker is sensitive enough to boost the sound signal with enough headroom.

System headroom is the difference between the audio system's typical operation level and the maximum level the system can attain. If a sound system usually operates at +4dBu but could operate as high as 23dBu, then it has 19dB of headroom. It is important to have enough headroom to handle transient performance peaks without distorting the signal.

As noted earlier, for a speech-only sound reinforcement system, 10dB of headroom is appropriate. For a music reinforcement system, as much as 20dB of headroom is needed to avoid clipping musical peaks.

Loudspeaker Sensitivity

Like microphones, loudspeakers are rated based on their ability to convert one energy form into another. Loudspeaker sensitivity is a measure of the efficiency at which the loudspeaker converts electrical energy to acoustic energy. It's the acoustic output resulting from the input of a known amount of electrical energy.

Loudspeaker sensitivity is expressed as an SPL at a specific power input and measurement distance. For example, a measurement for a loudspeaker might be 90dB SPL, 1 watt @ 1 meter.

Given the same reference electrical input level into two different loudspeakers, a more sensitive loudspeaker would provide a higher acoustical energy output than a less sensitive loudspeaker.

Loudspeakers vary substantially in efficiency. This does not mean that lower-sensitivity loudspeakers are of lesser quality. Like microphones, loudspeakers are designed and chosen for specific applications.

Power Amplifiers

Now that you know the power required at the listener position, you can specify the power amplifier. A power amplifier boosts the audio signal enough to drive the loudspeakers.

Power amplifiers are designed to be connected to a specific load (impedance)—either a low-impedance load (typically 2 to 8 ohms) or a high-impedance load, such as with a distributed or constant-voltage loudspeaker system. The power amplifier's specifications should state what impedance load it is designed to be connected to. Some power amplifiers have a switch, or different output terminals, that allow them to connect to various impedance loads. Other power amplifiers may require an internal or external transformer to function with a constant-voltage 70V or 100V load.

TIP The term *constant voltage* implies that an amplifier configured for a 70V line, for example, will never output more than about 70V RMS regardless of the number of loudspeakers connected to the output of the power amplifier. However, the actual number of loudspeakers you can connect will be limited by the power available from the power amplifier.

Specifying a Power Amplifier for Directly Connected Systems

To specify a power amplifier for directly connected loudspeakers:

1. Determine the required SPL at the listener position.
2. Determine your required headroom (10dB for voice or 20dB for music).
3. Determine the power (watts) required at the loudspeaker using the EPR formula (or an online calculator).
4. Round the result up to an amplifier value that can be readily purchased.

Specifying a Power Amplifier for Distributed Audio

This is a process for specifying a power amplifier for a constant-voltage distributed audio system. In this type of system, you need to specify your transformer tap settings before you can determine your amplifier need.

To specify the power amplifier required for a transformer-connected, constant-voltage loudspeaker system:

1. Determine the required SPL at the selected listener position.
2. Determine your required headroom (10dB for voice or 20dB for music).
3. Determine the power (watts) required at the loudspeaker using the EPR formula (or an online calculator).

4. Select and record the appropriate tap setting for the loudspeaker based on the power required.
5. Repeat steps 1–4 for each loudspeaker.
6. Sum the tap setting power from each loudspeaker.
7. Increase the total tap setting power by a factor of 1.5.
8. Round the result up to an amplifier value that can be readily purchased.

Microphones

If you're designing an audio system, chances are it will include microphones. Clients use microphones to be heard—in presentations, conferences, or performances. In this section, you will learn about the special qualities of microphones and how they factor into AV systems.

Microphones vary by type of transducer and directional characteristics. The selection of a microphone (mic) for any particular application is based on several factors, including directional or pickup patterns, sensitivity, impedance, and frequency response. Installers should also be aware of the mic's physical design and how the mounting surface will affect the overall audio performance. The basic types of microphones are discussed in the following sections.

Dynamic Microphones

In a dynamic microphone, you will find a coil of wire (a *conductor*) attached to a diaphragm and placed in a permanent magnetic field. Sound pressure waves cause the diaphragm to move back and forth, thus moving the coil of wire attached to it.

As the diaphragm-and-coil assembly moves, it cuts across the lines of flux of the magnetic field, inducing a voltage into the coil of wire, as shown in Figure 6-13. The voltage induced into the coil is proportional to the sound pressure and produces an electrical audio signal. The strength of this signal is very small and is called a *mic-level signal*.

Dynamic microphones are used in many situations because they are economical and durable, and they will handle high sound pressure levels. These microphones are very versatile because they do not require a power source.

Condenser Microphones

In the study of electricity, you found that if you have two oppositely charged (*polarized*) conductors separated by an insulator, an electric field exists between the two conductors. The amount of potential charge (*voltage*) that is stored between the conductors will

Figure 6-13 The workings of a dynamic microphone

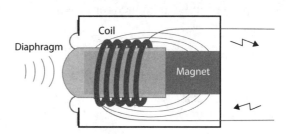

Figure 6-14
The workings of a condenser microphone

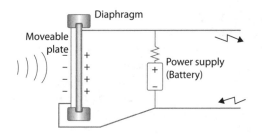

change depending on the distance between the conductors, the surface area of the conductors, and the dielectric strength of the insulating material between the two conductors. An electronic component that uses this principle is called a *capacitor* or *condenser*.

A condenser microphone contains a conductive diaphragm and a conductive backplate, as shown in Figure 6-14. Air is used as the insulator to separate the diaphragm and backplate. Electrical energy is required to polarize, or apply, the positive and negative charges to create the electric field between the diaphragm and backplate.

Sound pressure waves cause the diaphragm to move back and forth, subsequently changing the distance (spacing) between the diaphragm and backplate. As the distance changes, the amount of charge, or *capacitance,* stored between the diaphragm and backplate changes. This change in capacitance produces an electrical signal.

The strength of the signal from a condenser microphone is not as strong as the mic-level signal from a typical dynamic microphone. To increase the signal, a condenser microphone includes a preamplifier powered by the same power supply used to charge the plates in the microphone. This preamplifier amplifies the signal in the condenser microphone to a mic-level signal but is not to be confused with a microphone preamplifier found in a mixing console.

The power supply to charge the capacitor elements and drive the preamplifier may be a battery included in the microphone body, an external mains-powered supply device, or come from an external *phantom* power supply system.

As the diaphragm used in a condenser microphone is usually lower in mass than other microphone types, the condenser microphone tends to be more sensitive than other types and responds better to higher frequencies, with a wider overall frequency response. Very high-pressure acoustic waves can damage the microphone.

Electret Microphones

An *electret microphone* is a type of condenser microphone. The electret microphone gets its name from the pre-polarized material, or the *electret,* applied to the microphone's diaphragm or backplate.

The electret provides a permanent, fixed charge for one side of the capacitor configuration. This permanent charge eliminates the need for the higher voltage required for powering the typical condenser microphone. This allows the electret microphone to be powered using small batteries, or normal phantom power. Electret microphones are physically small, lending themselves to a variety of applications and quality levels such as in lavalier and head-worn microphones.

MEMS Microphones

A *microelectromechanical system* (MEMS) microphone is a member of a group of microscopic mechanical devices that are built directly onto silicon chips using the same deposition and etching processes used to construct microprocessors and memory systems. They are minute mechanical devices that can be directly integrated with the pure electronic circuitry of the chip. The best known of these in the AV world are the digital micromirror devices (DMDs) used for light switching in digital light processing (DLP) projectors. Many other MEMS devices are in wide commercial use as accelerometers, air pressure sensors, gyroscopes, optical switches, inkjet pumps, and even microphones.

Microphones built using MEMS technologies are generally variations on either condenser microphones or piezo-electric (electrical signals produced by the mechanical movement of a crystal) microphones. Although their tiny size means that MEMS microphones are not particularly sensitive, especially at bass frequencies, the huge advantage of MEMS microphones is that they can be constructed in arrays to increase sensitivity, and they can be placed on the same chip as the amplifiers and signal processors that manage both gain and frequency compensation. The on-chip processing can include analog-to-digital conversion, resulting in a microphone with a direct digital output. In addition to dedicated MEMS microphones, smartphones, smart speakers, tablet computers, virtual reality headsets, and wearable devices can incorporate chips that include a range of MEMS devices, including microphones, accelerometers, and gyroscopes and all of their associated processing circuitry.

Defining Phantom Power

Phantom power is the remote power used to power a range of audio devices, including condenser microphones. The supply voltage typically ranges from 12V to 48V DC, with 48V being the most common. Positive voltage is applied equally to the two signal conductors of a balanced audio circuit, with the power circuit being completed by current returning through the cable's shield. Because the voltage is applied equally on both signal conductors, it has no impact on the audio signal being carried and does not cause damage to dynamic microphones. However, an audio attenuator/pad device inserted in a phantom-powered line may form a bridge between the signal lines and the ground and cause unexpected and unwelcome currents to flow.

Phantom power is frequently available from audio mixers. It may be switched on or off at each individual microphone input, enabled for groups of microphone channels, or enabled from a single switch on the audio mixer that makes phantom power available on all the microphone inputs at once. If phantom power is not available from an audio mixer, a separate phantom power supply that sits in line with the microphone may be used.

Microphone Physical Design and Placement

Whether dynamic, condenser, electret, or otherwise, microphones come in an assortment of configurations to meet a variety of applications. The following are some common microphone configurations:

- **Handheld** This type is used mainly for speech or singing. Because it is constantly moved around, a handheld microphone includes internal shock mounting to reduce handling noise. Handheld mics are available in both wired and wireless configurations.

- **Surface mount or boundary** This type of microphone, also known as a pressure zone microphone (PZM), is designed to be mounted directly against a hard boundary or surface, such as a conference table, a stage floor, a wall, or sometimes a ceiling. The acoustically reflective properties of the mounting surface affect the microphone's performance. Mounting a microphone on the ceiling typically yields the poorest performance because the sound source is much farther away from the intended source (for example, conference participants) and much closer to other noise sources, such as ceiling-mounted projectors and heating. ventilation, air conditioning (HVAC) diffusers.

- **Gooseneck** Used most often on lecterns and sometimes conference tables, this type of microphone is attached to a flexible or bendable stem. Stems come in varying lengths. Shock mounts are available to isolate the microphone from table or lectern vibrations.

- **Shotgun** Named for its physical shape and long and narrow polar pattern, this type of microphone is most often used in film, television, and field-production work. You can attach a shotgun microphone to a long boom pole (fishpole), to a studio boom used by a boom operator, or to the top of a camera.

- **Instrument** This family of microphones is designed to pick up the sounds of musical instruments, either directly from acoustic instruments or from the loudspeaker cabinet of an amplified instrument. This type of microphone is usually either a condenser or a dynamic device, depending on the loudness, dynamic range, and frequencies to be picked up from the instrument. Some specialized instrument transducers use direct mechanical or magnetic pickups.

- **Lavalier and headmic** These microphones are worn by a user, often in television and theater productions. A lavalier (also called a *lav* or *lapel mic*) is most often attached directly to clothing, such as a necktie or lapel. In the case of a headmic, the microphone is attached to a small, thin boom and fitted around the ear. As size, appearance, and color are critical, lavaliers and headmics are most often electret microphones.

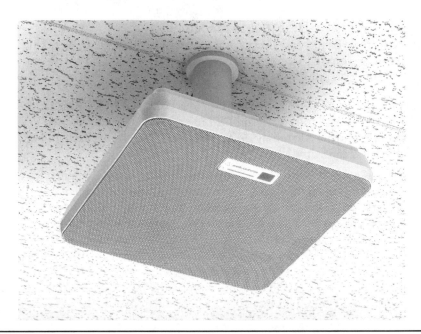

Figure 6-15 A ceiling-mounted multimicrophone array used for meeting rooms (Courtesy of Audio-Technica)

- **Beamforming array** Beamforming arrays have multiple microphone elements, usually condenser microphone capsules or MEMS microphones. These elements are configured in arrays of varying shapes and linked together through a digital signal processing system to form narrow beam patterns that can be electronically steered to pick up the desired sounds while rejecting ambient noise. Array microphones can be placed in convenient locations such as on a wall, on a tabletop, or on the ceiling, as shown in Figure 6-15. Microphone array devices are primarily used in meeting rooms, in conferencing spaces, on desks, and on lecterns.

Microphone Polar Response

One of the characteristics to consider when selecting a microphone is its polar response or pickup pattern. The pickup pattern describes the microphone's directional capabilities—its ability to pick up the desired sound from a certain direction while rejecting unwanted sounds from other directions.

Pickup patterns are defined by the directions in which the microphone is most sensitive. These patterns help you determine which microphone type to use for a given purpose.

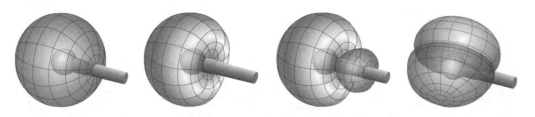

Figure 6-16 Microphone polar patterns (L to R): omnidirectional, cardioid, supercardioid, and bidirectional

There will be occasions when you want a single microphone to pick up sound from all directions (such as for an interview), and there will be occasions when you do not want a microphone to pick up sounds from nearby sources surrounding it (such as people talking or papers rustling). The pickup pattern is also known as the *polar pattern*, or the microphone's directionality.

Different types of pickup patterns include the following:

- **Omnidirectional**   Sound pickup is uniform in all directions.
- **Cardioid (unidirectional)**   Pickup is primarily from the front of the microphone only (one direction) in a cardioid pattern. It rejects sounds coming from the side but mostly rejects sound from the rear of the microphone. The term *cardioid* refers to the heart-shaped polar plot.
- **Hypercardioid**   A variant of cardioid, this type is more directional than a regular cardioid mic because it rejects more sound from the sides. The trade-off is that it will pick up some sound from directly at the rear of the microphone.
- **Supercardioid**   Provides better directionality than hypercardioid, with less rear pickup.
- **Bidirectional**   Pickup is equal in opposite directions with little or no pickup from the sides. This is sometimes also referred to as a *figure-eight* pattern, referring to the shape of its polar plot.

Figure 6-16 shows some of these microphone polar patterns.

Polar Plot

The polar plot is a graphical representation of a microphone's directionality and electrical response characteristics (see Figure 6-17).

Although a frequency-response graph shows the sensitivity in decibels at a given frequency, the polar plot shows the sensitivity in decibels by angle, or the pickup pattern of the microphone. Typically, microphone specifications will include several polar plots, each showing the pickup pattern at a different frequency.

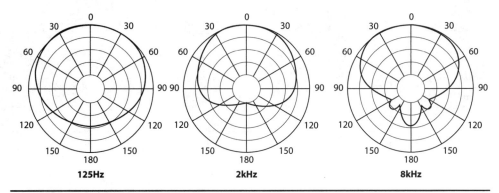

Figure 6-17 Polar plots of one microphone at 125Hz, 2kHz, and 8kHz

Imagine a microphone in the center of the plot, pointing straight at 0 degrees. If you stood at the 0-degree point, you'd be standing right in front of the microphone. At the 180-degree point on the plot, you would be directly behind the microphone.

The polar plot shows the angles from which the microphone picks up and transduces sound at the highest voltage. It extends farthest toward the 0-degree point, which shows that the microphone is best at detecting sounds coming from directly in front of it. To the left and right of the 0-degree mark, the curve falls away. This means as you move farther to the left or right of this microphone, it picks up less of the sound. In Figure 6-17, the microphone picks up no sound at all from directly behind at 2kHz, though it picks up some sound from behind at 125Hz and 8kHz.

Polar plots help you select and position a microphone. For example, say you need a microphone to pick up audio in a conference room. You want people on all sides of the table to be heard, so you would choose a mic whose polar pattern displays even pickup in all directions in the frequency range of the human voice.

The Right Mic for the Job

Have you ever handed a presenter a lavalier microphone and watched them pin it to their lapel—pointing sideways? If the microphone had an omnidirectional pickup pattern, that's probably OK. However, if the pickup pattern is directional, they just pointed it 90 degrees away from its optimum position.

As one AVIXA training expert puts it, "I always ask, 'Who's going to be pinning the microphone to the presenter? Does that person know how to use a directional mic?' If a sound technician is going to pin the lav to the presenter, I'll give the customer a directional mic. If the presenter's going to pin it to themselves, I'll probably use an omni."

Figure 6-18 Microphone frequency response

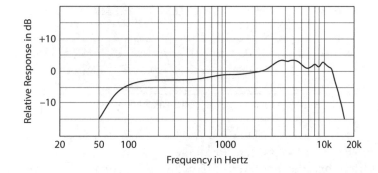

Microphone Frequency Response

The frequency-response specification is an important measure of a microphone's performance. It defines the microphone's electrical output level over the audible frequency spectrum, which in turn helps to determine how a microphone "sounds."

Frequency response is a way of expressing a device's amplitude response versus frequency characteristic, as shown in Figure 6-18. A frequency response is usually presented as a graph or plot of a device's output on the vertical axis versus the frequency on the horizontal axis.

A microphone's frequency response gives the range of frequencies, from lowest to highest, that the microphone can transduce. It is often shown as a plot on a two-dimensional frequency response graph. A microphone's frequency response graph shows the voltage of its output signal relative to the frequency of the sounds it picks up.

With directional microphones, the overall frequency response will be best directly into the front of the microphone. As you move off-axis from the front of the microphone, not only will the sound be reduced but the frequency response will change.

Microphone Signal Levels

A microphone, regardless of the type, produces a signal level called *mic level*. Mic level is a low-level signal—only a few millivolts. Table 6-2 shows the voltages of different audio signal types.

Because mic level operates at only a few millivolts, it is prone to interference. A microphone preamplifier amplifies the mic level signal to line level for routing and processing. Line level is the voltage of a regular audio signal and is used for transporting signals

Description	Voltage
Mic level	1 to 3 millivolts (–60 to –50dBu)
Line level (professional)	1.23 volts (+4dBu)
Line level (consumer)	316 millivolts (–10dBV)
Loudspeaker level	4 to ~100 volts

Table 6-2 Comparison of Signal Voltages

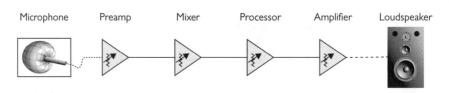

Figure 6-19 Mic level to line level to loudspeaker level

between system components. In a professional audio system, line level is about 1.23 volts (+4dBu); consumer line level is 316 millivolts (–10dBV). When you see a 3.5mm (1/8in.) jack plug or phono (RCA) connector, it often indicates a consumer-level signal.

Once the audio system has routed and processed the signal, it is sent to the power amplifier for final amplification up to loudspeaker level. The loudspeaker takes that amplified electrical signal and transduces the electrical energy into acoustical energy (see Figure 6-19).

Microphone Sensitivity

Another important performance criterion that characterizes a microphone is its sensitivity specification. Sensitivity defines how efficiently a microphone converts acoustic energy into electrical energy. It is expressed as decibels of voltage per pascal of sound pressure (dBV/Pa).

For example, a microphone may have a sensitivity specification that reads as follows: *–54.5dBV/Pa (1.85mV)*. This means that with a reference level of 1Pa (94dB SPL) into the microphone, it will produce a voltage level of –54.5dBV (1.85mV) out.

As 1Pa is equal to 94dB SPL, if a microphone with the previous specification receives less than 94dB of sound pressure, it will output less than –54.5dBV

If you expose two different microphones to an identical sound input level, a more sensitive microphone provides a higher electrical output than a less sensitive microphone. Condenser microphones are generally more sensitive than dynamic microphones.

Does this mean that lower-sensitivity microphones are of lesser quality? Not at all. Microphones are designed and chosen for specific uses. A professional singer using a microphone up close can produce a high SPL. In contrast, a presenter speaking behind a lectern and half a meter away from the microphone produces much less sound pressure. The singer needs a less sensitive microphone than the person talking.

For the singer, a dynamic microphone may be the best choice because it will typically handle the higher SPL produced by the singer without distortion while still providing more than adequate electrical output. The presenter would benefit from a more sensitive microphone, such as a condenser mic.

Another consideration is the physical robustness of the microphones that make them suitable for specific applications. Condenser microphones tend to be more delicate than dynamic microphones and may not stand up well to the sometimes-brutal treatment that handheld microphones are subjected to during a performance or presentation.

When determining microphone sensitivity, you will need to consider three factors:

- **Sound pressure level** The acoustic energy at the microphone.
- **Electrical signal level** The goal is to have a line-level signal after the preamplification.
- **Matching levels** Can the signal level from the microphone/preamp combination be amplified to the line-level signal that the audio system (mixer) requires?

Microphone Pre-Amp Gain

Let's say you need to choose a microphone for an auditorium. Your sound source is a presenter located 600mm (2ft) away from the microphone, with a measured SPL of 72dB. To route and process that signal, you need to amplify the microphone-level signal to line level (0dBu). Most microphone preamplifiers will provide around 60dB of gain.

You have a choice between two microphones:

- Dynamic microphone A, with an equivalent voltage specification of –54.5dBV/Pa (1.85 mV)
- Condenser microphone B, with an equivalent voltage specification of –35.0dBV/Pa (17.8 mV)

In this scenario, your microphone specification sheets tell you that if you put 94dB SPL into each microphone, –54.5dBV and –35.0dBV, respectively, will be produced. You need to select a microphone that will provide an adequate signal level for the application. To do this, you need to know what the required microphone pre-amp gain is for each microphone.

Mic output = SPLMeasured – SPLRef + Sensitivity + 2.21
Pre-amp gain required = Mic output – Required level

Where:

- Pre-amp gain required = the gain in dBu required to amplify the mic output to the required system level
- Mic output = microphone output in dBu
- Required level = pre-amp output in dBu required to meet system input specifications
- SPLMeasured = microphone input in dB SPL measured at the microphone position
- SPLRef = microphone reference level in dB SPL – from the specification sheet (usually 94dB SPL = 1Pa)
- Sensitivity = sensitivity of the microphone in dbV – from the specification sheet
- 2.21dB = conversion factor between dBV (0dbV =1V) and dBu (0dBu = 775mV)

In this example, the pre-amp gain required for dynamic microphone A is –74.29dBu. The pre-amp gain required for condenser microphone B is –54.79dBu.

Assuming you have a 60dB gain in your microphone pre-amp, dynamic microphone A is not sensitive enough for this application. The closer you can get to 0dBu, the better the microphone will be for the application. You need 83.5dB, but you have only 74.29dB, leaving you 9.21dB short.

Does this make dynamic microphone A a bad mic? No. It is designed for a different application—speaking or singing at a much closer range than 600mm and with much higher acoustic input. In this case, the mixer's preamplifier would normally be more than adequate to amplify dynamic microphone A's signal to line level.

With a pre-amp gain of −54.79dBu, condenser microphone B is sensitive enough for this application. The combination of the more sensitive microphone and the gain available from the microphone preamplifier in the mixer can take the mic level signal to the 0dBu level needed for signal routing and processing.

Microphone Mixing and Routing

Aside from microphones amplified to line level, your audio system will include other sources originating at line level. For professional audio line sources, little, if any, amplification will be required for routing and processing. Consumer line level may require some amplification.

Automatic Microphone Mixers

An automatic microphone mixer can control the number of open microphones (NOM) in the system. The user can set NOM on the mixer, but generally it's limited to one. The following are some of the situations where using an auto-mixer may be useful:

- Conference rooms
- Courtrooms
- Meeting spaces
- Live spoken-word events with handheld microphones

In these settings, you have multiple people speaking and need a way to control the different sources. The big question is, "How many people need to talk at any one time?"

If multiple people need to speak, you can assign a NOM that reflects that number. However, keep in mind that the number of open microphones affects the chances of feedback in the system. The fewer microphones open, the fewer chances of feedback or extraneous noise coming through the system. On the other hand, some collaborative and discussion sessions may actually be hindered by automated mixing systems locking out critical contributors or even the session moderator.

There are two ways an auto-mixer can limit the NOM in a design: gated or gain sharing. With *gated sharing,* each channel is set to an adjustable sound threshold that a microphone needs to surpass to be turned on. If it falls below that threshold, it is muted.

The threshold must be set high enough to keep the channel from being opened by background noise, yet low enough to open when someone is speaking. In simple systems this creates a binary effect—each microphone is either on or off. But there is a chance that the system can't pick up the first-spoken syllables of a low talker if not set correctly, and varying levels from conference participants (soft to loud) can lead to choppy audio. More sophisticated gate implementations can modify the attack and release characteristics of the gate to allow a gentler transition between talkers.

With *gain sharing*, the available gain is shared among all of the channels, and microphones with more signal get more gain. (Those with less receive less.) Because all the microphones are splitting gain depending on activity, pickup is gradual, resulting in smoother operation and greater sophistication than a simple gated design.

Microphone Placement: A Conference Table

Let's say you need to place microphones around a conference table. Here are two possible options:

- Six cardioid mics—one microphone per every two participants—with one each for the participants at the ends of the table (see Figure 6-20).
- Two omnidirectional mics (see Figure 6-21). Although two omni microphones would seem to cover the participants, you must also consider the environment. Omnidirectional mics are equally sensitive to sounds coming from all directions, so ask yourself, for example, "Will there be a projector above the table?" and "How loud will background noise from the HVAC system be?"

A more contemporary approach would be to use a single (or perhaps two—depending on table size) multielement, beam-steered array either on the conference table or hung from the ceiling above it. With appropriate programming, the effective NOM is selectable, and priority locations can be allocated to critical talkers such as the meeting host or moderator.

Figure 6-20
A conference table with a cardioid microphone coverage pattern

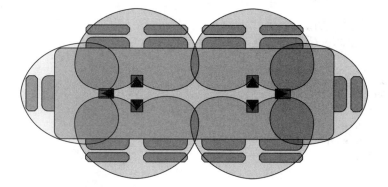

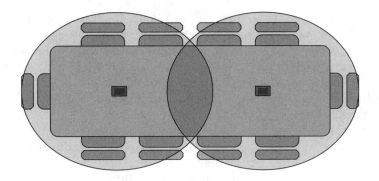

Figure 6-21
A conference table with an omnidirectional microphone coverage pattern

Microphone Placement: The 3:1 Rule

When using multiple microphones, consider how the sound from a presenter will reach the microphones.

In Figure 6-22, each person has been given their own microphone. When a presenter speaks, their voice is picked up by three microphones. The microphone directly in front picks up the sound louder than the off-axis microphones do. In addition, because they are farther away, the sound that these two microphones pick up may be delayed and thus out of phase.

Let's say the mics are separated by 300mm (1ft) and are 300mm (1ft) away from the person talking. The front mic picks up the talker at 65dB SPL. What are the two mics picking up?

Using Pythagoras's theorem, we can calculate how far the outside mics are from the person talking:

Distance = $\sqrt{(300^2 + 300^2)}$
Distance = $\sqrt{(90{,}000 + 90{,}000)}$
Distance = $\sqrt{(180{,}000)}$
Distance = 424mm

The loss over distance is as follows:

dB = 20 × log(424 ÷ 300)
dB = 3.01 or about 3dB less (not very much)

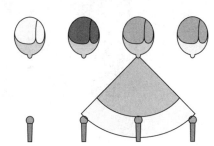

Figure 6-22
Each mic is 300mm (1ft) away from one presenter and 300mm (1ft) away from all other mics.

Figure 6-23
Each mic is 900mm (3ft) away from the other and 300mm (1ft) away from any one presenter.

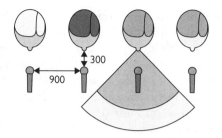

In an analog mixer, the three microphone signals would be combined. The phase differences between the microphones will create a comb-filter effect in the combined signal, which gives it a thin, hollow, tone quality.

In Figure 6-23, the microphones are separated at a greater distance, this time at a ratio of 3:1. That is, the distance between each microphone is three times the distance between the microphone and a presenter.

By separating the microphones using the 3:1 rule, we can use Pythagoras's theorem to calculate the distance from the outside microphones to the person talking:

Distance = $\sqrt{(300^2 + 900^2)}$
Distance = $\sqrt{(90,000 + 810,000)}$
Distance = $\sqrt{(900,000)}$
Distance = 949mm

The loss over distance is as follows:

dB = 20 × log(949 ÷ 300)
dB = 10dB loss, which sounds half as loud

When the three microphones are mixed together in an analog mixer, the phase differences between the microphones will still affect the sound. However, the levels being mixed are greatly reduced because of the distance, which makes the comb filtering effect less prominent in the mix.

NOTE   In signal processing, a comb filter adds a delayed version of a signal to itself, causing constructive and destructive interference. On a graph of frequency versus amplitude, a comb filter displays as a series of regularly spaced spikes, giving the appearance of a comb.

Reinforcing a Presenter

If you need to mic only one specific person, you have a few options for reinforcing their voice.

Lavalier microphones are effective for video or training presentations where the microphone needs to be unobtrusive. Despite a cable connection being significantly more reliable, they are often connected to a body-worn wireless transmitter pack to facilitate freedom of movement for the talent (presenter). The best position for a lavalier will be in the center of the chest, just below the hollow of the throat. You should get the microphone as close to the talent's mouth as possible, but take care to avoid any positions that might cause unwanted contact. As lavalier microphones are highly susceptible to picking up unwanted noises from the talent's clothing, jewelry, and cabling to earpieces or headsets, it is essential to check for such problems before going live with the microphone.

An ear-worn microphone is another relatively unobtrusive microphone solution, provided that its color is a reasonably close match to the skin tone of the talent. Ear-worn microphones should be placed about 10mm (1/2in.) away from the corner of the performer's mouth. Do not position the element directly in front of the mouth because it will pick up popping noise from plosives (*P*s, *D*s, and *T*s) and breath sounds. Place the element so it does not directly contact or rub the cheek of the presenter.

If an ear-worn microphone is not available, a lavalier placed along the center of the chest, as close as possible to the mouth, is best. Ask the talent to tilt their chin down as low to the chest as they possibly can, then place the microphone just below that point.

Handheld microphones will always give better gain before feedback when held as close to the mouth as possible. Advise inexperienced presenters to hold the mic just below their chin (then watch them forget to do so after about 10 minutes).

TIP   If you need to place microphones at a lectern, only one microphone should be active. Additional microphones won't give you more gain and should be used only for backup or redundancy. If you need more gain, try using a more sensitive microphone or one with a tighter pickup pattern. This solution is cheaper and won't introduce comb filtering issues. A mechanically isolating shock mount for the microphone will help to reduce noise from the presenter's hand gestures and shuffled papers.

Microphones and Clothing

When reinforcing (micing) a presenter, you need to consider the talent's clothing:

- If the talent is wearing multiple layers of clothing, such as a jacket over a shirt or blouse, care should be taken not to place the microphone where it will be rubbed by the relative movement of layers as the talent moves or makes hand and arm gestures.
- If the talent is wearing a knotted tie, you may be able to hide the microphone inside the knot.

- Avoid placing the microphone near silk. Silk blouses and ties can sound scratchy if they move against the microphone.
- Protective mounts are available that can be taped to the talent's chest and prevent rustling noises from clothing.
- Finding an unobtrusive and low-rustle cable route between the microphone pickup and its mic-cable interface or transmitter pack can be fraught with difficulties and issues of intimate physical contact with the talent, as can finding a suitable place to hide a transmitter pack.
- If your project involves a wardrobe manager or stage manager, working closely with them may facilitate microphone placement.

Polar Plots for Reinforcing a Presenter

When reinforcing a presenter with a lavalier microphone, you need to consider microphone placement and polar patterns. A lavalier microphone with a cardioid polar pattern will capture the presenter's voice without capturing background noise from other directions (see Figure 6-24).

This directional microphone is an excellent choice, but only if you know that the end users are all trained in how to place the microphone. For example, what if the user places the cardioid microphone off-center, on their lapel? What if they place the microphone upside down?

Instead of specifying a microphone with a cardioid polar pattern, consider specifying one with an omnidirectional polar pattern. An omnidirectional microphone will pick up the presenter's voice (among other things), even if it is placed off-center or upside down (see Figure 6-24).

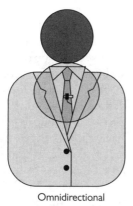
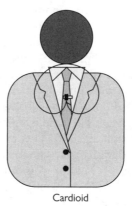

Omnidirectional Cardioid

Figure 6-24 Proper placement of omnidirectional (L) and cardioid (R) lavalier microphones

Audio System Quality

Now let's consider what you need to do to ensure the quality of the audio system and how it functions in the space. For one thing, you need to determine whether the audio system will be stable by comparing a couple of calculations, namely, PAG and NAG. You also need to check your sound-reinforcement design to make sure it will be effective based on the client's audio system needs.

Sound reinforcement is the combination of microphones, replay devices, audio mixers, signal processors, power amplifiers, and loudspeakers that are used to electronically capture, enhance, amplify, and distribute sound.

To provide adequate audio coverage for listeners, a sound system must accomplish three things:

- It must be loud enough.
- It must be intelligible.
- It must remain stable.

Before we delve into PAG-NAG equations and a fundamental goal of audio system design—eliminating acoustic feedback—let's look further at these three measures of audio system quality.

It Must Be Loud Enough

Can the listener hear the intended audio? If the answer is no, you have a problem. This may be a simple matter of changing the listener or loudspeaker location or increasing the volume. It may also be a sign of a much greater issue, such as unfriendly acoustics.

"Loud enough" depends on the application. Although sound systems encompass everything from conference rooms to concert stages, for now, at least, we'll focus on the context of a typical boardroom, conference room, or lecture hall.

For each audio system, you need a target sound pressure level. For a typical speech reinforcement system, that level is around 60 to 65dB SPL. That is the speech level of a human about 600mm (2ft) away. Through the audio system you're designing, your goal is to replicate that conversational level with the audience. One key consideration in achieving this is the signal-to-noise (S/N) ratio and its effect on audio loudness.

The signal-to-noise ratio is the ratio between the desired signal and the noise accompanying that signal. The ratio is usually measured in decibels. In your design, look for an acoustic S/N ratio of 25dB. This means you should have 25dB between the level of the sound system output (your signal) and the level of the room's background noise.

Using a 25dB S/N ratio, you can arrive at the required loudness of the audio system two different ways. If you've measured a background noise level of 55dB SPL, your system target level then needs to be 25dB above that—80dB SPL. But 80dB SPL would be a loud level for conversation.

Coming at it from a different direction, start with your target. Say you want to achieve 65 dB SPL. Your background noise level should be 25dB below that—40dB SPL. Think of this as the signal-to-noise level of the acoustic environment and design your acoustic space with that in mind.

 NOTE You may have heard that an audio system should have a 60dB S/N ratio. This is the *electronic* S/N ratio of the electronic components. The signal level in the system should be at least 60dB above the combined noise inherent in all the electronics in the signal path. A properly adjusted system should have no problem meeting a minimum 60dB S/N ratio. But don't confuse the *electronic* S/N ratio of 60dB with the *acoustic* S/N ratio of 25dB.

It Must Be Intelligible

Intelligibility describes an audio system's ability to produce a meaningful reproduction of sound. For example, an intelligible sound-reinforcement system can reproduce accurately the vowels and consonants of a source, which helps listeners identify words and sentence structure, giving the sound meaning.

In audio system design, intelligibility deals with the intensity and time arrival of the indirect sound. Reflected and reverberated sound energy arrives at the listener's ear after direct sound, which comes directly from the talent or loudspeaker. Late reflections sound like distinct echoes, and excessive reverberation masks intelligible speech.

Whether in amplified or unamplified speech presentation, the requirements of speech intelligibility are very much the same:

- **Speech loudness level** Average speech levels should be in the 60 to 65dB range if people with normal hearing are to understand it without effort. Under noisy conditions, the speech level should be raised accordingly.

- **Speech signal-to-noise ratio** For ease in listening, average speech levels should be at least 25dB higher than the prevailing noise level. However, at elevated speech levels, in noisy environments, a somewhat lesser speech-to-noise ratio is possible. Typically, a 15dB S/N ratio may suffice in situations where considerable noise is expected, such as in busy transport terminals and in live sports venues where there are spectators.

Normal face-to-face speech communication is about 60 to 65dB SPL, but most speech reinforcement systems operate between 70 and 75dB SPL. If the level of amplified speech increases beyond that range, to about 85 to 90dB SPL, overall intelligibility won't increase much, and most listeners will complain of excessive levels. Any louder, and intelligibility will diminish because listeners will feel oppressed by the high levels.

The direct-to-reflected sound ratio is a measure of the different energy in a space. It takes the level of direct sound—from source to listener—and compares it to the level of reflected sound, all of which arrives at the listener by an indirect pathway and determines the intelligibility.

Whether reflected sound is a problem or not comes down to two things:

- How late did it arrive at the listener in relation to the direct sound?
- At what energy level did that reflection arrive at the listener?

If the reflection arrives at the listener 50 to 80 milliseconds or later than the direct sound and at a sufficient energy level, it is considered an echo, and it adversely affects intelligibility.

TIP Harmonic distortion can also diminish audio quality. Harmonics are multiples of a fundamental signal frequency, and harmonic distortion describes energy that wasn't in the original signal. Harmonic distortion can occur in both an audio system's electronic components and loudspeakers, which makes it one of the most common types of distortion. Analyzing all the harmonics together and comparing them to the fundamental frequency gives you a measure of total harmonic distortion (THD), usually stated as a percentage of the original signal. The lower the THD, the better the audio signal quality.

It Must Remain Stable

Stability applies to audio systems that employ microphones and loudspeakers in the same space. Can the client turn up the volume loud enough to be heard without causing feedback? If someone must make constant adjustments to eliminate feedback, chances are the system wasn't designed or executed properly.

To determine whether an audio system is stable, a designer calculates *gain before feedback*—how loud the loudspeakers get before the microphones pick them up. Gain before feedback is frequently referred to as *potential acoustic gain* (PAG), which you will learn about later in this chapter.

Gain before feedback is largely a design issue. It has to do with distance relationships between the source (such as a presenter) and the microphone, the microphone and the loudspeakers, and the loudspeakers and the listeners.

The designer must also determine the amount of gain (amplification) required from the audio system to produce an adequate sound level. How loud do the loudspeakers need to be for listeners to hear the intended audio? This is known as *needed acoustic gain* (NAG), which you will also learn about later.

By comparing PAG to NAG, you can determine whether a system will be stable. If the system exhibits more potential acoustic gain than needed acoustic gain, it will be stable. But if its needs exceed its potential, either the client won't be able to turn the volume up loud enough or feedback may occur.

TIP You can perform calculations that help predict whether an audio system will create feedback, but as general rules, loudspeakers should be as close to listeners and as far from microphones as possible, and microphones should be as close to the presenter and as far from loudspeakers as possible.

PAG/NAG

To reiterate, audio system stability refers to the system's ability to amplify sound from a microphone without feedback or distortion. Proper equipment selection and placement are significant contributors to system stability (think acoustic treatment, equalization, feedback suppression, and mix-minus), but controlling acoustic gain is also an important method. A sufficient PAG/NAG ratio will minimize the potential for feedback.

NAG is the gain the audio system requires to achieve an equivalent acoustic level at the farthest listener equal to what the nearest listener would hear without sound reinforcement. It tells the designer how much you need to increase the amplitude of a sound wave on a microphone for it to be equivalent to some closer location, where no amplification would be required. Again, NAG is relevant only for systems incorporating microphones that are being reinforced in the same space.

Before you can solve for NAG, you need to understand *equivalent acoustic distance* (EAD). EAD is the farthest distance one can go from the source without needing sound amplification or reinforcement to maintain good speech intelligibility. It is a design parameter dependent on the level of the presenter and the noise level in the room. A simple illustration of this principle is two people talking as one person backs away until they can no longer communicate clearly. The farthest distance at which they can still communicate clearly is the EAD.

In calculations, the farthest listener in a space is at a position identified as D_0 which should be in the same ambient noise conditions as the EAD listener—the listener who can hear without amplification. Using the NAG formula, the resultant level indicates how many additional decibels of sound pressure are required for the farthest listener to hear as well as the EAD listener.

TIP EAD will be very different depending on whether you're working with music or voice. An EAD of 1.2 meters (4 feet) is often a safe bet for voice. This assumes a 25dB acoustic S/N ratio.

The purpose of the NAG calculation is to determine how much gain is needed to deliver the same level of sound to all the listeners in the space. The reference for the necessary sound level is determined by the EAD. Logically, excessive background noise level will cause the EAD to decrease.

The formula for NAG is as follows:

$NAG = 20 \times log(D_0 \div EAD)$

The answer is in dB and indicates the amount of acoustical loss from the EAD position to the farthest listener. Designers use this formula to decide whether any equipment is necessary to overcome the loss over distance. Because you calculate NAG before equipment is employed, no equipment is represented in the NAG formula.

Figure 6-25 shows an EAD chart, which allows you to determine the maximum physical distance that a talker and a listener could be separated and easily be heard and understood without a sound system. For example, let's say you are talking at a comfortable volume of 70 dBA at 600 millimeters (2 feet). If you look at the EAD chart where

Figure 6-25
Equivalent acoustical distance and ambient noise level

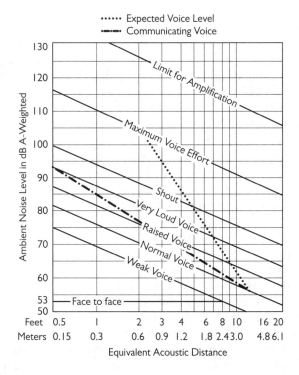

600 millimeters (2 feet) on the x-axis meets 70dBA on the y-axis, your plot would land within the "normal voice" range on the chart. Let's say that you change the level to 53dB. By following the normal voice line along the chart, the EAD would need to be 4.8 meters (16 feet) to stay within the normal voice range.

PAG describes the ability of a system to amplify live sound without creating feedback. You will usually compare PAG to NAG for a given space. The potential of the system needs to exceed the needs of listeners.

Whereas NAG is theoretical, PAG deals with actual equipment and comprises four distance factors in its calculation, as shown in Figure 6-26. The distance variables are:

- D_0 is the distance between the talker and the farthest listener
- D_1 is the distance between the microphone and the closest loudspeaker to it
- D_2 is the distance between the farthest listener and the loudspeaker closest to them
- D_S is the distance between the sound source (talker) and the microphone

The formula for PAG is:

$$PAG = 20 \times log[(D_0 \times D_1) \div (D_2 \times D_S)]$$

It is important to note that the PAG formula assumes the use of an omnidirectional microphone and an omnidirectional loudspeaker. A directional microphone could

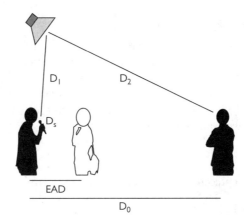

Figure 6-26
Distance measurements used in the NAG-PAG equations

increase PAG by picking up the presenter's voice while rejecting more of the loudspeaker's output than an omnidirectional microphone. A directional loudspeaker could increase PAG by delivering more of the power to the listener and away from the microphone. A directional loudspeaker could also reduce feedback potential by producing less room reverberation and thus reducing the reverberant sound level at the microphone.

However, directional equipment isn't always a good thing. As long as directional microphones and loudspeakers point away from each other, they are helpful. But if they point toward each other, it may become a catastrophe.

More Variables: NOM and FSM

Now that you know about PAG and NAG, you need to consider two other factors when ensuring audio system stability: *number of open microphones* (NOM) and *feedback stability margin* (FSM).

In speech-reinforcement or conferencing applications, NOM becomes an issue for a variety of reasons. For example, each time the number of open microphones increases, the system gain also increases. In other words, each decibel of gain is a decibel closer to feedback. Similarly, every time the number of open microphones doubles, acoustic power doubles, so when you double the number of open microphones (one to two, two to four, etc.), you lose 3dB before feedback. In conferencing applications, the level increases with each open microphone, which could introduce signal distortion.

FSM refers to how close the system is to actual feedback. Any system on the edge of feedback will produce a ringing behavior prior to actual feedback. The FSM in a carefully equalized system is typically a minimum of 6dB.

Putting NOM and FSM on the PAG side of a formula makes sense, because microphones and equalizers are pieces of equipment and PAG is relevant to equipment. Ultimately, you will compare PAG and NAG to determine whether an audio system will be stable and whether it can be built as designed. PAG must be greater than or equal to NAG.

PAG/NAG in Action

Now that you know about the NAG formula, the PAG formula, and how NOM and FSM affect them, you can combine them in one formula.

Remember, for a sound system to be stable, PAG must be greater than or equal to NAG. Conversely, NAG must be less than or equal to PAG. Although this chapter presents NOM and FSM on the PAG side of the formula, you could also factor them into the NAG side. For our purposes, the test for stability is

$$NAG < PAG$$

This is also represented by the complete test formula:

$$20 \times \log(D_0 \div EAD) < 20 \times \log[(D_0 \times D_1) \div (D_2 \times D_s)] - 10 \times \log(NOM) - FSM$$

Here's how to complete the test:

1. Ascertain the EAD and the distances D_0, D_1, D_2, and D_s for the presenter and farthest listener positions.
2. Ascertain the number of active microphones (NOM).
3. Ascertain the feedback stability margin (FSM).
4. Perform the NAG calculation using the formula given earlier.
5. Perform the PAG calculation using the formula given earlier.
6. Account for NOM and FSM by subtracting *10 × log(NOM) – FSM* from the PAG side of the equation.
7. Compare NAG to PAG. To have a stable sound system, NAG must be less than PAG.

Chapter Review

In this chapter, you learned how people perceive sound. You learned how sound is generated, moves through space, and interacts with the environment. Most importantly, you learned how to receive and process sound, as well as control its propagation to meet a client's needs.

Upon completion of this chapter, you should be able to do the following:

- Calculate the change in a sound and signal level using decibel equations
- Calculate power and distance using the 10-log and 20-log functions
- Relate a given sound pressure level measurement to human perception
- Identify three qualities of an effective sound reinforcement system
- Plot loudspeaker coverage in a room so that the SPL level is consistent within a range of 3dB throughout a defined listener area
- Specify a power amplifier for a sound reinforcement system by calculating the wattage needed to achieve a target SPL at the listener position

- Explain the purpose of the microphone preamplifier and identify the signal level used for routing and processing
- Select the appropriate microphone for a given application by evaluating the environment
- Determine whether a given audio system will be stable by comparing NAG and PAG calculations

Review Questions

The following questions are based on the content covered in this chapter and are intended to help reinforce the knowledge you have assimilated. These questions are similar to the questions presented on the CTS-D exam. See Appendix E for more information on how to access the free online sample questions.

1. When creating a schematic diagram of audio signal flow for a project, what must be included in the diagram?

    A. Microphones, mixers, switchers, routers, and processors

    B. Inputs, outputs, equipment rack locations, ceiling venting

    C. Conduit runs, pull boxes, bends

    D. Construction materials, wall panels, acoustic tiles

2. Sound pressure level should always fall between _____.

    A. −10 and 100dB SPL

    B. 0 and 140dB SPL

    C. 0 and 200dB SPL

    D. 10 and 140dB SPL

3. Using a 20-log decibel calculation and assuming a loudspeaker is generating 80dB SPL at a distance 7 meters from the source, what would the level be at 28 meters away?

    A. 40dB SPL

    B. 52dB SPL

    C. 68dB SPL

    D. 72dB SPL

4. A presenter is speaking to a large audience. Listener #1 is 2 meters away from the presenter, and listener #2 is 15 meters away from the presenter. What is the expected gain in SPL at the listener #1 position? Hint: This is a 20-log problem.

    A. −1.7dB SPL

    B. −7dB SPL

    C. −17.5 dB SPL

    D. 17.5dB SPL

5. The threshold of human hearing is specified as 0dB SPL at what frequency?
   A. 0kHz
   B. 1kHz
   C. 3kHz
   D. 10kHz

6. Calculating for loudspeaker coverage takes into account which of the following?
   A. Ceiling height, listeners' ear height, and the speaker's angle of coverage
   B. Dispersal patterns, room dimensions, and number of speakers
   C. Number of speakers, type of speakers, acoustic treatments
   D. Number of listeners, type of audio, speaker power

7. Assuming a loudspeaker coverage angle of 90 degrees, a mounted loudspeaker height of 3.6 meters, and listeners who are seated, what is the diameter of coverage?
   A. 2.4 meters
   B. 4.7 meters
   C. 4.8 meters
   D. 5 meters

8. How many watts are required at the loudspeaker with a sound pressure requirement of 70dB SPL, a loudspeaker with a sensitivity of 88dB SPL, 1w@1m, 10dB SPL of headroom, and a listener position of 14 feet (4.26 meters) from the loudspeaker? Hint: You'll need the EPR formula.
   A. 2.89 watts
   B. 4.0 watts
   C. 4.33 watts
   D. 5.12 watts

9. A(n) _____ microphone picks up sound uniformly from all directions.
   A. Bidirectional
   B. Cardioid
   C. Hypercardioid
   D. Omnidirectional

10. PAG and NAG are important calculations for ensuring an audio system is:
    A. Loud enough
    B. Intelligible
    C. Stable
    D. All of the above

Answers

1. **A.** When creating a schematic diagram of audio signal flow, you should include microphones, mixers, switchers, routers, and processors.
2. **B.** Sound pressure level should always fall between 0 and 140dB SPL.
3. **C.** 68dB SPL. To calculate the loss, because you're calculating for a distance farther away, dB = $20 \times log(D_1 \div D_2)$ = $20 \times log(7 \div 28)$ = 12dB. Therefore, 80dB – 12dB = 68dB SPL.
4. **D.** The gain is 17.5dB SPL. dB = $20 \times log(D_1 \div D_2)$ = $20 \times log(15 \div 2)$ = $20 \times log(7.5)$ = 17.5.
5. **B.** The threshold of human hearing is specified as 0dB SPL at 1kHz.
6. **A.** Calculating for loudspeaker coverage takes into account ceiling height, the listeners' ear height, and the speaker's angle of coverage.
7. **C.** 4.8 meters

   $D = 2 \times (H - h) \times tan(C\angle \div 2) = 2 \times (3.6 - 1.2) \times tan(90 \div 2) = 2 \times 2.4 \times 1 = 4.8$

8. **A.** 2.89 watts

$$EPR = 10^{\left(\frac{[L_p + H - L_s + (20 \times Log(D_2 \div D_r))]}{10}\right)} \times W_{ref}$$

   First, calculate the exponent part of the equation.

   $L_p + H - L_S + 20 \times log(D_2 \div D_r)$
   $70 + 10 - 88 + 20 \times log(14 \div 3.28)$
   $(80 - 88) + 12.6$
   $(-8 + 12.26) = 4.6$

   Second, divide by 10.

   $4.6 \div 10 = 0.46$

   Third, complete the exponent calculation.

   EPR = $10^{0.46}$
   EPR = 2.89

   Finally, multiply 2.89 by the reference value, which is 1 in this problem.

   EPR = 2.89×1
   EPR = 2.89

9. **D.** An omnidirectional microphone picks up sound uniformly from all directions.
10. **C.** PAG and NAG are important calculations for ensuring an audio system is stable.

PART III

Infrastructure

- **Chapter 7** Communicating with Allied Trades
- **Chapter 8** Lighting Specifications
- **Chapter 9** Structural and Mechanical Consideration
- **Chapter 10** Specifying Electrical Infrastructure
- **Chapter 11** Elements of Acoustics

CHAPTER 7

Communicating with Allied Trades

In this chapter, you will learn about
- The related professionals you work with on an AV project
- Three common organizational tools used to track an AV project
- The purpose of industry standards and the role they play in working with allied trades
- The importance of detail in specifying your design

The infrastructure that supports an AV design is handled by architects; electricians; lighting designers; structural, electrical, network, and mechanical engineers; artisans; builders; interior designers; acousticians; and many others. As you specify what will go into your design, you will need to communicate those elements to your allied trade partners.

Each trade requires certain information from you, which we will discuss in Chapters 8 through 11. But before you learn about working with individual allied trades, we will discuss briefly the general fabric of communicating with these professionals.

Duty Check
This chapter relates directly to the following tasks on the CTS-D Exam Content Outline:

- Duty A, Task 1: Identify Decision-makers and Stakeholders
- Duty B, Task 2: Coordinate with Architectural/Interior Design Professionals
- Duty B, Task 3: Coordinate with Mechanical Professionals
- Duty B, Task 4: Coordinate with Structural Professionals
- Duty B, Task 5: Coordinate with Electrical Professionals

(continued)

- Duty B, Task 6: Coordinate with Lighting Professionals
- Duty B, Task 7: Coordinate with IT and Network Security Professionals
- Duty B, Task 8: Coordinate with Acoustical Professionals
- Duty B, Task 9: Coordinate with Life Safety and Security Professionals
- Duty D, Task 1: Participate in Project Implementation Communication

In addition to skills related to these tasks, this chapter may also relate to other tasks.

Communicating with Stakeholders

All AV professionals must work with people from outside their organization to serve their customers' needs. This particularly includes the primary stakeholder—the project owners—via the owner representatives on the project.

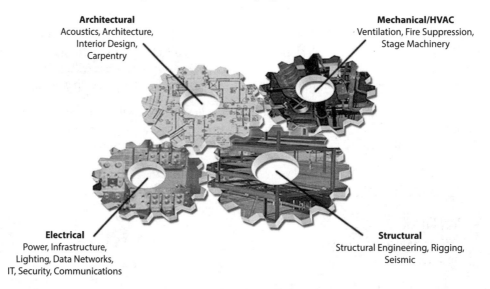

Architectural
Acoustics, Architecture, Interior Design, Carpentry

Mechanical/HVAC
Ventilation, Fire Suppression, Stage Machinery

Electrical
Power, Infrastructure, Lighting, Data Networks, IT, Security, Communications

Structural
Structural Engineering, Rigging, Seismic

Allied trades are the businesses that collaborate with AV professionals to complete integrated solutions for customers. Each trade has its own priorities and areas of expertise, but they must all work together to satisfy the customer. Typically, you will need to work with these other professionals in the design phase of the project to create proper spaces for AV systems. Cooperating and identifying issues early will produce the best results.

Table 7-1 shows some of the AV industry's allied trades and the areas where they might collaborate with AV professionals. This table is not exhaustive. A more complex or unusual AV project might involve even more allied trades.

Allied Trade	Collaborate with AV Professionals On...
Architects	Window placement in display environments; acoustics; space requirements
Building professionals	Building processes; construction schedules; project management; construction waste management; site access; building information management
Interior designers	Furniture; equipment positioning; wall, floor, ceiling, and window treatments
Electrical professionals	Wiring installation; power requirements; energy management; RF spectrum management
IT professionals	Network connectivity; media storage; equipment control; communications; Internet access; control system programming; physical and logical system security
Security professionals	Alarms; surveillance; entry and access control; presence detection; fire detection; Emergency Warning and Intercommunication Systems (EWIS)
Heating, ventilation, and air conditioning (HVAC) professionals	Equipment mounting; HVAC noise; pipe and duct placement; atmospheric special effects; equipment ventilation and cooling
Live production professionals	Production and effects lighting; musical performance audio; stage monitoring; communications and cueing; pyrotechnics; video relay; performance video and projection systems; production control; stage management
Structural engineering professionals	Equipment mounting and suspension; flying systems; stage machinery; construction methods and materials
Content developers	Digital signage, video wall, and soundscape content designers; UI/UX designers; videographers; post-production teams
Subject-matter experts	Topics specific to a particular project or vertical market; for example, a teacher might advise an AV professional on the requirements for a classroom, or a museum director might advise on the specific requirements for a gallery or display hall

Table 7-1 Areas of Collaboration with Allied Trades

Tracking the Project

Engaging with allied trades and understanding each other's needs are crucial to figuring out how you're going to design an AV system. In the process, it is important to develop a feasible timeline with your client and team so that everyone is aware of the dates by which certain milestones should be accomplished. Although this is more of a project management function, it's important that AV designers understand how it works.

Project coordination covers many different aspects of a project, not just the AV portion. Sometimes, other parts of a project must be completed before you can begin or finish certain AV elements. It is common for the AV team to wait until construction is finished before scheduling the installation of sensitive equipment.

The AV project manager must also monitor other parts of the project to ensure that considerations for AV equipment or specifications are met, such as monitoring where, specifically, an HVAC or sprinkler system is installed in a space to keep it from being collocated with a projector or cable run. These matters are the subject of the weekly onsite progress meetings between all trades on the project.

The most common way to coordinate the completion of such tasks and to track the overall project is to create an organizational chart. There are three basic types:

- **Work breakdown structure** A work breakdown structure (WBS) presents deliverables on products or services.
- **Gantt chart** A Gantt chart provides a timeline for all activities that are scheduled.
- **Logic network diagram** A logic network diagram shows which tasks have to be completed before you can begin the next task. Similar to a program evaluation and review technique (PERT) network chart.

These charts provide a view of the project at different stages. Let's look at what is involved with each.

Work Breakdown Structure

Figure 7-1 shows an example of a WBS. The purpose of this document is to make certain that all those in a group working together on a project know exactly what results are required by the rest of the project team and the client. The WBS defines aspects of the project as products or services. These are always stated as nouns and do not specify who will accomplish something, when it will be done, or how much of a specific product or service will be delivered.

The WBS divides the project's work into clearly understandable chunks. It also shows the relationship between these elements and to the entire project or system. Every lower level of the WBS shows a more specific component of the total statement and traces that specific component's relationship to a more general task.

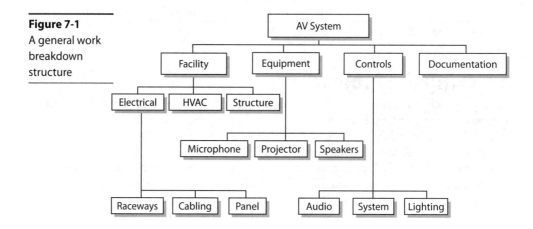

Figure 7-1 A general work breakdown structure

The WBS can serve as an outline for estimates, and it also makes it easier to assign clearly defined responsibilities. Additionally, the WBS can create a stable basis for recognizing and managing changes in the project plan and even for evaluating performance.

Gantt Chart for Project Schedules

A general project schedule is typically established during the design and bid phases. The owner's project manager finalizes it before work begins and will bring it to planning meetings and the regular onsite meetings. This way, every team project manager can see their project deliverables. Each trade or discipline must have a corresponding schedule of work (developed by each team's project manager), which is incorporated into the main project schedule. The AV project team needs to be represented on this construction schedule for the entire project. This responsibility often falls to the AV designer or consultant.

The most common way to present a project schedule is a type of bar chart known as a *Gantt chart* (named after Henry Gantt), which depicts the timeline for tasks and subtasks as horizontal bars or lines. It shows the sequence in which tasks should be performed and any project milestones, such as the completion of major categories of tasks. Project management software is often used to create Gantt charts, like the example shown in Figure 7-2.

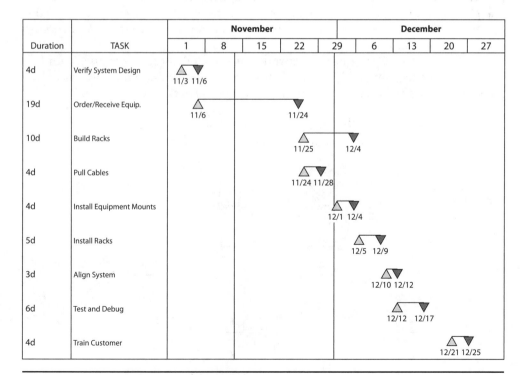

Figure 7-2 A Gantt chart showing a project schedule (dates in mm/dd format)

Because some tasks must be completed before others may start, the Gantt chart identifies these types of dependencies in a manner that clearly shows their sequence. As you know, walls within a room must be finished and painted prior to mounting some AV components, so the start dates of these AV tasks will be identified as dependent on the end dates of the room-preparation tasks. In Figure 7-2, the second task (ordering and receiving equipment) is dependent on the first task (verifying the system design).

Logic Network Diagram

Resources such as tools, test equipment, vehicles, and staff time are always limited, even in large organizations. Since it is likely that the installation company will simultaneously be working on more than one project, it is essential to plan for the allocation of resources for each project. This will help avoid conflicts and shortages when projects have overlapping needs. An effective resource plan involves creating a clear schedule, showing what resources will be required and when they will be needed, and highlighting possible shortages and related problems. Once these challenges are identified, you can initiate actions to assure the availability of more people, tools, or other resources needed to properly complete the job on schedule.

After the WBS has been completed, specific activities and milestones can be linked to the project's deliverables. A *deliverable* is an object or a service that is required to be delivered to a customer as a component of a complete project. The deliverable could be a single device, a complex functional system, a document, a piece of software, or a training session. These deliverables and their requirements form the basis for a *logic network diagram,* such as the one shown in Figure 7-3.

Also known as a project schedule network diagram (and not to be confused with IT's logical network diagram), a logic network diagram helps organize, and finally schedule, a project's activities and milestones. The uses of a logic network diagram include

- Listing activities and their sequential relationships
- Identifying dependencies and their impacts

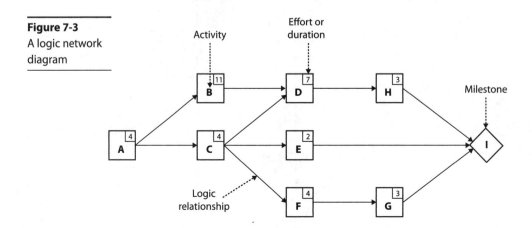

Figure 7-3 A logic network diagram

- Introducing activities into the network according to their dependencies, not on the basis of time constraints
- Identifying successors
- Identifying predecessors

The logic network diagram includes the activities and tasks that are required to create the deliverable as they consume time and resources. The functions of a logic network are expressed as verbs, such as *build, test, fabricate,* and *develop*. Each activity should have an associated deliverable.

In Figure 7-3, *effort* refers to the amount of time it will take to complete an *activity* or task in its entirety. Many companies calculate an effort value based on best-case scenarios. This optimistic practice usually increases risk and can lead to underbidding the labor component of an installation. For example, it may take only a few minutes to replace a faulty component, but if a technician must travel to a distant location to get the component, those few minutes can turn into hours.

Mature organizations look at resource requirements across all of their current projects before establishing timelines. They prefer to estimate *duration,* which refers to the amount of time it will actually take to complete a task, not just the time devoted to the task. Duration factors into several components, such as the number of work periods, resource availability based on other projects, and personal and organizational calendars (weekends, holidays).

Milestones are key events in the project, such as the completion of a major deliverable or the occurrence of an important event. They can be associated with scheduled payments, client approvals, and similar events. Examples include *materials delivered, racks programmed,* and *projectors calibrated*. As shown in Figure 7-3, milestones are typically shown as diamonds or triangles in the logic network diagram.

NOTE Gantt charts and logic network diagrams can be as simple or as complex as you want to make them.

Industry Standards as Common Language

AV projects are complex. Out of necessity, they're guided by standards, both in the AV industry and in those of allied trades.

A standard is a document that provides requirements, specifications, guidelines, or characteristics that can be used consistently to ensure that materials, products, processes, and services are fit for their purpose. Standards serve as a common platform for understanding and teamwork; they're a language for conducting day-to-day business. Standards are often voluntary, but they can also become regulation or code.

There are many kinds of standards: technical, management, performance, measurement, methodology, and even de facto standards. De facto standards are not necessarily written down or approved by some standards organization, but become a standard because they are overwhelmingly accepted by the public or a particular industry. In the AV and technology-related industries, technical standards help ensure that diverse products from different companies work together, which creates efficiency, consistency, competition, and interoperability. Performance standards help ensure that a system operates optimally, or as intended, so that it meets a client's expectations.

Standards are developed by specific organizations to document agreed-upon rules, regulations, specifications, measurements, and protocols. These organizations often create standards that meet or call upon other standards, which gives them the added weight of authority and encourages implementations.

Standards groups that are important to the AV and IT industries include the International Organization for Standards (ISO) (www.iso.org); the International Electrotechnical Commission (IEC) (www.iec.ch); the Institute of Electrical and Electronics Engineers Standards Association (IEEE SA) (standards.ieee.org); and regional standards coordinators such as Standards New Zealand (standards.govt.nz), Germany's Deutsches Institut für Normung (DIN) (din.de), and the American National Standards Institute (ANSI) (ansi.org) in the United States.

Of particular relevance to AV technologies are such professional groups as the Entertainment Services and Technology Association (ESTA) (esta.org), the Society of Motion Picture and Television Engineers (SMPTE) (www.smpte.org), the Audio Engineering Society (AES) (www.aes.org), VESA (formerly the Video Electronics Standards Association) (vesa.org), the Motion Picture Experts Group (MPEG) (mpeg.org), the Professional Lighting and Sound Association (PLASA) (plasa.org), the USB Implementers Forum (USB IF) (www.usb.org), the 8K Association (8KA) (8kassociation.com), the HDBaseT Alliance (hdbaset.org), and the Joint Photographic Experts Group (JPEG) (jpeg.org).

AVIXA became an ANSI-accredited standards developer to assist in bringing performance standards to the AV industry. Since developing its first standard in 2008, AVIXA's standards development processes have produced standards and recommended practices in a wide variety of areas ranging from audio systems performance and energy management practices to drawing symbols, project documentation, equipment rack construction, video display systems, and system verification methodologies. A more complete listing of these can be found in Appendix B, and a fully up-to-date listing, including standards currently being developed and revised and information on how to acquire your own copies of the AVIXA standards, can be found on the AVIXA website at avixa.org/standards.

NOTE Standards differ throughout the world. If you are interested in discovering AV standards relevant to your region, start your search at the ANSI Standards Portal (www.standardsportal.org) for information on standards in countries including the United States, the People's Republic of China, the Republic of India, the Republic of Korea, and the Federative Republic of Brazil, or try the ISO (www.iso.org).

Hierarchy of Design Consultation

Although AVIXA and allied trades offer standards that lend uniformity and professionalism to each project, there are a variety of contributing influences on design consultation that go beyond standards (see Figure 7-4).

As you consider which references you'll use for basic project information or as the basis for key decisions, always start with regulatory requirements. From there, you can work your way through the available references to identify the sources that will help you make and defend your decisions. For example, you may be working with an allied trade partner to resolve a conflict between lighting positions in a room. Although the lighting designer might refer to a manufacturer's recommendation for fixture placement, you might refer to a national standard that recommends a different solution. Instead of going with a personal preference, you can use your knowledge of the standard to resolve the problem with the lighting designer. When you consult a higher voice of authority, you can be more confident in the decision-making process.

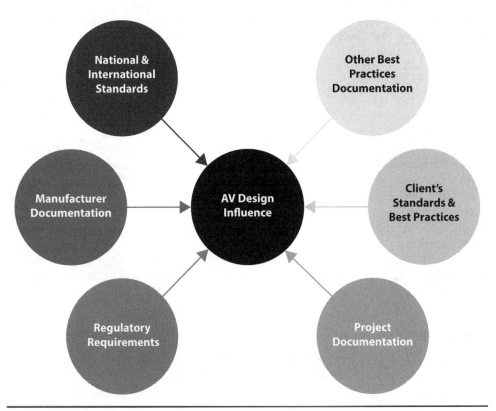

Figure 7-4 Contributing references for design consultation

Ultimately, your goal is to reference the proper source and use this knowledge to justify your design solutions. Don't overlook that some of your allied-trade colleagues may have a vast body of knowledge and experience from working on previous projects and should be carefully listened to during the decision-making process.

Showing Workmanship

As you progress through the AV design and documentation process, your most important ally is the integrator. The specifications you create for electricians, lighting designers, and other trades comprise the same documentation that an integrator will use to verify that the design you developed is the one the client sees. Integrators will be the people who ultimately install, set up, and verify the AV design based on your specifications.

With that in mind, don't assume anything. Your specifications are the lens through which allied trades and integrators picture the project; therefore, you don't want that vision to be anything but clear. For example, if the designer fails to clearly specify important details, such as the display contrast ratio or loudspeaker placement, the AV design intent may or may not be achieved. Even small details, such as how items should be labeled on input wall plates or the color of user-accessible input cables, can affect the overall success of a system.

To assist you in these processes, the ANSI/AVIXA 2M-2010 *Standard Guide for Audiovisual Systems Design and Coordination Processes* lays out a procedure for determining what project communication (documentation, meetings, and so on) needs to take place on a given project. At the time of publication, this standard was undergoing a major revision as ANSI/AVIXA D401.01:202X.

In addition, the ANSI/AVIXA 10:202X *Audiovisual Systems Performance Verification* standard is intended to help you evaluate a system's performance against its documented goals and provides some invaluable guidance on how to communicate your design intent and verify that it has been correctly executed. It includes a list of performance verification tests by system function. This list is divided into tests that should be performed before, during, and after system installation. At the time of publication, this standard was undergoing a major revision.

Specify the details, document the math, and use professional guidelines such as AVIXA standards in your design. Maintaining this mind-set will safeguard against unnecessary confusion and limit misconceptions.

Chapter Review

In this chapter, you reviewed the basic tools, resources, and strategies that are the foundation for communicating with allied trades during an AV design. You should now be ready to study each trade individually to learn exactly what they'll need from you as you design the AV system.

Review Questions

The following questions are based on the content covered in this chapter and are intended to help reinforce the knowledge you have assimilated. These questions are similar to the questions presented on the CTS-D exam. See Appendix E for more information on how to access the free online sample questions.

1. What is the first design reference you need to consider in your design consultation?
   A. Project documentation
   B. Regulatory requirements
   C. Manufacturer documentation
   D. Client's standards and best practices

2. What is a common way to track a project?
   A. Spreadsheet
   B. Note taking
   C. Work breakdown structure
   D. Statement of work

3. Who must monitor other areas of the project to ensure that considerations for AV equipment or specifications are met, such as monitoring where an HVAC and sprinkler system are installed?
   A. HVAC installer
   B. AV project manager
   C. Facilities manager
   D. Building owner

4. In which phase does a project manager need to work with allied trades to create proper workspaces?
   A. Training phase
   B. Installation phase
   C. Maintenance phase
   D. Design phase

5. Which document provides a common language for communicating with teams from different industries?
   A. Standards
   B. Project documentation
   C. Manufacturer documentation
   D. Meeting notes

Answers

1. **B.** In the hierarchy of design consultation, regulatory requirements are your first point of reference when making key decisions.
2. **C.** A work breakdown structure is one common way to track an AV project.
3. **B.** An AV project manager should monitor all aspects of a project to ensure that considerations for AV equipment are met.
4. **D.** A project manager should work with allied trades during the design phase to create proper workspaces.
5. **A.** Standards provide a common language for communicating with teams from different industries.

CHAPTER 8
Lighting Specifications

In this chapter, you will learn about
- The characteristics of light that impact human perception
- Selecting luminaires for an AV system based on a room's application
- Creating zoning plans for a room's lighting based on application
- Determining lighting-level presets for each zone in an AV space
- Addressing factors specific to maintaining lighting quality for videoconferencing
- Safety expectations in your design

An AV designer must specify lighting requirements for the space in which an AV system will be installed and operated. The goal of this chapter isn't to turn you into a lighting engineer. It's to make you aware of lighting specifications that can improve the quality of your AV design. Ultimately, you will likely coordinate with a lighting consultant to execute your design.

Light can come from many sources, such as natural light from windows or artificial light from luminaires and AV equipment (projectors, displays, etc.). All light sources, whether natural or artificial, must be considered in an AV design to ensure that it meets the client's needs and allows end users to operate an AV system effectively.

Keep in mind, users need enough ambient light to move safely around a space, view collateral materials (objects, signage, printed pages), and see presenters well enough to comprehend the intended message. Such ambient light, while necessary, can impact the effectiveness of display systems and will affect the design. Users also need enough light coming from the AV displays to perceive displayed images as faithful reproductions. Moreover, in situations where video cameras are required, you need to specify lighting that illuminates participants and the space so that they can be seen clearly and communicate well.

> **Duty Check**
>
> This chapter relates directly to the following tasks on the CTS-D Exam Content Outline:
>
> - Duty A, Task 6: Identify Scope of Work
> - Duty B, Task 2: Coordinate with Architectural/Interior Design Professionals
> - Duty B, Task 6: Coordinate with Lighting Professionals
> - Duty C, Task 1: Create Draft AV Design
> - Duty C, Task 2: Confirm Site Conditions
>
> In addition to skills related to these tasks, this chapter may also relate to other tasks.

Basics of Lighting

When considering lighting for an AV design, it is important to understand human perception. How humans perceive brightness, contrast, and color will help determine whether your design serves its intended purpose, as identified during the needs analysis, and will help users communicate effectively.

Visual perception in an AV space requires a light source (the luminaire), receptor (the human eye and/or the camera), and analyzer (the brain). It's up to the designer to control the light within a space so that users can accurately perceive, receive, and analyze information.

The human eye is stimulated by light that comes directly from a source, as well as light that reflects off surfaces before entering the eye. The brain interprets the various light waves it receives through the eye, causing people to see color, contrast, and shape. Everything humans see is the product of light absorbed by, transmitted through, and reflected back from the surfaces of objects. The important characteristics of light, which designers must consider, are brightness, contrast, color temperature, and color rendering.

Brightness

For the eye to recognize color and contrast, there needs to be enough light reflecting off or emitted from the surfaces in a space. As discussed in detail in the section about the display environment in Chapter 5, the intensity of light is measured using two types of meters:

- Incident meters measure illuminance, the light coming directly from a source such as a lamp, projector, luminaire, digital sign, video wall, or monitor. Illuminance is measured in lux (lx), or in the United States, in foot-candle (fc).

- Reflected meters, or spot meters, measure luminance—the light that bounces off an object like a projection screen, a wall, a participant in a videoconference, or a work surface. Luminance is measured in candela per square meter (cd/m^2), sometimes still called the *nit* (nt), or in the United States, in foot-lambert (fL).

Calculating Luminance from Illuminance

Incident light measurements can be converted into a reasonable approximation of the luminance of an illuminated matte surface if the light reflectance value of that surface is known. Light reflectance value (LRV), or reflectance, is the amount of light reflected from an object as a percentage of the light falling on it. Pure white surfaces have a theoretical LRV of 100 percent.

The conversion formula is:

$$L = E \times LRV \times \pi \div 100$$

where:

- L = luminance in cd/m^2
- E = illuminance at the surface, in lux
- LRV = the LRV of the illuminated surface (0 to 100)
- π = pi ~3.14

NOTE This formula only works for matte surfaces. If a surface has any specularity (shininess or sheen), this relationship is not valid.

Color Temperature

The types of light sources you choose for an AV space will be a determining factor in what wavelengths of light will reach users' eyes. White light is produced when a light source produces a mix of the main colors in the visible spectrum. All types of light source produce different distributions of the visible colors, so many types of white light have a different tint—some are actually reddish white, and others are bluish white. This difference in balance of the spectral colors is referred to as a light source's *color temperature*.

Color temperature is the scientific measurement for expressing the distribution of the spectral colors radiating from a light source, expressed in Kelvin (K). As the Kelvin measure is a reference to the colors emitted from a theoretical object under defined conditions, the measurement is expressed in Kelvin, never degrees. Known as the Planckian locus, the curved line in the middle of the chromaticity diagram shown in Figure 8-1, tracks the color of the theoretical black body radiator as it's heated to the temperature indicated in Kelvin along the line. The higher the color temperature, the bluer the light; the lower the color temperature, the more orange the light.

The color temperature assigned to a light source with chromaticity coordinates that do not fall along the Planckian locus is an approximation based on the coordinates'

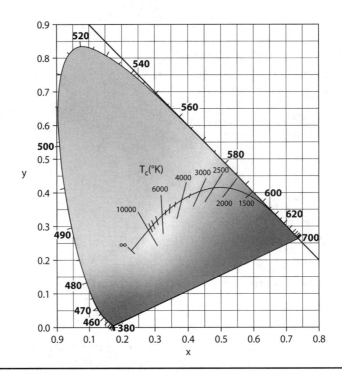

Figure 8-1 A chromaticity diagram

proximity to the locus. This is known as a *correlated color temperature* and is the figure quoted for most nonincandescent sources. Table 8-1 lists the correlated color temperatures of some common light sources.

Temperature	Source
1,900K	Candle light
2,400–3,000K	Tungsten filament lamps
2,200–3,000K	"Warm white" LEDs
2,200K	High-pressure sodium lamp
3,200K–3,400K	Tungsten Halogen filament lamps for film and video production
4,200K	"Cool-white" fluorescent lamp
4,000–6,000K	"Daylight" LEDs
5,400K	Daylight
5,500K	Noon, direct daylight, cloudless sky
6,500K	SMPTE reference white
3,000–9,500K	Flat-panel screen range

Table 8-1 Correlated Color Temperature of Some Common Light Sources

Color temperature is important to visual display. Here are some examples of how you will see color temperature applied:

- Electronic cameras are substantially more sensitive to changes in light's color temperature than the human eye. Setting the correct white balance is a critical step in producing good-quality images from any electronic imaging device. Color temperature deviations greater than ±200K within a video image should be avoided.
- Most direct-view displays and projectors have color temperature selection and adjustment capabilities. These allow you to alter the white balance of an image for the lighting conditions in the environment or to match projected images and monitor displays in a presentation.
- To maintain the correct color balance in camera images that include video displays in the field of view, the color temperature of the lighting for the image should be within the ±200K tolerance of the color temperature of the video displays.

TIP As virtually every meeting room, classroom, office, or presentation space is likely to involve some form of video recording, video relay, or videoconferencing as part of its AV functionality, it is advisable to plan the AV and lighting systems so that the color temperatures across the entire space lie within the ±200K tolerance of video imaging devices.

Color Rendering Index

The *color rendering index* (CRI) of a light source indicates how accurately it renders the colors in the objects it illuminates. All light sources produce different quantities of the range of spectral colors—the CRI is a measure of whether the full spectrum is present and in what relative proportions. By definition, the full spectrum light sources that produce light through incandescence have a perfect CRI of 100. Such sources include the tungsten filament in a lamp, the carbon in a candle flame, and the photosphere of the sun.

Most sources that produce light by other means—such as passing a current through a gas (sodium vapor), exciting a phosphor to glow (fluorescent), or passing a current through a semiconductor junction (LED)—have a lower CRI depending on the design and construction of the light source, as shown in Table 8-2. Special-purpose high CRI (but lower luminous efficiency) versions of some of these sources are available at higher prices.

Electronic cameras may have problems producing accurate images under light sources with a CRI under 60, even after a full white-balance. The closer the CRI is to 100, the more accurate the color quality of the image. It is important to render colors accurately when skin tones are in frame or if a client or a product has specific colors as part of their identity or brand.

A color checker chart (sometimes known as a Macbeth chart) may be helpful for a quick visual evaluation of a camera's response to a light source, but without access to a professional waveform monitor or calibration software, the results are not definitive.

Light Source	CRI/R_a
Incandescent lamp	100
Sunlight	100
High-pressure sodium vapor street light (orange)	24
Mercury vapor street light (blue/green)	20
Warm white tri-phosphor fluorescent	80
Cool white fluorescent	62
Kino Flo True Match fluorescent (film and video production)	95
HMI discharge lamp (film and video production)	94
LED torch	65
LED general-purpose lamp	85
LED special-purpose high CRI lamp	98

Table 8-2 Color Rendering Index/R_a of Some Common Light Sources

CIE (1995) CRI

CRI values measured using the International Commission on Illumination (CIE) method are usually quoted in light source specifications using notation R_a. Originally developed in 1965 (long before the microprocessor or the first visible-light LED), the CIE (1995) method for calculating the R_a of a light source uses a spectrophotometric comparison with eight specified low-saturation color test samples illuminated by a standardized D65 (daylight 6,500K) source. These eight specified colors, however, do not reveal some of the common shortfalls in the output spectrum of many recent lamp technologies, particularly LEDs, and consequently a number of newer methodologies for assessing CRI have been developed. Nevertheless, the dubiously useful R_a remains the most commonly quoted CRI for commercial light sources. See Table 8-2.

NIST Color Quality Scale

The U.S. National Institute of Standards and Technology developed the Color Quality Scale (CQS) CRI, which uses 15 more-saturated color samples than the CIE (1995) test. Instead of spectrophotometric tests of the light reflected from the color samples, this measure computes the CRI by comparing the full *spectral-power distribution* (SPD) curve of the light source under test with the SPD of a reference source. The SPD of a source is captured using a spectroradiometer, which measures the power output of the source at wide range of wavelengths across the visible spectrum.

Television Lighting Consistency Index (2012)

The Television Lighting Consistency Index (TLCI) was developed by the European Broadcasting Union (Eurovision) in 2012, and revised in 2018, to take the SPD of a light source and generate the image that would be produced for 24 color samples viewed on a standard three-chip video camera. The result of a TLCI calculation is a digital image of 24 color rectangles, where the center of each rectangle is the computed response for the light source, while the remainder of the rectangle is the reference color.

A wide selection of computed TLCIs for common film and video luminaires is in the public domain. At the time of publication, the current version of the TLCI did not accurately compute the response of single-chip video cameras.

CIE 224:2017 / ANSI/IES TM-30-20

Now adopted by the CIE as an international standard, the ANSI/IES TM-30-20 (TM-30) was developed by the Illuminating Engineering Society (IES) of North America to take the SPD of a light source and generate an image of a set of 99 reference color evaluation samples which take into account human color perception and a reference light source. The calculated TM-30 fidelity index (R_f), with a range of 1 to 100, is a comparison to the SPD of a D50 (5,000K) reference source.

TIP The rule of thumb regarding the CRI of light sources to use in projects is to completely avoid sources with a CRI under 60 and select the sources with the highest CRI you can afford.

Energy Consumption

Another factor you'll have to consider in a lighting design is a light source's energy consumption. The energy efficiency of a light source, known as its *efficacy*, is measured in lumens per watt (lm/W). Efficacy is calculated by dividing the initial lumens for the source by the nominal power it consumes.

Sustainability efforts, along with local and national energy efficiency regulations, seek to limit the waste of energy from light sources, so many manufacturers have been investing in luminaire designs and light sources with better efficacy. Figure 8-2 shows the range of efficacies for various lamps. As the efficacy of light sources increases, a secondary bonus accrues, as less cooling power is required to remove the waste heat generated by the lighting system.

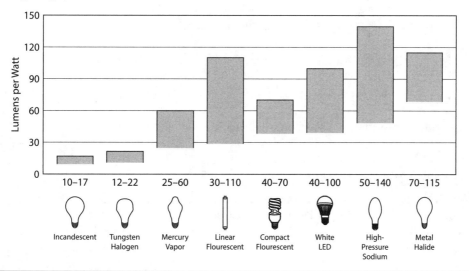

Figure 8-2 Lamp efficacy ranges

Keep in mind, however, that the highest-efficacy light source may not always be the right one for your AV design, regardless of sustainability efforts. Efficacy is only one measure among many you need to consider when determining the proper lighting. You must also take into account intensity, beam pattern, color, lamp life, and control capabilities.

Lighting the Space

Planning for the various luminaires and other lighting-related equipment in your AV design requires information gathered during the needs analysis. In that analysis, you and the client, together with the representatives of allied trades, should have discussed the different tasks that users will undertake in a space. Given space applications, you can start thinking about task lighting, window shades and blackout drapes, and light sources and luminaires.

 TIP It is important to be aware of the difference between a lamp and a luminaire. The term *lamp* describes a light-producing device such as a bulb, tube, or chip array. A *luminaire* is a complete lighting device. You may also hear the terms *lighting instrument*, *light fixture*, *lighting fixture*, and sometimes just *fixture* used by various trades to describe a luminaire, but luminaire is the internationally preferred, standard term.

Task Lighting

Task lighting concerns the amount of light you want to direct *toward* people for things such as note-taking, reading manuals, and operating equipment. It will need to be balanced against the amount of light you want to direct *away* from things such as a front-projection screen or videoconferencing camera. Many international standards recommend task-lighting light levels in AV areas to be in the range from a minimum of 150lux (14fc) to a preferred level of 320lux (30fc).

Ultimately, in planning the lighting, you need to specify the following in your design:

- Light you want
- Light you don't want
- Light you need to control

For more information about task lighting, see the section about task light levels in Chapter 5.

Shades and Blackout Drapes

Not all the light in a room comes from luminaires and display systems. Nor is all light desirable at every location of the room. Because there is usually ambient light—or general light filling a space—you will need to consider how to control it so that it doesn't negatively affect your design. Ambient light may strike the screens or the walls in the room and reflect into the viewing area, competing with the displayed image by reducing contrast and washing out the picture.

The ANSI/AVIXA standard V201.01:2021 *Image System Contrast Ratio* sets a minimum contrast ratio of 15:1 for basic decision-making applications, including information displays, presentations containing detailed images, classrooms, boardrooms, multi-purpose rooms, and product illustrations. Achieving compliance with the standard starts with limiting the amount of ambient light falling on screens. AV designers should specify the ambient light limits for these locations.

Although natural light can help set a mood or meet energy-savings requirements, the resulting level of ambience may overwhelm the image display system. One way to control the ambient light at a display is to specify opaque window shades, blinds, and blackout drapes for rooms with windows, glass walls, or doors that allow the passage of excessive natural light. If the ambient light is simply not sufficiently controllable for architectural, operational, or client-specified reasons, consideration of a more expensive, daylight-viewable display system may be the solution.

In some situations, a client may require a more naturalistic viewing environment than an isolated black box. To meet these requirements, the use of partially opaque and translucent window and door treatments may need to be considered. By working in collaboration with the architectural and interior design teams, it is possible to specify window, door, and transparent-wall treatments of appropriate transmission and color-shift values to create an ambient-light balance suitable for a display environment.

In other circumstances, the display system may be located in a space where a view through windows into a brightly illuminated operational space or to the outdoors is necessary. To achieve a lighting balance to meet the contrast requirements for visual display systems, it is usually necessary to control both the brightness and color temperature of the incoming light, together with the selection of high-brightness displays. Glass treatments such as tinted film or an external horticultural shade mesh can be used to reduce light transmission by a known factor to achieve an acceptable light balance (a 50% reduction is 1 f-stop; a 90% reduction is just over 3 f-stops). Color temperature correction may also be required to match the interior lighting, particularly if the space is being used for any video application such as recording, streaming, or videoconferencing.

While shade mesh is an inexpensive means of reducing light transmission, if the treated window will be in the camera's field of view, there is a risk of moiré effects arising between the window mesh and the camera's sensor array. The extent of the effect is dependent on the depth of field of the camera lens and the distance between the camera and the mesh.

Choosing Light Sources

Once you've specified your task light levels and your ambient light limits, you can consider the types of light sources and luminaires you'll want in the room. Light sources can be broken into four main groups:

- Incandescent
- Discharge
- Fluorescent
- Solid-state

Incandescent lamps produce light by passing a current through a tungsten wire filament, which causes the tungsten wire to glow white hot. To prevent the tungsten from boiling off the filament and blackening the lamp's envelope, the envelope is often filled with halogen gas. Although these light sources are relatively inexpensive and have a perfect CRI of 100, they are very energy-inefficient and have short lives. They can easily be dimmed by limiting the current flowing through the filament, although as the light output falls, there is a corresponding downward shift in color temperature. In many jurisdictions incandescent light sources do not meet the mandated minimum energy efficiency requirements and may no longer be used as a source for general illumination.

Discharge lamps produce light by passing a current between a pair of metal electrodes immersed in a gas or mixture of metal vapors and gases. This causes the gases to ionize and discharge light. There are a wide variety of gas mixes, electrode configurations, and envelope shapes available in a broad range of outputs and physical sizes. Some gas mixtures such as metal-halides output a sufficient mixture of wavelengths to be considered suitable for general illumination, with some tailored specifically for use in projectors, moving lights for production, and luminaires for film and video production. Discharge lamps are significantly more efficient than incandescent lamps and tend to have longer operational lives; however, they are quite difficult to dim electronically, and some have a very poor CRI.

Fluorescent lamps are a special type of discharge source, filled with a mix of gases that includes mercury vapor, which discharges ultraviolet light. The fluorescent lamp is coated with a mix of phosphors, which fluoresce in the presence of ultraviolet to produce visible light. The color of the light emitted is a result of the mix of phosphor materials used in the coating. As the green phosphor is the most efficient at producing light, many fluorescent lamps have a strong green cast to maximize their energy efficiency. Special-purpose fluorescent lamps are available tailored to a range of applications, including those designed to enhance the color of food products. There are also ranges of full-spectrum (but lower-efficacy) lamps at carefully matched color temperatures and high CRI for use in film and video production. Fluorescent lamps are higher efficacy than incandescent lamps, and while they can be electronically dimmed, they become much less efficient during the process. There is also significant trade-off between lamp efficacy and CRI.

Solid-state light sources include a mix of direct-output LED sources and phosphor materials excited by LEDs. In some lamp configurations, the sources are a mix of red, green, and blue LEDs (plus violet, lime, amber, and warm-white in some very-high CRI sources) combined to produce white light. In others, some of the light comes directly from LEDs (usually blue), and the remainder of the spectrum comes from phosphors excited by the LED. In a variation on the theme of the fluorescent discharge lamp, some solid-state sources produce ultraviolet/short blue light, which is used to excite phosphors coated on the LED. Solid-state sources have very high efficacy, are readily dimmed electronically, and have operational lives in the tens of thousands of hours. However, they require good cooling for stable operation and have CRIs that range from totally unusable to nearly perfect.

As all light sources produce waste heat, designers must account for that load in addition to the waste heat from other AV equipment when specifying the HVAC demands of the entire AV system.

Choosing Luminaires

Luminaires include not only the light source but also the physical features that control and distribute light. A luminaire is a complete lighting unit consisting of a light source; its power supply, ballast, or transformer; optical components such as reflectors, lenses, beam control and light-filtering systems; the housing and support structures; and the wiring, automation, communications, plumbing, ventilation, and power-control components.

Luminaires can be grouped based on several characteristics, but the following are the most common:

- By the type of light source they employ, such as incandescent, discharge, fluorescent, or LED. Discharge light sources are also often identified by their gas filling, such as sodium vapor, metal-halide, mercury vapor, etc.
- By mounting style, such as surface-mounted, recessed, semi-recessed, wall-mounted, track, lighting bar, or suspended.
- By optical configuration, such as plano-convex and Fresnel spotlights, beam lights, flood lights, soft lights, and profile/ellipsoidal/framing spotlights.
- By light distribution, classified by the IES as direct, indirect, semi-direct, semi-indirect, direct-indirect, and general diffuse. Such descriptions are based on the percentage of light output above and below a horizontal plane.
- By control interface, if any. Most directly controllable luminaires have interfaces only for a specific control protocol such as DMX512A, ZigBee, DALI, 0–10V DC, or an Ethernet-hosted protocol such as streaming ACN or ArtNet.

NOTE The *efficiency* of a luminaire describes the difference in percentage lumen output between the naked light source and the complete luminaire containing it, including all optical, electrical, and mechanical systems.

Luminaire Specifications

When selecting luminaires, it is important to assess the specifications of the luminaires you are considering. The data available for luminaires usually includes their light source type and mounting socket, voltage and current requirements, physical dimensions, weight, operational temperature range, available accessories, noise output, beam angle (1:2 variation), field angle (1:10 variation), Ingress Protection (IP) rating, output lumens, and photometric distribution plots.

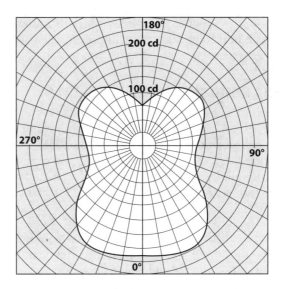

Figure 8-3
A photometric distribution plot with intensities in candela

Photometric distribution plots are usually displayed as polar plots that look similar to microphone and loudspeaker plots (see Figure 8-3). These plots show the distribution of light in the horizontal plane and provide an indication of the pattern of light spread. In the plot shown in Figure 8-3, the center of the plot is the center of the output from the luminaire and the 0° point is the direction in which the luminaire is focused.

With this data about the prospective luminaires in hand, you can set about the process of designing a lighting system.

Lighting Design

Lighting design is the process of getting the right amount of light, of the right color balance, coming from the right direction, to the places that you need it and keeping light away from everywhere else.

The Right Amount of Light

The *right amount* of light is the amount of light needed for both humans and video systems to work comfortably for the duration of the required task. Recommendations are available from a range of organizations as to the suitable light levels for humans to comfortably undertake a variety of tasks.

Sitting passively to view a video presentation or watching a presenter at a lectern requires very different light levels from taking notes about the presentation or operating a control console or a computer system interface for long hours based on the information being displayed on an array of screens. An emergency command and control center is a very different environment from a videoconference room, a classroom, a trading floor, or a presentation hall, although all require comfortable task lighting that does not compete with the visual display systems.

The right amount of light is usually not an even distribution light across all surfaces of an object or the even coverage of a surface. Shadows are critically important to our ability to perceive the shape and depth of human faces and other three-dimensional objects, particularly if the objects are more than 10 meters (30 feet) away from the viewer or are being viewed in a two-dimensional medium such as a photograph or a video display. The technique of portraiture—the practice of lighting a face to reveal its shape and the expressions it portrays—is critical in the lighting of AV spaces for in-person presentations and for capturing on video. The depth and contrast in the shadows cast on the human face by the light sources that reveal it have been the subject of artists, photographers, and filmmakers for centuries and are worth some consideration by anyone who considers lighting design.

If video cameras will be recording or relaying the activities in the space, the exposure requirements and dynamic range of the cameras must be taken into account when determining what is the right amount of light. If the light levels and contrast balance work well for the cameras, they will almost certainly be suitable for human vision, which is not so finicky.

Methods for measuring and documenting the amount of light incident on a surface, or being reflected from that surface, are described in detail in ANSI/IES/AVIXA RP-38-17 *Recommended Practice for Lighting Performance for Small to Medium Sized Videoconferencing Rooms*, which is available from the Standards section of the AVIXA website at www.avixa.org/standards.

The Right Color of Light

The *right color* of light is the balance of spectral colors that allows all of the elements in the visual environment to appear in their intended colors without causing visual confusion. Among the most important color references in AV environments are human skin tones and video display systems. The undistorted rendering of skin tones requires a moderately high (80+) CRI for human-vision environments where moderate color accuracy is tolerable, but a very high (90+) CRI for video imaging with its much tighter tolerances.

Most other objects and surfaces in an AV environment do not necessarily require high color rendering accuracy in their lighting, but care should be taken with well-known or commercially important objects to ensure that the colors are accurately represented. If a client's logo or product is on display, they will expect the colors to be accurately portrayed. Interior designers and architects can be very sensitive to having the carefully chosen colors of their designs distorted by the lighting.

The color temperature range of the light sources is critical in AV environments that will be captured by cameras. While human vision has a number of compensatory mechanisms that tolerate variation in color temperature and color rendering across the visual field, at the time of writing no available video capture system was equipped with artificial intelligence that could reproduce such flexibility. Although most professional video camera systems can be white-balanced to almost any light source with chromaticity coordinates within vague proximity of the Planckian locus, video pickup systems have very narrow tolerances of variation of color *within* an image, typically around ±200K from the point of color balance.

To avoid noticeable color differences in video images that include a video display, the other elements of the image need to be lit by light sources close in color temperature and brightness to the white point of the video display. As most video and television displays have a high-color temperature white point (nominally 6,500K/D65, but it may fall in the range of 4,000K to 9,000K+), "daylight" balance light sources are a good starting point. With the ready availability of high-efficiency, controllable light sources in the 4,000K to 6,000K "daylight" color temperature range, there is little reason to consider lighting designs for video capture environments based around the 2,000K to 3,400K color range of legacy (and low-efficacy) incandescent light sources.

Light from the Right Direction

Light coming from the right direction is critical to task lighting, environmental lighting, and especially to portraiture.

Task lighting should reveal the working area for the task without reflecting off work surfaces to cause glare. The light on the task area should not be obscured by the very process of undertaking the task. For example, writing notes on a page should not cast shadows so deep that the page is obscured: a problem with direct overhead lighting from a small number of compact light sources. The standard approach seen in many work environments is to employ arrays of large-area diffused overhead luminaires, often fitted with baffles to control glare. The selection, placement, and alignment of overlapping coverage from multiple luminaires is a similar process to the one described in Chapter 6 for designing loudspeaker coverage. This popular solution has a major drawback for AV environments due to the difficulties in keeping ambient light away from projection screens and video walls and in focusing visual attention on a presenter.

Light from steep overhead sources may cast shadows in a task area, but light from lower-angle sources may strike viewers directly in the eyes, causing disability glare. As low-angle glare is often due to spill light from distant luminaires, the application of eggcrate baffles and other optical control tools, such as zoom optics, barndoors, shutters, masks, flags, and snoots, will generally resolve the glare by removing coverage from affected areas.

Using light to reveal the shape and texture of objects is accomplished by selective illumination from different horizontal and vertical directions to create shadows of the features of the object or person. Although near-vertical light creates the smallest shadowed areas, has the least spill, and brings out horizontal features and textures, it essentially only illuminates the top of a human head with some highlights on the brow and the top of the nose. This unfortunately leaves the eyes and mouth mostly in darkness. Near-horizontal light casts long prominent shadows across an area, spills into places it isn't wanted, and produces substantial glare, but it brings out vertical textures and features. On human faces it illuminates eye sockets, cheeks, and mouths, allowing the expressions on a face to be clearly discerned. The art of portraiture for video and presentations is about selecting the right balance of horizontal and vertical angles from multiple light sources to reveal enough of the subject's features without causing excessive glare or discomfort.

The emitting area of a light source dictates the type of shadows that it casts. Compact light sources such as metal-halide projector lamps, strobe flash tubes, or spotlight lamps, produce sharp, well-defined shadows and are mostly employed in projectors and luminaires where precise optical paths are critical. Large-area light sources such as fluorescent tubes, softlights, illuminated ceilings, softbanks, and LED arrays cast soft-edge shadows because the light striking an object is coming from many points on the light source, allowing light to "wrap around" the object and fill in some of the shadowed areas. The larger the effective emitting area of a light source, the softer the shadows it creates, but the more difficult it is to use optics to control the area being illuminated. Eggcrate baffles and louvers are widely used to control the beam shape of large area sources, despite the loss of light output from absorbing the straying light.

Despite the often-voiced complaints from presenters, it is simply not possible to have an audience or camera see the expressions on the presenter's face without some glare. The lighting designer's challenge, be it on a stage, in a conference hall, in a desktop videoconference, in a meeting room, in a video studio, or in a classroom, is to strike a balance between presenter comfort and the ability of the audience to understand the gestures and expressions that the presenter is there to communicate.

Creating a Zoning Plan

Different room functions and tasks require different lighting, so your design specifications will need to include a way to alter the areas being illuminated and the level of illumination. During a video presentation when note-taking is required, your system is likely to require a different selection of luminaires at different levels from what is required to achieve the proper contrast ratios during a videoconference.

To facilitate changing between differing lighting requirements in a space, the control of luminaires with similar functions is often grouped together. These luminaire groups are usually associated by the area they illuminate, not necessarily the area where they are actually located. These groups are often referred to as *zones,* a term also used for groups of loudspeakers or groups of video displays that service a specific zone, or area, of a facility.

A zone can be as simple as a single row of luminaires focused on a display wall or as complex as a multitude of individually controlled luminaires. The control for each zone can be as simple as an on/off switch or as complex as remote, full-range dimming.

Figure 8-4 shows a room's lighting zones organized by luminaire function and labeled with letters.

Determining Zones

Determining which luminaires go in which zone depends on the needs analysis. Keeping in mind how the space should operate will help you determine the complexity of the zoning plan.

If you have a room that will be used only for video presentations and general meetings, you might create three separate zones: the lighting above the meeting space, the lighting for a possible presenter, and the lighting closest to the screen. Such a simple zoning plan would allow users to adjust the lights manually based on the room's usage. Notice in Figure 8-5 that groups of lights are linked together so they can be manipulated to meet the needs of various tasks in the room.

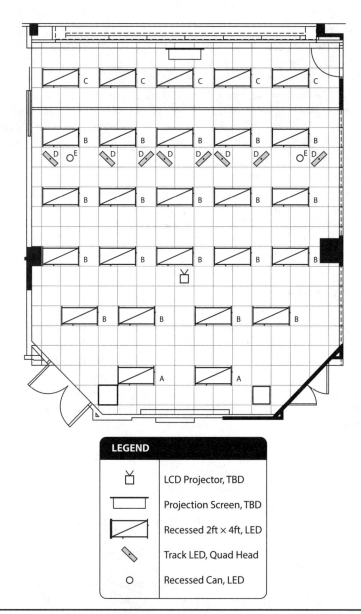

Figure 8-4 Lighting zones organized by letter

Chapter 8: Lighting Specifications

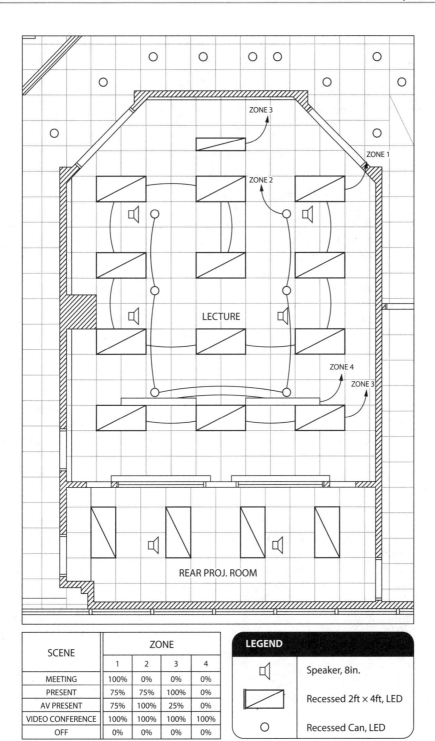

Figure 8-5 Lighting zones based on tasks

If you are designing for a space that will host videoconferences, you should consider assigning the lighting into the following zones:

- Ambient, or general wash illumination
- Key and fill lighting on the conference participants
- Backlighting on the conference participants
- Task lighting on table and work surfaces
- Lighting of vertical and perimeter surfaces
- Additional zones for accent and display lighting

The more complex the AV space, the more complex the zoning required to meet various needs. It may be more appropriate to go beyond the concept of fixed zone groups and allocate each luminaire to an individual control channel, using the lighting control system to create zone groups during the plotting/commissioning process. This allows the end users to create flexible zone groups as required during the operation life of the facility.

You should document your lighting zones to show which luminaires will be controlled together. As you document the zoning plan, be logical. You can color-code them, if that helps, or you can label them using symbols and a key.

Documenting Luminaires

When you have considered the quantity and quality of light, its color, destination, and direction for the space and selected the types and locations of the luminaires required, you will need to document your decisions for communication with the project's lighting designer and/or lighting engineer and/or electrical engineer.

Although documenting luminaires for an AV design might seem straightforward, it's often a process that's better accomplished using computer-modeling software. Such software can produce artful and scientifically calculated models, based on your lighting design information. More importantly, by using software, the luminaires you specify for your design can be imported into the lighting designer's modeling software, putting the AV needs front and center.

Your lighting documentation also needs to take into account other possible systems, such as HVAC and sprinklers, which compete for space in the ceiling. Furthermore, you will have to consider your own loudspeaker coverage pattern, if applicable, and your sightline study. This is where you will want to review reflected-ceiling plans (see Figure 8-6). You cannot put a luminaire where a loudspeaker is, and you cannot suspend luminaires that impede sightlines, that cast shadows of suspended equipment, or that interrupt projector beams.

The bottom line is, be thoughtful and specific. Your fellow tradespeople can't meet your AV design needs if you don't communicate them through documentation.

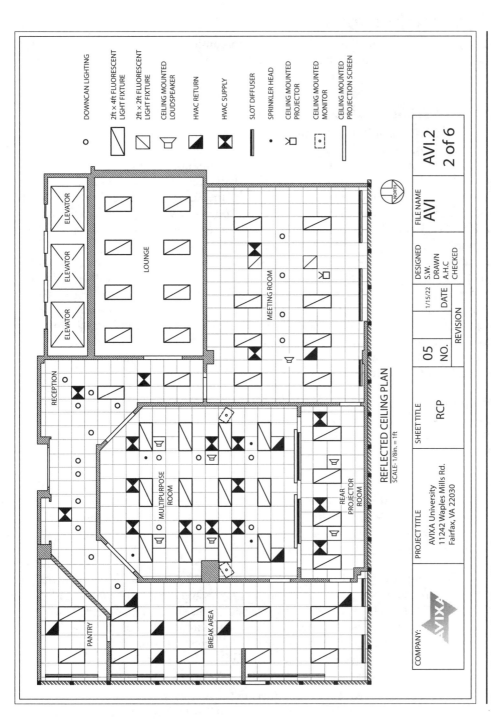

Figure 8-6 Reflected-ceiling plan

Lighting Control

After zoning your lights, you will then need to consider how the end user will interact with them. This means determining how lights will be controlled, setting up preset scenes, and incorporating control interfaces.

On/Off vs. Dimmable

The simplest way of adjusting lighting levels is a switch. Switches give you the option of two conditions: on and off. They can control a single light or an entire zone.

The most basic lighting design for an AV room would be two zones of switched luminaires. Typically, the luminaires in front of display screen(s) can be turned off separately from the rest of the room. Switches are rated by the amount of current they can safely pass. When the number of luminaires grouped onto one zone approaches the current limit of the switches being specified for the installation, it is common practice to control them using high-current controllers such as relays or *contactors*.

Another, more complex way of controlling a lighting system is through dimmers, which give users greater flexibility to set light levels based on the task at hand. Dimmers are devices that can reduce the output from a light source. This is usually achieved by electronically varying the voltage (or current) of its power supply. They can be simple, such as a dimmer/switch on the wall for manual control, or as complex as a computer-based controller. Most discharge light sources cannot be electronically dimmed over a useful range of their output power and must be dimmed by mechanically operated shutter, louver, or iris systems.

Dimming

Some luminaires include integrated dimming, requiring only a power supply and a control signal such as analog 0–10V DC or digital DALI or DMX512. Low-current fixtures powered via Ethernet PoE generally include integrated dimming controlled by an Ethernet-hosted control protocol such as ArtNet or Streaming ACN.

Electronic dimmers are rated according to their current-handling capability and the shape of their output waveform (see Figure 8-7), and therefore are different for LED/solid-state, incandescent, discharge, and fluorescent lamps. Fluorescent lamps can only be dimmed when fitted with appropriate dimmable ballasts.

Common types of electronic dimmers include

- **Phase control** AC dimmers use a switching device such as a *Triac* or a pair of silicon-controlled rectifiers (*SCRs*) to progressively remove the *leading* edge of each half-cycle of the sine wave–shaped mains supply to a device. This is a relatively inexpensive, widely available, and mature technology suitable for dimming incandescent and some modified fluorescent lamp loads. Some acoustic noise and radio frequency interference (RFI) are generated in the load wiring and from dimmed loads.

- **Reverse phase control** AC dimmers use switching devices such as power metal oxide silicon field-effect transistors (*power MOSFETs*) and insulated-gate bipolar transistors (*IGBTs*) to progressively and smoothly remove the *trailing* edge of each half-cycle of the sine wave–shaped mains supply to a device. This is a more

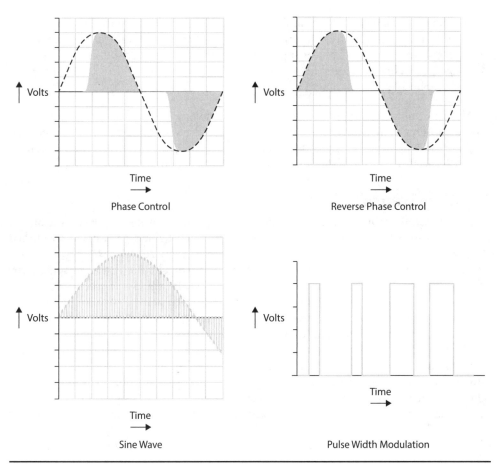

Figure 8-7 Dimmer output waveforms

expensive and recent technology suitable for dimming incandescent, compact-fluorescent, mains-voltage LED, and some modified fluorescent lamp loads. Some acoustic noise and RFI are generated in the load wiring and from dimmed loads, but less than from forward phase control.

- **Sine wave** AC dimmers use IGBTs switching at 400 to 600 times the mains power frequency to control the amplitude of the sine wave–shaped power cycle by pulse-width modulation (PWM). This is similar in operating principle to switched-mode power supplies and class-D audio amplifiers. This is an expensive and relatively recent technology suitable for dimming incandescent, compact-fluorescent, mains-voltage LED, and some modified fluorescent and other discharge lamp loads. It is adored by acousticians, musicians, and recording engineers for the very low levels of acoustic noise and RFI emitted by dimmed loads.

- **Constant voltage** DC dimmers use IGBTs or power MOSFETs to control the DC current by PWM. This is a tried-and-true technology traditionally used for dimming DC-powered LED devices that require a fixed voltage.

- **Constant current**   DC dimmers use power MOSFETs to control the DC current by simple current limiting—the equivalent of adding a variable resistance in series with the DC power supply. This is a well-proven, traditional current-control technology suitable for dimming DC-powered LED devices that tolerate variable voltages.

The selection of the appropriate dimming technology and control protocol for each type of luminaire in your lighting design may require consultation with an electrical engineer or a lighting designer. An extensive list of lighting control protocols can be found in the IES Technical Memorandum TM-23-11 *Lighting Control Protocols*.

Lighting Scenes

A *scene, preset,* or lighting state is a recallable configuration of lighting levels for one or more zones or areas. If your design specifies dimmable lighting, it can be controlled through scenes or presets that dictate the light levels in every zone.

Thinking back to the needs analysis for the room, you should be able to specify the light levels you need from each zone and preset to create the proper scene. There are many relevant standards and recommended practice documents that recommend light levels and contrast ranges for specific identified tasks. However, if your project, such as a theme park ride, a haunted house, or an interactive museum experience, does not readily fit into the common applications specified in standards or recommended practices, you may need to consult with a lighting engineer or lighting designer to establish which elements of your project may be required to meet specific illumination standards. The levels recorded in the preset scenes specified in the lighting design should be set and verified during the commissioning and verification stages of a project, before closeout and sign-off.

For any given specified scene, every zone or area is adjusted to produce the correct lighting for that particular task. As an example, in a scene/preset for "presentations,"

- The zone covering the screen area will need to be lowered or turned off to keep light away from the projection screen
- The zone(s) covering the general areas of the room for movement and task lighting will need to be dimmed to working or note-taking levels—possibly those listed in a standard or a recommended practice
- The zone(s) covering decorative or accent lighting will need to be adjusted to balance the appearance of the room against the ambient light levels that will affect viewing the presentation
- The zone(s) covering the presenter area will need to be raised to provide a balance between the brightness of the screen(s) and good portrait modeling on the presenter, plus coverage of the presenter's notes

When a user selects the "presentations" function on the control interface, the lighting in all of these zones is automatically set to the levels to support that task.

Figure 8-8
A programmable control panel with an LCD display and six buttons (Courtesy of Philips Dynalite)

Figure 8-8 shows a programmable lighting control panel with an LCD screen, four scenes, and two control options. Each scene selection function will recall the stored preset levels for the lighting in the room for a different task. As a designer, you should be thinking about how your users will interact with your AV system. Your responsibility is to specify a panel design and layout that labels scenes by task using unambiguous words or clear pictographs/symbols/icons that are readily understood by the end users (not necessarily the client, the architect, or the AV/IT technicians, none of whom are likely to have to operate an unfamiliar panel in a state of anxiety or duress).

 TIP Be sure to specify control-override features and keep window treatments and lighting independently addressable. Override features should allow end users to operate the lighting when the need arises for unscheduled or unusual events. Being able to take control of shades and drapes independently from lighting presets allows end users to adjust the window treatments without needing to cycle through lighting scenes.

Lighting a Videoconference

Videoconferencing requires a particular set of design criteria to address both the viewers in the space and the camera(s) capturing images. The ANSI/IES/AVIXA RP-38-17 *Recommended Practice for Lighting Performance for Small to Medium Sized Videoconferencing Rooms* includes some very helpful explanations, recommendations, and techniques

Participant Need	Camera Need
Visually comfortable room	Proper portrait lighting on faces
Views without glare on video displays and tasks	Appropriate contrasts between subjects and room surfaces
Proper brightness and contrast of displays	Clear views of participants at a natural eye height
Unobstructed view of other displays that might be used (whiteboards, confidence monitors, document and replay feeds, etc.)	No direct views of luminaires or of far-end video displays; specific seating arrangement

Table 8-3 Lighting Requirements in Videoconferencing Spaces

for approaching the lighting design for these spaces and, if possible, should be read in conjunction with this chapter.

Although the people in a videoconferencing space are capable of capturing and processing images using their eyes and their brains, a video camera simply captures the levels of light falling on the pickup sensors. Humans have the ability to adjust to a far greater range of image contrasts, movement speeds, color differences, and luminance variations than even the smartest camera.

With this in mind, when designing a room to support videoconferencing, you must think of the people in the room and the camera. Your lighting requirements need to strike the right balance between the two, as described in Table 8-3.

Glare

Any light that causes discomfort or a reduction in visibility is called *glare* and detracts from the effectiveness of your AV design.

Glare is the sensation produced by luminance within the visual field that is sufficiently greater than the luminance to which the eyes are adapted and causes annoyance, discomfort, or loss in visual performance or visibility. There are two types of glare:

- *Direct glare,* or any excessive bright light coming directly into the eyes or camera
- *Indirect glare,* or any excessive bright light reflecting off a surface to enter the eyes or the camera lens

When it comes to videoconferencing, your goal is to avoid direct glare from light sources entering the camera, as well as indirect glare, called *veiling reflections,* from light reflecting off shiny surfaces and washing out critical text and details. Some ways of addressing glare include the following:

- Specifying the location of the camera and the use of lens hoods or matte boxes
- Specifying the alignment, pan, and tilt angles of screen(s)
- Specifying the luminaire distribution, focus/alignment, and beam masking with shutters, barndoors, eggcrate baffles, snoots, or flags
- Specifying the room and furniture finishes

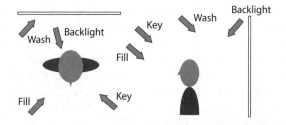

Figure 8-9 Horizontal and vertical angles for the functional groups of videoconference lighting

Light Balance

Videoconferencing systems require proper light balance to produce high-quality images. In general, three-point lighting of a presenter will create suitable lighting in many AV designs, but it's especially true of videoconferencing. This is explained in more detail in the recommended practice RP-38-17. Additional light on the walls behind the participants helps to control the contrast balance (see Figure 8-9). Four functional groups of lighting are usually deployed in videoconference setups:

- **Key light**   This is the main shadow-casting (form-revealing) light on the subject.
- **Fill light**   This helps to control the depth of shadows created by the key light.
- **Backlight**   This adds highlights to the subject's shoulders and hair/scalp/headwear to aid separation between the subject and the background.
- **Wash or dressing light**   This adds light to the background of the image to help control the contrast between the subject and the background.

The primary goal of lighting for video is to ensure that the contrast and color temperatures in the captured image remain within the exposure range and color tolerance range of the imaging system. Once these primary requirements have been met, the lighting designer/lighting director/video cinematographer can then go about the business of communicating the intent of the design, including such matters as good portraiture on people, covering the specified image content, and aesthetically pleasing image composition.

Automatic Exposure Controls

If the contrast range within an image exceeds the capabilities of the imaging system, signal processors in the image-capture chain automatically limit the video signal: some of the brighter image content will be compressed into undifferentiated areas of white (white clipping) and/or darker image content will be compressed into undifferentiated areas of black (black crushing). To assist with bringing image content within the dynamic range of the camera's imaging system, some video cameras include automatic exposure correction.

The best known of these mechanisms is the auto-iris system, which seeks to both prevent overexposure and bring the central brightness range of image content within a preset mid-range value by adjusting the aperture of the iris in the camera's imaging chain. Auto-iris is quite effective if the content of the image lies within the preset average range, but content found in everyday videoconferencing images can confuse the automation into producing unacceptable images. This content includes skin tones outside the range of average European complexions; bright, dark, and heavily patterned clothing on the participants; and dark or bright surfaces behind the participants. In some camera systems the auto-iris may be adjustable. Independent control over the light levels on the background of videoconferencing images allows some adjustment of the image contrast to prevent the auto-iris from spoiling perfectly good pictures.

Automatic gain control (AGC) systems are included in many cameras to compensate for situations where insufficient light is reaching the camera's imaging device. If the AGC detects an average signal level below a preset threshold, additional gain is applied to the video output until that threshold is achieved. As with any electronic system, increasing signal gain also increases the electronic noise in the signal, which in video cameras usually manifests as random sparkles in the darker areas of the image. If the lighting design provides sufficient levels for good overall exposure of the imaging device, the AGC remains inactive and the dark noise is reduced. Just like the auto-iris system, the AGC can be triggered into making unnecessary and unhelpful contributions to videoconferencing images by content which doesn't meet the preset levels of the system. Such content includes skin tones outside the range of average European complexions, dark or light clothing on the participants, and dark or bright surfaces behind the participants. Independent control over the light levels for each of the components of videoconference lighting allows adjustment of the exposure and contrast in the image to prevent the AGC from interfering with the pictures. In some camera systems, the AGC threshold may be adjustable.

There is an only *slightly* paranoid view held by many lighting designers, directors of photography, and cinematographers that auto-iris and AGC systems were created solely to mess with the quality of their images.

Color Balance

Some video cameras have manual settings for color temperature selection, which should be set to the color temperature of the light sources being used. If the camera has an auto-white balance capability, care should be taken that when the auto–white balance function is activated, the majority of the content of the image is both true white and illuminated by the same light sources that will be lighting the captured image. Some camera systems perform a white-balance during startup, while others have a button or control interface option to activate the white balance. If substantial areas of the image are not true white or not illuminated by the correct lighting at the time of the auto–white balance, there is a strong probability that the camera will not be correctly balanced and image colors will be distorted.

Wall and Table Finishes

When designing a space to support videoconferencing, remember that the materials, colors, and reflections within the camera's field of view can significantly impact the camera's ability to capture the intended images. There are many surfaces that affect a camera, from the materials on the walls, ceiling, floor, tables, chairs, and furniture to the colors and properties of surfaces. When light encounters such materials, the characteristics of light will change depending on their properties. Here are some examples:

- Polished material, such as glass, marble, acrylic, or mirror, creates specular (mirrorlike) reflections of light sources and display screens.

Finishes/Furniture	Suggested Reflectance Values
Walls	40 to 60 percent reflectance; no small patterns/stripes; no specular finishes; avoid black, white, high-saturation shades of orange, yellow, green, and red. Grays and low-saturation blues and mauves are good.
Floor	Less than 60 percent reflectance, low sheen.
Ceiling	70 to 90 percent reflectance, low-saturation colors.
Windows	Blackout or opaque shades may be required for windows; interior finish of shades as per wall finishes.
Whiteboards	40 to 60 percent reflectance gray, not white.
Tables	20 to 60 percent reflectance; monolithic, neutral colors such as gray, buff, taupe, and lighter; no specular finishes.
Chairs	As per wall finishes.

Table 8-4 Suggested Reflectance Values of Room Surfaces

- Irregular material, such as etched metal, creates spread reflections similar to specular but more widely distributed.
- Semigloss materials, such as painted walls and cupboards and benches finished with gloss and semigloss laminates, exhibit shiny hot-spot reflections of light sources and display screens.
- Matte materials, such as plaster and matte paint, create diffuse reflections that reflect light without direction and achieve a wide distribution of light.

As you coordinate with allied trades, you will want to specify the reflectance values of various surfaces throughout the space to achieve a good distribution of light and contrast balance within the space. See Table 8-4 for guidance.

NOTE During the design process, you need to ensure that all of the stakeholders take the time to coordinate their approach to the interior design. Getting the lighting designer and the interior designer and/or architect together is important in ensuring that furniture and interior finishes meet the functional requirements of the space.

Illuminated Exit Signs

AV spaces may be required to include illuminated exit signs to guide emergency egress. While essential for public safety, these signs may interfere with the functioning of your AV design. Illuminated exit signs may not only cast ambient light into the video display environment, but they may also be reflected by display panels and furniture finishes and may also provide a visual distraction if they are within direct audience or camera sightlines.

Figure 8-10
Illuminated emergency exit sign with black background (Courtesy of Clevertronics)

As exit sign placement is rarely a matter for negotiation, it is the AV designer's responsibility to take account of the exit signs when choosing the locations of light-sensitive elements of the system design. For example, it may be a bad idea to locate a projection screen close to an exit door which will be marked with an illuminated sign. In some jurisdictions, lower-brightness, or black-background (theater) versions of illuminated exit signs (see Figure 8-10) may be approved for use if a space is used for events such as stage presentations or screenings, where the house lights are dimmed or extinguished when an audience is present.

An important step in the initial site survey and examination of the construction drawings is to ascertain the locations of the exit signage and all other life safety equipment in the space. In addition to exit signage, safety equipment such as sprinkler plumbing, alarm button panels, combustion and thermal detectors, and evacuation warning/information loudspeakers all require wall and ceiling real estate that may limit your choices for locating elements of your AV system.

Chapter Review

In this chapter, you learned about important lighting specifications you will need to consider as you design the lights within your space. You will use this information when communicating with the lighting consultant, electrical engineer, interior designer, and architect about your design, and ultimately you will hand these specifications over to your integrator as a way for them to verify the work.

Review Questions

The following questions are based on the content covered in this chapter and are intended to help reinforce the knowledge you have assimilated. These questions are similar to the questions presented on the CTS-D exam. See Appendix E for more information on how to access the free online sample questions.

1. For light coming directly from a source such as a bare lamp, what measure and instrument should be used?

    A. Foot-lambert with an STL

    B. Candela with an incident meter

    C. Lux with an incident meter

    D. Foot-candle with a light impedance meter

2. What type of electronic dimmer is suitable for controlling a mains-voltage LED light source?

    A. Constant current

    B. Phase control

    C. Reverse phase control

    D. Constant voltage

3. What is the minimum recommended color rendering index (CRI) for light sources that will be used for videoconferencing or video capture?

    A. 60

    B. 80

    C. 100

    D. 40

4. What other type of light, besides light coming directly from a source, must be measured?

    A. Refracted

    B. Distorted

    C. Fluorescent

    D. Ambient

5. When designing a room for videoconferencing, what is the suggested range of suitable colors and light reflective values for whiteboards that are in a camera's field of view?

    A. Less than 60 percent reflectance, low sheen

    B. 40 to 60 percent reflectance gray

    C. 70 to 90 percent reflectance, low-saturation colors

    D. 20 to 60 percent reflectance; neutral colors such as gray, buff, and taupe

Answers

1. **C.** Use lux with an incident meter to measure illuminance, the light coming directly from a source.

2. **C.** Reverse phase control (trailing-edge) electronic dimming is suitable for mains-voltage LED light sources. Constant current and constant voltage dimming are suitable for DC-powered LED sources.

3. **A.** The minimum recommended CRI for light sources that will be used for videoconferencing or video capture is 60, but 80 is recommended, and >90 is preferred.

4. **D.** When considering lighting for an AV design, you should measure ambient light as well as light coming directly from a source.

5. **B.** When designing a room for videoconferencing, whiteboards that are in a camera's field of view are recommended to be 40 to 60 percent reflectance gray, not white.

CHAPTER 9

Structural and Mechanical Considerations

In this chapter, you will learn about
- Building codes and issues that have an impact on AV design
- Equipment mounting requirements
- Mounting requirements in racks and how to calculate power consumption and heat loads
- Heating, ventilation, and air conditioning (HVAC) and fire safety issues that impact AV systems
- The *Audiovisual Systems Energy Management* standard and how you can use it to monitor and conserve energy consumption in your AV systems

In a building, there is a limited amount of space for infrastructure, and AV systems aren't the only types of systems that need to be installed. There are HVAC, fire suppression, security, electrical, lighting, communications, and numerous other systems. And like AV systems, these others comprise not only the actual equipment but also power, cabling, cable pathways, and more. To ensure the systems integrate seamlessly, coordinating the AV design with the members of the architectural, engineering, and construction (AEC) team is paramount. Moreover, the standards of all applicable trade groups must be carefully considered and effectively communicated.

This chapter—divided into mechanical and structural topic areas—will discuss various infrastructure elements that are important to your design so you can better communicate with others on the design team. We start by delving into rules and regulations governing the design of the structural and mechanical elements of an AV design.

> **Duty Check**
>
> This chapter relates directly to the following tasks on the CTS-D Exam Content Outline:
>
> - Duty B, Task 2: Coordinate with Architectural/Interior Design Professionals
> - Duty B, Task 3: Coordinate with Mechanical Professionals
> - Duty B, Task 4: Coordinate with Structural Professionals
> - Duty C, Task 3: Produce AV Infrastructure Drawings
>
> In addition to skills related to these tasks, this chapter may also relate to other tasks.

Codes and Regulations

Building codes exist to protect life and safety. They define standards for the design and construction of electrical systems, fire protection, plumbing, and structural systems. The regulatory bodies tasked with overseeing the construction of public spaces want to know the answers to the following questions:

- How many people should we permit within a space?
- How much room is required for exiting?
- How should we protect the people during an emergency situation?
- How should people be protected from accidental electrical shock?
- What construction materials are appropriate (and inappropriate)?

There are many code considerations an AV professional should learn. Most codes are logical and serve a good purpose. Even if they are not enforced in your area, you may want to consider codes as best practices. The following are some examples:

- The Health and Safety Executive in the United Kingdom has a regulation that limits the exposure of employees to certain sound pressure levels. The *Control of Noise at Work Regulations* state that employees should not be subjected to more than an average of 85 A-weighted decibels (dBA) over an 8-hour period without some specific protective action being taken.
- Part 4 of the *National Building Code of India* provides guidance on the width of aisles for typical seating, the width of exit rows, and furniture arrangements, which are all critical to AV design.
- The United Kingdom's *Disability Discrimination Act* offers guidelines to ensure all people, regardless of disability, have the same experience in a space. New Zealand's *Public Health and Disability Act* and the U.S. *Americans with Disabilities Act* (ADA) are similar to the UK guidelines.

First, we'll go into more depth on specific design elements that consider the needs of people with disabilities. Then we'll discuss some of the electrical and construction codes that affect AV design.

Designing for Equal Access

Many regions have laws or codes that prohibit discrimination based on disability. These laws often impact structural considerations in new construction. Australia's *National Construction Code* identifies a range of other construction standards that AV designers must be aware of when laying out a room. For example, everyone should have equal access to a space, and passageways must be kept clear of protruding elements that could interfere with a person's passage through the space.

When it comes to designing for equal access, you will need to refer to the different authorities having jurisdiction in the location of your project. Governing bodies provide building code provisions for equal access to people with disabilities. For example, the U.S. ADA states that the same experience must be provided to all people. However, it does not detail the specific solutions required to meet that requirement.

In the case of AV systems, guidelines dictate that every presenter or end user must have access to the same control functions. This typically means that operator controls should be between 380mm (15in.) and 1370mm (54in.) above the floor, depending on whether someone in a wheelchair is directly in front of the equipment or approaching from the side. This allows a person in a wheelchair to insert media or operate controls. If the equipment must be approached from the front because a wheelchair cannot come beside it, then the access points should be between 380mm (15in.) and 1220mm (48in.) high (see Figure 9-1). See the "Ergonomics" section in this chapter for more information.

The guidelines also dictate that every presenter must have access to all sources, including slides, video, and recordings. For example, if there is fixed seating and an audio system, an assistive listening system should be provided for the hearing impaired.

Figure 9-1
ADA guidelines for access to control functions

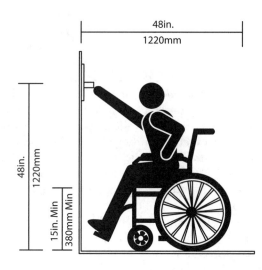

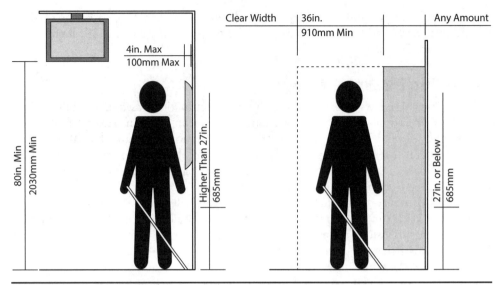

Figure 9-2 ADA guidelines for head and aisle clearance

The ADA also recommends certain design guidelines related to head and aisle clearance (Figure 9-2). For example, equipment, such as a display, should protrude no farther than 100mm (4in.) from a wall. And when hanging screens, projectors, or other devices from a ceiling, it's important to note that ADA guidance states that 2030mm (80in.) is the minimum height for any passageway.

Electric and Building Codes

Codes are an extensive compilation of minimum standards for construction purposes. They protect the health and safety of people who use or occupy buildings and structures. Typically, the authors of codes have no enforcement power. They are merely experts offering their opinions. A governing or ruling body takes these suggestions and adopts them as their own.

Some examples include the U.S. *National Fire Protection Association* (NFPA) and the *European Committee for Standardization* (CEN). The NFPA authors the U.S. *National Electrical Code* (NEC), a source of electrical installation and usage guidelines for the United States. The CEN develops a wide range of construction standards that are adopted in various forms by jurisdictions throughout the European Union.

A country may adopt uniform rules that cover any or all of these items. For instance, the Australian Building Codes Board is a joint initiative of all levels of government in Australia. Its mission is to provide guidance for design, construction, and use of buildings through nationally consistent building codes, standards, regulatory requirements, and regulatory systems. Australia and its neighbor New Zealand share a jointly developed electrical standard known as *Wiring Rules*.

A country may also choose to let smaller governing bodies make the rules. This is the case in the United States. The individual states can develop their own codes, and regions and cities may add their own. They typically reference and adopt portions of the NEC or material from the U.S. *International Building Code* (IBC). You will revisit electric and building codes in Chapter 10.

NOTE Make sure you are aware of the codes at work on your project as prescribed by the *authority having jurisdiction* (AHJ) or regional regulatory authority. Ignorance is never an excuse. In some cases, several conflicting codes from jurisdictions may overlap. In a case such as this, it is important that you follow the most restrictive code.

Mounting Considerations

The building structure is the part of a building that is capable of supporting other materials. This may be structural steel, concrete, or wood trusses. Keep in mind, there are portions of the building structure that you can see but may not be true structural components.

You can have the perfect design with all the right equipment to fulfill a client's needs, but if the building structure can't sustain the weight of mounted AV components, the installation will be unsafe. To create a plan for mounting AV devices as necessary, ask the following questions at the beginning of the project:

- Will the building structure hold each piece of equipment?
- Will the mounted support system hold each piece of equipment?
- Will any existing hazardous materials (asbestos, radioactive, lead paint, etc.) be disturbed during mounting?

To answer these questions, you should determine what type of building material is behind finished walls and ceilings. Looks can be deceiving. What looks like a solid wall may in fact be old plaster with no structural integrity. Verify by inspection how the building components such as structural beams and wall frames are fastened in areas where you will be mounting AV components. All mounting decisions for heavy equipment should be taken in consultation with either the owner's representative, an approved structural engineer, or both.

Mounting Options

You have a few options for mounting AV equipment in a space—on the floor, wall, or overhead. This allows you to offer alternatives depending on the structure of the room or building.

Floor Mounting

Floor mounting is the simplest form of mounting because gravity is working in your favor. It is typically used for mounting equipment racks, kiosks, projection pedestals, and other AV components. Mounting to the floor (see Figure 9-3) ensures that the equipment is securely attached and protected from theft or mistreatment.

Figure 9-3
Bolts hold rear-projection mirror mounts in place.

Alignment of a rear-projection mirror assembly or a multiprojector array takes many hours of labor. It needs to be performed in a way that prevents accidental movement and resulting damage or the need for realignment. Once the alignment is complete, the assembly can be attached to the floor permanently. However, this task should not be performed without the building owner's permission.

Wall Mounting

Audio and video components are routinely mounted to walls. Typical wall-mounted devices include projection screens, flat-panel displays, videowalls, touch screens, monitors, video cameras, loudspeakers, control panels, and small equipment racks. When it comes to mounting equipment to a wall, an AV designer has to consider several structural elements in addition to the equipment weight.

Suppose you intend to mount a remote-controlled video camera to the wall of a training room. The camera assembly weighs 15 kilograms (33 pounds), and the mounting bracket gives a footprint of 150 millimeters by 150 millimeters (6 inches by 6 inches) for support.

It may seem like attaching the mounting bracket to the wall is a simple solution, but there are other considerations:

- Can the wall withstand the load of this camera assembly? You should investigate the construction of the wall.
- Will the weight of the camera on the end of the mounting bracket pull the assembly out of the wall? Suspending a camera from the wall on a bracket creates a lever.

Avoid wall mounting heavy objects wherever possible. Heavy components should preferably be mounted overhead or to the floor.

Proper blocking is critical to wall mounting. Blocking is the support system or construction material that is added to the wall, typically before the wall finish is applied. As shown in Figure 9-4, blocking consists of pieces of wood or other load-bearing materials

Figure 9-4
Blocking provides a secure mounting point.

that have been inserted between structural building elements to provide a secure mounting point for finish materials or products.

The building contractor/prime contractor/general contractor may include reinforcement material into the wall assembly across three or more studs prior to the application of the plaster wall-sheeting. Once the finish is applied to the wall, the blocking is invisible to the eye. This can be done only with previous planning and coordination between the AV designer, the project manager, the architect, and the builder.

Overhead Mounting

Audiovisual components such as projection screens, projectors, flat-panel displays, monitors, luminaires, and loudspeakers are often mounted overhead, typically from the ceiling. Anytime you suspend or hang something overhead, think about what needs to be done to support the weight. You must attach a mounting support to the structure of the building and not the ceiling itself. Therefore, your concern is not the ceiling type, but the structure of the building.

You will want to know the type of materials used in the construction of the building. If the building is made of concrete, find out whether it is poured two-way joist slab (waffle-like concrete), reinforced construction (smooth concrete with steel reinforcing mesh placed before pouring), or post-tension construction, which is also smooth concrete, but with high-tension cables running through it for support. In both reinforced and post-tension construction, the enclosed reinforcements act as a mesh wrapped in concrete.

Construction factors are important because if someone on your projects drills through a cable in a post-tension ceiling/flooring, it can snap with such great force that it could even tear off the outside of the building. Anything in its way is in danger of damage or serious injury.

Always verify the depth and location of these cables with the building owner or building contractor before attempting to mount anything to the ceiling or floor. If this information is not available for the building, then an x-ray of the proposed drilling or trenching site must be conducted and interpreted by specialists.

A building's beams have a weight load they can handle. It is possible that the beams are just strong enough to hold the roof and that any additional weight could cause structural damage. Overhead mountings should never be drilled into a universal beam/rolled steel joist (RSJ)/I-beam. It is better to weld or clamp a piece of steel-channel strut (for example, Unistrut, Kindorf, or EzyStrut) to it.

When you're planning for mounting to a wall or overhead, check the load limit in the mounting equipment specifications to ascertain the weight at which the item will structurally fail.

Load Limit

Load limit is an important part of AV equipment's specifications. One element of the load limit is the Safe Working Load (SWL) or a similar rating called the Working Load Limit (WLL), which is the weight that must not be exceeded. The SWL is determined by dividing the load limit by 5. Five is the *safety factor,* sometimes referred to as *load factor, design factor,* or *safety ratio.* For example, if the rating for structural failure is 500 kilograms, then the safe mounting load is no more than 100 kilograms (in other words, 500 ÷ 5 = 100).

Likewise, when assessing the weight of equipment to be mounted, it is best practice to multiply the published design weight by 5. For example, if a loudspeaker with mounting hardware weighs 50 kilograms (110 pounds), you should use mounting methods and devices that will handle 250 kilograms (550 pounds).

Note that the load factor specified for the mount refers only to the mount and not to the bolts or any other components used in mounting. Each individual component has its own load factor, and you should be able to identify the weakest link.

If you exceed the load limit, your device won't just fall; it could also cause damage to other equipment and the space itself, not to mention possibly injure people nearby. When considering load limits and possible mounting malfunctions, you need to guard against the following:

- **Improper calculation of loads**   This occurs when only the device load is taken into consideration. Load calculation must also include the weight of all mounting hardware.
- **Shearing**   The load or weight of the mount and component can result in gravity pulling the mount down and shearing off the bolt (see Figure 9-5).

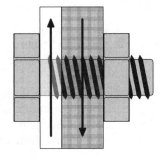

Figure 9-5
Shearing occurs when the weight of the component is exerted across the bolt.

Figure 9-6
Tensile load pushes a bolt.

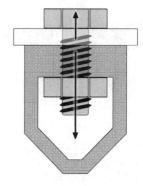

Figure 9-7
A bolt pulling out of a wall

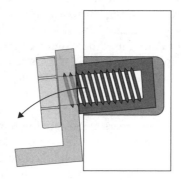

- **Tensile strength**   This refers to the maximum amount of tension the mount hardware, such as fasteners, can withstand before it fails (see Figure 9-6). Inferior-quality bolts could stretch, twist, or break because of the load.
- **Pull-out**   This occurs when force caused by the load pulls a bolt from the wall or ceiling to which the mount is attached (see Figure 9-7).
- **Placement of bolts**   Bolts placed in the top of the mount carry more stress than bolts placed lower. Follow the manufacturer's instructions on bolt placement.

Mounting Hardware

When it comes time to mount AV equipment, installers will need the following:

- The product to be mounted
- The mount
- The appropriate mounting hardware, including backing plates, locking devices, the connection between the two such as a metal pipe or threaded rod, and brackets
- Cable dressing, as needed

Figure 9-8 ISO property classification markings on bolt heads

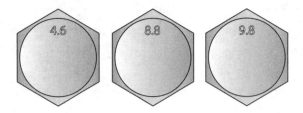

The choice of mounting hardware, or *fasteners,* depends on the type of mounting surface. For example, your options for mounting in concrete will not necessarily work for hollow-block bricks or plaster wall-sheeting. Hardware specifications will be categorized by type and manufacturer and the method of sinking into the material.

To select the correct mounting hardware, consider the holes or slots in the mounting (and note that holes are stronger than slots), as well as their location. The number of bolts used in a mount determines the structural strength per attachment point. Understand the size, type, and length of the bolts, as well as the number and size of threads. This will all help AV installers successfully execute your design.

Standards organizations, such as the International Organization for Standardization (ISO), SAE International (formerly the Society of Automotive Engineers), and ASTM International (formerly the American Society for Testing and Materials), have developed ways to categorize mounting hardware by composition, quality, and durability.

The ISO has a numerical marking system called a *property classification,* as shown in Figure 9-8. The property class numbers relate to the strength of the material that the hardware is made from. The higher the number, the stronger the bolt.

The ASTM International and SAE International use a different marking technique for bolt heads. As shown in Figure 9-9, the raised bumps on the head of the bolt indicate the bolt's grade. If there are none, then it is Grade 2 (see the first image on the left of Figure 9-9), and the metal is usually softer than in higher grades. One raised bump is Grade 3 (not shown). The more bumps there are, the stronger the bolt. Selecting the proper hardware for mounting is critical for a sturdy and safe installation.

 TIP When designing for mounted equipment, specify bolts that are no less than ISO-rated 8.8 or ASTM and SAE Grade 5. It is best to procure your fasteners from a specialty hardware vendor, as a general hardware supplier may not have ISO 8.8 or Grade 5 hardware in stock.

Figure 9-9 ASTM and SAE grade markings on bolt heads

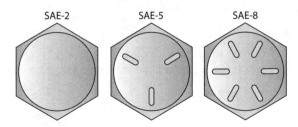

Factors to consider when selecting mounting hardware include size, couplings, length and thickness of pipe or threaded rod, and gauge of steel. Usually, a pipe or threaded rod is used to suspend a device. Black iron (mild steel) provides substantial strength, but there are other material choices.

Pipes are usually specified by material type and wall thickness, but in the United States the term *schedule* when associated with pipes refers to the material thickness, similar to the U.S. wire gauge system. For long runs, specify a single piece of pipe without couplings. Each coupling gives the mounting two additional weak spots or points of failure. A threaded rod should be used only for vertical mounts.

 TIP When using metal pipe, any threaded junction should have some form of locking function, such as a set screw or an alignment hole with a through bolt. This will prevent the assembly from unscrewing over time and with use.

Designing the Rack

As an AV professional, you know that some equipment gets mounted and some goes in a rack. What does a rack do? The most obvious purpose is to group together systems and subsystems. But racks also protect the elements of your AV design and provide security after installation. They can also prolong the life of the design by maintaining system integrity.

In addition to the material covered in this chapter, it is advisable to take advantage of the vast body of knowledge assembled by AXIVA's panels of industry experts in the standards developed as guidelines for good practice in working with equipment racks:

- AVIXA F502.02:2020 *Rack Design for Audiovisual Systems*
- AVIXA F502.01:2018 *Rack Building for Audiovisual Systems*

Racks come in many different styles, including the following:

- **Wall-mounted**   These racks attach to any type of wall when used with the proper hardware. As shown in Figure 9-10, the swing-frame mechanism allows rear access to the equipment for service.
- **Floor standing**   These stand-alone racks have mounting rails that hold the AV equipment in place and are used typically at fixed installations. They usually have enclosed sides and may also have doors at the front and/or rear. They may be fitted with rear mounting rails to carry deep and heavy equipment and to support cable management and power distribution equipment.
- **Open frame**   As shown in Figure 9-11, these fixed racks have an open mounting-frame and can be combined side by side to accommodate more devices. Open-frame racks may be fitted with side panels and front and/or rear doors. Like floor-standing enclosed racks, they may be fitted with rear mounting rails to carry deep and heavy equipment and to support cable management and power distribution equipment.

Figure 9-10
The swing frame in a wall-mounted rack allows rear access to the equipment for servicing.

- **Two post** Also known as relay racks or telco racks, two-post racks are used where floor space is tight and where the equipment mounted is not very deep or heavy at the rear. They are primarily used for mounting patch panels and network devices.
- **Slide-out** These racks are used where there is only limited access to service the installed equipment. As shown in Figure 9-12, the enclosed equipment rack can slide forward out of its housing for installation or service. This type of rack may also have the ability to rotate or spin in place for ease of access during installation, wiring, and service.

Figure 9-11
Open-frame racks can be combined to accommodate more devices.

Figure 9-12
Slide-out racks may also rotate by 45 or 90 degrees.

- **Mobile/portable** These racks could be metal, wood, or plastic with casters and handles for portability. A large selection is available for components such as mixing consoles, communications systems, radio microphones, amplifiers, outside broadcast systems, media servers, and lighting control. Portable racks may take the form of trunks, cases, or custom solutions for AV equipment and maintenance gear that requires protection and storage during transport.
- **Portable shock-mounted** These portable racks for delicate equipment are vibration isolated in dense foam or suspended by shock-absorbing mounts. They are designed to protect the enclosed equipment from rough handling and transport by road, rail, or air.

Racks of all types come in various sizes to match the width and height of specific components. Many of the sizes are industry standard to make designing a rack easier.

Rack Sizes

Racks vary in the width of equipment that can be mounted and are selected to match the width of specific components. The inside of the standard equipment racks used for audiovisual equipment is 19 inches wide (482.6 millimeters).

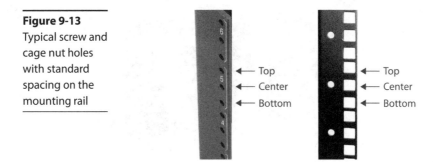

Figure 9-13
Typical screw and cage nut holes with standard spacing on the mounting rail

Many of the technical specifications for a rack, including size and equipment height, are determined by standards that have been established by numerous standards-setting organizations. The external width of the rack varies from 533 to 635 millimeters (21 to 25 inches). Racks are also classified by their vertical mounting height. The height of the usable mounting space in a rack is measured in *rack units*. One rack unit (RU) is equal to 1.75 inches (44.45 millimeters) in height. A 300-millimeter (1-foot)–high rack would be considered to be 6 RUs in height, while a 2.1-meter (7-foot)–high rack would be considered 48 RUs high. When talking about rack height, people often drop the *R* in RU and describe, for example, a rack enclosure with 48 units as a 48U rack.

Like racks themselves, equipment height is measured in RUs, with most rack-mountable equipment designed to fit into a whole number of RUs. A simple audio mixer might be 1 RU high, while an amplifier might be 3 RUs high. Some equipment, like switchers, can be 10 RUs high or even taller. While 48 RUs is a widely used rack height, and all that rack space is handy, some taller racks may not fit through standard doorways or be transportable in a passenger elevator/lift.

Hole spacing on standard rack units is designed to match RU dimensions and provide stability for the mounted equipment. On each side of the mounting rail, there are typically three screw holes or cage nut slots for each RU. Figure 9-13 shows the standard spacing for mounting holes; notice that the holes are not evenly spaced. Equipment manufactured for rack mounting has "ears," or extensions, on both sides of the faceplate that align with the holes in the mounting rail. Sometimes the ears are ordered separately and attached to the sides of equipment before mounting. Rack screws through the ear holes secure the equipment in the rack.

Among other considerations noted earlier, the selection of a rack's size is based on the size of the space into which the rack will be installed. The rack must be able to fit into the space and able to pass through the entries and building passages to reach the installation site. Available options vary in depth from 300 to 800 millimeters (12 to 32 inches) or more.

Ergonomics

When designing a rack layout, consider how the end users will interact with specific components; what do they need to see, use, or adjust? Also, consider how the field technician servicing the rack will access various components. Is it difficult to remove a piece

Figure 9-14
User line of sight at standing and seated positions

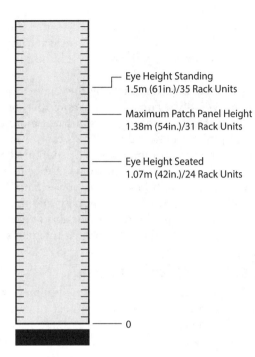

of equipment for repair? Can the technician get to terminals for troubleshooting? Should you add test points to allow for quick and easy calibration and recalibration? Is there space for additional equipment to be added to the system?

Equipment with user-interface components, such as optical-disc and media players, should be placed within easy reach of standing and sitting users. For example, you want the player at a convenient height for the end user to insert the media. (See Figure 9-1 in "Designing for Equal Access.")

Media players and control panels usually go between eye height for a standing and sitting user, as shown in Figure 9-14. Devices that are not necessary for a user to interact with, such as a wireless microphone receiver or a patch panel, can be located at the top or bottom of the rack.

Weight Distribution

Not only must the equipment in a rack be placed where it's accessible, its weight should be distributed strategically. If a freestanding rack has heavy equipment near the top, the whole unit could become unstable and tip over. To avoid this, place the heaviest equipment in the bottom of the rack. For example, a local uninterrupted power supply (UPS), a power line conditioner, or a powerful analog audio amplifier may be the heaviest devices that you would put in an equipment rack. UPSs may weigh more than 30 kilograms (66 pounds). When equipment racks are mounted to a wall, or even bolted to the floor, the added stability may make it acceptable to install some heavier components at the top of a rack.

For the stability of racks that are not fixed to the structure of a building, AVIXA F502.02:2020 *Rack Design for Audiovisual Systems* recommends that no less than half of the weight of the equipment in the rack should be contained in the bottom third of the rack.

Most equipment is supported entirely by the rack screws in the front mounting rail. Heavier equipment typically occupies two or more RUs, providing more support. Some equipment may need the additional support of a rear mounting rail.

Professional equipment will have integral rack ears for mounting or rack ears that may be removable. Either way, the equipment will be supported by mounting the rack ears to the rack mounting rails, using either rack screws or cage nuts and screws. Heavier equipment, such as UPSs, line conditioners, and power amplifiers, will include additional mounting provisions at the rear of the equipment and need to be secured to additional rack mounting rails located in the rear of the rack. Always check the manufacturer's equipment manual for mounting and spacing requirements.

Signal Separation Within a Rack

In rack layout design, it is common to group equipment according to their function. For instance, you may want to install and connect all video components in one area of the rack to avoid electromagnetic interference, resulting in noise.

Signal separation protects against crosstalk between cables. As shown in Figure 9-15, cables carrying mic level signals (at about 1 to 10mV) are kept separate from cables carrying power (100 to 240V). The difference in voltage between a mic level signal and a power cable can be seen as a ratio of about 50,000 to 1 (47dB). Keeping the cables separate helps to maintain the integrity of the signals.

Figure 9-15 Signal separation in an equipment rack

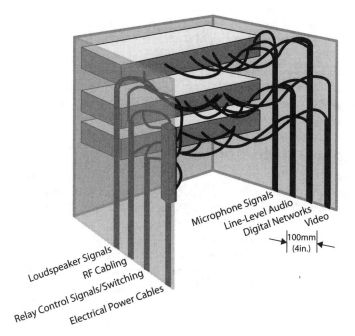

Signal Type	Mic Level	Audio Line Level	Video	Data Twisted Pair	RF Coax	Speaker	AC Power
Mic Level	–	Separate Bundles	Separate Bundles	100mm (≈4in.)	100mm (≈4in.)	100mm (≈4in.)	300mm (≈12in.)
Audio Line Level	Separate Bundles	–	Separate Bundles	Separate Bundles	50mm (≈2in.)	50mm (≈2in.)	100mm (≈4in.)
Video	Separate Bundles	Separate Bundles	–	Separate Bundles	50mm (≈2in.)	50mm (≈2in.)	50mm (≈2in.)
Data Twisted-Pair	100mm (≈4in.)	Separate Bundles	Separate Bundles	–	Separate Bundles	Separate Bundles	50mm (≈2in.)
RF Coax	100mm (≈4in.)	50mm (≈2in.)	50mm (≈2in.)	Separate Bundles	–	Separate Bundles	50mm (≈2in.)
Speaker	100mm (≈4in.)	50mm (≈2in.)	50mm (≈2in.)	Separate Bundles	Separate Bundles	–	50mm (≈2in.)
AC Power	300mm (≈12in.)	50mm (≈2in.)	50mm (≈2in.)	50mm (≈2in.)	50mm (≈2in.)	50mm (≈2in.)	–

Table 9-1 Recommended Minimum Separation of Cables from AVIXA Rack Standards

Signal separation allows audio wiring to remain shorter because most of that cable will be between the components themselves. Shorter cable runs mean less of an opportunity for the cables to pick up induced noise. Table 9-1 shows the recommended minimum cable separations listed in the AVIXA rack-related standards F502.01:2018 and F502.02:2020.

Block Diagrams

Before installers can start wiring everything together in a rack, they have to be able to visualize the signal flow among the components within a system. To help them do that, you should create block diagrams.

A *block diagram* illustrates the signal path through a system. It is drawn with simple icons (triangles, circles, or boxes) representing the system's devices, as shown in Figure 9-16.

Interconnecting lines (sometimes with arrows) indicate the signal path. They are drawn or read from left to right and from top to bottom.

A designer typically provides block diagrams to installers. A good diagram will have well-marked icon labels because several common symbol variations are used in block diagrams. For example, an amplifier is usually, but not always, symbolized as a triangle.

The goal of a diagram is to visualize the signal flow. Analyze two connected icons at a time and ask yourself, "Will this work?" Then work your way into greater detail. This process will reveal any potential problem areas.

Heat Load

In your AV design, you have to calculate how much heat is produced during normal operation by all equipment within an AV rack. Any piece of equipment that is powered on generates heat, whether it is idle or doing work. The resulting *heat load* is measured in kilojoules (kJ), while strangely, the U.S. customary unit is British thermal units (Btu). 1kJ = 0.95Btu.

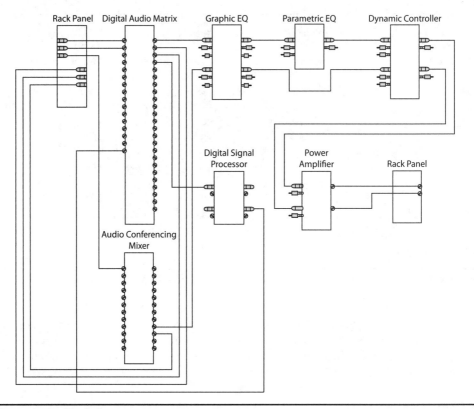

Figure 9-16 Block diagram

An AV designer typically calculates the total heat load from a system and provides that information to the HVAC engineer. HVAC system designers can then design appropriate cooling to prevent equipment failure and balance HVAC systems so people can be comfortable in the space. Consult the owner's manual or device labels for the power consumption of each device in the AV rack.

The formulas for calculating heat load are:

$$Total\ kJ = W_E \times 3.6$$

or

$$Total\ Btu = W_E \times 3.41$$

where:

- W_E is the total watts of all equipment used in the room.
- 3.6 is the SI conversion factor, where 1 watt of power generates 3.6 kilojoules of heat per hour.

- 3.41 is the U.S. customary conversion factor, where 1 watt of power generates 3.41 Btu of heat per hour. (Some older material may still use a less accurate conversion factor of 3.4.)

These formulas do not account for the heat load generated by amplifiers.

Calculating Heat Load from Power Amplifiers

Power amplifiers are different from most other components in the amount of heat they dissipate. They will typically have the highest power consumption of all the equipment you are handling, with the only exception perhaps being the projector.

A multichannel power amplifier that is rated at 200W per channel may also have a power consumption rating of 600W maximum. From a heat load perspective, designers are less interested in the wattage per channel rating than the consumption rating, but the maximum consumption of 600W does not tell the whole story. Some factors need to be considered with power amplifiers:

- The efficiency of the amplifier. How much of the power that is consumed is used up within the mounted case compared with the amount transferred to the loudspeakers? Most digital (switched mode) power amplifiers are more energy efficient than analog amplifiers.
- Voice paging, background music, rock music, video presentation, and pink noise have drastically different demands on the amplifier. This is referred to as the *duty cycle*—how much of the time it is being driven and how much it is at rest. What kind of program material is being amplified?
- How much power will be demanded of the amplifier? Just because it is capable of producing 200W per channel, are you designing it to do so, or have you allowed a lot of headroom for transient peaks?
- How often is the power amplifier powered up? Is it always on and operational or only running for 1 hour per week?

It would be improper to simply take a specified 600W consumption and multiply that by 3.6 to come up with 2.16MJ/hr (2160kJ/hr), as this ignores all the previously mentioned factors.

The amount of heat the equipment generates is directly related to its power consumption. For example, a 20W equalizer produces less heat than a 40W equalizer. If electronic equipment gets too hot, it may perform badly or even cease to operate. Therefore, manufacturers incorporate methods to circulate cooler air in and hot air out of electronic equipment.

Cooling a Rack

With all the heat-generating equipment contained close together in a rack, it is up to the AV designers and technicians to direct a flow of cool air into the rack and hot air out.

There are two main methods of cooling a rack:

- **Evacuation** The evacuation cooling method uses fans to draw air *out* of the rack, usually through the top vents (see the top left of Figure 9-17). This is much like sucking on a straw. Cool air is drawn into the rack from the side and bottom vents and through vent plates. Since there may not be adequate filtering over every vent or slot, dirt can infiltrate the equipment rack. It is important to keep the equipment rack dust and dirt free with periodic maintenance.
- **Pressurization** Cooler air is blown *into* the rack by fans in the bottom of the rack (see the lower right of Figure 9-17). The fans may have a filter to prevent dust and dirt from entering the rack. Vents on the vent plates and sides and top of the rack provide an escape for the hot air. The positive air pressure in the rack reduces the ingress of dust. In larger equipment rooms, pressurized, chilled air may be ducted into the bottom of the racks, often from under the floor.

Most electronic devices have vents through which the hot air can escape, typically in the back or top of the unit, and slots or vents where cool air can enter. Pay attention to the heat flow between the vents on adjacent equipment. Beware of mounting equipment with opposite heat flow immediately adjacent to each other because it will cause circulation of only the hot air.

Separate pieces of equipment that produce a lot of heat by placing blank or vented panels between them. Vent plates in the front of the rack may allow cool air to escape before reaching the equipment farther up the rack and may need to be replaced with blank plates. Equipment producing a greater amount of heat should go near the top of the rack.

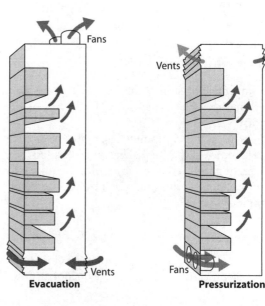

Figure 9-17 Evacuation (left) and pressurization (right) methods to cool an AV rack

The installation of heavy analog audio amplifiers in a rack can create a dilemma. They can cause the rack to tip over if placed higher up in the rack, and they also create a lot of heat, which is best vented if it is placed near the top of the rack. Only after the rack is securely mounted in place to the wall or floor should such an amp be installed at a higher rack elevation.

You know about removing the heat from the equipment rack, but you should now consider where that heat should go. How do you remove it from the room? HVAC systems are the physical systems that control the environmental (temperature and humidity) conditions of a workspace. Let's look at AV-related HVAC requirements.

HVAC Considerations

AV designers are primarily concerned with the relationship between HVAC systems and the maintenance of temperature and humidity in the AV equipment, as well as the associated noises and vibrations created in presentation spaces.

AV designers are not responsible for, and in many cases are barred from, the design of HVAC systems. However, an AV designer should coordinate HVAC requirements with an HVAC designer and make the HVAC designer aware of the AV system's impact on an HVAC system.

All HVAC systems generally have the same basic components. A system consists of components that do the following:

- Extract heat to and from the external environment
- Extract heat and humidity to and from the conditioned spaces
- Move air throughout the conditioned spaces

HVAC Issues That Impact Design

A designer should understand how HVAC systems affect AV systems. *Background noise* within a space can affect speech intelligibility, which creates additional challenges for AV designers. Other than noise from sources external to the AV space, HVAC is usually the main source of noise that a designer will have to address. Being aware of the issues can help an AV designer collaborate with a mechanical engineer.

HVAC system components that can generate noise include the following:

- **Compressors and pumps**   Low-frequency vibrations and acoustic rumble.
- **Fans and impellers**   Motor vibrations and air turbulence.
- **Diffusers**   Passage of supply or return air through the diffuser vents. HVAC diffusion vents can generate a substantial amount of noise and unwanted air flow if they are not designed and balanced to fit the space's function. Oversized diffusers may eliminate some of these problems.
- **Ducts**   Passage of air. The ductwork also can carry noise generated by other sources throughout the building. This is especially problematic in meeting, presentation, and conferencing spaces.

As discussed in Chapter 6 in the section "It Must Be Loud Enough," sound systems typically need to perform 25dB above the ambient noise level. For example, if the HVAC ambient noise measures 45dB SPL, the loudspeaker system would need to produce at least 70dB SPL at the listener's ear.

The AV designer, architect, and mechanical engineer should work together to ensure large mechanical units are acoustically well isolated and not placed directly adjacent to the spaces that require a low-noise floor.

Mechanical vibration and noise created by vibration are often major problems in buildings. The low rumble of a chiller, large fan, or impeller can resonate throughout a building if it is not designed properly and given the appropriate vibration isolation. Low-frequency vibration should also be considered because it can affect the stability of display devices rigidly attached to the building structure.

Many components can be used to control vibration. Larger pieces of equipment are often provided with spring isolators, air mounts, or rubber pads. Piping and ducting systems can also be isolated with the use of flexible connectors between sections to control vibrations traveling along the pipe or duct walls.

The cooling diffusers should be placed in locations that ensure cooler air is available to the presenter area. You should not rely on diffusers from the audience area to cool the presenter area, or the audience may complain about being too cold. The air flow from a diffuser can also cause a projection screen to flutter or sway.

Current construction techniques conserve space by using the open space (plenum) above the visible ceiling as a pathway for ventilation. Return air is returned via this space to the ventilation system. When utilized for this function, the space is called a *return air plenum*.

If fire breaks out in the return air plenum space, the smoke will circulate throughout the building. Any toxic smoke and gases from the combustion of materials located in this space can cause serious physical harm to people. All material should be rated for safe use within the return air plenum. In many locations, it is a violation of rules set forth by the AHJ to use materials that are not rated for plenum space use.

AV designers must be aware of the locations and installation requirements of HVAC system components to create a compatible overall system design.

Fire and Life Safety Protection

In case of a fire or other life-threatening emergency, an alarm system will attempt to alert the occupants so they can be evacuated from the space. The AV system may need to be integrated with the fire alarm system to do any of the following:

- Mute the sound from an ongoing AV system to allow the alarm to be heard
- Interrupt display and replay sequences and return any mechanical systems to a safe state
- Interrupt any atmospheric or pyrotechnic effects
- Relay emergency warning and information audio

- Send visual messages to displays
- Activate a lighting exit state to allow egress
- Notify AV system operators of the emergency situation

Fire suppression systems may also be installed and employed by the life safety professionals. You should be aware of the following:

- Wet systems are fire protection systems that use pressurized water. Water and electricity do not play well together. These systems damage interiors and destroy electronic equipment.
- Dry systems do not have water constantly present in the system—they are pressurized by fire fighters in the event of a fire. These systems don't leak and are much harder to trigger by accident than pressurized water systems, but still produce the same amount of damage when activated.
- Chemical systems replace the room air with fire-inhibiting gas such as carbon dioxide or halomethanes (chemical agents that contain halogenated methane). The chemicals used in these systems are greenhouse and/or ozone-depleting gases and can asphyxiate any human occupants of the space.
- Hypoxic air systems use inert gas to reduce the ambient oxygen concentration in a protected space to between 10 and 14 percent, a level that is breathable but will not support flaming combustion. These systems use gases such as argon and nitrogen, either as mixtures or alone. They are greenhouse and ozone layer neutral.

The AV designer must coordinate their work with the mechanical engineer so that sprinkler systems are not located over ceiling-mounted AV equipment, which would hinder the flow of water to a fire. Care should be taken that when projection screens and masking drapes are deployed, they do not obscure control panels, lighting controls, alarm activation panels, or alarm indicators.

AV systems for experiences that include atmospheric effects (smoke, haze, fog, etc.), flame projectors, or pyrotechnics should be designed in conjunction with fire safety engineers to enable the required levels of fire safety without the risk of falsely triggering fire detection systems.

Fire Isolation

While drilling through a firewall to run a cable is sometimes unavoidable, there are ways to do so safely. For instance, always use *firestop* materials to refill the hole you have made. Firestop comes in many forms, such as foam, blocks, or plugs, but it all has the same basic function. It acts as an impenetrable, nonflammable barrier.

When mounting AV equipment in the ceiling, you may find firestop materials in the way of your installation. Remove as little firestop as necessary, and make sure you replace it immediately.

Energy Management

Before we end this chapter, we need to examine one more element of structural and mechanical systems: energy management. You will find the standard ANSI/AVIXA S601.01–2021: *Audiovisual Systems Energy Management* invaluable in your design. The purpose of the standard is to provide requirements for energy-efficient design, implementation, use, and management of energy-consuming AV systems.

When an AV system is managed in conformance with this standard, results from real-time monitoring, recording, and reporting will lead to more informed decision making. AV systems designers can provide AV solutions that incorporate optimum energy efficiency while adhering to the specified operational requirements.

Energy conservation management of an AV system is accomplished by

- **Documentation**   An energy management plan and a system connection diagram provide necessary documentation for communicating the energy management goal to project stakeholders.
- **Automation**   By using control system interfaces such as vacancy sensors, occupancy sensors, and actuators, the AV system can maintain the lowest possible energy-consuming state for the intended AV functionality.
- **Monitoring**   Measurement systems gather and record data on an ongoing basis. This entails ascertaining the power efficiency of all devices in the system, then monitoring and recording their power consumption over time.
- **Reporting**   The analysis of the recorded data provides usage and trending information that can be reviewed and analyzed to determine if the efforts to maximize energy reduction have been successful or require modifications.

NOTE   Some organizations take power conservation so seriously that they turn it into an internal competition. To assist with this, a designer could set up an intranet page for the competing spaces, or a display outside each space, to reveal how much energy has been consumed over a given period.

Chapter Review

Mechanical systems, including the HVAC and fire suppression systems, have a significant impact on the design and layout of an AV system. The AV designer must carefully consider these elements to make sure the end users are comfortable and the equipment functions as expected.

The designer must also consider how the AV system will be integrated into the building's structure. This includes making sure the equipment is mounted securely, accessible to the end user, and in accordance with relevant codes. Once again, the structural considerations include issues of both functionality and safety. Once you complete this chapter review, you'll be ready to tackle the electrical infrastructure requirements of your design.

Review Questions

The following questions are based on the content covered in this chapter and are intended to help reinforce the knowledge you have assimilated. These questions are similar to the questions presented on the CTS-D exam. See Appendix E for more information on how to access the free online sample questions.

1. If an amplifier draws 7A at 230VAC, without considering amplifier efficiency, what is the heat load it generates?
   A. 5796kJ/hr
   B. 2737kJ/hr
   C. 5474kJ/hr
   D. 3024kJ/hr

2. What factors need to be considered when designing and drawing rack elevation diagrams?
   A. User mobility, competence of the technician, quality of the rack material
   B. Equipment manufacturer, model year of equipment, number of USB ports
   C. Ergonomics, weight distribution, RF and IR reception, heat loads, signal separation
   D. Number of RJ-45 connectors used, amount of conduit used

3. According to best practices, what is the Safe Working Load (SWL) requirement for overhead mounting of a 35kg (77lb) projector?
   A. 140kg (309lb)
   B. 175kg (386lb)
   C. 350kg (772lb)
   D. 210kg (463lb)

4. When working on a project where there are several applicable accessibility codes relating to assistive hearing systems in the space, which code should you follow?
   A. The code specified by the architect
   B. The code that will cost the least to implement
   C. The most restrictive code
   D. The code that will be easiest to implement

5. What is the best place to install a 27kg (60lb) audio amplifier in a freestanding equipment rack?
   A. Top of the rack
   B. Middle of the rack
   C. Bottom of the rack
   D. The location is irrelevant

Answers

1. **A.** If an amplifier draws 7A at 230VAC, without considering amplifier efficiency, it generates 7 × 230 × 3.6 = 5796 kJ/hr (5.8MJ/hr).

2. **C.** When designing and drawing rack elevation diagrams, you need to consider ergonomics, weight distribution, RF and IR reception, heat loads, and signal separation, among many other things.

3. **B.** The SWL requirement for overhead mounting of a 35kg (77lb) projector is 5 × 35 = 175kg (386lb).

4. **C.** When there are multiple overlapping codes on a project, it is important that you follow the most restrictive code.

5. **C.** The best place to install a 27kg (60lb) audio amplifier in a freestanding equipment rack is at the bottom of the rack.

Specifying Electrical Infrastructure

In this chapter, you will learn about
- How current, voltage, resistance, impedance, and power interact so you can apply them to the power and earthing/grounding required for an AV system
- Specifying the circuits needed to support AV equipment
- Differentiating among system and equipment earthing/grounding schemes so that you can identify and specify them within a given AV system
- Specifying infrastructure that will protect AV equipment from magnetic and electric interference
- Specifying AV cable support, conduits, and calculating conduit jam ratio
- Common power and earthing/grounding issues

As an AV professional, you are responsible for requesting the appropriate power and accompanying infrastructure for your system's needs. This will require you to calculate those needs and physically specify the location of outlets in the design space. You will have to integrate your AV system with existing power and grounding infrastructure and work with the electrical trade to verify that your power needs are met.

Power and earthing/grounding may seem simple, but it is one of the most misunderstood topics in the AV industry. There are many electrical terms that sound similar but often mean different things. When specifying electrical requirements, you need to apply safe principles. A lack of foundational knowledge can lead to improper recommendations and serious consequences.

Moreover, correctly applying power and earthing/grounding principles will help prevent common analog signal problems. As you will learn, some symptoms in analog audio and video system performance can indicate problems with the electrical system. You will learn about the sources of these problems so you can avoid or troubleshoot them. A solid foundation in power and earthing/grounding theory will help you identify the source of an issue and then formulate and implement the necessary corrective measures.

> **Duty Check**
>
> This chapter relates directly to the following tasks on the CTS-D Exam Content Outline:
>
> - Duty B, Task 5: Coordinate with Electrical Professionals
> - Duty C, Task 3: Produce AV Infrastructure Drawings
>
> In addition to skills related to these tasks, this chapter may also relate to other tasks.

Circuit Theory

Working with power for AV systems requires an understanding of the fundamentals of electronic circuit theory. From the work you undertook to obtain your CTS certification, together with the material revised in the section on electrical basics in Chapter 2 and the sections on transformers and connecting loudspeakers in Chapter 6, you should by now have a basic understanding of the following electronic concepts:

- Voltage
- Current
- Power
- Resistance
- Impedance
- Ohm's law and the power equation
- Transformers
- Series and parallel circuits
- Alternating current (AC) and direct current (DC)

If you are uncertain about how well you understand these ideas, it is advisable to review your CTS notes or CTS Exam Guide and to check back over Chapters 2 and 6 before proceeding with the more detailed sections that follow.

There are two types of electrical current: direct current (DC) and alternating current (AC), as shown in Figure 10-1.

Direct Current

Direct current flows in one direction. Batteries, plugpacks, and USB chargers are examples of DC devices; they have both a positive electrode and a negative electrode. In a DC circuit, the negative electrode releases electrons into the circuit, where they travel to the positive electrode, which attracts electrons.

Figure 10-1
The difference in voltage over time in DC (left) and AC (right)

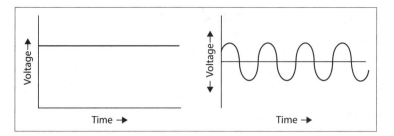

In a DC circuit

- The current always flows in a single direction (negative to positive).
- The current may not be continuous and may vary in level.
- Most electronic equipment runs on a DC power supply, which comes from batteries, an external power supply device, or a mains-powered DC power supply section within the equipment.
- Most digital data signaling is pulsed DC (which has many of the characteristics of AC).

Alternating Current

Alternating current refers to an electrical current that periodically alternates (reverses) its polarity and its direction of flow. The frequency of the reversals is measured in hertz (cycles per second). Alternating currents found in the AV world include all analog audio and video signals, the feeds to and from radio frequency (RF) transmitters and receivers, and the mains supply that provides the power to most AV equipment.

The voltage of the AC mains power from general-purpose outlets is typically 220 to 240V at a frequency of 50Hz in many parts of the world, including Europe, Africa, and most of Asia, and 110 to 120V at a frequency of 60Hz in North America (Mexico to Alaska), parts of the Caribbean, and some parts of South America. In other places, mains voltages may be 100V, 120V, or 220V to 240V at frequencies of either 50Hz or 60Hz.

Although definitions vary between regulatory jurisdictions, to electricians and electrical engineers, low voltage (LV) is usually considered as ranging between approximately 50V and 1000VAC. Extra low voltage (ELV) generally includes anything under 50VAC. Each jurisdiction has rules or codes on which type of workers are allowed to install or work with differing voltage levels.

The values quoted for AC currents, power, and voltages are generally not the peak positive or negative values, but *root mean square* (RMS) values. RMS is an algorithm that calculates an average effective value, taking into account both positive and negative peak values. For a regular sinewave signal, such as the AC mains, the RMS value of the waveform is the peak value divided by the square root of two (1.414). For example:

$$V_{RMS} = \frac{V_{Peak}}{\sqrt{2}}$$

$$A_{RMS} = \frac{A_{Peak}}{\sqrt{2}}$$

Circuits: Impedance and Resistance

The amount of current that flows in a circuit is limited by the opposition to current flow in the circuit. In a continuous-DC circuit, resistance is the only form of opposition to current flow. In AC circuits and pulsed-DC circuits, impedance is also a factor in the opposition to current flow.

Resistance

The resistance (R) of a conductor is a property of the material it is made from, its length, and its cross-sectional area. For a given material, resistance decreases as conductor size increases. Conductor size is measured in cross-sectional area (or sometimes by wire gauge).

One way to reduce circuit resistance is to use a cable with a larger cross-sectional area. Resistance increases with cable length; therefore, a longer cable run will result in a lower voltage at the end of the cable (voltage drop). When resistance is too high, it will degrade the signal or even prevent an effective current from flowing.

Impedance

Impedance (Z) is the total opposition to current flow in a circuit. It includes the resistance of the materials in the circuit, plus their *reactance* (X). Reactance is the opposition to current flow in the circuit caused by reactions to the changes in the voltage and current, and is directly related to the frequency of the changes in the current. In a continuous-DC circuit, where there are no changes, the frequency is zero, and hence the reactance is zero, so the impedance is equal to the resistance.

Reactance has two components: *capacitive reactance* (X_C) due to the capacitance in the circuit and *inductive reactance* (X_L) due to the inductance in the circuit.

The formula for calculating impedance is:

$$Z = \sqrt{R^2 + (X_L - X_C)^2}$$

Although audio signals are AC, for loudspeaker calculations it is simpler (and a reasonable approximation) to treat impedance as resistance when making loudspeaker impedance calculations.

Capacitance

Capacitance occurs when two electrically charged conductors are separated by a non-conductive material. The electrostatic field set up between the conductors stores some charge. The amount of charge stored is governed by the area of the conductors, the distance between them, and the dielectric strength of the material separating them. An electronic component that uses this principle to store charge is called a *capacitor*. The construction of a capacitor usually employs two large-area conducting electrodes, known as plates, to maximize the amount of charge that can be stored.

A charged capacitor discharges when a conducting path is present between the plates. Figure 10-2 shows a capacitor in series with a load, shown as a resistor.

Figure 10-2
A series circuit with an AC signal source, capacitor, and resistor

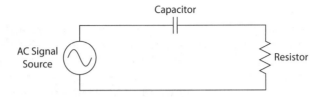

The capacitor affects the flow of current through this circuit. You can see this impact on the graph in Figure 10-3. The current flowing in the circuit is displayed on the y-axis, and frequency of the AC is shown on the x-axis. As frequency increases in this circuit, so does current flow. Capacitors have less effect at high frequencies and more effect at lower frequencies.

The capacitance of an electric circuit acts to oppose changes of voltage in the circuit by storing some of the energy. This effect is called capacitive reactance (X_C). The higher the frequency of the signal in a circuit, the lower the capacitive reactance losses, as can be seen in the formula:

$$X_C = \frac{1}{2\pi fC}$$

Where:

X_C = Capacitive reactance in ohms (Ω)

π = 3.142

f = Frequency in hertz (Hz)

C = Capacitance in farads (F)

Significant capacitance between the conductors in a cable can attenuate and distort the signal carried in the cable. More cable creates more capacitance, resulting in a greater loss of high frequencies. One means of reducing the capacitance in a cable is to keep the conductors well separated, such as in the coaxial cables used for analog video, Serial Digital Interface (SDI) video, and RF applications.

Figure 10-3
The impact of a capacitor on an AC circuit. As frequency of the AC increases, the current will also increase.

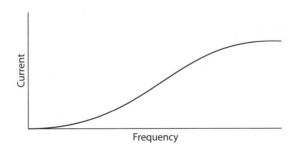

Inductance

When current flows through a conductor, it produces a magnetic field around the conductor. Conversely, when a conductor is placed in a changing magnetic field, a current is *induced* to flow in the conductor. However, the induced current flows in the opposite direction to the current that produced it and therefore acts to reduce to the original current flow. This opposition to current flow is inductive reactance (X_L). The higher the frequency of the signal in a circuit, the higher the opposing current from inductive reactance, as can be seen from the formula:

$$X_L = 2\pi f L$$

Where:

X_L = Inductive reactance in ohms (Ω)

π = 3.142

f = Frequency in hertz (Hz)

L = Inductance in henry (H)

An electronic component that uses the principle of inductance to create a magnetic field is called an inductor. As the magnetic effect around a conductor is quite small, most practical inductors are constructed from a long length of conductive material formed into a coil to concentrate the magnetic field. Many inductors also incorporate an iron (or other ferrous material) core at the center of the coil to further concentrate and intensify the induced magnetic field.

As the induction effect occurs in a changing magnetic field, induction can be produced by an alternating current (or pulsed direct current) in a stationary conductor or by the relative physical movement of a conductor and a magnetic field. Electric motors, alternators, and generators are devices where a conductor and a magnetic field are in relative motion. Figure 10-4 shows a simple circuit with an AC signal source, an inductor, and a resistor.

Figure 10-5 shows the relationship of the current and frequency in an AC circuit. In the graph, the current flowing in the circuit is displayed on the y-axis, and frequency of the AC is on the x-axis. As the frequency of the AC in this circuit increases, the current flow decreases.

Figure 10-4
A series circuit with an AC signal source, an inductor, and a resistor

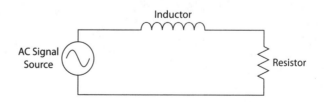

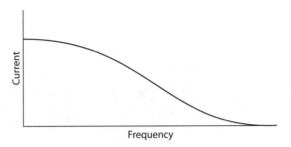

Figure 10-5
The impact of an inductor on an AC circuit. As the frequency of the AC increases, the current flow will decrease.

The changing magnetic fields around a conductor carrying an alternating (or pulsed direct) current induce currents with similar waveforms in nearby conductors, even if there is no direct electrical connection. The substantial induced currents from mains power cables can result in unwanted mains-frequency signals (and their many harmonics) being induced into signal cables carrying data, control, video, and audio. This manifests as hums and/or buzzes in analog signals or as random corruption in digital signals.

Induced currents from adjacent signals in multicore cables can cause signal distortion and other forms of interference: the phenomenon known as *crosstalk*. Methods of reducing induced crosstalk in multicore cables include using grounded shields around signal pairs to carry the induced current away to ground, as found in audio multicores and Category 7+ twisted-pair data cables. Another crosstalk reduction approach is to physically separate the conductor pairs by including spacer material between the signal pairs, as found in Category 6A+ twisted-pair cables.

Induced currents are the operating principle for the transformer, where the current flowing through a primary coil produces a magnetic field that induces a current in a secondary coil, without any direct electrical connection. Transformers are used to electrically (*galvanically*) isolate one part of a circuit from another or to change voltages between circuits. In the section on constant-voltage loudspeakers in Chapter 6 you will find more detailed material about transformers and about their application in drive circuits for loudspeaker systems.

Inductors, Capacitors, and Resistors in a Series Circuit

In a series circuit, AC current can flow only where the frequency inductor and capacitor cross currents over. Figure 10-6 shows a series circuit in which the capacitor blocks lower-frequency current, while the inductor blocks higher-frequency current.

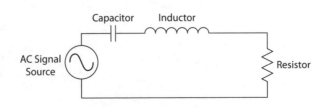

Figure 10-6
A series circuit with an AC signal source, a resistor, a capacitor, and an inductor

Figure 10-7
Where inductor current flow and capacitor current flow intersect, you have a bandpass.

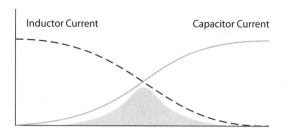

In Figure 10-7, the circuit's current is displayed on the y-axis and its frequency on the x-axis. The dashed curve represents the inductor current flow; the continuous curve is the capacitor current flow. The shaded area beneath is the resultant current flow.

The intersection of the two curves creates an *x*, which is known as the *crossover* or *x-over*. This is where the effects of the capacitor and inductor cross. Current can flow in this circuit only in this frequency region. The result is that only a portion of the frequency range can pass current. This is a *bandpass filter*.

Inductors, Capacitors, and Resistors in a Parallel Circuit

In a parallel circuit, AC current flows *except* at the frequency crossover point. In the parallel circuit shown in Figure 10-8, the capacitor passes higher-frequency current, while the inductor passes lower-frequency current. This type of circuit can be used as either a high-pass filter to remove low frequencies from a signal or a low-pass filter to remove high frequencies from a signal. These filters are used in the crossover filter system for multispeaker cabinets to ensure that each speaker receives only the frequencies that it reproduces well.

In Figure 10-9, the circuit's current is displayed on the y-axis and its frequency along the x-axis. The dashed curve represents the inductor current flow; the continuous curve is the capacitor current flow. The shaded area is the resultant current flow. The dip at the crossover represents a reduction in current flow. This is a *notch filter*. You can find these types of circuits in audio equalizers.

Figure 10-8
A parallel circuit with an AC signal source, a resistor, an inductor, and a capacitor

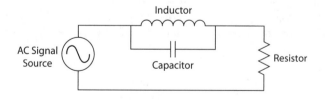

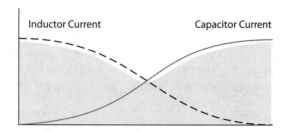

Figure 10-9
A notch filter, where the current flows intersect, marks a reduction in total circuit current flow.

Specifying Electrical Power

When a building is built or a room remodeled, an electrical engineer plans for supplying power throughout the project. These general plans may not include requirements for AV systems, in part because each AV system is different—ranging from a single screen and replay audio facility; to a multiscreen, multichannel incident response room; or an interactive museum exhibit with dozens of displays, dozens of audio channels, and atmospheric and mechanical effects.

AV designers need to specify everything from power outlet/receptacle types and ratings to power locations. They need to calculate expected current loads for every item of AV equipment. You may also have some specific electrical power and earthing/grounding requirements based upon the size and signal-to-noise requirements of an AV system.

In addition to specifying power locations and requirements, AV designers will usually be required to

- Specify or design interface and connection plates
- Review electrical plans and schedules to confirm they meet AV requirements

They may also be required to

- Size raceways and cable trays
- Specify junction-box sizes and locations
- Create raceway riser diagrams
- Perform conduit-related calculations

To do all this properly, an AV designer must know local codes and regulations so that the infrastructure is safe, conforms to standards and best practices, and can be understood by electrical engineers and contractors.

Established Terms

You may come across slang terms while working with power and earthing/grounding. The IEEE Standard 1100-2005, *Recommended Practice for Powering and Grounding Electronic Equipment,* identifies a list of terms to be avoided because they're just not clear. Still, several of them are commonly used in the AV industry, such as the following:

- Clean ground
- Clean power
- Computer-grade ground
- Dedicated ground
- Dirty ground
- Dirty power
- Equipment grounding safety conductor
- Frame ground
- Shared circuits
- Shared ground

Don't use these terms! Electricians or other project team members might misinterpret your meaning. For example, what exactly is clean power? Because this term does not have a clear definition, your idea of clean power may be different from your co-worker's or the electrical engineer's. Because AV integration relies so heavily on an electrician's work, avoid ambiguous terms to ensure your message gets across. Learn the electrician's power and earthing/grounding vocabulary to ensure your safety as well. When in doubt on terminology, check with the regional authority having jurisdiction (AHJ) for your project.

Codes and Regulations

AV professionals must use and reference standards and codes in their AV designs. Most regions have standardized methods and procedures, which they will publish as a code for all electricians and electrical engineers to follow. Examples of such codes include

- European Committee for Electrotechnical Standardization (CENELEC) HD 60364: *Electrical Installations for Buildings*
- India IS 732: *Code of Practice for Electrical Wiring Installation*
- U.S. *National Electrical Code* (NEC)
- United Kingdom BS 7671: *Requirements for Electrical Installations. IET Wiring Regulations*
- Canadian CSA C22.1: *Canadian Electrical Code* (CE Code)
- Australia and New Zealand AS/NZS 3000: *Wiring Rules*

It's important that you understand the relevant codes because you need to verify that your electrical infrastructure specifications and AV systems installation are compliant. You may work on a project that crosses jurisdictions, and their codes may conflict. In such cases, it is always best practice to follow the more restrictive or stringent code requirements. When in doubt, consult with the AHJ.

Electrical Distribution Systems

In all but the simplest systems, current draw, voltage, and the locations and types of receptacles required for AV equipment must be calculated, specified, and located on drawings. For a new facility or major renovation, this information will go to the electrical engineer and contractor for inclusion in their drawings and specifications. For AV integration in an existing facility, the required electrical circuits and receptacles may be coordinated through the owner, a building engineer, or an electrical contractor. In some cases, an AV firm may have a contracted electrician or electrical engineer who can evaluate the situation and add the necessary distribution infrastructure, circuits, and receptacles.

In any case, you or the installation team should perform simple tests on the electrical outlets/receptacles in a space to check for proper voltage and earthing/grounding prior to connecting AV equipment. In addition, you should locate overcurrent-protective and ground/earth-fault–protective devices, such as circuit breakers, residual current devices (RCD)/earth leakage circuit breakers (ELCB)/ground-fault circuit interrupters (GFCI)/core balance relays (CBR), for AV circuits. In larger installations, you may have to inspect the electrical system further to ensure that it meets your specifications.

Electrical engineers and electrical contractors need to know how much power your AV system requires so they can plan accordingly. To communicate effectively, you need an intermediate understanding of power distribution.

So, where does the power for your AV system come from? At a power-generation station, the generator puts out extremely high-voltage AC (66kV to 765kV), which travels to a transmission transformer for the long trip down the power lines. Because transmission at high voltage reduces the amount of current, there is less loss from resistance. This means long-distance transmission can serve a broad geographic area. Due to the presence of three separate induction coils/windings in mechanical alternators (generators), current is produced as three equal, alternating currents, all running at the same frequency, but each separated by a 120-degree phase angle. In large-scale DC-based power generation, such as storage-battery and photovoltaic (solar) systems, the DC current is converted to three-phase AC current through an inverter system before being fed into the supply and distribution network. Power transmission systems have feeds in multiples of three lines.

These primary transmission lines deliver power to electrical substations, which are located strategically around population areas or industrial centers. The substations have transformers that reduce the voltage before local distribution. The electricity is then distributed to a local transformer, where it is further reduced to the voltage required for your application—whether residential or commercial.

When the power from the substation's transformer reaches your local transformer, it is normally reduced to the supply voltage required for your AV systems. This supply-voltage power can be distributed to your AV systems in two ways: *single-phase* or *three-phase*. In a single-phase system there is a single line/live/active/phase/hot conductor connected to only one of the phases in the supply system. In a three-phase system, there are three line/live/active/phase/hot conductors that carry current from all three phases of the generation system. It's important to know which type of power you need so you can communicate with the electrical engineer and electrical contractor to ensure distribution of sufficient power for the installation you design.

Power Distribution Basics

Most equipment used in either domestic or AV electronic systems operates on single-phase mains power. The mains supply to a single-phase device requires two conductors:

- **Supply** A line/live/active/phase/hot conductor that connects to the power network grid or a local power source such as a generator set or a standby power source. All switches, disconnection, and current protection devices are inserted in this supply conductor. In most large installations there are three sets of live/phase conductors to distribute the load more evenly across the power generation network.

- **Return** A neutral/return/ground conductor that provides a current return path from connected devices to the power source. Regardless of which supply line/active/live/phase they are connected to, all devices are connected to the same neutral or return path. While usually being connected to the ground (planet Earth) at the same point as the safety earth, the return conductor is part of the distribution and supply system and serves no safety function. In double-pole circuit breakers the return conductor is disconnected at the same time as the supply conductor.

Protective Earth/Safety Ground/EGC

The protective-earth/safety-ground/equipment-grounding conductor (EGC) connection in a power distribution system serves the solitary purpose of providing an uninterrupted low-impedance path to ground/earth from any conductive parts of electrical systems and devices that can be accessed by an end user. The function of the protective-earth/safety-ground/EGC is to carry away to the earth/ground (planet Earth) any voltage that may arise on conductive parts of the installation that may otherwise cause harm to the end user. Figure 10-10 shows the earth/ground connection point for an installation.

The protective-earth/safety-ground path must not include any switching or current protection devices. In a plugged electrical supply connection, the protective-earth/safety-ground pin must connect first on insertion and disconnect last on removal. Most electrical regulations require that the protective-earth/safety-ground pin on an equipment connector be longer than any line/phase/active or neutral/return pin.

The exposed conductive parts of electronic equipment should be earthed/grounded. Electronic equipment with conductive cases; conductive knobs, buttons, or sliders; or exposed conductive screws may become energized if they accidentally become connected

Figure 10-10
An earth spike/grounding electrode outside a building

with a line/phase/active conductor due to an electrical fault. Equipment earthing/grounding will prevent the exposure to voltage that these exposed conductive parts might carry in such conditions by providing a direct low-impedance path to the earth/ground. This will allow a very large current to flow, which will open the fuse or trip the circuit breaker in the supply line.

Components of an electrical supply system that must have a protective-earth/safety-ground/EGC connection include

- Electrical equipment enclosures
- Distribution boards/panelboards and switchboards
- Junction boxes and backboxes
- Power outlets/receptacles and audiovisual plates
- Equipment racks
- Luminaires
- Conductive raceways, ducts, conduit, and cable trays
- Dimmer racks
- Electrical load patch systems
- Any other metallic or conductive housing that has the potential to become energized in a fault situation

Figure 10-11 shows an example of equipment earthing/grounding. The electronic device could be any component of an AV system.

The dotted conductor in Figure 10-11 shows a continuous pathway to earth/ground (planet Earth) beginning with the conductive enclosure of the electronic device, through the long pin of the power connector, through the power outlet/receptacle, back to the distribution board, and through the earth bus, which is connected to the earth at the main electrical distribution board.

The dashed conductor in Figure 10-11 shows the return/neutral pathway for current from the electronic device to earth/ground (planet Earth) through the return/neutral pin of the power connector, through the power outlet/receptacle, back to the distribution

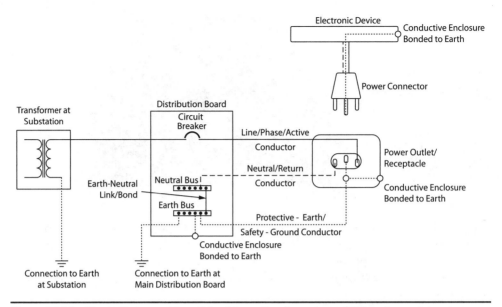

Figure 10-11 Neutral/return and earthing/grounding for a piece of AV equipment (the electronic device)

board, and through the neutral bus, which is then linked/bonded to the earth bus, which is connected to the earth at the main electrical distribution board for current return to the generation and distribution system.

Earthing/Grounding Conductors

Although there is a diverse range of standards specifying the insulation colors on phase/active/line conductors and neutral/return conductors, there is a broad consistency between jurisdictions that protective-earth/safety-ground conductors have either all-green or green-and-yellow insulation. In a few jurisdictions and in some older, legacy installations, the protective-earth/safety-ground conductor is a bare (uninsulated) conductor.

Some jurisdictions still allow the protective-earth/safety-ground path to be via conducting components on the cable support and protection raceways, ducts, and conduits, but due to the risk of accidental (and potentially fatal) discontinuities arising along such a path, the practice has generally been eliminated.

Grounding/earthing the chassis of equipment rated as IEC Class II protection (double-insulated) or Class III (supplied from a separated extra-low voltage [SELV] power source) is not required in most jurisdictions and is strictly forbidden as an additional risk factor in some others. IEC Class II– and Class III–protected equipment is identified with the symbols shown in Figure 10-12, usually located near the power input point on the device.

Figure 10-12
IEC Class II and III protection symbols

IEC protection Class II
Double Insulated

IEC protection Class III
Separated Extra-Low Voltage
Power Source

Ground/Earth Faults

The purpose of a protective-earth/safety-ground/EGC is to protect people. Perhaps an integrator has installed an AV device but the power supply becomes defective. The insulation may have melted away or some internal component may have come loose or failed. The result is an accidental connection, called a *ground* or *earth fault*, between the line/phase/active conductor and the conductive enclosure of the device. This fault brings the conductive enclosure of the AV device up to *mains voltage*.

The human body has approximately 1kΩ of resistance, depending upon where you measure the conductivity of your skin. If a person came into contact with a device energized by a ground/earth fault, as much as 230mA (120mA in 120V systems) could flow through that person's body—more than enough to cause muscle contraction, heart failure, nerve damage, or death (see Table 10-1).

In a normally functioning device, all current flows along the line/phase/active and neutral/return conductors. In our hypothetical fault situation, the line/phase/active conductor of the electrical circuit has come into contact with the conductive enclosure of

Current Flow	Possible Outcomes
1mA	Perceivable to humans with a slight tingling sensation; under some conditions, a threat can exist
5mA	A slight shock can be felt; no pain yet, but disturbing; an average person can let go; however, sudden and strong involuntary movements could lead to injury
6mA to 16mA	Painful shock and loss of muscular control; the "let go" range
17mA to 100mA	Excruciating pain, respiratory arrest with severe muscular contractions occurs; the ability to let go is lost and death is possible
> 50mA	Respiratory arrest occurs
101mA to 200mA	Ventricular fibrillation occurs along with muscular contraction and nerve damage; death is likely

Table 10-1 Effects of Current on the Human Body

an electronic device, and the current flows from the now-live enclosure directly to the ground/earth (planet Earth) via the device's protective-earth/safety-ground/EGC conductor. Due to the low impedance of the path to earth/ground, a very high current will flow through earth/ground/EGC conductor, causing the supply circuit's overload protection device (such as a fuse or a circuit breaker) to disconnect the circuit from the power supply and render the faulty electronic device now safe.

Earth-Leakage Protection Devices

An overload circuit breaker or fuse is designed to interrupt the supply to a circuit when the current exceeds a preset limit. With a thermally activated circuit breaker, a small over-current (due to overloading) can be tolerated for a prolonged period before tripping, although a large over-current due to an earth/ground fault will trip fairly quickly. A magnetically activated circuit breaker will trip quickly once its threshold current has been reached. As you can see from Table 10-1, the current necessary for fatal electrocution is quite small, so an overload breaker offers virtually no timely protection from electrocution.

Earth-leakage protection devices such as a *residual current device* (RCD), a *core balance relay* (CBR), a *ground-fault circuit interrupter* (GFCI), an *earth-leakage circuit breaker* (ELCB), or a *residual current breaker with over-current protection* (RCBO) are based on the principle that the amount of current entering a device should be exactly the same as the amount of current leaving the device and that any discrepancy is due to current flowing somewhere that it shouldn't, which is a bad thing. If there is an earth/ground fault that allows the body of the device to become connected to the active/phase/line, then when a person touches the device, they provide a path to earth/ground for the current, which is not via the neutral/return. Unfortunately, an overload circuit breaker or fuse will not detect that a small, albeit fatal, additional current is now flowing: this is the situation for which an earth-leakage protection device is employed.

In an earth-leakage protection device, any earth-leakage current is sensed by comparing the current in the active/phase/line with the neutral/return currents flowing through a circuit. Under normal conditions, with identical currents in both conductors but flowing in opposite directions, the effects of the currents balance each other out. Any residual current detected is due to leakage to earth/ground via a fault and triggers the protection device to rapidly disconnect the circuit before a lethal current can flow.

Earth-leakage protection devices are available in a range of sensitivities for fault currents between 10 and 100mA and with interrupt times ranging between 5 and 100 milliseconds after a fault current is sensed. Some devices such as fluorescent tube ballasts, discharge-lamped projectors, large switched-mode devices, and arc welders have small inherent leakage currents, and others produce small transient leakage currents during normal operations, so it's important to seek expert advice on the most appropriate earth-leakage protection devices that will both keep users safe and avoid nuisance tripping at power-up, lamp-strike, motor-start, or shutdown.

A Ground/Earth Fault Case Study

An AV-specific example of why you must pay close attention to the earthing/grounding in an AV system occurred in the United States in 2005. A pastor in a church in Waco, Texas, was performing a baptism during a Sunday morning service. While standing waist-deep in water, he reached for a microphone so his congregation could hear him. The moment he touched the microphone, he was hit with an electric shock that killed him on the spot.

There is a possibility this tragedy could have been avoided had someone recognized the dangerous ground-fault condition in the water heater. The pool was energized, so when the pastor touched the microphone, the microphone cable's shield completed the circuit path.

These types of incidents are more likely to occur where power receptacles are not equipped with a protective-earth/safety-ground connection, such as in some legacy installations.

The Dangers of Three-to-Two-Pin Adapters

To attempt to solve hum and buzz problems in analog systems, some AV technicians have been known to troubleshoot signal ground faults using a three-to-two-pin AC adapter with the protective-earth/safety-ground disconnected or removed. These are also known as a *ground-lifting* adapter or *widowmaker*. However, such a process removes the protective-earth/safety-ground path for fault currents, which could bring the entire connected system to mains supply level, with the only available earth path being through the equipment users.

If you have a hum problem due to a potential difference in signal grounds (i.e., a ground loop), instead of removing the protective-earth/safety-ground pin or using a three-to-two-pin adapter, you should follow proper troubleshooting procedures to identify the cause of the problem.

TIP The number-one cause of hum and buzz in analog systems (or subsystems) is a poorly connected or disconnected shield/screen on a signal cable.

Grounds for Confusion

AV, electronic, and electrical terminology share some words that have very different meanings, but confusingly, are often used interchangeably. The term *ground* has at least three mutually exclusive meanings:

- **Signal ground**   This is the common system zero-reference point for electronic signals such as analog audio, video, and control signals and for the analog pulses that underlie digital signaling systems. Signal ground is normally connected to a single point and may or may not also be connected to the ground/earth (planet Earth). It is not part of either the power distribution or the protective-earth system.

- **Supply ground** This is one of several names, including *return* and *neutral*, used for the current return system, where current supplied from the mains power generation and distribution system is returned to the generation source via a connection to the ground (planet Earth). Such ground-return connections are located at all points of power generation and consumption. Some stand-alone power generation and supply systems use a wired return path instead of ground-return. The ground-return connection point to the ground/earth (planet Earth) is usually shared with the protective-earth/safety-ground system.

- **Safety ground** This is also known as the *protective earth* and the *equipment grounding conductor* (EGC). The protective-earth/safety-ground system provides an uninterrupted low-impedance path to ground/earth (planet Earth) from any exposed conductive parts of electrical systems to protect equipment and users from the possibility of harm from those exposed parts accidently becoming live. The safety-ground/protective-earth system usually links to the earth/ground (planet Earth) at the same point as the supply neutral/return system.

Power Distribution Systems

The most common electrical-distribution systems operate at 230V single phase and 230/400V three phase at a frequency of 50Hz, but 240/415V and 220/380V at 50Hz are also used. North America, some parts of the Caribbean, and some parts of South America use 60Hz systems operating at 120V and 240V single phase and 120/208V and 277/480V three phase. In Japan, while the voltage is 100V single phase and 100/200V three phase, some regions run at 60Hz and others at 50Hz.

A 230V single-phase system is employed in both domestic and smaller-scale industrial applications, including most AV equipment. As shown in Figure 10-13, this system uses three conductors. The first conductor is the 230V phase/line/live/active, the second conductor is the system neutral/return, and the third conductor is the protective-earth/safety-ground.

The 230/400V three-phase system (Figure 10-14) uses four conductors. The first three conductors carry the individual phases/lines/actives/lines with a potential of 400V between any two phases (230V between any phase and the neutral/return); the fourth conductor is the neutral/return conductor, which is usually at earth/ground potential.

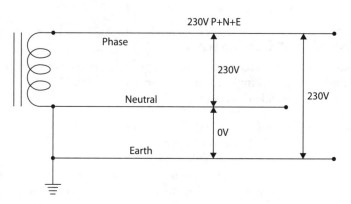

Figure 10-13 A 230V single-phase system

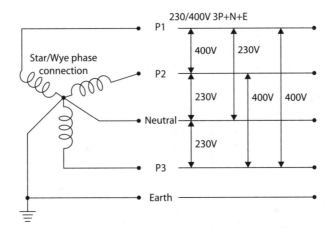

Figure 10-14
A 230/400V three-phase system

A fifth protective-earth/safety-ground/EGC conductor is usually part of the system, although it is not part of the energy supply. This system is commonly found in commercial and industrial applications that have greater power requirements, such as larger buildings, manufacturing facilities, and larger-scale AV installations and events.

In a perfect world, when working with three-phase systems, you want all the power for AV equipment to come from only one of the phases to avoid the possibility of higher phase-to-phase voltages being present in adjacent equipment. For this reason, standards in most jurisdictions require equipment and wiring to be insulated against phase-to-phase voltages rather than phase-to-neutral.

TIP With any three-phase electrical distribution system, the electrical engineer will attempt to design a balanced load across the phases. While balanced phase loads are desirable, three-phase systems are designed to work reliably even when completely out of load balance. Keep in mind that a large, infrequently used AV system would not practically be placed on one phase of the building's power system. If it is, the system will be unbalanced, either when running or when idle. The AV designer's preference for loads on the same phase and the engineer's competing preference for balanced loads should be discussed as early as possible in the project.

The Neutral/Return in Three-Phase Distribution

The neutral/return conductor in most three-phase power distributions connects from the common termination of each secondary transformer winding on the mains power source to the earth/ground (usually planet Earth). This type of interconnection of the three supply phases (see Figure 10-14, Figure 10-17, and Figure 10-18) is usually referred to as a *star* or *Wye* configuration due to its resemblance to the shape of a star or the uppercase letter "Y."

As the common neutral/return conductor for the three-phase system is carrying the return current for all three phases, the current flow will be the sum of the return currents for all phases. Because the three supply phases are delayed by 120 degrees from each other (see Figure 10-15), the currents in the neutral will also be delayed by 120 degrees, so while some phases are drawing current *from* the neutral conductor, others may be supplying current *to* the same neutral conductor.

Neutral Currents

If the currents from all three phases are equal in magnitude, the sum of the currents in the return/neutral conductor will actually be zero. If any phase is drawing less than the other two, the phases will be out of balance and there will be current flowing in the neutral conductor. In the most extreme case of an out-of-balance system, the maximum current will be flowing from a single supply phase and back through the neutral/return. As soon as current flows in either of the other phases, the neutral/return current will be reduced. The result of the current cancellation in three-phase systems is that the size of the neutral/return conductor is only required to be the same as any one phase conductor.

There is one exception to that rule: neutral current cancellation only applies for loads that draw *complete cycles* of current. A device such as a switched-mode power supply or a phase-control/reverse phase-control lighting dimmer distorts the waveform of the current being drawn, preventing complete phase cancellation. Contrary to expectations, this may result in neutral/return currents *greater* than the current for any single phase. Allowance for such conditions is specified in most electrical wiring standards, which generally specify words similar to the U.S. National Electrical Code in 310.15 (B)(4)(c), which requires "systems with high harmonic content to treat the neutral in the same way as a phase conductor and calculate its size based on actual (rather than assumed) load." The common practice in high harmonic-load installations is to specify neutral/return conductors with double the current capacity of any single phase. As many AV devices such as digital amplifiers, signal processing equipment, solid-state light sources, network devices, and projectors with solid-state or discharge light sources use switched-mode power supplies, they are all contributors to the harmonic content of the power distribution system.

TIP It is important that you inform the electrical power engineers working on your project about the number of dimmed luminaires and switched-mode devices in your system design.

Figure 10-15 Three-phase AC waveforms

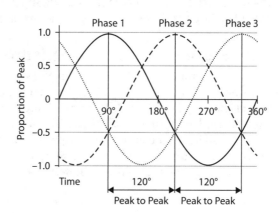

Figure 10-16
North American center-tapped single-phase power distribution system

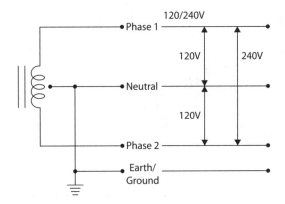

Power Distribution Systems in North America

The most common single-phase system in North America is a 120/240V center-tapped configuration (Figure 10-16). The 120/240V single-phase system uses three conductors. Two phase conductors connected to the ends of a center-tapped transformer winding carry the supply voltage, while a conductor connected to the center-tap point on the transformer winding is the neutral/return. The voltage between either phase conductor and the neutral will be 120V, while voltage between the two phase conductors will be 240V. This unusual system provides a 120V single-phase supply for standard devices and a 240V single-phase supply for higher-power devices such as heating and cooking appliances.

The 120/208V three-phase system uses four conductors, as shown in Figure 10-17. The three phase conductors carry the supply, while the fourth is the grounded neutral/return conductor. Voltage between a phase conductor and ground will be 120V.

Figure 10-17
A 120/208V three-phase system

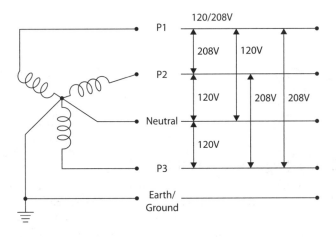

Figure 10-18
A 277/480V three-phase system

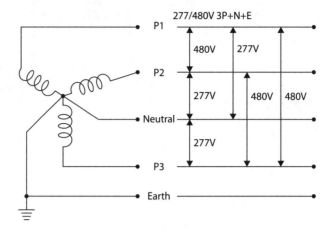

Voltage between any two of the three phase conductors will be 208V. Because of its flexible nature, this is the most popular system used in commercial, institutional, and light industrial buildings.

The 277/480V three-phase system (Figure 10-18) is similar to the 120/208V system. It is used in large commercial and industrial buildings. It also employs four conductors and is important to be aware of because it is used for broadcast, stage and event lighting, atmospheric effects devices, equipment winches, large LED video walls, and stage machinery, but not for general AV and IT electronic equipment.

Three-Phase vs. Single-Phase Voltages

The voltage between any one phase and the system neutral/ground is equal to the voltage between any two phases divided by the square root of 3:

$$V_{P-N} = \frac{V_{P-P}}{\sqrt{3}}$$

Where:

V_{P-N} = The voltage between any phase and the system neutral/return

V_{P-P} = The voltage between any two phases

$\sqrt{3}$ = 1.732

Specifying AV Circuits

AV designers may be responsible for calculating the number of circuits needed to support their AV systems. Projectors, screens, displays, conference room tables, lecterns, mixers, radio mics, control consoles, in-ear monitors, processors, monitors, credenzas, bring-your-own (BYO) devices, luminaires, active loudspeakers, effects equipment, mechanical effects, network equipment, autocues, hearing assistance systems, and more may all require power.

Begin by reviewing your project documentation for the location of all equipment, plus other locations that require power, such as

- Conference tables
- Lecterns
- Control consoles
- Floor-level wall outlets for
  - Mobile lecterns
  - Mobile control consoles
  - Temporary videowalls
  - Mobile facilities trolleys
  - Temporary video monitors and foldbacks
- Overhead rigging positions for temporary/transient
  - Lighting
  - Projection
  - Video monitors
  - Active loudspeakers

Match the equipment to power receptacle locations. Receptacles may be single- or multiple-outlet and rated for standard or heavy-duty loads. They can be located in benches, desks, fittings, ceilings, walls, wall plates, wall boxes, equipment racks, and floor boxes and in projection, equipment, and control rooms.

Branch Circuit Loads

A branch circuit can only safely handle the amount of current specified by the regional wiring regulations. These limits are defined according to construction of the cable and the conditions in which the cables are installed. Each branch circuit is protected against currents that exceed the rated load of the wiring by the presence of devices such as fuses and circuit breakers. These protective devices disconnect all loads from the circuit when their current limits are exceeded.

Most electronics that you plug into an outlet—a PA amplifier, a projector, a Blu-ray rewinder, a signal processor—require less than the maximum rated current for a branch circuit. However, designers must calculate the total current drawn when the branch circuit is in typical use. In other words, can the circuit handle the PA amplifier, signal processor, Blu-ray rewinder, *and* projector operating at the same time?

To do this, you need to know which outlets and directly wired devices are connected to a specific branch circuit. Each piece of equipment has a power-consumption rating, normally stated in amps and usually found on an information panel, plate, or label located near the power-input point of the device. The power consumption information is usually stated as a voltage or voltage range and a current load, for example, "230V 2.4A" or "120V 5A" for fixed-voltage devices or "110-240V 2.2A" for devices that operate across a range of supply voltages.

Totaling the current demand for all devices on a circuit will enable you to ascertain if the load is within the capacity of the circuit. It is good practice to allow some spare capacity in any circuit with outlet receptacles to provide for additional equipment being added to the system or for replacement devices being installed during the operational life of the system. The long-term trend, however, is that succeeding generations of AV devices tend to be more energy efficient, so demand may actually decrease after upgrades.

NOTE The U.S. NEC requires that loads do not exceed 80 percent of the available current in a branch circuit (i.e., a maximum of 16A on a circuit rated for 20A).

Calculating the Number of Circuits: An Example

In this example, consider you are designing a classroom for a university. The architect and electrical engineer have provided a floor plan (Figure 10-19). At this point in your design, you've already placed equipment in the room—two flat-panel monitors and two projectors for a visual display system. You've placed a wall-mounted video camera and an AV rack with audio, video, networking, and streaming equipment in the back of the room. The lectern at the front has a document camera plus provision to power a presenter's notebook or tablet computer.

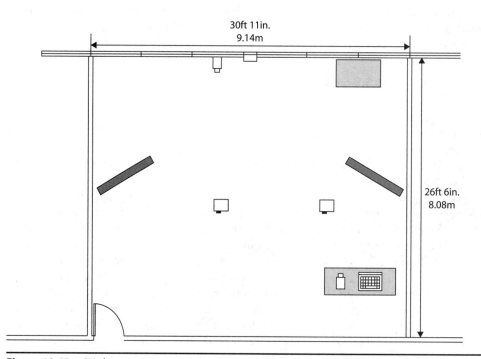

Figure 10-19 AV elements in a university classroom design

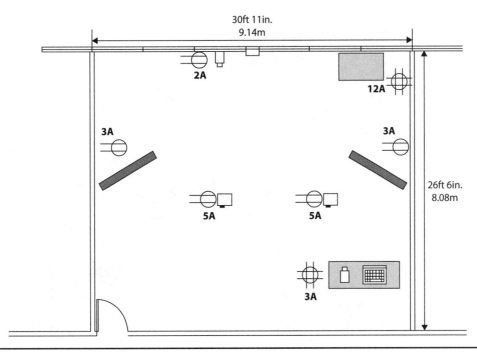

Figure 10-20 Receptacle/outlet locations in an AV installation

All the AV gear requires power, so you need to ask the electrical engineer to place receptacles at these locations. Based on this drawing, you request five double/duplex outlets/receptacles, one at the rear camera, one by each projector, and one by each monitor. You also require two four-way/quadruplex outlet/receptacle boxes, one by the rack and one at the lectern. Each piece of AV gear requires a certain current. The rack requires 12A, the rear camera requires 2A, the monitors each require 3A, the document camera plus BYO devices require 3A, and the projectors each require 5A. See Figure 10-20.

You have looked up standard circuit rating information from your AHJ and determined that you will use 16A-rated circuits. When you add up the current for the devices you plan to put on an individual branch circuit, you will need to add a circuit every time you reach 16A.

Figure 10-21 shows one solution for this room. The receptacles have been divided into three groups. Circuit 1 has the rear video camera and the rack for a total of 14A. Circuit 2 has two projectors for a total of 10A. Circuit 3 has two monitors and the lectern devices for a total of 9A. There are, of course, many other solutions that meet the necessary requirements.

Once you've documented how many circuits your system will require, you will need to include some general notes on the drawing. The notes may read something like this:

- Circuits for AV use only. Do not share with any non-AV loads.
- All AV circuits must originate from the same phase via the same sub-distribution board.

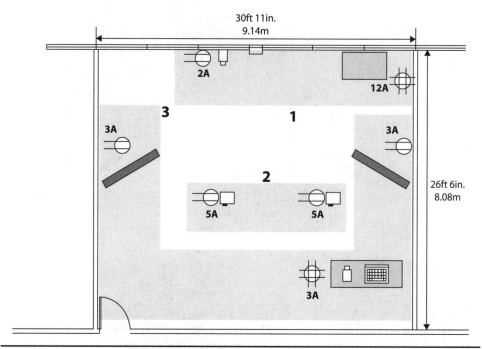

Figure 10-21 Sample circuit arrangement in an AV installation

- All AV circuits must include a protective-earth/safety-ground conductor. (This note is only necessary in jurisdictions where circuits without such conductors may be permissible.)

At a minimum, the circuits specified for the AV system should be labeled for AV system use only. In other words, no non-AV loads should be shared with the circuits designated for the electronic elements of the AV system or IT equipment associated with the AV system.

Power Strips and Leads

When you have several devices that require power at a single location, you could use power strips, variously known as power boards, outlet strips, multi-outlet boards, electric portable outlet devices, socket outlet units, relocatable power taps, and many other titles. These families of devices come with regional standards requirements in the matters of current ratings, captive inlet cables, mains-available indicators, the requirement for a main switch and/or a switch per outlet, requirements for overload protection devices, and the allowability of internal surge-protection devices.

In many jurisdictions it is not recommended, and sometimes even not permissible, to connect one power strip to an outlet on another power strip (daisy-chain), although the technical basis for these recommendations/regulations are sometimes obscure and lack sound technical explanations. In the United States, for instance, it is permissible to plug a *rack-mounted* power distribution strip into an outlet on a power strip, but not to plug one free-standing power strip into another.

If a power distribution system has outlets rated for a specified maximum current, it makes no electrical or safety difference if the total current is drawn through just one socket or through a multitude of similarly rated sockets, provided that the sum of the currents is within the ratings of the circuit.

And then there are power cables—key components for delivering power to electrical or electronic devices. In AV, power cable management is important. Sometimes power cables are too long, and they must be wrapped and secured neatly. Let Figure 10-22 be your guide to best practices.

You should consider neatness and accessibility when placing a cable within a rack or equipment area. The first example, on the left of the figure, shows a power cable wrapped as a coil—an acceptable way to tie a power cable. The example of a folded cable shown in the center of the figure is also acceptable. The example of the two cables wrapped together on the right of the figure is unacceptable because it would be hard to access the individual cables for service. Removing the cable for one piece of equipment would mean unbundling several cables. It also has the potential for heat buildup in the bundled cables. In high-current-draw situations, cables wrapped together without sufficient ventilation can overheat, producing cable damage and potentially resulting in a fire.

 TIP Longer power cables often take up more space in a rack and require extra attention. For equipment with detachable power leads, neater cable management can be achieved by replacing longer cables with shorter ones.

Figure 10-22 Best practices in power cable management

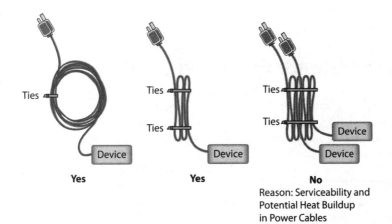

Isolated Ground

An isolated-ground system is used in North America to minimize interference problems, such as hum and buzz, caused by problematic *ground loops*. It also provides a more stable reference for the audiovisual circuits and can possibly reduce the noise level getting into the audiovisual system. The wide adoption of technologies such as optical fiber cabling, wireless connections, and digital signaling systems has reduced the likelihood of such ground loops occurring.

In most other regions, the *mesh-earth* system serves a similar function. In this system all protective-earths are in a separate mesh to the neutral/return system until the two are finally linked up at the main distribution board.

In a North American isolated-ground system, the equipment safety-ground conductors for the AV system power are isolated from the neutral/return conductors. AV equipment racks and other AV enclosures also need to be isolated and insulated from contact with grounded objects, such as conduit, building steel, and conductive concrete floors.

Isolated ground conductors are only isolated from the neutral/return up to the point where the system neutral/return and safety-ground conductors come together in the main distribution board. This ensures that any fault current has a low-impedance pathway back to the source.

Some AV systems may require a technical power system. A technical power system is a power system designed especially for AV systems and associated computers and utilizes an isolated ground. Such systems are useful in analog critical-listening environments that require a low noise floor, such as broadcast studios, postproduction facilities, performing arts centers, recording studios, and other analog critical-listening environments.

Residual noise from interference, problematic ground loops, and more will be evident in environments with low noise floors. Assuming good design practices and properly adjusted gain structure, the normal acoustic noise floor in noncritical listening spaces will mask any system noise. The widespread adoption of technologies such as optical fiber cabling, wireless connections, and digital signaling systems has reduced the likelihood of such ground loops occurring.

Interference Prevention and Noise Defense

Interference, or *noise*, is any electrical signal in a circuit that's not the desired electrical signal. The most common evidence of interference in analog systems is a hum or buzz. In digital systems, it usually manifests as random corrupted data.

In analog audio systems, *hum* is identifiable as an undesirable 50 or 60Hz noise in a signal. It manifests in analog video systems as a rolling horizontal band or "hum bar" on the display. *Buzz* is a hum with additional harmonic energy.

How does interference find its way into a signal? There has to be a source of interference and a device that's sensitive to the interference. You may hear the two referred to as *source and receiver, source and victim, source and sink*, or *source and receptor*.

There must also be a means or pathway by which the interference traverses from source to receiver. *Field theory* explains the transfer of energy from one circuit to another via an electrical or magnetic field. There is no physical contact or electron flow between circuits during this type of energy transfer. The transfer usually occurs as a result of either *magnetic-field coupling* or *electric-field coupling* between the circuits.

Magnetic-Field Coupling

As current flows through a conductor, a magnetic field develops around that conductor. This magnetic field is often depicted as invisible lines of force (see Figure 10-23), referred to as *magnetic lines of flux*. If the current remains constant, the lines of flux remain stationary. This explains what happens with DC, because the current is flowing in only one direction, so the field doesn't change over time.

In the case of an alternating current, the magnetic field expands, contracts, and changes direction in response to changes in the magnitude and direction of the current. With increased current flow in the conductor, the density of the magnetic lines of flux increases in direct proportion to the current flow. Similar to other physical phenomena, such as light, gravity, and sound pressure waves, the lines of flux diminish in intensity as they get farther from the conductor, decreasing as the inverse of the square of the distance ($\frac{1}{d^2}$).

Current can be induced to flow through a conductor when there is relative motion between a conductor and a magnetic field. This means that current can be induced either by moving a conductor through a magnetic field (thereby cutting across the lines of magnetic flux) or by making the magnetic field (the lines of flux) expand and contract around a stationary conductor by varying the current.

Let us consider the case of a mains conductor carrying an alternating current at 50Hz. As the current changes in intensity and direction, the surrounding magnetic field expands, contracts, and changes direction in response to the current flowing in the conductor.

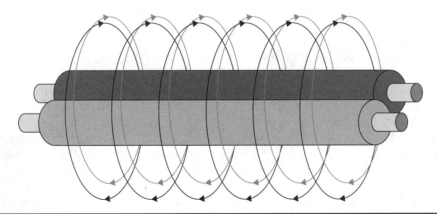

Figure 10-23 Magnetic field lines around adjacent conductors

Now consider a second conductor placed within the varying magnetic field of the first conductor. As the magnetic field from the first conductor expands, contracts, and changes direction, the lines of flux cut across the second conductor, thereby inducing an electromagnetic field (EMF) in the second conductor.

Although this is a highly desirable effect in a transformer, it is not necessarily desirable in most other situations. In other words, the varying 50 hertz current in the AC conductor is induced into the second conductor. This is *magnetic-field coupling*.

Electric-Field Coupling

Positively or negatively charged bodies (containing either a shortage or an excess of negatively charged electrons) exert influence on the space surrounding them. For example, a positively charged conductor (lacking electrons) creates an electric field in the surrounding region, attracting electrons. This attraction—the electric field—creates a negative charge in a nearby conductor through what is called *electric-field coupling*. This difference in charge creates an EMF or potential difference between the two conductors. The movement of electrons that creates this voltage potential is *current flow*. As long as there is no physical connection between the two conductors, the voltage potential will remain.

When two conductors are separated by an insulator (dielectric), along with a difference in potential between them (EMF), there's an electric field between the two conductors. The strength of the electric field increases in relation to the difference in voltage that exists between the two conductors. Similar to other physical phenomena, such as light, gravity, and magnetic fields, the strength of the electric field diminishes in intensity as the conductors are separated, decreasing as the inverse of the square of the distance ($\frac{1}{d^2}$).

With continuous DC, the field doesn't change over time. Current is flowing in only one direction; therefore, the electric field extends out, and the lines of force are stationary. With an alternating current, the polarity and magnitude of the field changes over time, because the electric field expands and contracts. Therefore, the polarity of the electric field changes with the changes in voltage. Pulsed or variable DC produces changes in the strength of the electric field, just like an alternating current, except that the field doesn't change direction.

The capacitance (ability to store a charge) of the electric field depends on the distance between the conductors, the surface area of the conductors, and the dielectric strength of the insulating material between the conductors. *Dielectric strength* refers to the maximum possible potential difference across the insulating material between two bodies before the dielectric breaks down and an arc occurs, thus equalizing the charge between the bodies. For example, air is the dielectric (and a good insulator) between charged storm clouds and the earth. A lightning strike is a breakdown of the dielectric—the air in this case—allowing a current to flow and thus equalizing the charge between cloud and earth.

Although electronic components specifically designed to have high capacitance are known as *capacitors,* an electric field will exist between any two conductors with a potential difference (voltage) between them. As insulators are not perfect, some current will flow through the insulator (dielectric) between conductors.

Electromagnetic Shielding

An electromagnetic shield is a grounded conductive partition placed between two regions of space to control the propagation of electric and magnetic fields between them. Shields are used to contain electric and magnetic fields at the source or to protect a receiver from electric and magnetic fields. A shield can be the conductive enclosure that houses an electronic device or the enclosure (aluminum foil, conductive polymer, copper braid, conductive cable duct, or metallic conduit) that surrounds a conductor.

Shielding works bi-directionally in that it prevents signals from getting out or getting in. The effectiveness of a shield is determined by the type of field, the material used for shielding, and the distance from the source of the interference.

Cable shielding can be implemented in a variety of ways. It can enclose a single insulated conductor or enclose individual insulated conductors in a multiconductor cable. Shields are also used to enclose multiple insulated conductors. Shields may even be used as a return path for current that originates at sources of interference.

There are three basic types of cable shielding:

- **Foil shield**   This type of shielding uses aluminum foil or conductive polymer wrapped around an insulated conductor or conductors. For termination purposes, an additional uninsulated conductor in continuous contact with the foil shield is used. The bare conductor provides an electrical *drain* or *ground* connection.
- **Braided shield**   This type uses fine, uninsulated wires to form an interwoven cover over an insulated conductor or conductors. Some braided shields are woven in multiple layers to improve shielding. Ground or drain termination is made directly with the braided shield.
- **Combination shield**   This type uses both foil and braid shielding, with the possible addition of a drain wire.

The main considerations in selecting a shield are coverage, flexibility, and frequency range. Coverage is expressed as a percentage and indicates how much of the inner cable will be covered. Flexibility is a subjective measurement, which directly correlates to a cable's flex life. Not all shielding can protect the enclosed conductor from all sources of EM interference. Because of this, a shield's effectiveness is narrowed to a certain frequency range.

Shielding is also used to protect other devices and cables from the EM radiation generated by the signals in the enclosed circuit.

Shielding from Magnetic Fields

Magnetic shielding employs any magnetically permeable material to absorb and conduct magnetic lines of flux to redirect them away from an unintended victim, such as a circuit or a transducer.

Permeability is a material's ability to concentrate magnetic lines of flux. A magnetically permeable material will concentrate the magnetic lines of flux within itself, rather than let the magnetic lines of flux pass through it.

Magnetically permeable materials include iron, nickel alloys, ferrite ceramics, and steel, which vary in their ability to conduct magnetic fields. A design that puts magnetically sensitive AV cabling in steel conduit or fully enclosed steel ducts yields excellent magnetic-field shielding.

Nonmagnetically permeable materials include aluminum, brass, and copper, which do not have the ability to absorb or redirect a magnetic field. This means that AV cabling with aluminum foil/conductive polymer shielding or braided-copper shields will *not* attenuate interference from magnetic fields.

Shielding from Electric Fields

The purpose of *electric-field shielding* is to lead current to the system earth/ground. Electric-field shielding employs electrically conductive materials to absorb and conduct electrical fields away from an unintended victim.

Foil or braided shielding does an excellent job of absorbing energy from a nearby electric field, but that energy will need somewhere to go. As discussed earlier, electricity seeks to return to the source, and such a pathway is provided when the shield is connected to the system earth/ground. This pathway must be a low-impedance connection.

Ground Loops

In a theoretical ideal AV system, all equipment could be connected to a single AC outlet so that all shields for the interconnecting cables would be ground-referenced to the protective earth/safety ground of that single outlet and eventually to the earth/ground (planet Earth). Usually this will not be the case. You may have multiple devices plugged into different outlets originating from different circuits on different subdistribution boards/panelboards. Although all AC circuits may be referenced to the earth/ground (planet Earth) connection at the main distribution board, the various protective-earth/safety-ground conductors may take different pathways.

Those different pathways—even when well designed and implemented—may have slightly different impedances, which can create potential differences between points, and hence create small current flows in the cable shields. These currents are known as *ground loops*.

The low-impedance mesh-earthing system used in most power distribution schemes reduces the risk and magnitude of these ground loops.

As the use of long analog signal runs in AV systems diminishes with the adoption of digital balanced-line twisted-pair and optical-fiber signal paths, the problem of potential differences on cable shields and signal return paths also diminishes.

Balanced and Unbalanced Circuits

All electrical circuits—and the cabling used to connect them—generate electromagnetic fields that interact with other electrical circuits and cabling, including AV circuitry and cabling. This interference and noise degrade the quality of the signal and, as discussed previously, may introduce hum, buzz, crackles, and splats.

Figure 10-24
A balanced circuit. The signal spikes are canceled out by inversion.

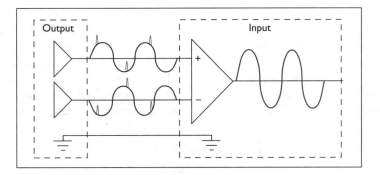

One way to reduce the noise in a circuit is to employ an actively balanced electrical circuit, as shown in Figure 10-24, which offers a defense mechanism against noise. In a balanced circuit, the signal is transmitted over two identical parallel conductors, with the signal on one of the conductors inverted. These signals are said to be *in balance*. At the receiver, the signals are combined, with the inverted signal reversed in a differential mode amplifier. This produces a higher-strength signal but, more importantly, cancels out any external signals picked up by both wires. This cancellation process is known as *common mode rejection* (CMR). To ensure that any unwanted signal or noise is induced equally in both conductors, the two conductors are usually twisted tightly around one another so that both are equally exposed to the offending noise source, a configuration known as a *twisted-pair*. For further noise immunity, the twisted-pair may then be surrounded by a shield that is tied to system ground. This configuration of cable, well-known for its use for analog audio signals, is known as *shielded twisted-pair cable*.

In an unbalanced circuit, as shown in Figure 10-25, the signal is transmitted down a single signal conductor. Without the second inverted signal to provide common mode rejection, all external signals picked up by the signal conductor would be present in the system output. The single conductor in an unbalanced cable is usually surrounded by a cable shield that also acts as the return electrical path for the circuit. Unbalanced circuits are also known as *single-ended circuits*.

Figure 10-25
An unbalanced circuit

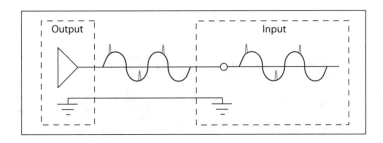

With either a balanced or unbalanced design, the longer the cable run, the more noise the cabling is subjected to. Therefore, unbalanced lines are extremely limited in the distance they can cover and their ability to transfer a usable signal.

The ability of a specific AV device to reject common-mode noise is described by its *common-mode rejection ratio* (CMRR). The CMRR is expressed in decibels, with higher numbers indicating better rejection, usually 70 to 120dB, depending on need. Typical component CMRRs may vary as much as 5 percent from their stated values.

Transformer-Balanced Circuits

One of the best defenses against common mode noise is to specify transformer-balanced circuits. Electronically balanced circuitry may be less expensive, but it is typically not as effective at rejecting common-mode noise. An untapped transformer is inherently balanced, which leads to its excellent common-mode rejection.

Common-mode noise will arrive on both ends of the primary side of the transformer. Because it's common-mode noise, there will not be any potential difference between the ends of the primary winding, and therefore, noise current will not flow through the primary side of the transformer and be transferred to the secondary side.

Cable Support Systems

A *raceway* or cable support is a device or mechanism that supports cables by equalizing the tension throughout the length of the cable. Raceways vary in construction from full cable enclosures to partial enclosures and cable suspension systems. Some common cable raceway systems are listed in the following sections.

Cable Tray

A cable tray, as illustrated in Figure 10-26, is an assembly or grouping of sections of noncombustible materials (usually metal) to provide rigid support for cables. Cable trays are usually open-topped structures and frequently include slots or holes that allow cables to be anchored to the structure after being run out. Cable trays for vertical cable runs often take the form of a cable ladder. They provide no EMI or RFI shielding for the supported cables.

Figure 10-26
Slotted cable tray (left) and cable ladder (right)
(Image: vav63/Getty Images)

Figure 10-27
Various types of conduit

Conduit

A conduit is a tube, usually circular in cross-section, that completely encloses the cables that are run through it, supporting them and protecting them from damage. A conduit, as shown in Figure 10-27, may be constructed of metal or a range of polymer materials that may be weatherproof, suitable for burial in the ground or in concrete, or even ultraviolet-resistant for use in direct sunlight. Steel conduits may act as EMI and RFI shields for cables carrying low-level signals. Conduits are used for permanent installations or may be left unpopulated for temporary use in production or exhibition spaces.

Cable Duct

Cable ducts are structures that enclose and support cables on at least three sides. They may be rigid or partially flexible and may be constructed of metal or a range of plastic materials. Many ducts have a capping mechanism that allows them to be closed after completing cable installation and reopened for service access. Some duct systems have internal dividers to separate the different services being carried. In enclosed metallic ducting systems, the cables are isolated from external EMI and RFI sources. In internally divided metal ducting systems, the separate sections offer EMI and RFI isolation between sections.

Hook Suspension Systems

A wide range of hook-type suspension systems have been developed for use in situations where other cable support systems may be unnecessary or difficult to install. They consist of a series of hooks (as shown in Figure 10-28) or rings suspended from the building structure. The bundles of cables are then looped between hooks. While generally lower in cost and easier to run cables through, hook suspensions offer little mechanical support and no physical or signal protection to the cables they carry. This is not usually an issue for cables run through spaces that are infrequently accessed.

Figure 10-28
A hook cable suspension

Conduit Capacity

To maintain cable and signal integrity, conduits should not be overfilled. As shown in Figure 10-29, when cables are inserted into a conduit, the remaining space, known as *permissible area,* ensures that the cables will not be jammed together and damaged. Permissible area rules are usually identified in electrical codes.

Typical examples of permissible area percentages for cables are listed in the U.S. National Electrical Code:

- One cable may occupy 53 percent of the conduit's inside area.
- Two cables may occupy 31 percent of the conduit's inside area.
- Three or more cables may occupy 40 percent of the conduit's inside area.

Figure 10-29
Conduit showing space for cables and the permissible area

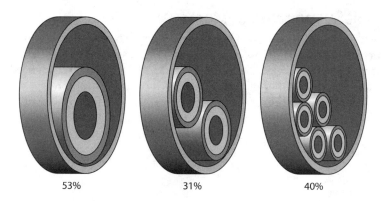

For example, if you want to run three cables in a piece of conduit, you will first have to calculate the capacity using the conduit capacity formula to know whether it will meet that 40 percent permissible area. Permissible area in any given region is determined by the AHJ.

Calculating conduit capacity requires

- Inner diameter of the conduit: ID
- Outer diameter of each conductor: $od_1, od_2, \ldots od_n$
- Permissible area of the conduit

Use the appropriate formula for calculating conduit capacity based on the number of cables, as shown next.

Conduit Capacity Formula for One Cable

$$ID > \sqrt{\frac{od^2}{0.53}}$$

Conduit Capacity Formula for Two Cables

$$ID > \sqrt{\frac{od_1^2 + od_2^2}{0.31}}$$

Conduit Capacity Formula for Three or More Cables

$$ID > \sqrt{\frac{od_1^2 + od_2^2 + od_3^2 \ldots}{0.40}}$$

Jam Ratio

In addition to conduit fill percentage calculations, you need to calculate a jam ratio. You will need to use a jam ratio to ensure that when three cables are installed in a conduit, they do not slip or jam. When three cables are installed in a conduit, one of the cables may slip in between the other two, especially at a bend where the conduit may be slightly oval because of the bending process. If the third cable slips between the other two, the three cables may line up and jam at the exit of a conduit bend. (See Figure 10-30.)

To find the jam ratio, calculate the ratio of average cable outer diameter to inner conduit diameter. If the result is between 2.8 and 3.2, the cables are at risk of jamming. If the calculation falls within that range, the next larger size conduit should be specified to avoid any potential jam.

A jam ratio of 3.0, for example, would mean that the value of the combined diameters of the three lined-up cables equals the inner diameter of the conduit.

$$Jam\ ratio = ID \div \left(\frac{od_1 + od_2 + od_3}{3}\right)$$

Figure 10-30
Cable jamming in conduit

Chapter Review

You just learned a lot about a subject that factors relatively little in the CTS-D Exam Content Outline. You've learned about circuits, electrical power and distribution, grounding, preventing interference, specifying conduit, and more. The reason for all the detail is that electrical infrastructure is critical to the performance of AV systems and can be one of the most misunderstood areas of communication among project team members. It's not uncommon for an AV professional to use one term when trying to describe electrical infrastructure and an electrical engineer to understand something different. Think of this chapter not only as test prep but also as an important reference for professional collaboration.

Review Questions

The following questions are based on the content covered in this chapter and are intended to help reinforce the knowledge you have assimilated. These questions are similar to the questions presented on the CTS-D exam. See Appendix E for more information on how to access the free online sample questions.

1. For branch circuit loads covered by the U.S. National Electric Code, you should plan to use not more than _____ of the rated current.
    A. 60 percent
    B. 100 percent
    C. 75 percent
    D. 80 percent

2. In a three-phase power distribution system for an AV installation that includes no devices with switched-mode power supplies, what is the maximum current likely to flow through the neutral/return conductor?
    A. The same maximum current that could flow through any one of the phase conductors
    B. $\sqrt{3}$ times the maximum current that could flow through any one of the phase conductors
    C. Three times the maximum that could flow through any one of the phase conductors
    D. No current flow due to phase cancellation

3. The main considerations when selecting a cable shield are:
   A. Materials, permeability, and conductors
   B. Coverage, flexibility, and frequency range
   C. Impedance, flexibility, and materials
   D. Cable types, gauge, and interference
4. Which of the following electrical properties is measured in ohms?
   A. Impedance (Z)
   B. Capacitive reactance (X_C)
   C. They are all measured in ohms
   D. Inductive reactance (X_L)
5. The first step when determining conduit capacity for cable should be to determine the:
   A. Cable resistance and multiply by a factor of 5
   B. Allowable fill percentage based upon local codes and regulations
   C. Wire dimensions exclusive of insulation multiplied by 3.1459
   D. Bend radius for all cables utilized in the installation
6. Where should a three-to-two-pin (ground-lifting) adapter be used in an AV installation?
   A. On audio equipment exhibiting mains-frequency hum
   B. On analog video monitors exhibiting rolling bars
   C. Nowhere—ever
   D. On Ethernet network devices exhibiting randomly corrupted data

Answers

1. **D.** For branch circuit loads covered by the U.S. National Electric Code, you should plan to use no more than 80 percent of the available rated current. (In most other jurisdictions, the entire rated capacity of the branch circuit is considered to be available for use.)

2. **A.** In a three-phase power distribution system for an AV installation that includes no devices with switched-mode power supplies, the maximum current likely to flow through the neutral/return conductor is the same as the maximum current that could flow through any one of the phase conductors. (Where switched-mode devices are used, the neutral/return current may exceed the maximum current that could flow through any one of the phase conductors.)

3. **B.** The main considerations when selecting a cable shield are coverage, flexibility, and frequency range.

4. C. Resistance (R), impedance (Z), capacitive reactance (X_C), and inductive reactance (X_L) are all measured in ohms.

5. B. The first step when determining conduit capacity for cable should be to determine the allowable fill percentage based upon local codes and regulations.

6. C. Devices such as three-to-two-pin, or ground-lifting, adapters that interrupt the protective-earth/safety-ground connection to any device must never be used. Ever. They are known as *widowmakers* for a good reason.

Elements of Acoustics

In this chapter, you will learn about
- How sound is produced
- Acoustics-related intensity and pressure
- Sound reflection, reverberation, diffusion, absorption, and transmission
- Measuring background noise and specifying maximum allowable background noise levels for a given environment and application

The first question designers consider when planning a sound system is whether such a system is even required. If a space is small, the acoustics are good, and the background noise level is low enough, there may not be a need for any system that seeks to reinforce the sound.

If, however, the sound is from an electronic source, or the audience can't adequately hear a presentation, performance, videoconference call, etc., then using microphones, audio mixers, signal processors, power amplifiers, and loudspeakers to amplify a sound source will probably be necessary. The sound can also be distributed to a larger or more distant audience.

The second question is how much privacy a space requires, both from sound originating outside the space and from sound leaking out. Depending on the answers, infrastructure decisions will play a pivotal role in sound system design.

This chapter assumes knowledge of sound reinforcement needs assessment and specifications, which were discussed in detail in Chapter 6 of this guide. This chapter does not provide the level of knowledge required of an acoustician or sound engineer but will help you to make informed infrastructure decisions and design for a comfortable listening experience.

> **Duty Check**
>
> This chapter relates directly to the following tasks on the CTS-D Exam Content Outline:
>
> - Duty B, Task 1: Review A/E (Architectural and Engineering) Drawings
> - Duty B, Task 8: Coordinate with Acoustical Professionals
> - Duty C, Task 1: Create Draft AV Design
> - Duty D, Task 1: Participate in Project Implementation Communication
>
> In addition to skills related to these tasks, this chapter may also relate to other tasks.

Acoustic Engineering

Acoustic engineering is an important allied trade to the AV industry. Acoustics is the science and technology of sound in all its aspects. When acousticians look at a room, they consider the following properties of sound:

- **Production**   How the energy is generated; in other words, the source of the sound
- **Propagation**   The pathway of the energy
- **Control**   How (sound) energy is generated and subsequently propagated
- **Interaction**   How material responds to the sound energy imposed upon it
- **Reception**   How hearers' ears and brains will respond to the stimuli placed upon them

Ideally, any application where audio quality is crucial will require an acoustic consultant's expertise during the programming phase. As an AV professional, you should be aware of your design environment's acoustic properties and be prepared to make recommendations regarding acoustic criteria.

NOTE   This chapter will cover each of the listed sound properties, but you'll find control-related information in discussions of sound transmission class (STC) and impact insulation class (IIC).

Sound Production

When you think about where sound comes from, it's probably easiest to envision vibrating objects, such as a moving loudspeaker cone or a vibrating string, producing waves in the air. However, sound is produced by any disturbance that creates changes in pressure or velocity in an elastic medium such as air. For example, air escaping from an air-conditioning outlet comes out of a vent that's much smaller than the duct.

This increases the velocity of the air, resulting in a whoosh of air. The change in velocity and pressure, rather than mechanical vibration, produces the sound.

Not only is mechanical vibration not required to produce sound, you don't even really need air. Sound waves do need a physical medium in which to propagate, but that medium can be water, other liquids, or solids. The fact that sound can be carried by so many different media is part of what makes acoustics such a complex field.

In his book *Music, Physics, and Engineering* (Dover Publications, 1967), Harry F. Olson wrote:

> Sound is an alteration in pressure, particle displacement, or particle velocity which is propagated in an elastic medium, or the superposition of such propagated alterations.... Sound is produced when the air or other medium is set into motion by any means whatsoever. Sound may be produced by a vibrating body, for example, the sounding board of a piano, the body of a violin, or the diaphragm of a loudspeaker. Sound may be produced by the intermittent throttling of an air stream as, for example, the siren, the human voice, the trumpet or other lip-reed instruments, and the clarinet and other reed instruments. Sound may also be produced by the explosion of an inflammable-gas mixture or by the sudden release of a compressed gas from bursting tanks or balloons.

Indeed, any disturbance that creates changes in air pressure or air-particle velocity will be processed by our ear/brain system as sound.

Sound Propagation

Sound waves have physical length, or *wavelength*. Wavelength is the distance between two corresponding points in consecutive cycles. Knowing how a sound wave of a certain size interacts when it comes into contact with surfaces of different sizes and materials helps us to understand sound wave behavior.

You may be wondering about the effect altitude has in relation to the velocity of sound. It turns out that temperature has a much more pronounced effect on the velocity of sound than atmospheric pressure.

The velocity of sound depends on the temperature and molecular weight of atmospheric gases. It does not depend on pressure changes. The pressure decreases with an increase in altitude and therefore only slightly affects the velocity of sound. The velocity of sound in air, at sea level and at 20°C (68°F), is 343 meters per second (1125 feet per second).

You can obtain the wavelength value if you divide the velocity of sound by frequency, as shown in the following formula:

$$\lambda = \frac{v}{f}$$

where:

- λ is the wavelength measured in meters (feet).
- v is the velocity of sound in meters per second (feet per second).
- f is the frequency in hertz.

Using this formula, you can calculate the wavelengths at the limits of the frequencies most humans can hear—20Hz to 20kHz.

The wavelength at 20Hz = 343 ÷ 20 = 17.15m (56.25ft)
The wavelength at 20kHz = 343 ÷ 20,000 = 17.15mm (0.67in.)

Considering such a wide range of wavelengths, it's important that you have a solid understanding of how waves of certain physical dimensions behave in space.

Although the sounds you hear every day are complex waveforms, every sound can be broken down into a series of individual simple sine waves at different frequencies. *Fourier analysis* is the mathematical process used to calculate what is known as the Fourier series of component frequencies for a waveform. Figure 11-1 is an oscilloscope display of the complex waveforms from the left and right channels of a music track.

The calculation and mathematical manipulation of the components of a waveform's Fourier series underlies digital signal processing (DSP) and analysis applications.

From the formula you can see that frequency and wavelength are inversely proportional. The lowest frequencies have the longest wavelengths, and the highest frequencies have the shortest wavelengths.

Sound Intensity

When you talk about how "loud" a sound is, you are referring to its power. This is the amount of sound power falling on (or passing through, or crossing) an area. Power is measured in watts, so the unit for sound power is watts per square meter (W/m^2).

In practice sound is not usually measured using a power meter because these are complex and expensive instruments. Instead, it is typically measured in terms of pressure, which is the amplitude of sound over an area, rather than power over an area. You can measure pressure with a simple, familiar device such as a microphone, because, like the human eardrum, microphones detect changes in air pressure.

Figure 11-1 Complex waveforms

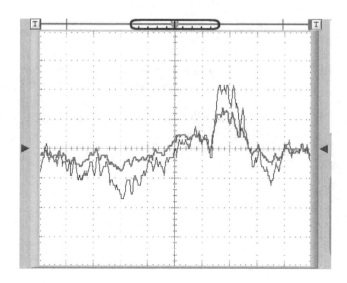

The use of decibels to measure sound intensity (amplitude) by comparing a measured sound pressure to a reference pressure, such as the threshold of human hearing, is discussed in detail in Chapter 6.

Particle Displacement

Particle displacement, or *displacement amplitude* (measured in meters), is the distance of a particle's movement from its equilibrium position in a medium as it transmits a sound wave by compression and rarefication. When a sound wave is traveling through air, the air molecules are displaced at the particle speed of the oscillations of the compression wave being transmitted traveling through the air, while the sound wave itself moves at the velocity of sound.

Sound Interaction

Sound energy radiates from the source in all directions. Unless the sound is generated in a completely free space, the energy will encounter a boundary or surface. If you are outdoors, the only likely boundaries may be the ground or nearby buildings. There are many boundaries and surfaces when you are indoors. Besides the walls, ceilings, and floors that you may be thinking of, furniture, objects, and people also affect what happens with the sound energy in the environment.

So, what happens to the sound energy produced by either a sound reinforcement system or other source of generated sound? The Law of Conservation of Energy tells us that energy can be transformed from one form to another and transferred from one body to another, but the total amount of energy remains constant.

When sound energy encounters a surface or room boundary, one or more of the following three things occur (see Figure 11-2):

- Reflection
- Absorption
- Transmission

Before getting into the specifics of what happens to sound within a space, we'll give you a good overall view of what is occurring in that space.

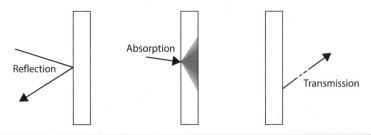

Figure 11-2 Sound interaction

The separate "sound events" that occur over time have a significant effect on the quality of sound as experienced by the listener. An acoustical consultant, understanding how a venue is to be utilized, can predict each of these events by evaluating and making recommendations regarding the venue's size, optimum shape, finishes and materials to be used, and so on.

Let's discuss each event in detail.

Reflection

Figure 11-3 shows the types of sound events that occur in a space. Of such events, you will always have *direct sound*—sound that arrives directly from the source to the listener.

If a space has boundaries (such as walls, ceilings, floors, furniture, and so on), there will also be *reflections*—sound that arrives to the listener after the direct sound as reflected energy. This sound takes an indirect path, taking more travel time before arriving at the listener.

If the sound energy is not absorbed by some material or transmitted into an adjoining space, it will be reflected back into the originating space. Reflections can be direct, diffuse, or somewhere in between.

Specular Reflections

Specular, or *direct reflections,* as shown on the left of Figure 11-4, bounce directly off a surface like light bouncing off a mirror. Like light, the incoming angle (the angle of incidence) will equal the outgoing angle (the angle of reflection).

Diffuse Reflections

The second type of reflection pattern is *diffused,* or *scattered reflection,* as shown on the right of Figure 11-4. Diffusion is the random scattering of sound waves from an uneven surface, like light scattered from a matte-white projection screen. It occurs when surfaces are at least as long as the incident sound wavelengths but not more than four times as long.

Figure 11-3
Reflection

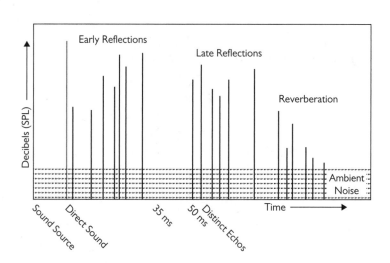

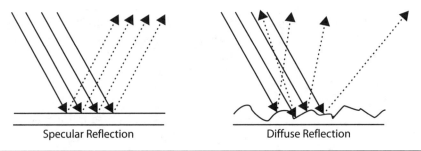

Figure 11-4 Specular and diffuse reflections

When the sound energy is diffused, it is dispersed and therefore less noticeable. Dispersed sound can have a tendency to do some self-canceling, in that as it is scattered, the hearer will receive diffused energy from all directions, causing some canceling.

With a matte-white screen, all incident light that strikes the screen is scattered evenly in all directions. This even distribution in all directions is easily accomplished with visible light energy because the light visible to humans covers less than an octave. Audible sound energy covers ten octaves, making diffusion, or any other acoustical control, over a wide frequency range difficult to implement. For reflecting surfaces to function as diffusers, they must be heavily textured and irregular—the dimensions of irregularities should be *nearly equal to the wavelength of sound*.

Diffusers   Diffusers (pictured), as the name implies, are a good complement to sound absorption because they reduce echoes and reflections without removing sound energy entirely. Their various shapes, sizes, and materials provide a surface for sound to scatter or disperse in different directions. With diffusion, the energy is reflected back at different angles rather than the majority of the energy being reflected at a single specific angle. They are useful for breaking up dead spots from phase cancellation, but they also smooth out the peaks or the response of the room. Additionally, diffusion, like any other acoustical treatment, is effective only over a limited frequency range.

A space without diffusion can feel cavernous, so many acousticians include diffusers to improve speech intelligibility and the overall experience for listeners.

NOTE Diffusers should not be confused with *bass traps*, which are large-sized sound absorbers for lower frequencies. Bass traps are used mainly for reducing low-frequency standing waves.

Flutter Echoes

A *flutter echo* is a type of hard reflection. A flutter echo is a series of reflections that continue to bounce back and forth between hard parallel surfaces such as large walls, ceilings, windows, and floors. Flutter echoes are distracting and should be avoided.

VIDEO Watch a video from AVIXA's online video library demonstrating flutter echoes. Note how the sound bounces back and forth off the walls of the room. Check Appendix D for the link to the AVIXA video library.

Other types of flutter echoes are "pitched-roof" flutter echoes and flutter echoes created by concave surfaces, but these concepts are beyond the scope of this guide.

Room Modes

A *mode* (or standing wave or room resonance) is the amount of energy at fixed positions within a room. Room modes significantly affect the perceived low-frequency performance in smaller rooms such as boardrooms, conference rooms, classrooms, edit suites, home theaters, control rooms, huddle rooms, music practice rooms, and small studios—really, any relatively small space.

A *standing wave* occurs when a sound wave travels between two reflecting surfaces, such as two parallel walls. At some frequency, the room's dimension equals exactly one-half a wavelength at that particular frequency, as well as multiples of that one-half wavelength. The wave reflects back on itself, producing fixed (standing) locations of high and low pressure due to wave interference.

For example, let's take a room that is 7m (23ft) deep. In air (at 343m/s), a wavelength of 7m corresponds to a frequency of 49Hz. As the resonant frequency of the room is half this wavelength, the room resonates at 24.5Hz. (The same result would be obtained using 1125ft/s and 23ft.)

The following are standing waves based on the example:

- **7m (23ft)**   One-half wavelength at 24.5Hz
- **14m (46ft)**   One wavelength at 49Hz
- **10.5m (34.5ft)**   One-and-a-half wavelengths at 73.5Hz
- **7m (23ft)**   Two wavelengths at 98Hz

In practice, the nulls (minimal pressure zones) can cancel well enough that they could be indiscernible from the room's acoustic noise floor.

NOTE This chapter takes into account only axial modes (involving one pair of surfaces). There are also tangential modes (involving two pairs of surfaces) and oblique modes (involving all three room-surface pairs), which are not discussed here. Critical room design would include consideration for tangential and oblique modes.

Note in Figure 11-5 that the modes would cover the full width and length of the room. It also illustrates only a slice of the modal distribution. The floor-to-ceiling dimension is almost always the room's smallest dimension (RSD) and sets the frequency at which model behavior becomes dominant.

At what point do room modes dominate the low-frequency performance of the room? A simple formula is used to approximate this critical (or *Schroeder*) frequency.

$$F_c = 3 \times \frac{C_{air}}{RSD}$$

where:

- F_c is critical frequency.
- C_{air} is the velocity of sound in air (343m/s or 1125 ft/s).
- *RSD* is the room's smallest dimension (in meters or feet to match C_{air}).

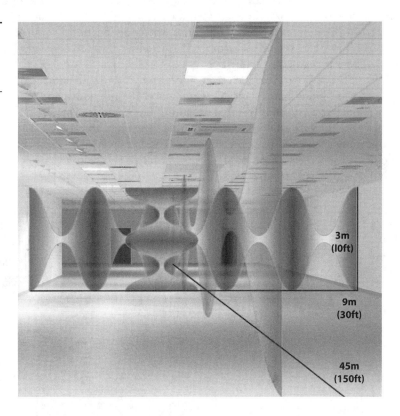

Figure 11-5 Some fixed energy locations in a room

As we've already noted, the room's smallest dimension will invariably be the distance from the floor to the ceiling. Also, F_c is not to be considered a single frequency, but a transition area. Below the F_c, you would expect room modes to dominate the low-frequency performance of the space.

For example, take a meeting room that is 6m wide and 10m deep, with a floor-to-ceiling height of 3m. The distance from the floor to the ceiling is the variable used in the calculation.

$$F_c = 3 \times 343 \div 3$$
$$F_c = 343\text{Hz}$$

In this example, the transition area into where room modes dominate would be from around 343Hz and below.

During system equalization, placing a measurement microphone in specific locations of either high or low energy will result in an erroneous representation of the actual sound system response.

These areas of high and low energy distribution will also be revealed by the levels shown on an audio spectrum analyzer. With room modes, the energy level shown by the analyzer is completely dependent on the relationship between the analyzer microphone's position and the mode. If the microphone is placed in the general location of a peak (maximum energy), the analyzer would show an excessive energy level that one might think needs to be compensated for in equalization. If the microphone is placed in the general location of a null, the sound energy might not show at all but would appear as part of the noise floor. And, of course, there are locations for all the energy levels in between those two states. Unfortunately, there isn't a way to equalize electronically based on a room's physical characteristics.

Figure 11-5 shows fixed energy locations in the room, illustrating only a slice of modal distribution. Note that the modes would cover the full width and length of the room.

Rooms having dimensions that are multiples of one another are not very desirable—a cube being the absolutely least desirable room shape. Table 11-1 lists the optimum room proportions suggested by some acoustical researchers using a range of research methodologies. Using these recommended ratios will likely yield a reasonable distribution of low-frequency axial room modes.

Table 11-1 Recommended Room Proportions		Ceiling Height	Room Width	Room Length
	L.W. Sepmeyer	1.00	1.14	1.39
		1.00	1.28	1.54
		1.00	1.60	2.33
	M.M. Louden	1.00	1.40	1.90
		1.00	1.30	1.90
		1.00	1.50	2.50
	J.E. Volkman	1.00	1.50	2.50
	C.P. Boner	1.00	1.26	1.59

Generally speaking, tangential modes will be −3dB of the axial mode energy, and oblique modes will be −6dB less than the axial mode energy. This is because tangential and oblique modes involve greater distances and more surfaces. The difference in levels between these types of modes will be more pronounced in a room with acoustic treatment.

Room modes will dominate the low-frequency performance of the room. Because these areas of maximum and minimum pressure (peaks and nulls) are created by the relationship between wavelength and room dimensions and are location specific, these locations will be evident and can be displayed on a spectrum analyzer. Determining the frequency below which room modes dominate helps us to know where our analyzer provides unreliable data. In other words, the data displayed by the analyzer completely depends on microphone placement at those room mode frequencies. It is important to verify that the measurement microphone was not placed in an area of maximum or minimum pressure.

Because the analyzer data is unreliable, it is best to set low frequencies on an equalizer by ear. After equalizing an audio system effectively, there may still be unavoidable issues caused by the physical shape and size of a space. In other words, there is no way to "equalize a room." Different room dimensions cause sound to react in different ways.

Ultimately, the AV professional needs to know when the data seen on the analyzer is completely unreliable. Rooms with dimensions that are multiples of one another reinforce modal frequencies. This can be mitigated by spreading out the room modes—using room dimensions that are not multiples of one another. Those involved with small studio or home theater design might know these as the so-called *golden room* ratios.

Although smaller rooms, such as conference rooms, huddle rooms, and boardrooms, are characteristically dominated by room modes, reverberation is an issue in larger rooms.

Small Rooms

The "sound" of a small room is impacted not only by the modal behavior at lower frequencies (around 300Hz and below—the bottom four octaves of our hearing spectrum) but also by early reflections. Direct sound combines with the energy from early reflections (10ms or less), and the result is severe tonality shifts. In other words, these severe tonal shifts in the sound occur when early reflected energy combines with direct energy. This doesn't even require a sound system to occur and helps to explain why a small room "sounds like a small room." Small rooms sound poorly because the dimensions create this direct and early reflection problem.

Reverberation

Reverberation is the sound that persists in a room after the energy that created it is stopped. Adding audio equipment cannot compensate for a room with undesirable acoustic properties.

Reverberation is the combination of many acoustic reflections, which are dense enough that they don't sound like reflections, but rather act as a sonic decay "tail" to the sounds in a room.

True reverberation is really only a large-room phenomenon and can be a significant issue in large rooms used for reinforced speech or contemporary (amplified) music performances.

Figure 11-6 Measuring RT_{60}

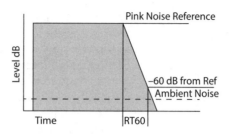

Examples include gymnasiums used for assembly purposes; enclosed sports stadiums such those used for swimming, basketball, gymnastics, and boxing; velodromes; an armory used as a reception hall; or a cathedral-style house of worship introducing contemporary-style (amplified) music. Longer reverberation times are more appropriate for liturgical houses of worship, cathedrals, chamber music, concert halls, and other venues used for acoustic (nonreinforced) performances.

You can think of a reverberant room as being a sound energy storage tank. The room cannot readily drain off the energy being introduced into it either by absorption or by transmission.

A statistical reverberant field is reverberation that is well enough spatially diffused that it is essentially the same at all points in a room. Many of the formulas used in acoustics and sound systems assume the existence of a *statistical reverberant field*, but this does not always exist. A reverberant field doesn't exist in a small room because the sound energy is not able to develop a diffuse, uniform, random distribution.

When we talk about the reverberation time (RT_{60}) of a room, we assume the room has a statistical reverberant field.

Reverberation Time Persistence of sound energy (reverberation) is measured in terms of time and level. It is also frequency-specific. The *reverberation time* of a space (Figure 11-6) is said to be the time it takes for sound to decay by 60dB (to one-thousandth of its original level). Hence the term RT_{60}.

The RT_{60} value will vary with frequency and environment. In general, spaces primarily intended for speech or more contemporary music require shorter RT_{60} levels than music, depending on the style. An acoustician can quantify acceptable RT_{60} times and design the venue accordingly.

Figure 11-7 illustrates some examples of reverberation times in certain types of venues. Note how the reverberation time directly relates to the volume of a room.

Whether the reflections are either useful or distracting depends on how late the reflection arrives to the listener after the direct sound, as well as the energy level of the reflection in comparison to the direct sound energy level. See the "Sound Reception" section in this chapter for more information about sound integration.

Absorption

Sound can also be absorbed. Absorption is about slowing down particle velocity (power) using various surfaces, such as porous absorbers. It is the friction between the air molecules and a material.

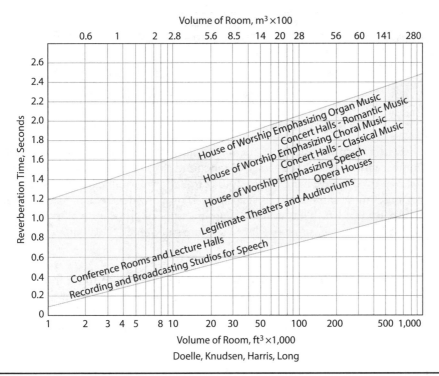

Figure 11-7 Reverberation times of various venues

Sabin

The effectiveness of different types of absorbers, such as porous or resonant absorbers, is frequency (wavelength) dependent. An absorber effective for a wavelength of 150mm (6in.), approximately 2kHz, will be of little use for a wavelength that is more than 1.2m (4ft), or about 280Hz. Such effectiveness is quantified as a coefficient (*sabin*) over a single octave band.

By definition, an empty space, such as an open window, has 100 percent sound absorption, or 1.00 sabins of absorption in a given octave, while a material that absorbs 40 percent of the incident sound over that octave has an absorption of 0.4 sabins.

NRC

A widely used measure of absorption was the now-outdated *noise reduction coefficient* (NRC), with values ranging from 0.00 to 1.00 (rounded to the nearest 0.05). NRC used a measure of absorption averaged over four test frequencies (250Hz, 500Hz, 1kHz, and 2kHz). Generally speaking, materials with NRC values smaller than 0.20 are considered to be reflective, while those with values greater than 0.40 are considered to be absorptive.

While absorption reduces the amount of sound-level energy within a room, absorption does not prevent sound energy from being transmitted into an adjoining room. In other words, absorption is not an acoustic barrier.

Table 11-2 Air Attenuation Coefficient

Frequency	Air Attenuation Coefficient (sabins/m)	Air Attenuation Coefficient (sabins/ft)
2kHz	0.009	0.003
4kHz	0.025	0.008
8kHz	0.080	0.025

SAA

A more-recent absorption metric is the *sound absorption average* (SAA). The SAA is a single-number rating that is the average of the sound absorption coefficients of a material for the 12 one-third octave bands from 200Hz to 2.5kHz (rounded to the nearest 0.01). As the SAA uses more sample points (12 versus 4) over a wider frequency range than the NRC, it is considered to be a more accurate measure of absorption. Despite the SAA replacing the NRC rating, some product literature still quotes NRC values.

Air Absorption

While air absorption will not be a factor in a typical conference room, boardroom, or meeting room, it is a factor in larger spaces. As you get farther away from the source, the sound energy spreads out and some of the energy is absorbed by the air.

Table 11-2 shows air attenuation coefficient values at some frequencies.

Absorption Coefficient

The absorption coefficient of a surface is the ratio of the energy absorbed by the surface to the energy incident. It typically lies between 0 (nonabsorbing) and 1 (totally absorbing surfaces). The absorption coefficient can be defined for a specific angle of incidence or random incidence as required. Absorption coefficients are usually measured at 125Hz, 250Hz, 500Hz, 1kHz, 2kHz, and 4kHz. The amount of absorption varies with frequency, and porous absorbers have limited effectiveness below about 250Hz.

Although theoretically impossible, values greater than 1 are often found when taking random incident-sound measurements. This usually occurs because of diffraction/edge effects.

Porous Absorbers

As the displaced air molecules pass through a porous absorber, the friction between the molecules and the material of the absorber slows down the molecules. While there may still be some reflected sound back into the room, a much greater portion of the sound has otherwise been absorbed.

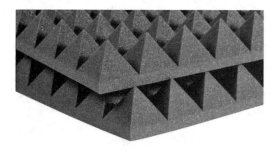

Typical porous absorbers include carpets, acoustic tiles, acoustical foams (pictured), curtains, drapes, upholstered furniture, people, and their clothing. The effectiveness of different types of absorbers, such as porous or resonant absorbers, is frequency (wavelength) dependent.

Placing porous absorbers directly on a room boundary (such as a wall) may be the most convenient location, but unfortunately not the most effective, as the particle velocity at a boundary is near zero. Porous absorbers are actually most effective when placed at a quarter wavelength from a room boundary. For absorbing a 100Hz tone, this would be approximately 1m (3ft) from the boundary.

Porous absorbers are primarily effective at middle and high frequencies. Fittingly, this range is where the ear is most sensitive and where noise control is most needed in many environments.

Resonant Absorbers

Resonant absorbers are used to manage frequencies where the size or thickness of porous absorbers in those cases would be too expensive or difficult to manage. Additionally, resonant absorbers are typically placed in room boundaries or wall surfaces where porous absorbers fail to affect low-velocity sound waves. Typical resonant absorber construction consists of a gypsum/plaster board/wallboard panel and spacing filled with absorbent, insulating material.

 NOTE The absorber must be compliant with local regulations for fire-rated materials. Foam materials that are not clearly fire-rated should never be used in acoustical engineering applications.

Transmission

We know sound can be transmitted through walls and other building materials. Some materials and construction methods can be rated on their ability to attenuate the sound passing through them. The greater a material's mass, the higher the sound transmission loss.

Acoustic Mass Law

You can predict the sound transmission loss of single-layer, impermeable materials using a formula known as the *acoustic mass law*:

$$TL = 20 \times \log(m_s \times f) - 48$$

where:

- TL is the transmission loss in dB.
- m_s is the mass per unit area of the material in kg/m².
- f is the frequency in Hz.

The equation shows that each time the frequency measured or the mass per unit area of a single-layer wall is doubled, the transmission loss is increased by approximately 6dB.

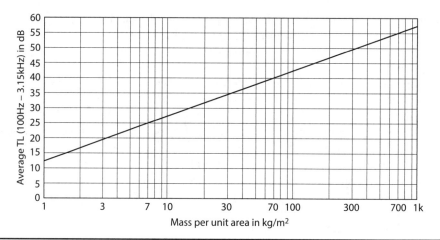

Figure 11-8 Acoustic mass law for single-layer partitions

To increase the sound transmission loss of a partition by 12dB at all frequencies, the mass per unit area must therefore be increased by a factor of four. A sound transmission loss increase of 18dB requires an eight-times increase in mass per unit area.

According to the formula, the greater a material's mass, the higher the transmission loss of any given frequency. In addition, the higher the frequency, the higher the transmission loss of any given mass of material. This explains why you can hear bass thumping outside a music venue but not the vocals.

Figure 11-8 illustrates the relationship between mass per unit area and transmission loss for single-layer partitions, but unfortunately, it doesn't translate directly to the real world.

As an AV designer, you won't typically be expected to perform acoustical calculations such as the total absorption of a space or the transmission loss of a partition. These are the purview of an acoustician, an acoustic consultant, or an acoustical engineer.

However, knowing the basics of how sound wave behavior can be predicted and manipulated can help you make recommendations that prevent adjacency issues and improve the performance of the AV system.

As can be seen from Figure 11-9, where transmission loss is plotted against frequency, the acoustic mass law has important shortcomings:

- It doesn't account for acoustic effects that occur at very low frequencies (below 50Hz) where transmission loss is more dependent on resonance effects and material stiffness than mass.

- It doesn't account for acoustic effects that decrease transmission loss at high frequencies (above 5kHz). Transmission loss also drops at the critical frequency—the frequency at which incident sound waves graze the surface because they are parallel to it.

- Finally, the mass law cannot account for any leakage that allows sound to be transmitted through the air around the material.

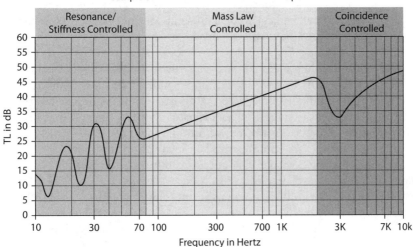

Figure 11-9 Limits to application of the acoustic mass law

Still, the mass law remains a good general guideline—denser materials provide better transmission loss, reflecting more sound than less dense materials.

Sound Transmission Class

Sound transmission class (STC) rates transmission loss at speech frequencies from 125Hz to 4kHz plotted against a standard contour as the reference. STC does not evaluate performance at frequencies below 125Hz, where music and mechanical equipment noise levels can be high.

Table 11-3 shows typical STC ranges associated with certain expected levels of privacy, assuming background noise levels of approximately 35dB SPL A-weighted. Note that these ranges will not provide isolation from low-frequency energy.

Also note that these ratings assume no significant flanking paths or partition openings. Slight breaches, such as gaps around conduit, cableways, and plumbing, or unsealed cracks at the ceiling or floor allow sounds to come through walls that are supposed to be barriers. They are undesirable and defeat the purpose of the barrier.

	STC Range	Speech Privacy
Table 11-3 Typical Speech Privacy Associated with STC Ratings	0–20	No privacy. Voices clearly heard between rooms.
	20–40	Some privacy. Voices will be heard.
	40–55	Adequate privacy. Only raised voices will be heard.
	55–65	Complete privacy. Only high-level noise will be heard. Even this level may not be sufficient for some secure facilities.

TIP Because STC doesn't account for the low frequencies produced by music and mechanical noise, it is useful only for measuring the transmission properties of a material with respect to speech.

Selecting proper building materials and construction techniques can help to keep the conference room quiet. A concrete block wall where each block is 200mm (8in.) thick will yield an STC 45 rating. If there were a 92dB SPL sound on one side of the wall, the sound could penetrate but be reduced to a level of about 47dB SPL (barring any other means for the sound to get over, around, or through the wall).

This same block wall can become an STC 56 by applying plaster sheeting to both sides. Additional construction materials applied can further increase the STC rating.

Figure 11-10 shows the STC ratings of various wall constructions. These are all based on using 16mm (5/8in.) industrial-grade gypsum/plaster board/drywall and assume no penetrations or flanking paths.

The STC 36 example utilizes wooden wall studs and absorptive material. Both the STC 39 and STC 44 examples use metal studs, but the STC 44 includes absorptive material. Metal studs are often more effective because their flexible nature provides certain isolation from vibration. They decouple one side from another and so reduce noise transmission through the structure.

The STC 46 example uses staggered studs that also can support further mounting of shelves and hardware. Typically, wood is chosen, as it is difficult to stagger metal studs because of the need to use a continuous runner at the top and bottom plates. Staggered studs represent a compromise between single-stud and double-stud construction.

The STC 57 example implements a double-stud technique with absorptive material on both sides. In both the STC 46 and STC 57 cases, the layers have an air gap of 25mm (1in.), providing more airspace and less possibility that vibrations will resonate through the wall via the studs.

Impact Insulation Class

Impact insulation class (IIC) is a rating used to quantify impact sound absorption. It is an average of the attenuation in decibels that occurs at frequencies ranging from 100Hz to 3.15kHz. The higher the IIC rating is, the better insulation from impact noise the material provides. An IIC rating of 50 is considered the minimum for flooring in residential buildings like multifloor apartment blocks.

Think about how the spaces above and below your AV environment will be used. How much impact noise will result from the activities in the space above yours? What about in the AV space itself? Ensure that the IIC of the floor above is sufficient to keep listeners

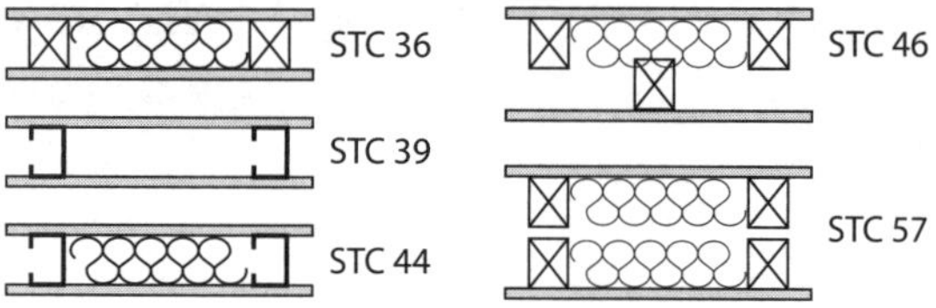

Figure 11-10 Construction for various sound transmission coefficient ratings

Table 11-4 Some Impact Insulation Class Ratings	Topping	IIC Rating
	None	28
	Vinyl flooring	35–40
	Hardwood flooring	30–35
	Hardwood flooring with resilient layer	45–50
	Carpet and underlay	75–85

from being distracted by impact noise. You should also make sure that the IIC of your own floor keeps the occupants of the space below from being bothered by activities in the AV space.

Table 11-4 shows IIC ratings for a 150mm (6in.) concrete slab with various toppings.

NOTE Like STC or NRC, IIC is a single-number metric that is an average of performance across a frequency range. It doesn't tell the whole story, especially since the impact of a high-heel shoe on a tile floor produces sound at a different frequency from a boot on a wooden floorboard.

Sound Reception

Many things can distract from the intended message in a meeting space. The lights may emit a buzzing sound. The water cooler in the corner may turn on and off. The ventilation system may rattle as it heats or cools the room. Noise from outside the space, such as traffic, other events, or nearby construction, can intrude into the space. Each of these items adds to the overall background noise in a space.

Because excessive noise levels interfere with the message being communicated, ideally background noise-level limits will be specified by an acoustician, audiovisual consultant, or designer appropriate to the type of room and its designed purpose. In other words, the criteria and limits for background noise levels for a gymnasium will be much different from those of a conference room. The HVAC system, partitions, and any necessary acoustical treatment will be designed and applied so that the background noise-level criteria is not exceeded.

A room's acoustical properties, such as reflections and their types, and the amount of transmission allowed, together with background noise levels, are significant contributors to a sound system's overall effectiveness. We'll discuss the effects of each and how they can be handled in design.

The Integration Process

Within certain time frames, our ear/brain system *integrates* direct and reflected sound energy to be perceived as a single sound event. In some time frames these reflections can be useful, adding apparent "fullness" to the direct sound. This integration of direct and reflected sound is similar to the human visual system integrating a series of still pictures in film and video presentations that result in apparent motion.

An important effect of integration is an increase in loudness, which occurs when the reflected sound is within what is often called the *integration interval*—about 30ms for speech and 50ms for music, depending on the temporal structure of the sound.

Small conference rooms often have reflections arriving 10ms or less after the direct sound. These short time frames often require room treatments to preserve tonal balance and improve speech intelligibility, especially in rooms with audioconferencing or videoconferencing capabilities. Large fan-shaped auditoriums often have late reflections arriving later than 50ms after the direct sound.

While possibly acceptable for acoustic music (but not for amplified presentations), reflections arriving later than 50ms are not acceptable in a speech application, because late reflections hinder speech intelligibility.

Some "early" reflections are actually useful in that if they arrive close enough in time behind the direct sound, your ear/brain system actually integrates both the direct and delayed reflection and perceives only one sound. The perception of integration will depend somewhat on what you are listening to. Your brain can integrate a greater delay with music than it can with speech.

Our brain uses the comparison between direct and reflected sound to determine the direction of the sound's origin, a psychoacoustic effect called the *precedence* (or Haas) *effect*. Our brains use the direct sound—the sound arriving first at the listener—for directional cues. This has important implications for sound system design in presentation environments.

In situations where auxiliary loudspeakers are required to fill in coverage under an auditorium balcony or to cover the more distant parts of a large listening space, the auxiliary loudspeakers are usually fed via a delay system to ensure that arrival times of the direct sound and the auxiliary sound are time-aligned. If you want listeners' attention to be drawn to a presentation area at the front of the room, the sound from loudspeakers at the front must reach them before the sound from loudspeakers in other parts of the room. Also, sound from the auxiliary loudspeakers can't be allowed to drown out sound from the former. Delay and sound pressure level of loudspeakers must be set accordingly.

If sound reflections arrive after the integration times indicated for music and speech, we will perceive these reflections as discrete echoes, and these later reflections can diminish intelligibility.

What about reflections arriving within 10ms of the direct sound? Knowing that sound travels about 343 meters per second in air, you can determine the wavelength of various frequencies by using the wavelength formula. For a frequency of 1kHz, the wavelength is 343mm (1.13ft). A complete cancellation will occur at one-half of that wavelength. So, if a reflection takes an additional path length of 172mm (6.8in.) over the direct sound, cancellation, or a *notch* in frequency response, will occur at 1kHz and subsequently every multiple of 1kHz, creating an overall comb filter response.

You can discover the frequency of the first notch by using the following formula:

$$f = \frac{1}{2t}$$

where:

- f is the frequency.
- t is the time in seconds.

So, calculating for a 5ms delay:

$f = 1 \div 2 \times 0.005 = 100\text{Hz}$

The first notch of the comb filter would be at 100Hz and create additional notches at every multiple of 100Hz. The first notch of a 10ms delay would be at 50Hz, with its subsequent multiples (100Hz, 150Hz, 200Hz, etc.). These notches take out large chucks of the audible spectrum, especially in the speech region. Once you get out to about 15ms or 33Hz, the subsequent multiples become much denser, and the notches created by the delayed wave energy become less noticeable.

In small rooms, the additional path lengths taken by reflections are quite short when compared to larger rooms. This explains why smaller rooms devoid of acoustic treatment and with lots of hard, highly reflective surfaces sound the way they do. The listening environment in a small room is not always pleasant.

Not all reflections are bad. Figures 11-11, 11-12, and 11-13 show how an acoustician may direct reflections in an auditorium in a useful way. Bear in mind that these illustrations show only energy reflected off the ceiling. Reflections from the floor and walls should also be taken into consideration. This is an example of where an acoustician can provide recommendations for room shapes and materials.

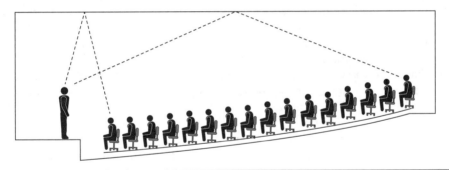

Figure 11-11 Section through an auditorium with a flat ceiling

Figure 11-12 Section through an auditorium with a segmented ceiling

Figure 11-13 Section through an auditorium with a stepped flat ceiling

Background Noise

Background noise can be a large factor in the effectiveness of an AV system. Excessive ambient noise levels result in less effective communication because of listener fatigue. Ambient noise is the background noise that originates from all sources other than the desired source. It is any sound other than the desired signal.

Background noise can come from outside of the room and could include sounds from adjacent rooms, foot traffic from spaces above, structure-borne vibrations, and traffic on the street. While an electronic sound system has inherent noise in the electronic components, rooms also have noise associated with them. Background noise can come from sources within the room as well, particularly from the HVAC system and from AV equipment with active heat-management systems, such as luminaires, atmospheric effects devices, projectors, displays, power amplifiers, dimmers, uninterruptible power supplies (UPSs), and IT equipment.

An acoustician, AV consultant, or AV designer will specify the maximum background noise limits to the architect, other consultants on the project, and the owner. These limits will inform the other parties in regard to certain decisions about construction techniques, devices, and implementation of various aspects of the building services.

Measuring Background Noise

There are two common methods for quantifying background noise levels:

- **Noise rating (NR)** The ISO *noise rating* method is commonly used by the building industry to specify or quantify the background noise levels from ventilation equipment or other sources in an occupied space. Noise rating is referred to as *noise criterion* (NC) in the United States. See Table 11-5 for the recommended NR ratings for some spaces.

- **Room criteria (RC)** This metric is helpful in quantifying the tonal characteristics of ventilation noise, such as rumble or hiss. The RC method takes all of the components (walls, ceilings, diffusers, etc.) in a room as a whole to determine the criteria levels for room design.

Maximum Noise Rating Level	Applications
NR 25	Concert halls, broadcasting and recording studios, churches
NR 30	Private dwellings, hospitals, theaters, cinemas, conference rooms
NR 35	Libraries, museums, courtrooms, schools, hospitals, operating theaters and wards, flats/apartments, hotels, executive offices
NR 40	Halls, corridors, cloakrooms, restaurants, night clubs, offices, shops
NR 45	Department stores, supermarkets, canteens, general offices
NR 50	Typing pools, offices with business machines
NR 60	Light engineering works
NR 70	Foundries, heavy engineering works

Table 11-5 Recommend Noise Rating Levels for a Range of Applications

NR and RC are single-number ratings that represent a complete spectrum of sound-pressure levels to meet a particular criterion. For example, to meet NR 35, the sound-pressure levels in the space cannot exceed the levels in any octave band on the NR 35 curve, as shown in Figure 11-14.

TIP Do not use A-weighted sound pressure levels for specification or noise analysis. dB SPL A-weighted measurements are only rough approximations and provide no information regarding spectrum shape and frequency content.

Figure 11-14
Noise rating (NR) curves

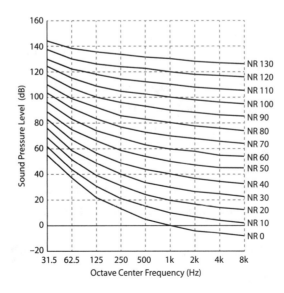

Chapter Review

You have just reviewed the impact that a facility's acoustical properties can have on audio system design. Sound is a huge factor when you're designing in a space. There are many important considerations that you as a designer need to incorporate into your plans. Now that you know what to look out for, you'll be able to design successfully your AV systems around the space you're working in.

Review Questions

The following questions are based on the content covered in this chapter and are intended to help reinforce the knowledge you have assimilated. These questions are similar to the questions presented on the CTS-D exam. See Appendix E for more information on how to access the free online sample questions.

1. What formula is used to predict the transmission loss of single-layer, impermeable materials?

    A. Ohm's law

    B. Mass law

    C. Frequency coefficient

    D. Inverse square law

2. What is the STC range for adequate privacy?

    A. 0–20

    B. 20–40

    C. 40–55

    D. 55–65

3. If sound transmission class measures a material's ability to halt sound traveling through the air, how is transmission of sound resulting from impacts, such as footfalls on the floor above, quantified?

    A. Noise reduction coefficient (NRC)

    B. Sound transmission class (STC)

    C. Impact insulation class (IIC)

    D. Noise criteria (NC)

4. What effect occurs when the reflected sound is within the integration interval—about 30ms for speech and 50ms or more for acoustic music?

    A. Scattering of the sound throughout the venue

    B. Echo

    C. Amplitude modulation

    D. An increase in loudness

5. Which device will reveal areas of high and low energy distribution?
   A. Peak level indicator
   B. Spectrum analyzer
   C. Oscilloscope
   D. Pink noise generator

Answers

1. **B.** Mass law allows designers to predict the transmission loss of single-layer, impermeable materials.
2. **C.** The STC range for adequate privacy is 40–55.
3. **C.** The transmission of sound resulting from impacts, such as footfalls, is quantified by impact insulation class (IIC).
4. **D.** When reflected sound is within what's known as the integration interval, you'll hear an increase in loudness.
5. **B.** A spectrum analyzer will show you areas of high and low energy distribution.

PART IV

Applied Design

- **Chapter 12** Digital Signals
- **Chapter 13** Digital Video Design
- **Chapter 14** Audio Design
- **Chapter 15** Control Requirements
- **Chapter 16** Networking for AV
- **Chapter 17** Streaming Design
- **Chapter 18** Security for Networked AV Applications
- **Chapter 19** Conducting Project Implementation Activities

CHAPTER 12

Digital Signals

In this chapter, you will learn
- How to calculate the required bandwidth in bits per second of an uncompressed digital audio or video signal
- How to differentiate between compression and encoding
- How to describe common digital video compression methods

In modern AV systems, audio and video are generally transported as digital signals. As an AV professional, you need to be familiar with the physical characteristics of digital signals, including how far you can transport them. You also need to be aware of digital media bandwidth requirements and the burden they can place on a network. Finally, you need to be conversant in common encoding and compression techniques so that you can specify AV systems with the right capabilities.

Before you begin this chapter, it is advisable to revise your CTS notes or CTS Exam Guide and confirm your understanding of the basics of digital signals, digital audio, and digital video, as the material in this chapter builds upon that knowledge.

Duty Check

This chapter relates directly to the following tasks on the CTS-D Exam Content Outline:

- Duty C, Task 1: Create Draft AV Design
- Duty C, Task 2: Confirm Site Conditions
- Duty D, Task 2: Conduct System Performance Verifications

In addition to skills related to these tasks, this chapter may also relate to other tasks.

Digital Signals

The concept of digital signaling is based on the idea that most objects, quantities, and parameters in the universe can be described or represented by a series of numbers. Those numbers may represent such properties as the frequency or amplitude of a sound, the amount of red light in a particular pixel of an image, the level of a fader on a control console, the angle of rotation of a shaft, the acceleration of a moving object, or whether or not a button is being pressed.

In our digital world those numbers are usually represented by a string of binary numbers where the presence of something such as a voltage, current, electromagnetic field, stream of photons, polarized magnet, etc., is used to represent the binary digit 1 and the absence or inverse of that property is used to represent the binary digit 0.

Being just a stream of numbers, digital signals are capable of carrying more than just one signal type at a time. Some digital connections carry video, audio, control, communications, and TCP/IP.

Digital Audio Bandwidth

An audio signal's bandwidth requirements are in direct relationship to the signal's sampling rate (measured in hertz) and bit depth (measured in bits). The Nyquist-Shannon sampling theorem tells us that an analog signal can be reconstructed if it is encoded using a sampling rate that is greater than twice the highest frequency being sampled. If, for example, we want to accurately sample audible sounds that occur in the frequency range of human hearing, from 20Hz to 20kHz, then the sampling rate should be higher than 40kHz. Many audio experts are of the view that very much higher sample rates are necessary to capture and reproduce the higher-order harmonics that add a sense of authenticity to the reproduced sounds, hence the high sample rates used for high-fidelity professional audio.

The formula for the required data throughput of an audio stream is:

$$Bit\ Rate\ (bps) = Sampling\ Rate \times Bit\ Depth \times Number\ of\ Channels + Overhead$$

Common audio sampling rates include

- **Telephone**   8kHz
- **Audio CD**   44.1kHz
- **DVD audio**   48kHz
- **Blu-ray Disc**   96 or 192kHz
- **Super Audio CD (SACD)**   2.82MHz

Bit depths as low as 8 bits per sample may be used in some low-quality voice communications applications, but more common bit depths include 16 bits for CD-quality audio replay and 24 and 32 bits per sample for more demanding applications. These increased bit depths are necessary for the higher dynamic range and reduced aliasing required for professional recording, mixing, mastering, post-production, and live production applications.

Digital Video Bandwidth

The bandwidth required for video signals is dictated by the

- Horizontal resolution of each frame
- Vertical resolution of each frame
- Bit depth of each pixel
- Number of channels of video signals
- Amount of overhead and/or metadata in each frame
- Number of frames per second

Table 12-1 lists the resolutions and frame rates for some common video formats.

Digital Video			
Format	Width (Pixels)	Height (Lines)	Frames/Sec
480i	720	486	29.97
576i	720	576	25
480p	720	480	29.97
575p	720	576	25
720p	1,280	720	25
1080i	1,920	1,080	29.97
1080p	1,920	1,080	25
DCI 2K	2,048	1,080	24
UHDTV1 (4K)	3,840	2,160	30
DCI 4K	4,096	2,160	24
UHDTV2 (8K)	7,680	4,320	60
Computer Graphics			
Format	Width (Pixels)	Height (Lines)	Frames/Sec
XGA	1,024	768	60
WXGA	1,280	768	60
SXGA	1,280	960	60
WSXGA	1,440	900	60
WUXGA	1,920	1,200	60
WQUXGA	3,840	2,400	120

Table 12-1 Some Video Format Characteristics

Bit Depth

Video sample bit depths vary between formats. Most standard dynamic range (SDR) formats use 8-bit samples for all three color channels—R, G, and B—yielding a total of 24 bits per pixel. Some computer color formats include an additional 8-bit transparency (alpha) channel for a total of 32 bits per pixel.

High dynamic range (HDR) video signals may be 10 bits per channel, requiring 30 bits per pixel; 12 bits per channel, requiring 36 bits per pixel; or 16 bits per channel, requiring 48 bits per pixel.

YUV Subsampling

YUV is a color video processing system that was originally developed to reduce the bandwidth requirements for broadcast television and color video recording. The YUV process converts RGB video signals into a luminance channel (Y), which contains the brightness information and most of the detail of an image, and two chrominance channels (U and V, otherwise known as Cb and Cr), which carry color difference information to generate the colors in the image. The YUV color space is slightly smaller than the RGB, so there are some color accuracy losses in conversion to this format.

Reducing the bit depth of the luminance channel of a YUV signal drastically reduces the perceived sharpness and detail in the resulting image. However, reducing the bit depth of the chrominance channels has a lower, and sometimes indiscernible, effect on how the human eye perceives the resulting image. Herein lies an opportunity to reduce the bandwidth of a video signal without drastically degrading the perceived picture quality.

In the process known as subsampling, each of the YUV channels is sampled at different rates to achieve the desired trade-off between video quality and bandwidth. The various subsampling compression schemes are identified by a triplet (X:X:X) indicating the number of pixels in a 4 × 2 block that are sampled. Because of the importance of luminance information, all compression schemes sample all pixels in the luminance channel. Table 12-2 lists the most common subsampling schemes.

Note that these subsampling methods may not be directly available for video devices such as computer graphics, which have RGB output.

Subsampling Scheme	Luminance Samples per Row	Top Row Chrominance Samples	Bottom Row Chrominance Samples	Pixel Chrominance Sample Pattern			
4:4:4	4	4	4	✓	✓	✓	✓
				✓	✓	✓	✓
4:2:2	4	2	2	✓		✓	
				✓		✓	
4:2:0	4	2	0	✓		✓	
4:1:1	4	1	1	✓			
				✓			

Table 12-2 Video Subsampling Schemes

4:4:4 Sampling
4:4:4 full bit-depth sampling samples all pixels in all three YUV channels in equal proportion (1 + 1 + 1 = 3 channels wide). The formula for three-channel, uncompressed digital video signal bandwidth is the same as for an RGB signal:

Bit Rate (bps) = Horizontal Pixels × Vertical Pixels × Bit Depth × Frame Rate × 3 Channels

4:2:2 Sampling
4:2:2 sampling includes all the luminance information, but only half the chrominance information (1 + 0.5 + 0.5 = 2 channels wide). As it maintains most of the vertical resolution in the color signals, the drop in quality is not particularly noticeable. 4:2:2 is used for broadcast television.

The formula for 4:2:2 digital video signal bandwidth is:

Bit Rate (bps) = Horizontal Pixels × Vertical Pixels × Bit Depth × Frame Rate × 2 Channels

4:2:0 Sampling
4:2:0 sampling includes all the luminance information, but only one quarter of the chrominance information (1 + 0.5 + 0 = 1.5 channels wide), resulting in blocky color patterns and only moderate color accuracy.

The formula for 4:2:0 digital video signal bandwidth is as follows:

Bit Rate (bps) = Horizontal Pixels × Vertical Pixels × Bit Depth × Frame Rate × 1.5 Channels

4:1:1 Sampling
4:1:1 sampling, like 4:2:0, includes all the luminance information, but only one quarter of the chrominance information (1 + 0.25 + 0.25 = 1.5 channels wide), resulting in blocky color patterns and only moderate color accuracy. Some digital video and some Moving Picture Experts Group (MPEG) video formats (see Chapter 17) sample at this rate.

The formula for 4:1:1 digital video signal bandwidth is as follows:

Bit Rate (bps) = Horizontal Pixels × Vertical Pixels × Bit Depth × Frame Rate × 1.5 Channels

Bandwidth: Determining Total Program Size
The total bandwidth required for a full video program stream includes the video stream; multiple audio streams, particularly for surround sound, audio description, and multilingual schemes; plus streams for metadata, control, time code, and subtitle data.

AV Stream Bit Rate (bps) = Video Bit Rate + Audio Bit Rate + Overhead Data Bit Rate

Once AV program bandwidth is calculated, you may find that you need to select lower sampling rates in some streams to fit within the available system bandwidth.

Some sophisticated video codecs can analyze the image content and select the optimum subsampling for detail versus bandwidth. Decoders may also interpolate the missing chrominance subsamples to produce video from a 4:2:2 subsampled stream that is virtually indistinguishable from a 4:4:4 stream with most program material.

Content Compression and Encoding

Compression is the process of reducing the data size of video and audio information before storing it or sending it over a network. The information is then decompressed for playback or display when it reaches its destination. Simple compression processes take advantage of repeated data patterns in a signal or file to reduce the number of times the pattern is stored or transmitted. The original signal can later be mathematically reconstructed to its original size and pattern without any loss of accuracy. This is known as *lossless compression* and can work well in compressing text, music, and video streams, although the amount of compression can be relatively low. ZIP, FLAC, MPEG-2, PNG, and some versions of JPEG 2000 and JPEG XS are examples of efficient lossless compression methods. Only lossless compression techniques are considered to be acceptable for critically accurate applications such as text transmission, telemetry, financial data, and some critical images.

The important steps in the compression process are analyzing and *encoding* the digital data to reduce its size and later *decoding* the compressed data into its original format. The devices and software processes that perform these functions are known as *codecs,* a witty combination of the terms *coder* and *decoder*.

Advanced compression processes are based on the human brain's inability to perceive all of the available detail in an image or a sound, and they use a variety of signal processing techniques to reduce the amount of unnoticeable detail transmitted or stored. This is known as *lossy compression* and can be almost undetectable at moderate compression levels, although the lack of fidelity becomes quite noticeable at high compression levels.

Common lossy methods of video compression include MPEG4, Motion JPEG (MJPEG), Motion JPEG 2000, JPEG XS, AOMedia Video 1 – AV1 (AOM AV1), Advanced Video Coding – AVC (H.264 or MPEG-4 Part 10), High Efficiency Video Coding – HEVC (H.265 or MPEG-H Part 2), Versatile Video Coding – VCC (H.266 or MPEG-I Part 3), and Essential Video Codec – EVC (MPEG-5 Part 1).

Common lossy methods of audio compression include G.722, MPEG-1 Audio Layer III – MP3, Advanced Audio Coding – AAC, Dolby AC-3, Windows Media Audio – WMA, and Ogg Vorbis. Even the highest-quality MP3 stream represents an approximately 75 percent reduction in bandwidth from CD audio's 1.4Mbps. The more commonly used 160kbps MP3 stream represents a 90 percent bandwidth reduction from uncompressed CD-quality audio.

For example, a single frame of uncompressed HD video at a resolution of 1920 × 1080 pixels, with 3 × 8-bit color samples per pixel, requires about 6.2MB of storage. At 30 frames per second, that's about 11GB per minute of storage, or a 1.5Gbps video stream. Using H.264 compression, Internet streaming services are delivering HD video over just 5Mbps of bandwidth, which represents a compression ratio of about 300:1. However, as this is lossy compression, it may not be as sharp or vivid as the original, but it is considered acceptable in less critical applications.

As the mathematical techniques for data compression are under constant development, you can expect compression ratios to continue to improve with little, if any, additional loss of fidelity, with H.266 being four times as efficient as H.264, at virtually identical picture quality.

To achieve an optimum level of data compression while also reducing the complexity of the decoding process, encoding often involves complex signal analysis, pattern detection, and intensive computation—a feat not always possible in real time. This can lead to substantial latency (signal delay) in a streaming system pipeline, or a pause between recording an uncompressed signal and replaying it in its compressed form. Specialized codec systems have been developed for audio and video streaming, videoconferencing, video recording, and post-production.

Digital Media Formats

A digital media file contains two elements:

- **Container**   The container (such as AVI, which stands for Audio Video Interleaved) is the structure of the file where the data is stored. The container holds the metadata describing how the data is arranged and which codecs are used.
- **Codec**   The codec provides the method for encoding (compressing) and decoding (decompressing) the file. Many video and audio codecs are in current use, with new ones constantly being created. In some cases, the codec must be installed in the device's operating system to decode the file.

Formats can be confusing because the term *codec* is frequently used interchangeably to describe the container and the codecs used within the container. In addition, some codec names describe both a codec and a container. An AVI container, for instance, could contain data encoded with an MPEG codec.

Codecs

Codecs come in a wide variety: one-way and two-way, hardware- and software-based, compressed and uncompressed, symmetrical and asymmetrical, specialized telephony and videoconferencing, and so on. Selecting the type of codec to use for a design requires considering many factors, including

- IT policies. The playback software that users currently have (or are allowed to have) determines the codecs their playback devices use, which in turn determines the codec you need to encode the streams.
- Licensing fees associated with the codec.
- The resolution and frame rate of the source material.
- The acceptable latency of the compression and decompression processes.
- The desired resolution and frame rate of the stream.
- The bandwidth required to achieve your desired streaming quality.

When it comes to matching the bandwidth you need with a codec's capabilities, you may not be able to find technical specifications to help you. Some testing may be in order.

Digital Audio Compression: MP3

MPEG-1 Audio Layer III – MP3 is one of the more common audio encoding formats. It was defined as part of a group of audio and video coding standards designed by the Moving Picture Experts Group (MPEG). MP3 uses a lossy compression algorithm that reduces or discards the auditory details that most people can't hear, drastically reducing overall file sizes.

When capturing and encoding MP3 audio, you can choose your bit depth and sampling rate. The MPEG-1 format includes bit rates ranging from 32 to 320 kbps, with available sampling rates of 32kHz, 44.1kHz, and 48kHz. MPEG-2 ranges from 8 to 160 kbps, with available sampling rates of 16kHz, 22.05kHz, and 24kHz.

MP3 encoding may be accomplished using constant bit rate (CBR) or variable bit rate (VBR) methods. You can predefine CBR encoding to match your bandwidth, especially where bandwidth is limited. VBR encoding uses a lower bit rate to encode portions of the audio that have less detail, such as silence, and uses a higher bit rate to encode more complex passages. CBR encoding results in more predictable file sizes, which helps in planning for bandwidth requirements. However, VBR can result in better perceived quality.

Digital Video Compression

Two types of compression can be applied to a video stream. *Intraframe compression* attempts to minimize the amount of information required to represent the contents of each individual frame of the video stream (see Figure 12-1) and is essentially the same process used to compress still images. Because it sends each individual frame, intraframe compression is preferable for editing video. Motion JPEG and DCI formats use intraframe compression. Not surprisingly, the resulting data files are quite large.

Interframe compression attempts to minimize the amount of information required to represent the contents of series of frames, usually by discarding the information that remains unchanged between successive frames in the stream. It detects how much information has changed between frames and sends a new frame only if there are significant differences (see Figure 12-2). It requires less bandwidth to stream and results in smaller files.

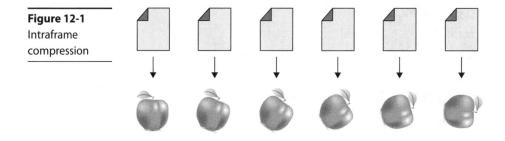

Figure 12-1 Intraframe compression

Figure 12-2 Interframe compression

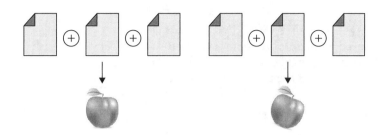

MPEG video algorithms use interframe compression, and so do most videoconferencing codecs. An interframe-compressed video stream contains three frame types.

- **Intraframes (I-Frames)**   The I-Frame is the reference point for the frames that follow. It's a complete independent frame of information, compressed using only intraframe methods.

- **Predictive Frames (P-Frames)**   A P-Frame consists only of the differences between the current frame and its reference I-Frame or the previous P-Frame—a process known as forward prediction. P-Frames can also serve as a prediction reference for B-Frames and future P-Frames. A P-Frame is much more compressed than an I-Frame.

- **Bi-directional Frames (B-Frames)**   Also known as Back Frames, B-Frames are derived using bi-directional prediction, which uses both a past and a future frame as references, only storing the differences between them. B-Frames show the highest level of compression.

The I-, B-, and P-Frames from a single shot form a block of information known as a group of pictures (GOP), as shown in Figure 12-3. A GOP can vary in length from three frames to several seconds.

> **NOTE**   An interframe codec sends a new frame every time it sees a shift in the moving moiré patterns produced by tightly patterned garments and image backgrounds, resulting in a far higher bandwidth data stream. Yet another good reason (aside from good taste) to avoid such patterns on garments and in room décor.

You will learn more about compression and its role in AV systems design in Chapter 17, where we discuss streaming applications.

Figure 12-3 A group of pictures

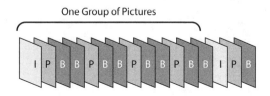

Chapter Review

You need to plan your design around the proper bandwidth requirements of the AV signals you will be distributing. This should be based on the client's needs and your understanding of the fundamentals of compression and encoding.

Review Questions

The following questions are based on the content covered in this chapter and are intended to help reinforce the knowledge you have assimilated. These questions are similar to the questions presented on the CTS-D exam. See Appendix E for more information on how to access the free online sample questions.

1. Your customer wants to stream CD-quality stereo music at 24 bits per sample in stereo to his lobby area. How much bandwidth will this application require?

    A. 754bps

    B. 1.5Kbps

    C. 2.1Mbps

    D. 1Gbps

2. You need to stream 10 channels of 96kHz audio. You have 25Mbps of bandwidth available. What is the highest bit depth you can use?

    A. 24 bit

    B. 26 bit

    C. 32 bit

    D. 48 bit

3. You currently have the bandwidth capacity to stream 30 channels of 48kHz, 24-bit audio. How many channels could you stream if you upgraded to 96kHz, 24-bit audio?

    A. 15 channels

    B. 25 channels

    C. 30 channels

    D. 60 channels

4. What is the required bandwidth for a 4:4:4, progressive digital video signal (1920 × 1080, 8 bits, at 30Hz)?

    A. 1.49Gbps

    B. 2.60Gbps

    C. 3.20Gbps

    D. 460.4Mbps

5. What is the required bandwidth for an RGB computer image (2560 × 1440, 8 bits, at 75Hz)?

   A. 8.85Gbps

   B. 6.64Gbps

   C. 4.42Gbps

   D. 3.32Gbps

6. What is the required bandwidth for 4:2:2 progressive digital video (1920 × 1080, 8 bits at 60 Hz)?

   A. 100Mbps

   B. 1.99Gbps

   C. 3.24Gbps

   D. 1Gbps

7. What is the required bandwidth for a 4:1:1 progressive digital video signal (1920 × 1080, 8 bits at 30 Hz)?

   A. 746.5Mbps

   B. 1.49Gbps

   C. 3.30Gbps

   D. 1.30Gbps

8. _____ compression is common for networked AV applications, such as streaming media and IP telephony.

   A. Lossless

   B. Lossy

   C. Intraframe

   D. Apple QuickTime

Answers

1. **C.** 44,100 × 24 × 2 = 2.1Mbps
2. **B.** 96,000 × X × 10 = 25Mbps; X = 26 bit
3. **A.** 48,000 × 24 × 30 = 34.6Mbps = 96,000 × 24 × X; X = 15 channels
4. **A.** 1920 × 1080 × 8 × 30 × 3 = 1.49Gbps
5. **B.** 2560 × 1440 × 8 × 75 × 3 = 6.64Gbps
6. **B.** 1920 × 1080 × 8 × 60 × 2 = 1.99Gbps
7. **A.** 1920 × 1080 × 8 × 30 × 1.5 = 746.5Mbps
8. **B.** Lossy compression is common for networked AV applications, such as streaming media and IP telephony.

CHAPTER 13

Digital Video Design

In this chapter, you will learn about
- Frame rates and resolution
- 720p, 1080i, 1080p, 4K, 8K, and other video formats
- Signal properties and uses of Serial Digital Interface, High-Definition Multimedia Interface, DisplayPort, Thunderbolt, HDBaseT, and Universal Serial Bus
- The purpose and method of DisplayID/Extended Display Information Data communication and designing video systems that account for Extended Display Information Data issues
- Managing High-bandwidth Digital Content Protection keys in a digital video system

AV designers need to be familiar with the common digital video signal types and their applications. In many ways, these digital video signals are closer to "plug and play" than the analog signals of the past. Still, challenges exist.

When designing for interdevice communication in digital video systems, you need to consider the following:

- The properties and capabilities of your video source
- The properties and capabilities of your destination devices
- How you'll get the signal from source to destinations
- How you'll ensure electrical compatibility—cabling, adapter, or interface
- An Extended Display Information Data (EDID) strategy for simple and complex designs
- Digital rights management, including incorporating High-bandwidth Digital Content Protection (HDCP) keys for securing content licenses as needed

This chapter assumes knowledge of needs analysis, video source identification and selection, and display design, which were covered in Part II.

> **Duty Check**
>
> This chapter relates directly to the following tasks on the CTS-D Exam Content Outline:
>
> - Duty A, Task 5: Identify Client Expectations
> - Duty C, Task 1: Create Draft AV Design
> - Duty C, Task 2: Confirm Site Conditions
> - Duty D, Task 2: Conduct System Performance Verifications
>
> In addition to skills related to these tasks, this chapter may also relate to other tasks.

Digital Video Basics

Digital video devices have processors that produce different resolutions and frame rates. It is up to the AV designer to select an output resolution that looks good on the video system's displays. If an image is larger or smaller than a display's native resolution, for example, it won't look as good. If the image is a different height or width than the display, a scaler for the display may stretch or create borders around the image.

Video processors and displays have a two-way relationship. Most systems allow users to change the image resolution within the abilities of the video processor and display to optimize the image and meet the client's needs. Whenever possible, it's best to design around the native resolution of the display.

The *native resolution* is the number of rows of horizontal and vertical pixels that create the picture. For example, if a display has a resolution of 2560 pixels horizontally and 1600 pixels vertically, it has a native resolution of 2560×1600. The native resolution describes the actual resolution of the display device—not the resolution of the signal delivering it. A higher-resolution display means more pixels, more detail, better image quality, and (most importantly) more digital information required to create the desired image.

Frame rate is the number of frames per second (fps) sent out from a video source. For instance, a 1080p high-definition video signal may have a frame rate of 60fps (often stated as 60Hz). The more different frames per second you display, the smoother the movement in the video will appear.

Frame rate is not the same thing as refresh rate. *Refresh rate* is the number of times per second a display will redraw the image sent to it, typically measured in hertz. Displays use buffering and scaling circuitry to match the frame rate and resolution of a source. A display's refresh rate should be equal to or greater than the frame rate of the signals sent to it.

High-Definition and Ultra High-Definition Video

The majority of video systems you will be required to design will be for one of the video formats known as high definition (HD). The most common high-definition video signal and display formats available are

- **720p** This is an HDTV signal format. The number 720 represents 720 horizontal lines, and the *p* indicates that it uses progressive scanning. The aspect ratio is 16:9 (1.77:1), and the resolution is 1280×720. It typically delivers approximately 50 frames per second (720/50p) in most regions and approximately 60 frames per second (720p60 or 720/60p) in Japan and North America. A 30 frame per second rate (720/30p) is found in some videoconferencing systems.

- **1080i** This has a resolution of 1920×1080 with an aspect ratio of 16:9 (1.77:1) and uses interlaced scanning at 25 or 30 frames per second. A single frame of 1080i video has two sequential fields of 540 lines of 1920 horizontal pixels. The first field contains all the odd-numbered lines of pixels, and the second field contains all the even-numbered lines. In areas such as Europe, Oceania, much of Asia, Africa, and part of Latin America, 1080i television signals are broadcast at 50 fields per second (1080/50i). In North America, Japan, and most of Latin America, this format is broadcast at 60 fields per second (1080i60 or 1080/60i).

- **1080p** This has a resolution of 1920×1080 with an aspect ratio of 16:9 (1.77:1) and uses progressive scanning at 60 frames per second. Typically, everything you install, from displays to switching equipment, should be at least 1080p compatible to simplify image setup and maximize display quality. Many sources, such as Blu-ray Disc players, media players, and computer systems, output at 1080p. It is sometimes referred to as *full HD*.

- **Ultra HD 4K** This has a minimum resolution of 3840×2160 pixels in a 16:9 (1.77:1) aspect ratio. Originally used in digital cinema, DCI 4K video resolution is 4096×2160 in a 1.9:1 full frame format, 3996×2160 for 1.85:1 conventional widescreen, and 4096×1716 for 2.39:1 CinemaScope screenings. The related video resolution consumer television format is known as 2160p, *4K Ultra HD*, and *UHDTV-1*, which specifies 3840×2160 pixels at a 16:9 aspect ratio.

- **Ultra HD 8K** This has a minimum resolution of 7680×4320 pixels in a 16:9 aspect ratio, exactly double the width and double the height of 4K. The related video resolution consumer television format is known as 4320p, *8K Ultra HD*, and *UHDTV-2*.

- **Ultra HD 16K** This has a minimum resolution of 15360×8640 pixels (132.7 megapixels) in a 16:9 aspect ratio, double the width and height of 8K.

Video formats usually include rounded frame numbers such as 60p. But in many cases, the precise frame rate is 59.94 (or 29.97) per second. In practice, this usually makes no difference because most common displays and other equipment tolerantly accept all typical rates. However, occasionally the transfer of video is interrupted because of small frame rate or related differences between equipment.

In some large-scale projects, the images you will be displaying may have resolutions that far exceed any of the standard HD formats and in aspect ratios that bear no resemblance to conventional television or cinematic displays. Modular LED displays may have arbitrarily large pixel arrays, and mapped-projection displays frequently require complex multiprojector arrays with complex image processing, masking, overlap, and blending, to wrap around objects. To make such projects possible, the video display systems for this class of projects will eventually need to be broken down into a series of standard high definition elements for composition, storage, delivery, and display.

It's important to understand that clients who want "4K" resolution video may not understand that "4K UHDTV" and "4K" are not necessarily the same thing. The term *4K* is generally used to refer to video signals with a horizontal resolution of about 4000 pixels. But to be accurate, 4K is a resolution of 4096 pixels horizontally by 2160 pixels vertically, or 8.8 megapixels. This is a cinematic standard for 4K film projection (Digital Cinema Initiatives 4K, or DCI 4K). What many clients may not realize is that DCI 4K carries a 17:9 aspect ratio, whereas most displays today are built for a 16:9 aspect ratio. Just to add to the confusion, as mentioned in the earlier description of 4K formats, there are also DCI specifications for widescreen (1.85:1) and CinemaScope (2.39:1) formats, which should also form part of your discussions with clients.

The International Telecommunication Union (ITU) has defined a UHD resolution that would fit a 4K-like image into the currently standard display ratio. Such an image has 7 percent fewer pixels in the horizontal aspect. On the left and right edges, 3.5 percent of the pixels are eliminated without compression or stretching so the image fits a 16:9 ratio.

When it comes to designing 4K, 8K, or 16K systems, the needs assessment is critical. Designers should ask:

- What is the goal for using 4K/8K/16K? What does the end user need to be able to see?
- What infrastructure is required to support this amount of data?
- Which type of display or projector does the end user want to use?
- Which type of receivers and processors are suitable for this system?

Consider, for example, a video device that accepts 4K. The specifications might state that it can support a UHD signal at 60fps with 4:2:0 chroma subsampling and 30fps with 4:4:4 chroma subsampling. Depending on the end user's needs, those specifications could factor heavily into the design.

In general, if a client informs you they want real 4K video, think to yourself "4096×2160 or 3840×2160 at 60fps with 4:4:4 chroma sampling and 10 bits per color (a 30-bit color depth)." That will require about 16Gbps of throughput. Then work backward to arrive at the version of 4K that fits their infrastructure or their budget. It may have to run at 30fps or use 4:2:0 subsampling or 8-bit color (a 24-bit color depth). You may need to demonstrate various implementations of 4K to ensure the experience matches what they desire from a 4K application.

If a client believes that they require 8K or 16K display resolution for their project, you should caution them from the outset that the technology required to deliver this resolution is substantially more complex and expensive than 4K.

The Cliff Effect

When designing a video system, it's critical that AV professionals consider the length of the cables that will carry the digital signals. Never forget that digital signals are simply analog pulses with a rapid rise time. In cables, those electronic pulses are subject to the same effects of cable resistance, capacitance, and EMI as every other analog signal. Similarly, in an optical fiber, the analog pulses of light are subject to optical losses, pulse stretching, and scattering along their journey. If a digital signal is carried too far on a cable or fiber, at some distance, the rise time of the pulses in the waveform will eventually become too long for the receiving circuitry to detect, and the signal will suddenly become indecipherable. This is identified as the *cliff effect*, because once the signal reaches that critical point, rather than slowly diminishing in quality, it simply becomes completely useless, as can be seen in the shape of the curve in Figure 13-1.

The distance to the edge of the cliff will vary based on the type and quality of the cable or fiber and the data rate of the signal. Signals with a higher data rate will generally reach the cliff sooner than signals at lower data rates. For example, a 1080p signal might travel only 20m (66ft) before falling off the cliff, while a 720p or 1080i signal might travel as far as 40m (132ft).

Before selecting a particular cable, check with the manufacturer to make sure it will carry the signal as far as you need it to travel. Allowable cable types and lengths are usually specified in a device's documentation.

When deciding on the cable you need for a feed, it is necessary to consider all the data being sent. Some video, which might seemingly use less bandwidth than a 1080p signal, may need a higher-bandwidth cable if the video entails extended data, such as high dynamic range (HDR), wide color gamut (WCG), or is sent at a high frame rate (HFR). To be safe, you may want to specify higher-quality cables than you might immediately need for a simple digital replay system. The cable you install to transport 1080p today may not work for HDR, WCG, HFR, 4K, 8K, or 16K next year.

Figure 13-1 The graph shows how, at a certain distance, a digital signal becomes indecipherable.

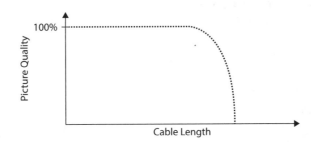

TIP To give your client the best longevity on the system you design for them, it is good practice to advocate for the highest-bandwidth cables and fibers they are prepared to install. Even with new advances in compression technologies, AV bandwidth requirements are unlikely to be decreasing in the foreseeable future.

You can test the stability of a signal by examining the bit error rate (BER), which is the total number of altered bits in a signal after traveling down a cable or fiber, caused by impedance, noise, interference, distortion, or synchronization errors. It's calculated by performing a BER test, which uses predetermined stress patterns consisting of test bit sequences generated by a test-pattern generator. The testers are often identified as BERTs (bit error rate testers).

Video Signal Types

Now that we have covered frame rates and resolutions, we will consider video signal types. When considering how video signals will be carried from source to sink (monitor, LED module, projector, or other type of display device), the connection format will be one of your first considerations. This includes both the signal format and the physical means of transporting that signal.

Uncompressed digital video signals exist in several forms; these forms can contain information in any one of a variety of formats and can be delivered via any one of several transport carriers.

AV over IP can be either a full-bandwidth or compressed transport stream, depending on the codec (encoder/decoder) employed. As TCP/IP is a serial data stream, all video must at the very least be serialized from the original RGB and sync channels before transmission. Many hardware and software systems are available to convert an uncompressed video format such as HDMI for transmission over IP, to then be decoded into the original uncompressed format for display or recording. The resolution and frame rate of the original video, together with the available bandwidth of the IP stream, will dictate how much (if any) compression is required by the codecs at the end points of the IP network. Some systems such as software-defined video over Ethernet (SDVoE) allow on-demand reservation of Ethernet network bandwidth and configuration of the transport codecs.

Common formats for digital display connections include

- Serial Digital Interface (SDI)
- Digital Visual Interface (DVI)
- High-Definition Multimedia Interface (HDMI)
- HDBaseT
- DisplayPort
- Thunderbolt
- Universal Serial Bus (USB)
- Video over IP networks

We will examine these formats in more detail in the sections that follow.

 NOTE DVI, HDMI, and DisplayPort are all licensed technologies and belong to different organizations. HDMI is most often used for home entertainment equipment, while DisplayPort is commonly used for computer displays. DVI was frequently used in computers but has been largely replaced by HDMI. You could potentially see any of the three on just about any piece of digital display equipment.

Serial Digital Interface

SDI is a family of video standards defined by the Society of Motion Picture and Television Engineers (SMPTE). It is an uncompressed, unencrypted digital video signal, characterized by serial, one-way communication over coaxial cable or optical fiber.

It is designed for local transport up to 100 meters (328 feet) of high-quality, uncompressed, unencrypted, and standardized digital video over a coaxial cable (75Ω) with a BNC connector. It is used in live production, broadcast, image magnification (IMAG), staging, and videoconferencing applications, although its lack of support for copy-protected content can be a significant problem. Many SDI video processors include the capability for audio channel swapping and signal control. SDI has low latency (delay) and is excellent where lip sync is an issue. SDI is often considered better suited than HDMI for connection directly to recording devices because it has a locking connector and will also transport time code. It is often used for infrastructure wiring in professional production and presentation environments.

The SMPTE has established standards for several SDI formats, as shown in Table 13-1.

Transition-Minimized Differential Signaling

Before delving further into digital video formats, let us explore the technology that underlies both HDMI and all digital variants of DVI. Transition-minimized differential signaling (TMDS) is a technology for transmitting high-speed serial video data at high, native, true-color resolutions. It was originally developed as a standardized digital video interface between a PC and a monitor. It employs differential signaling to help reduce EMI for faster, lossless video transmission. Unlike SDI, which is based around coaxial cables,

Format Name	Max Data Speed	SMPTE Standard	Application
SDI	270Mbps	SMPTE 259M	SD (480i, 576i)
HD-SDI	1.5Gbps	SMPTE 292M	HD (720p, 1080i)
3G-SDI	3Gbps	SMPTE 424M	HD (1080p60)
6G-SDI	6Gbps	SMPTE ST 2081	UHD (1080p120, 2160p30)
12G-SDI	12Gbps	SMPTE ST 2082	UHD (2160p60)
24G-SDI	24Gbps	SMPTE ST 2083	UHD (2160p120, 4320p120, 7680p30)

Table 13-1 Single-Link SDI Formats

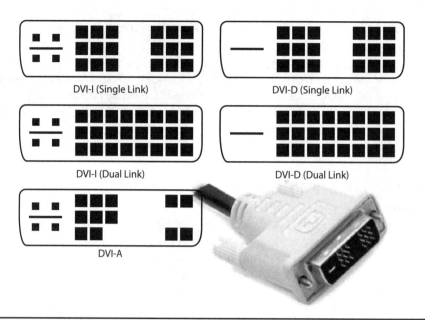

Figure 13-2 Many different connectors are available for DVI signals.

TMDS-based formats employ balanced twisted-pair cables for noise reduction during transmission. The signal is transmitted in a balanced differential mode to take advantage of common-mode noise rejection.

DVI

Once widely used for carrying data between computers and display devices, DVI was available in a bewildering variety of analog and digital variants. As shown in Figure 13-2, there are many different connectors available for DVI signals, including DVI, DVI-A, DVI-D, DVI-I, and mini DVI. The two multipin connectors for DVI signal transport are DVI-D for only digital information (no analog video information can be sent) and DVI-I for digital or analog information. Cable lengths for DVI connections are limited to about 5 meters (16 feet) for 1920×1200 and up to 15 meters (50 feet) for 1280×1024 video. Digital DVI does not support digital audio or control.

DVI has largely been replaced by formats such as HDMI, DisplayPort, and USB. DVI digital signals are directly compatible with HDMI, allowing simple, passive, DVI-to-HDMI adapters to be used to connect devices with DVI outputs to display on systems requiring HDMI inputs.

HDMI

HDMI is a point-to-point connection between video devices and has become a standard for high-quality, all-digital video and audio. HDMI signals include audio, control, and digital asset rights management information. It is a "plug-and-play" standard with video signals that are fully compatible with DVI.

Figure 13-3
HDMI transmission utilizing TMDS

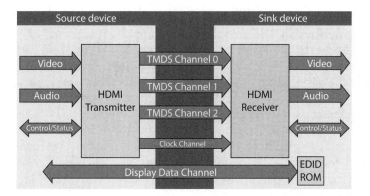

HDMI cables and connectors carry four differential pairs that make up the TMDS data and clock channels. These channels are used to carry video, audio, and auxiliary data. Figure 13-3 depicts the output of an HDMI source communicating with the input of an HDMI sink. The TMDS channels in the middle (0, 1, 2) carry the RGB video. The clock channel, just below the TMDS channels, carries the sync. The fifth connection on the bottom of the diagram carries Display Data Channel (DDC) information, which allows the source and sink to share resolutions and other parameters, including device identity information. There are also dedicated connections for 5V power and "hot plug detect" for monitoring a new connection or a device being powered up.

HDMI meets EIA/CEA-861-B standards for uncompressed, high-speed digital interfaces and is a widely adopted digital video format because it includes two communication channels in addition to TMDS. The DDC is used for configuration and status exchange between a single source and a single sink. It is the electrical channel that EDID and HDCP use for communication between a source device and a display.

In addition, HDMI sinks are expected to detect what are known as InfoFrames to process audio and video data appropriately. InfoFrames are structured packets of data that carry information regarding aspects of audio and video transmission, as defined by the EIA/CEA-861B standard. Using this structure, a frame-by-frame summary is sent to the display, permitting the automatic selection of appropriate display modes. InfoFrames typically include auxiliary video information, generic vendor-specific source product description, MPEG, and audio information.

An optional HDMI channel carries Consumer Electronics Control (CEC) data, which provides high-level control functions between all the various AV products in a user's environment. CEC is a single-wire, bi-directional serial bus that uses AV link protocols to perform remote-control functions for system-level automation—when all devices in an AV system support it.

Since HDMI 1.4, a previously reserved pair has become the HDMI ethernet and audio return channel (HEAC), which is a differential, bi-directional link that can carry an audio return channel (ARC) and 100Base-TX Ethernet at up to 100Mbps.

At the time of publication, HDMI 2.1a was the most recent version of this protocol available. Its 48Gbps bandwidth supports resolutions of 8K/50/60/100/120Hz and 4K/50/60/100/120Hz, plus some 5K(5120×2160)/50/60/100/120Hz and 10K(10240×4320)/50/60/100/120Hz resolutions. Dynamic HDR (10/12/16 bits per channel) is available on a frame-by-frame basis at most resolutions, but to conserve bandwidth, is the only form of HDR available for 4K/120 and 8K/60+.

Equipment manufacturers can choose to support either VESA DSC 1.2a compression or uncompressed video in most of the 4K and 8K modes. Color subsampling is supported using 4:2:0 at 16 bits per channel, 4:2:2 at 12 bits per channel, and 4:4:4 at 8 bits per channel.

HDBaseT

A variety of devices, including high-resolution displays, projectors, video source devices, switchers, and matrix switchers, can be connected to each other over distances of up to 100 meters (328 feet) via a single UTP cable (Cat 5e+) using HDBaseT devices and technology. HDBaseT is a proprietary protocol that has been licensed to a multitude of AV equipment manufacturers who have installed HDBaseT interfaces into their products. Adapter devices are readily available to connect to devices without built-in HDBaseT interfaces. HDBaseT carries HDMI-standard uncompressed 2K/4K video, as well as audio, USB, and control signals. It is also capable of carrying up to 100 watts of DC power, making it unnecessary to plug some HDBaseT displays and other end nodes into the wall. HDBaseT can also carry 100Mbps Ethernet, which allows HDBaseT devices to access content on connected computers. All devices in the HDBaseT signal chain must be capable of transmitting and/or receiving HDBaseT. HDBaseT has been designated as the IEEE 1911 standard.

DisplayPort

This Video Electronics Standards Association (VESA)–developed connection method utilizes a 20-pin, high-resolution, high-speed digital display, audio, and data interface with up to 77.4Gbps effective bandwidth in version 2.0. It is typically used to connect a computer to one, two, or three displays and is backward compatible with DVI, HDMI, and other display interfaces through the use of appropriate active or passive adapters. The Mini DisplayPort is a smaller variant of this interface, often found on thin form-factor portable devices, and uses the same connector as Thunderbolt 1 and 2.

DisplayPort has native support for fiber-optic cables and support for standard cables up to 15 meters (50 feet) in length at lower resolutions. It also supports HDCP 2.x signals and bi-directional communication. The interface uses low-voltage differential signaling (LVDS) and 3.3V for control signaling. It supports a dual-mode design that can send single-link HDMI and DVI signals through an adapter, which also converts the voltage from 3.3V to 5V. It also works with USB3.2 gen2x1 and later, with Type-C connectors supporting "DP Alt Mode" and USB4 via "DisplayPort Tunneling." Table 13-2 lists some of the higher-resolution transmission modes supported by DisplayPort 2.0,

Number of Displays	Image Resolution	Frame Rate	Channel Bit Depth	Bits per Pixel	Color Subsampling	Compression
Direct DisplayPort and via USB4 DisplayPort Tunneling						
1	1560×8640 (16K)	60Hz	10 (HDR 10)	30	4:4:4	VESA DSC
1	10240×4320 (10K)	80Hz	8 (SDR)	24	4:4:4	None
2	7680×4320 (8K)	120Hz	10 (HDR 10)	30	4:4:4	VESA DSC
2	3480×2160 (4K)	144Hz	8 (SDR)	24	4:4:4	None
3	10240×4320 (10K)	60Hz	10 (HDR 10)	30	4:4:4	VESA DSC
3	3480×2160 (4K)	90Hz	10 (HDR 10)	30	4:4:4	None
USB3.2 gen2x1+ with Type-C Connector Using DisplayPort Alt Mode						
1	7680×4320 (8K)	30Hz	10 (HDR 10)	30	4:4:4	None
3	3480×2160 (4K)	144Hz	10 (HDR 10)	30	4:4:4	VESA DSC
3	2560×1440 (QHD)	120Hz	8 (SDR)	24	4:4:4	None

Table 13-2 Some Higher-Resolution Modes Supported in DisplayPort 2.0

although a multitude of lower-resolution, lower-bit-depth and more compressed modes are also supported. DisplayPort 2.0 supports multichannel digital audio.

USB4, USB3.2, and USB Type-C Connectors

USB4 extends the capabilities of the USB ecosystem by using a substantially different communications architecture from previous USB generations, but maintains backward USB compatibility by tunneling USB3, together with VESA's DisplayPort and Intel's PCIe (hosting Thunderbolt), over its 40Gbps data fabric. USB2 signals are carried on separate dedicated channels.

The new architecture consists of three types of USB4 routers linked by USB Type-C cables:

- **USB4 Host** routers, consisting of
  - Inputs: DisplayPort, PCIe, USB 2.0, and USB3 adapters
  - Outputs: USB4
- **USB4 Hub** routers, consisting of
  - Inputs: DisplayPort, USB2, USB3, and PCIe adapters, USB4
  - USB3 and USB2 hubs and a PCIe switch
  - Outputs: DisplayPort, USB2, USB3, and PCIe adapters, USB4
- **USB4 Device** routers, consisting of
  - Input: a single USB4
  - Outputs: USB2, USB3, DisplayPort, and PCIe adapters

The minimum USB4 network (known as a USB4 domain) consists of

- A single Host router, with Connection Manager software running on the same platform
- Zero or more Hub routers
- Zero or more Device routers

A total of six routers can be present in a USB4 domain. Data can be routed to any adapter, on any router in the USB4 domain, and any router can contain up to 64 adapters. USB4 is only supported on 24-pin USB Type-C connectors and cables.

In summary, USB4 can transport DisplayPort and/or Thunderbolt video (via PCIe), USB4 data at up to 40Gbps, and USB2 and USB3 data. You can run your USB2 webcam, charge your tablet, hot-plug your USB memory key, and route a 4K video stream to multiple monitors.

The vertically reversible, male-to-male, Type-C cable and connectors have 24 pins and 24 wires. Four pairs are assigned for SuperSpeed differential data channels and two pairs for USB2 differential data. There are also two pairs allocated for the power bus, two pairs for ground, and one pair each for sideband use (out-of-channel signaling–SBU) and the configuration channel (CC).

Using the accompanying USB Power Delivery Specification revision 3.1, USB4 can negotiate using the adjustable voltage supply (AVS) to deliver 5A at up to 20V for a maximum 100W supply. Additionally, with extended power range (EPR)–qualified devices, a USB4 router can use the AVS to negotiate voltages up to 48V, which at 5A delivers up to 240W of power.

USB3

USB3.2 encompasses the USB versions previously known as USB 3.0 (now USB3.2 gen1), USB 3.1 (now USB3.2 gen2x1), and USB 3.2 (now USB3.2 gen2x2). Through Type-A and Type-B cables and connectors, it is a *dual-bus architecture* that is backward compatible with USB 2.0. These cables have nine primary conductors consisting of three twisted-signal pairs for USB data paths and power. There are two twisted-signal pairs (transmit and receive) for the 5Gbps data path in USB3.2 gen1 and 10Gbps in USB3.2 gen2x1 and gen2x2. There is also a pair for USB 2.0 backward compatibility and a pair for power. The ninth wire is a signal ground.

With the more recent USB Type-C cables and connectors, USB3.2 gen2x2 supports 20Gbps speeds and the "DP Alt Mode," which allows the routing of DisplayPort over the SuperSpeed twisted-pairs.

Table 13-3 shows several generations of USB for comparison.

Thunderbolt

Thunderbolt is a technology developed by Intel and Apple that can transfer high-speed data, video, audio, and DC power bi-directionally over one cable. It also permits a user to connect up to six displays, hard drives, and other compatible devices via a daisy-chain

Version	Speed Designation	Data Rate	Power
USB1.1	Low speed/full speed	1.5Mbps/12Mbps	2.5W (5V at 500mA)
USB2.0	High speed	480Mbps	2.5W (5V at 500mA)
USB3.2 gen1 (previously USB 3.0)	SuperSpeed USB 5Gbps	5Gbps	4.5W (5V at 900mA)
USB3.2 gen2x1 (previously USB 3.1)	SuperSpeed USB 10Gbps	10Gbps	4.5W (5V at 900mA) via Type-A and Type-B connectors. 7.5W (5V at 1.5A) or 15W (5V at 3A) via Type-C connector
USB3.2 gen2x2	SuperSpeed USB 20Gbps	20Gbps	4.5W (5V at 900mA) via Type-A and Type-B connectors. 7.5W (5V at 1.5A) or 15W (5V at 3A) via Type-C connector
USB4	USB4 40Gbps	40Gbps	100W (20V at 5A) via AVS or 240W (48V at 5A) via AVS + EPR

Table 13-3 Versions of USB

connection or hub. Most often found on Apple computer equipment, Thunderbolt combines DisplayPort and PCI Express (high-speed serial computer expansion bus, or PCIe) capabilities into one serial data signal. Thunderbolt versions 1 and 2 use a connector that is physically and electrically identical to Mini DisplayPort. Thunderbolt 3 and 4, which use a USB Type-C connector, have a maximum bandwidth of 40Gbps at up to 2 meters. The Thunderbolt 3 and 4 multistream transport (MST) capability allows USB3.2 or DisplayPort 1.4 signals to be embedded (tunneled) in its data stream.

Thunderbolt 4 supports four lanes of PCI Express Gen 3 serial data, DisplayPort 2.0, and resolutions up to 16K at 60Hz, 10 bits per channel, for a single native display, or two DP 1.4 displays of 4K at 60Hz. It is compatible with USB devices and cables; DVI, HDMI, and VGA displays via adapters; and previous generations of Thunderbolt.

NOTE Thunderbolt-to-Ethernet connections are possible via an adapter.

As you can see in Figure 13-4, a Thunderbolt 1 or 2 port is the same as a Mini DisplayPort socket. Mini DisplayPort–to–HDMI adapters will work with Thunderbolt-capable computers, but clients will not be able to utilize all that Thunderbolt offers without proper Thunderbolt cabling.

Figure 13-4
A Thunderbolt 1 or 2 socket—identical to a Mini DisplayPort socket (Getty Images)

Video over IP Networks

Video carried over TCP/IP-based Network-layer (layer 3) AV protocols can be routed across multiple networks, including the Internet, if sufficient network bandwidth is available. Some proprietary IP-based protocols are widely licensed to many manufacturers and have significant ecosystems that provide a huge range of solutions for control and system integration, while some other protocols are tightly limited to a particular manufacturer's range of products. As a system designer, you have a wide range of AV over IP network architectures to choose from. A selection of these protocols is examined in Chapter 16.

Introduction to EDID

As an AV professional, the AV systems you are designing are probably expensive, sleek, and modern. But what do you do when a user wants to connect an outdated digital device to your beautiful system? Managing Extended Display Identification Data (EDID) will help.

 VIDEO Watch an AVIXA video about EDID. Check Appendix D for the link to the AVIXA video library.

EDID was originally developed for use between analog displays and computer video devices, but has since made its way into DVI, HDMI, and DisplayPort. EDID has since been extended in capabilities to become Enhanced EDID (E-EDID), then further enhanced to become DisplayID.

Displays and video sources need to negotiate their highest common resolutions so they can display the image at the best available quality. EDID is a method for source and display (*sink*) devices to communicate this information, eliminating the need to configure the system manually. An EDID data exchange is the process where a sink device describes its capabilities such as native resolution, color space information, and audio type (mono or stereo) to a source device.

Each generation and version of EDID consists of a standard data structure defined by the VESA. Without the acknowledged handshake between display and source devices, the video system could provide unreliable or suboptimal video.

An EDID handshake can be between a source and the next device in line, whether it's a display, videowall, switcher, distribution amplifier, matrix, or scaler. However, some devices simply pass the EDID data through without responding. Without a successful handshake between source and display devices, the entire video solution could be unreliable. Therefore, creating an EDID strategy can be vital to your success as an AV designer.

EDID Packets

During the handshake process, the sink (display) device sends packets of EDID data to the source. The data packets carry the following information:

- Product information
- EDID version number
- Display parameters
- Color characteristics
- Audio characteristics
- Timing information for audio and video sync
- Extension flags

How EDID Works

Figure 13-5 shows how the EDID negotiation works, and the steps are detailed next. The process begins with the activation of a *hot plug* link. The hot plug signal is an always-on, 5V line from the video source device that triggers the EDID negotiation process when a sink device is connected.

The EDID sequence works as follows:

1. On startup, an EDID-enabled device will use *hot plug detection* (HPD) to see whether the device is on.
2. The sink device returns a signal alerting the source that it received the HPD signal.
3. The source sends a request on the DDC to the sink for its EDID information.
4. EDID is transmitted from sink to source over the DDC.
5. The source sends video in its nearest possible approximation to the sink's preferred resolution, refresh rate, and color space. The source's selection can be manually overridden in some cases.
6. If the sink's EDID contains extension blocks, the source will then request the blocks from the sink. Extension blocks can be compatible timings relevant to digital video, as well as supported audio formats, speaker allocation, color space, bit depth, gamma, and if present, lip-sync delay.

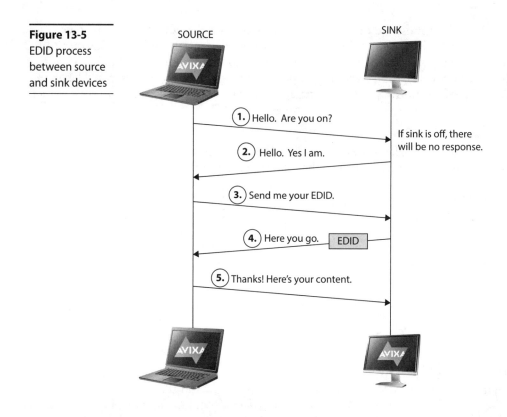

Figure 13-5 EDID process between source and sink devices

NOTE Not all digital video extension technologies handle the hot plug detection and Display Data Channel correctly. If EDID is required for a video display system to operate, it is important to verify the EDID compatibility of any video extension system specified in the installation.

EDID Table

An EDID table is a list of video resolutions and frame rates supported by a sink/display device. The DVI EDID data structure defines data in a 128-byte table; EDID for HDMI connections use 256 bytes followed by additional 128-byte blocks, while DisplayID may include multiple 256-byte blocks of capability data. Table 13-4 shows some of the display capabilities of a specific monitor.

As an AV professional, what should you do when a user attempts to connect their legacy device to a new system? After all, your job is to make that device work and look as good as possible with the new system you have installed. Managing EDID will help you accomplish this goal.

Display Parameters

Video Input Definition	Digital signal
DFP1X Compatible Interface	True
Max. Horizontal Image Size	600mm
Max. Vertical Image Size	340mm
Max. Display Size	27.2 inches

Gamma/Color and Established Timings

Display Gamma	2.2
Red	x = 0.653, y = 0.336
Green	x = 0.295, y = 0.64
Blue	x = 0.146, y = 0.042
White	x = 0.313, y = 0.329

Established Timings
800×600 @ 60Hz (VESA)
640×480 @ 75Hz (VESA)
640×480 @ 60Hz (IBM, VGA)
720×400 @ 70Hz (IBM, VGA)
1280×1024 @ 75Hz (VESA)
1024×768 @ 75Hz (VESA)
1024×768 @ 60Hz (VESA)
800×600 @ 75Hz (VESA)

Standard Timings

Standard Timings n°	4
X Resolution	1,152
Y Resolution	864
Vertical Frequency	75
Standard Timings n°	5
X Resolution	1,600
Y Resolution	1,200
Vertical Frequency	60
Standard Timings n°	6
X Resolution	1,280
Y Resolution	1,024
Vertical Frequency	60

Preferred Detailed Timings

Pixel Clock	241.5MHz
Horizontal Active	2,560 pixels
Horizontal Blanking	160 pixels
Horizontal Sync Offset	48 pixels
Horizontal Sync Pulse Width	32 pixels
Horizontal Border	0 pixels
Horizontal Size	597mm
Vertical Active	1,440 lines
Vertical Blanking	41 lines
Vertical Sync Offset	3 lines
Vertical Sync Pulse Width	5 lines
Vertical Border	0 lines
Vertical Size	336mm

Table 13-4 Capabilities of a Specific Display Screen Extracted from Its EDID Table

Developing an EDID Strategy

For example, suppose you are installing a new video system for a university teaching space. Your client has asked for a system and described it kind of like this:

> "Our teaching staff each have their own laptops or tablets, and they like to use PowerPoint slide shows, websites, and online videos in their lectures. We often have three or four different presenters in each room per day. Because the presenters do not have a lot of preparation time before class, we need a system where they can plug in their devices and the information shows up on the screen immediately. We want it to just happen."

Having an EDID strategy means that the projectors and/or screens will always be able to read the information from the presenters' devices. This makes the system easy to use, which will make your client happy. If the same signal is sent to multiple displays, EDID will allow you to control the source signal so it is consistent. The goal of an EDID strategy is to allow a display to present the signal at its native resolution without scaling. It also allows an installer to set up and configure a system based on the designer's goals. In some situations where the sink and source devices have substantially different capabilities (e.g., 720p versus 3840p), the EDID strategy may need to take an approach other than seeking native resolution to retain optimum image detail and feature resolution.

If the system has displays of different resolutions or aspect ratios, without a proper EDID strategy, the resolution and aspect of the output will be unreliable and may vary when switching between sources.

All display devices in a video system need to have the same aspect ratio. This will prevent many EDID problems. Computers and other devices tend to have either a 16:9 or 16:10 aspect ratio, so your projectors and other display devices should reflect this. Installed display devices tend not to get replaced as rapidly as personal computers and tablets do, so the unfortunate reality is that your display devices may quickly become outdated. If you have an EDID strategy, though, the system should still work for a long time.

One way to future-proof a system is to use the most common resolution between devices as a benchmark. That way, you can ensure all displays will look the same.

Even if your source and sink have an EDID strategy, sometimes non-EDID equipment such as switchers, expanders, and cables interrupt the EDID path and cause problems. Here are some strategies to make sure the system you're installing is EDID compliant.

An EDID emulator acts as a sink. It can be set to a specific aspect ratio and native resolution so that the source outputs a consistent aspect ratio and resolution. So, for example, a laptop will read the EDID information in the emulator instead of the EDID information from one of the multiple displays attached to the switcher. The laptop will then output the fixed resolution set in the emulator.

You can use an EDID emulator to make sure your system is EDID compliant. Sometimes, switchers will already have emulators built into them. Place the emulator as close to the source as possible, or in the switcher. This will give you the most accurate information about your system.

EDID Data Tables

Whether planning or executing a design, you will benefit from containing your EDID information in one place. Creating a data table will help you organize and manage all of your EDID information.

You may encounter disruptions in the EDID conversation. The best troubleshooting method is prevention. Compiling and maintaining an EDID data table for all devices in an installation can avoid many EDID problems. This table will help you track all of the expected resolutions and aspect ratios for every input and output. Currently there is no standard for documenting EDID information. Figure 13-6 shows an example of how an EDID data table might look. Note how the inputs and outputs are grouped together.

An EDID data table should contain the following information for the inputs on a switcher:

- Input number on the switcher
- Type of connected source device
- Preferred resolution
- Audio format
- Additional notes about the source or its settings

EDID/DisplayID Data Table											
Input	Device	Resolution/s	Colorspace	Audio	Notes	Output	Device	Resolution/s	Colorspace	Audio	Notes
1	Laptop	1080p	RGB	Stereo		1	Video wall	1080p	RGB	N/A	
2	Laptop	4Kp	RGB	Stereo		2	LCD panel 1	1080p	RGB/YUV	N/A	
3	Desktop	2560x1440	RGB	Stereo		3	LCD panel 2	1080p	RGB/YUV	N/A	
4	Tablet	1080p	RGB	Stereo		4	LCD panel 3	1080p	RGB/YUV	N/A	
5	Tablet	1080p	RGB	Stereo		5	LCD panel 4	1080p	RGB/YUV	N/A	
6	IPTV	720p	RGB/YUV	5.1		6	LCD panel 5	4Kp	RGB/YUV	Stereo	
7	Media player	1080p	RGB	5.1/Stereo		7	LCD panel 6	4Kp	RGB/YUV	Stereo	
8	Blu-Ray	1080p	RGB/YUV	5.1/Stereo		8	Projector 1	1920x1200	RGB/YUV	N/A	
9	PTZ Cam 1	1080p	RGB	N/A		9	Projector 2	1920x1200	RGB/YUV	N/A	
10	PTZ Cam 2	1080p	RGB	N/A		10	Wall plate 1	scaler-1080p	RGB	Stereo	
11	Doc camera	1080p	RGB	N/A		11	Wall plate 2	scaler-1080p	RGB	Stereo	
12	Wall plate 1	1080p-scaler	RGB	Stereo							
13	Wall plate 2	1080p-scaler	RGB	Stereo							
14	Aux	1080p-scaler	RGB	Stereo							
Input	Device	Resolution/s	Colorspace	Audio	Notes	Output	Device	Resolution/s	Colorspace	Audio	Notes
1	Laptop	720P	RGB	Stereo		1	Projector	1680x1050	RGB/YUV	Stereo	
2	Laptop	1920x1200	RGB	Stereo							
3	Desktop	1920x1200	RGB	Stereo							
4	Blu-Ray	1080p	RGB/YUV	Stereo							
5	Media player	1080p	RGB	5.1/Stereo							
6	Doc camera	1680x1050	RGB	N/A							
7	Wall plate 2	1680x1050	RGB	Stereo							
8	Aux	1680x1050	RGB	Stereo							

Figure 13-6 An EDID data table

It also should contain the following information for the outputs on a switcher:

- Output number on the switcher
- Type of connected sink device
- Device's native or support resolution
- Audio format
- Any additional notes about the sink device or its settings

Resolving EDID Issues

If you need to troubleshoot a problem that you think may be an EDID issue, follow the usual fault-locating principles:

1. **Identify the symptoms:**
    - Did the system ever work correctly?
    - When did the display system last work?
    - When did it fail?
    - Did any other equipment or anything related to this system change between the time it last worked and when it failed?
2. **Elaborate the symptoms:**
    - Confirm that every device is plugged in and powered on.
    - Make only one change at a time as you search for the problem.
    - Note each change and its effect.
3. **List probable faulty functions.** Identify potential sources of the problem.
4. **Localize the faulty function.** Simplify the system by eliminating the equipment that you know is not the source of the error.
5. **Analyze the failure.** Substitute the suspect devices or components with devices that you know work correctly. You can also use EDID test equipment that can emulate sources and sinks.

EDID Tools

There are a range of tools you can use to discover and troubleshoot EDID problems:

- **Software** Many software applications can be used to read, analyze, and modify EDID information. Some are available as tools to complement EDID hardware devices, but there is also a selection of freeware and public domain applications that run on Linux, Windows, or macOS platforms. With these you can connect the test computer up to a display and read the EDID from displays and/or sources to identify whether there may be a problem. Some of these applications will also allow editing of the EDID information to test devices or to resolve a problem.

As an example, the Linux "read-edid" command can extract a device's EDID information and help diagnose whether something is wrong.

- **EDID readers or extractors**   EDID readers and analyzers are available as special-purpose handheld devices or as additional functions on video test signal generators and analyzers.
- **Emulators**   An EDID emulator or processor may be a useful solution if you are unable to get the handshake process to function properly. It takes the place of the sink device's EDID output and forces one or only a few EDID choices.

Resolution Issues

If a computer cannot read the EDID from the sink, it may default to its standard resolution. If the user subsequently attempts to manually set the system resolution to match the display, some graphics cards may enforce the default lower resolution and create a misaligned (size, aspect, centering) output without actually changing the video resolution.

If a computer is connected to multiple displays but can read the EDID from only one display, it may send an output that is mismatched to the other displays.

Some devices in the signal path between the source and the sink, such as switchers, matrix switchers, distribution amplifiers, video signal processors, and signal extender systems, have factory-set default resolutions. If the equipment has such settings, you will need to configure the device to pass EDID information from the sink to the source and vice versa. This may involve setting the EDID to match preset information about the capabilities of a sink device.

Note that if you make the presets something the source does not recognize, you may not get any image, or the image will have a very low resolution. For source components such as Blu-ray Discs, be aware that some players will send a low-resolution 480p output that is compatible with many, but not all, older display devices.

No Handshake, No Picture

Many sources fail to output video if the handshake fails, but computer devices typically will send an output at a default lower resolution to ensure the user can still work with their computer. If this is the case, you may still see a picture from the PC source, but it will be of a lower-than-optimal resolution.

Some source devices will not output a video signal unless the display's EDID data gives confirmation that it can properly display the signal. If there is no signal output from the source, the problem could be that the display's EDID data is not being transferred to the source. If the hot plug detection pin cannot detect another device, the initiator of the conversation will interpret the sink as disconnected and cease the EDID communication. Potential problems with hot plug detection can involve the source not being able to supply sufficient voltage due to voltage drop in a long cable run, a bad (resistive) connection, or a digital video component such as a switcher or splitter intercepting the hot plug detect line.

Switching Sources

When switching sources, the changeover can be very slow, and there can be total picture loss during the switching process. This may be related to nonsynchronous (crash) switching between sources and the delay required to resynch the frame (vertical sync) signal on the new source, or it may be due to a delay in EDID negotiations with the new source.

When EDID sources are not receiving hot plug signals, they generally conclude that the sink device is disconnected. Some low-quality direct switchers disconnect and reconnect signals when switching between devices. If an EDID connection is broken, when it is reconnected, the negotiation begins anew. Another good reason to use the best switcher you can afford.

Managing EDID Solutions

Some approaches for dealing with EDID problems include the following:

- Always test the continuity of all circuits in cables.
- Only use cable lengths that fall within manufacturer guidelines.
- Determine whether a device has a default setting. If that setting is not optimal for the installation, research what it is optimal and how to change it.
- Always use EDID-capable signal extension, distribution, and switching devices.
- If the AV design specifies displays with different aspect ratios and resolutions connected to one source, select the highest common resolution on the EDID emulator.

EDID and Displays

Multiple displays with many resolutions will make EDID complicated. If you are responsible for designing and specifying a new video system, specify the same resolution and aspect ratio for every display in the system. If you do this, you will not have to make any compromises on preferred resolutions for the displays since they will all look the same. This will simplify your system and the amount of work you need to do.

When designing an AV system that includes only fixed sources, such as a desktop PC installed within a rack, an IPTV box, a media player, or a Blu-ray Disc player, you have control when it comes to how the sources will look on the displays. Choose a resolution that will be good for all of those displays, such as 1080p. If you need to make a compromise, remember to find the most common resolution between them and design around it.

If your AV system will incorporate personal devices such as a bring your own device (BYOD) system, you need to consider how the user will want their content to be displayed. Some users may prefer the display to be treated as an extended monitor, but usually users will want the display to duplicate their device's screen. In this case, they will want the native resolution of their laptop or device to appear on the display. You will therefore need to plan for your EDID to cover as many resolutions as the display can handle.

Digital Rights Management

Your customers may want to share or distribute content that they did not create, such as clips from a media library, a Blu-ray Disc video, or music or video from a streaming service. Unlicensed distribution of these materials can violate copyright laws.

Both you and your customers must be aware of potential licensing issues related to the content they want to use. You may need to negotiate a bulk license with a content service provider, such as a cable or satellite television provider or a streaming service. If you fail to obtain the proper licenses to stream content, you are not just risking the legal repercussions of copyright infringement. You are risking the system's ability to function at all.

Publishers and copyright owners use digital rights management (DRM) technologies to control access to and usage of digital data or hardware. DRM protocols within devices determine whether content can be allowed to enter a piece of equipment. Copy protection, such as the Content Scrambling System (CSS) used in DVD players, is a subset of DRM.

High-Bandwidth Digital Content Protection

HDCP is a form of encryption developed by Intel to control digital audio and video content. If the content source requires HDCP, then all devices that want to receive that content must support it.

HDCP is merely a way of authorizing playback. The actual AV signals are carried on other wires in the cable. HDCP is used to authorize the transmission of encrypted or nonencrypted content across a wide range of sources of digital content to eliminate unauthorized copies being made.

For example, there is a pause when you power up a Blu-ray player while the AV system is verifying that every device in the system is HDCP compliant. The Blu-ray player first checks whether the disc content is HDCP compliant. If so, the player proceeds to conduct a series of handshake exchanges to verify that all other devices in the display chain are also HDCP compliant before making the content available for display.

HDCP Interfaces

HDCP 2.x is capable of working over the following interfaces:

- DVI
- HDMI
- DisplayPort
- HDBaseT
- Mobile High-Definition Link (MHL)
- USB3.2 gen2+ via Type-C connectors
- USB4
- TCP/IP

Some Apple devices may block content if HDCP is not present, even if the content is not HDCP-protected.

How HDCP Works

HDCP's authentication process determines whether all devices in the display system have been licensed to send, receive, or pass HDCP content. No content will be shared until this entire process is completed. If there is a failure at any point in the process, the whole process has to restart.

The authentication steps are

1. **Device authentication and key exchange**, where the source verifies that the sink is authorized to receive HDCP-protected content and then exchanges device encryption keys
2. **Locality check**, where the source and sink verify that they are within the same location by confirming that the round-trip time for messages is less than 20ms
3. **Session key exchange**, where the source and sink exchange session encryption keys so the content can be shown
4. **Authentication with repeaters**, which allows repeaters in the system to pass HDCP content

When a switcher, splitter, or repeater is placed between a source and sink/display device to route signals, the inserted device's input becomes a sink/display, and its output becomes a new source and continues the video signal chain.

HDCP Device Authentication and Key Exchange

The authentication process is designed so the source verifies that the sink is authorized to receive HDCP-protected content. Each device must go through this process.

1. The source initiates the authentication process by sending a signal requesting the receiver to return its unique ID key.
2. The receiver sends its unique key to the source. The source must receive the sink's unique key within 100ms, or the device is not compliant with HDCP 2.x specification.
3. The source checks that the receiver's key contains a specific identifier that is given only to authorized HDCP adopters. If the key is missing, the process is aborted.
4. The source then sends its master key information to the receiver.
5. The receiver's software verifies the master key and uses it to compute new values that are returned to the source.
6. The source then verifies the receiver's calculated values. If the calculated values are not received within the appropriate amount of time, the authentication process is aborted.

HDCP Locality Check

The locality check verifies that the source and receiver are in the same locality, for example, in a nearby meeting room, auditorium, or classroom, and not a couple of cities away.

1. The source sends a message to the receiver containing a random number and sets a timer for 7 milliseconds. Both the source and receiver will have the same algorithm that takes the random number and generates a new one.
2. The receiver gets the random number, generates the new one, and sends that back to the source before the 7-millisecond timer expires.
3. The source verifies that the number it calculated and the one the receiver returned are the same. If they match, the authentication process continues. If they do not match or the timer expires, the authentication process for the locality check begins again. The locality check will be repeated for a total of 1,024 tries before it aborts the authentication process and everything must start over again.

HDCP Session Key Exchange

Once the source and receiver have passed the device authentication and locality check, the process moves on to the individual session using the following steps:

1. The source device generates a session key and sends it to the receiver with a message to pause for at least 200 milliseconds before using it.
2. The source pauses and then begins sending content encrypted using the session key. Each HDCP session has its own key and therefore unique encryption.

Once the session commences, the HDCP devices must reauthenticate periodically as the content is transferred. Several system renewability messages (SRMs) must be exchanged. If during SRM exchange it is discovered that the system has been compromised, the source will stop sending content.

HDCP and Switchers

All of the HDCP processes explained so far have assumed that the AV system you are installing has a single video source for a single video sink/display. In these systems, the number of keys and how they are exchanged will be handled between the two devices. However, when multiple video sources, displays, processors, and a switcher are added, HDCP key management becomes more complex.

Some switchers maintain key exchange and encryption sessions continuously, so the communication will not need to be restarted. These switchers can act as a source or sink to pass the protected and encrypted HDCP data to its destination. You will need to verify that the switcher can handle HDCP authentication from the manufacturer's documentation.

HDCP Authentication with Repeaters

An HDCP repeater is a device that can receive HDCP signals and transmit them on to another device, such as a switcher, display device, processor, or distribution amplifier. In a system with repeaters, as shown in Figure 13-7, the HDCP authentication process occurs after the locality check and device authentication have taken place between all the devices in the system. The session key exchange has not begun.

The HDCP authentication process in a system with a repeater follows these steps:

1. The repeater compiles a list of IDs from all its connected downstream devices.
2. The repeater sends the list of IDs and number of devices to the source and sets a 200-millisecond timer.
3. The source reads the list and compares it to a list of revoked licenses in the media. Each new HDCP device or media has an updated list of revoked license numbers provided by Digital Content Protection, LLC. If any of the downstream devices are on the revoked list, the authentication process fails.
4. The source then counts how many devices are downstream. If the number of devices is less than the maximum of 32, the authentication process moves forward.

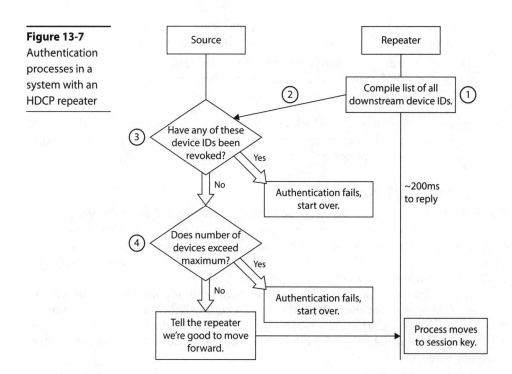

Figure 13-7 Authentication processes in a system with an HDCP repeater

HDCP Device Limits

HDCP 2.x supports up to 32 connected devices for each transmitter and a maximum of four repeater levels for each transmitter. Figure 13-8 illustrates an example of connection topology for HDCP devices.

Although HDCP 2.x supports up to 32 connected devices, in practice, the number of sink/display devices allowable from a single HDCP-protected media source is typically much more limited and based upon the number of keys allowed by the source.

HDCP Troubleshooting

Once your system has been designed and installed, you need a method for verifying that the HDCP keys are being managed correctly. You know that your keys are being managed correctly if the image appears on the sink and is stable over a period of time. If the HDCP keys are not managed correctly, an image constraint token (ICT) or a digital-only token (DOT) will be displayed on your sink.

An ICT is a digital flag built into some digital video sources. It prevents unauthorized copies of content from appearing on a sink device. This encryption scheme ensures that high-definition video can be viewed only on HDCP-enabled sinks. A DOT is a digital flag that is embedded into digital sources, such as Blu-ray Discs. Its purpose is to limit the availability or quality of HD content on any component output of a media player.

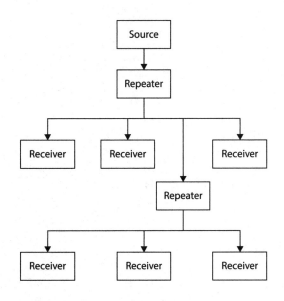

Figure 13-8 Depth of two repeater levels and device count of six in HDCP 2.x topology

There are devices on the market to help you troubleshoot HDCP problems. Some have feature sets that may include the following:

- Hot plug/5V presence detection
- DisplayID/EDID verification
- HDCP status indication
- Indication of the number of keys accepted by the source (in other words, the maximum number of devices supported)
- Cable verification

When HDCP authentication fails to initiate correctly, there can be a range of symptoms that do not directly implicate HDCP failure as their cause. These may include an image appearing for a few seconds at the start of a session and then disappearing, a green screen, white noise, or a blank screen, any of which could be caused by a broken cable, a bad connector, or dozens of other problems.

An HDMI signal analyzer or signal generator/analyzer is one of the few tools that will let you track and analyze the HDCP negotiation process, providing you insight into where the negotiations have gone wrong and a clue as to how you might remedy that issue. If such an analyzer is not available, then the signal flow troubleshooting strategy is likely to be the most productive. Start at the source end of the signal path with a known working display device and follow the signal path, testing at each point until you locate the fault.

There are professional AV situations, such as the capture of lectures and courtroom proceedings or feeds to videoconferencing systems and streaming sites, where HDCP on an output stream must be bypassed for the system to fulfill its required function. This requires the use of specialized equipment, such as mixers with inputs that behave exactly as an active fully compliant HDCP-complaint device, but then pass an HDCP-free stream through designated outputs.

Managing HDCP can be a challenge, but remember that if your client wants to bypass digital rights management, they shouldn't. Most jurisdictions have adopted some form of copyright law that criminalizes attempts to circumvent control access to copyrighted works. Be sure to make your clients aware of these laws when discussing their needs on a project.

Chapter Review

Digital video is often oversimplified—either the video appears at its destination or it doesn't. Having completed this chapter, you know some of the intervening issues that come with designing digital video systems and how to manage them.

Some of the questions you'll work through with clients when determining their digital video needs may include: "What transport solutions can I employ based on the needs and budget of the client?" "Will I face distance limitations? Bandwidth limitations?"

"If the client wants UHD, what implementation of UHD will satisfy their needs?" "Are the system's devices HDCP 2.x compliant? Do they need to be?" "What should the EDID strategy be in the case of multiple displays and differing display formats and resolutions?" With this information and more, your digital video design will reliably deliver what the client expects.

Review Questions

The following questions are based on the content covered in this chapter and are intended to help reinforce the knowledge you have assimilated. These questions are similar to the questions presented on the CTS-D exam. See Appendix E for more information on how to access the free online sample questions.

1. Ultra high-definition video (UHD) describes video formats with a minimum pixel resolution of:

   A. 1920×1080

   B. 3840×2160

   C. 4096×2160

   D. 7680×4160

2. DisplayPort via "DP Alt Mode" may be supported over which of the following versions of USB when using USB Type-C connectors?

   A. USB3.2gen1

   B. USB 2.0

   C. USB 1.1

   D. USB3.2gen2x2

3. A video system design to transmit 4K at 60 fps, with 4:4:4 chroma sampling and 10 bits per color, requires about _____ of throughput.

   A. 10Gbps

   B. 16Gbps

   C. 22Gbps

   D. 32Gbps

4. Both _____ use transition-minimized differential signaling to transmit high-speed serial data.

   A. DVI and HDMI

   B. HDMI and DisplayPort

   C. DVI and DisplayPort

   D. USB4 and DisplayPort

5. In a video system with multiple displays, problems with image quality in some, but not all, displays could indicate a problem with what?

   A. The video formats used

   B. The lengths of the video cables

   C. Incompatible HDCP keys

   D. EDID

6. An EDID data table should contain which of the following to describe the inputs of a switcher? (Choose all that apply.)

   A. The version of EDID used

   B. Input numbers

   C. Preferred resolution

   D. Color space support, such as RGB or a component

7. Clients who want to deploy 4K video over an integrated video system will likely need _____ devices to make it work properly.

   A. HDCP 2.x–compliant

   B. DisplayPort 2.0–compatible

   C. 8Gbps–capable

   D. 4:2:0-sampled

8. An image constraint token message on-screen indicates a problem with _____.

   A. HDMI

   B. UHDTV

   C. EDID

   D. HDCP

Answers

1. **B.** UHD describes video formats with a minimum pixel resolution of 3840×2160.

2. **D.** DisplayPort via "DP Alt Mode" is supported only under USB3.2gen2x2 when using USB Type-C connectors. DisplayPort over USB4 is tunneled via the USB4 data fabric.

3. **B.** A video system design to transmit 4K at 60 fps, with 4:4:4 chroma sampling and 10 bits per color, requires about 16Gbps of throughput (4096×2160×60×10×3 = 15,925,248,000bps).

4. **A.** DVI and HDMI use transition-minimized differential signaling to transmit high-speed serial data.

5. D. In a video system with multiple displays, problems with image quality in some, but not all, displays could indicate an EDID (Extended Display Information Data) problem.

6. B, C, D. An EDID truth table should include information about input numbers, preferred resolutions, and color spaces. The version of EDID/DisplayID in use is not required.

7. A. Clients who want to deploy 4K video over an integrated video system will likely need HDCP 2.x–compliant devices to make it work properly.

8. D. An image constraint token message on-screen indicates a problem with HDCP.

CHAPTER 14

Audio Design

In this chapter, you will learn about
- The difference between analog and digital audio systems
- How to compare and contrast different digital signal processor (DSP) architectures
- Different methods of signal metering and how to establish proper signal levels
- Comparing input and output impedances and implementing correct equalization practices
- Distinguishing between different types of audio processors and applying them to a design
- Graphic and parametric equalizers
- Applying crossover, feedback-suppression, and noise-reduction filters to solve problems within an audio system

For the foreseeable future audio systems will remain a hybrid of digital and analog components. Humans are analog, as are most microphones and loudspeakers. Communication starts with an analog sound into a transducer (a microphone), which converts it into an electrical signal. Eventually, that electrical signal is turned back into analog sound waves by an earpiece or loudspeaker, so it can be received by the ears of an analog human.

What happens between the microphone pickup element and the loudspeaker or earpiece diaphragm can be either analog or digital or some hybrid of the two. Depending on the quality of the analog-to-digital (A-to-D) conversion, digital-to-analog (D-to-A) conversion, sample depths, sample rates, bit rates, compression type, and so many other factors, there can be as wide a variety of sound quality found in the digital domain as there is in the analog world.

In Chapter 6, you began to determine the parameters of a sound system design. You started by asking whether the client even needed a sound system, and then you quantified sound pressure levels, analyzed background noise, and more. If indeed you found it necessary to install a sound system, you mapped out loudspeaker locations based on coverage patterns, calculated the amount of power required at the loudspeakers, determined when to apply direct-connect or constant-voltage power amplifiers and loudspeakers, matched microphone sensitivities to the application, and explored whether the locations you chose for your loudspeakers and microphones resulted in a stable sound system, with good spectral response and intelligibility.

Next, we will consider the processors and infrastructure between microphones and loudspeakers.

> **Duty Check**
>
> This chapter relates directly to the following tasks on the CTS-D Exam Content Outline:
>
> - Duty B, Task 2: Coordinate with Architectural/Interior Design Professionals
> - Duty C, Task 1: Create Draft AV Design
> - Duty C, Task 2: Confirm Site Conditions
> - Duty D, Task 2: Conduct System Performance Verifications
>
> In addition to skills related to these tasks, this chapter may also relate to other tasks.

Analog vs. Digital Audio

Where once part of planning the design for an audio system was to choose whether it should be assembled from either analog or digital components, such a choice is no longer either necessary or even practicable. Almost any system that you design is going to have some analog and some digital elements, even if it may not always be obvious which is which. The microphone inputs on digital devices almost always have an analog pre-amp (often proudly so), while even the simplest analog mixing console will likely include a basic digital delay, possibly some limiting, ducking, and perhaps a few simple digitally implemented effects.

If you choose, you can have a purely analog path for a simple reinforcement system—from microphone to mixer to processors to power amplifier to loudspeakers. While analog technology may be proven and reliable, it is subject to more system noise and signal degradation over distance than a digital one. It also requires the use of single-function components, which can be limiting by comparison to the flexibility possible with digital signal processing units.

If you choose to stay analog, your system will have more termination points and many more cables and therefore will need larger raceways. All those termination points will require additional labor. You will need more space for infrastructure, and if you need to split analog signals in a point-to-multipoint configuration, you'll need to specify transformers or distribution amplifiers.

Digital audio can travel over shielded or unshielded twisted-pair (S/UTP) copper, optical fiber, an existing data network, or even radio frequency (RF) Wi-Fi links, either on its own or as part of a digital AV signal such as HDMI, SDI, Q-SYS, or NDI. As digital audio is multiplexed, with hundreds of audio channels being carried on a single fiber, network, or cable, the amount of raceway real estate needed is much smaller.

Digital audio systems generally require less rack space, but may generate more heat. If your system design requires a point-to-multipoint setup, a digital system is easier to implement.

NOTE Regardless of whether an audio system is analog or digital, designers need to consider bandwidth. In the analog world, you would quantify bandwidth in terms of frequency, usually in megahertz. With digital, you would quantify bandwidth in terms of bits per second (bps), usually gigabits per second.

DSP Architectures

More often than not, even in the middle of an otherwise analog path, you will need a central digital signal processor. It could be all-inclusive, with the mixer, signal processors, and a small power amplifier. Or it could comprise just the mixer and signal processors, or even just the signal processors.

AV professionals configure DSP devices using either the manufacturer's software interface or front-panel controls. System configurations and settings can be saved, copied, and password-protected. Specific presets can be recalled through a control system interface.

Designing around a single DSP device rather than multiple analog devices reduces the number of connections required and simplifies installation. On the other hand, too many dedicated-task DSPs in the signal path runs the risk of introducing discernable latency due to the A-to-D and D-to-A conversions at each device. Moreover, using different sample rates and bit depths along the signal path can actually degrade audio quality and should be avoided. Once a signal has been converted into the digital domain, it is advisable to keep it digital until all processing has been performed before returning it to an analog signal.

Digital signal processors come in three basic varieties: flexible, fixed, and hybrid architecture.

- **Flexible**   *Flexible-architecture processors* are characterized by a drag-and-drop graphical user interface (GUI). In the GUI depicted in Figure 14-1, audio functions such as mixers, equalizers, filters, delays, and crossovers can be dragged from the processing library on the left and placed almost anywhere in any order along the signal chain on the right.

    Flexible-architecture processors give you more freedom in your design. You can manage many functions within a single box and develop complex signal paths. They are an excellent choice for systems where complexity or multiple applications may come into play, such as large-scale paging systems in airports. However, flexible-architecture processors may be more costly on a per-channel basis than fixed or hybrid systems. In addition, they require more DSP power because memory register stacks must be allocated for any eventuality, and code space cannot readily be optimized.

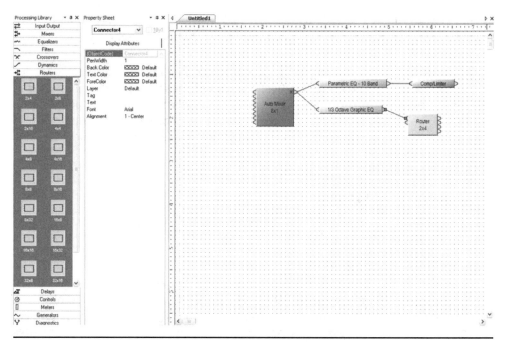

Figure 14-1 A drag-and-drop GUI for a digital signal processor

- **Fixed** *Fixed-architecture processors* handle one type of function. For example, they may handle compression/limiting, automatic mixing, equalization, or signal routing. They are often easy to set up and operate, and may be good for system upgrades because you configure them much as you would their analog counterparts. However, they are functionally limited and may not be scalable.
- **Hybrid** *Fixed-multifunction (hybrid) processors* (see Figure 14-2) allow you to adjust multiple functions. They operate along a predictable, known pathway and are fairly cost-effective, although some are limited in their routing and setup options. Generally, however, a hybrid architecture offers flexibility in signal routing, specifically with respect to which inputs connect to which outputs. It also optimizes DSP processing power because memory stacks and registers can be tightly packed.

Note in Figure 14-2 the level of control in pairing inputs and outputs while also providing for other adjustments along each path. If you look at input 1 on the left, the signal goes to an invert switch and then a mute switch, a gain controller, a delay, a noise reduction filter, a filtering setup, a feedback eliminator, and finally a compressor. From there, you see the matrix router, which allows you to route the inputs to the different mix buses. These mix buses route to the outputs on the right. The outputs have their own set of controls, including delay, filters, compressor, muting, gain, and limiting.

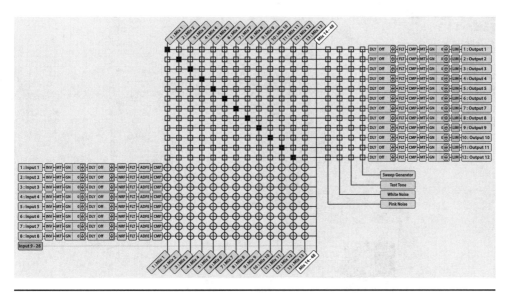

Figure 14-2 The interface for a hybrid DSP

Signal Monitoring

Signal-level monitoring helps ensure that an audio system doesn't clip the signal or add distortion. It also helps ensure that the signal level is high enough to achieve an adequate signal-to-noise ratio without actually adding noise. Signal levels that are too low decrease the system's signal-to-noise ratio and can result in background hiss.

Two types of meters are typically used for signal monitoring. The first—and most common indicator in general audio use—is the *volume unit (VU) meter* (Figure 14-3). The second is the *peak program meter (PPM)* (Figure 14-4).

A VU meter indicates program material levels for complex waveforms. In essence, it is a voltmeter calibrated in decibels and referenced to a specific voltage connected across a 600-ohm load. A VU meter meets stringent standards regarding scale, dynamic characteristics, frequency response, impedance, sensitivity, harmonic distortion, and overload. It is commonly used in public address (PA), background audio, and reinforcement applications.

 TIP When using a VU meter, set the average levels to read about –6 and set the peaks to about 0.

A PPM responds much faster than a VU, showing peak levels instantaneously. It is useful for recording, streaming, and broadcast or when the level must not exceed 0dB full scale (dBFS). Going above the full range of an A-to-D converter will cause clipping. PPMs are often found in audio mixers and post-production equipment. They may take the format of a calibrated analog meter with a fast attack and slow decay response; a dedicated LCD, vacuum fluorescent, or LED indicator column; or part of a video display.

Figure 14-3
A VU meter responds in a similar way to how humans respond to loudness.

Figure 14-4 A PPM shows instantaneous peak levels and is useful for broadcast, streaming, and recording. Left: Digital version. (Courtesy of spencerdare / Getting Images) Right: Analog version. (Courtesy of Tondose / Wikimedia (no changes) under the terms of: https://creativecommons.org/license/by-sa/4.0/legalcode and https://www.gnu.org/licenses/fdl-1.3.en.html)

Digital level meters may also include such features as peak hold indication, a color change on over-peak levels, and signal presence indication.

Analog vs. Digital Signal Monitoring

Professional analog audio line-level signals typically operate around 0dBu or +4dBu, but depending on the headroom available in the system, they can be as much as +18dBu to +28dBu before distortion (clipping) occurs. Digital signals, on the other hand, have a hard limit—0dBFS (full range on the A-to-D converter) represents the maximum level for a digital signal. At 0dBFS, all the bits in the digital audio signal are equal to 1. In 24-bit digital audio systems, this may be as high as +44dBu.

You can use the bit depth of an audio signal to calculate its dynamic range. The formula for the dynamic range of a digital audio signal is as follows:

$$dB = 20\log(2^{Bitdepth})$$

where:

- *dB* = The dynamic range of the signal
- *Bitdepth* = The bit depth of the signal samples

To calculate the theoretical dynamic range for the 16-bit samples used for many standard applications such as the compact disc:

- $dB = 20\log(2^{16})$
- $dB = 20\log(65,536)$
- $dB = 20(4.8165)$
- $dB = 96$

A 16-bit system has a dynamic range of 96dB.

Using the same formula, you can calculate that the 24-bit audio widely used for concert, broadcast, film, and recording work has a dynamic range of 144dB, while the 32-bit audio used in some precision applications has a dynamic range of 192dB.

Setting Up the System

When configuring a DSP, you may need to consider many common functions. We will go through some of these so you are clear on what they can do and how you can set them to the recommended settings.

- Input gain
- Filters (input and output)
- Equalization
- Feedback suppression
- Crossovers
- Noise reduction
- Dynamics
- Compressors
- Limiters
- Gates
- Delay
- Routing

Where to Set Gain

Regardless of whether you're working in the analog or digital domain, proper gain structure and signal monitoring are required. Setting gain correctly provides an optimal signal-to-noise ratio for an AV system and helps to avoid signal distortion. This enables the system to deliver the best performance so the user does not hear hiss, noise, or distortion.

There are two methods for setting system gain: unity gain and the system optimization method. Both methods are relatively simple, and neither requires expensive equipment.

- **Unity gain**   For a typical presentation room, conference room, or boardroom where you have installed professional audio components, unity gain will provide an adequate system electronic signal-to-noise ratio of around 60dB. Using the unity gain method, the strength of the output signal should equal the strength of the input. For example, if the first device you measure shows a signal-level output of 1.23V (+4dBu), you should be able to measure a 1.23V signal at the output of each device all the way to the power amplifier inputs.

- **System optimization**   For a more critical listening environment such as a studio, broadcast facility, performing arts facility, or lecture hall, the system optimization method will provide the optimal signal-to-noise ratio allowed by the equipment in the signal path. This optimization method provides uniform headroom throughout the signal path. It also maximizes the electronic signal-to-noise ratio of the audio system. However, this method requires more time and skill than the unity gain method.

Most mixers will produce +18 to +24dBu output levels without clipping; 10 to 20dB of headroom will support an emphatic talker or an especially loud section of program material. With preamplifiers properly set and all other adjustments set at unity, the mixer's output meter should be reading about 0 (analog) or about –18 to –20dBFS (digital) under normal operating conditions. Although clipping an analog signal produces an undesirable and unpleasant distortion, digital clipping is also annoying and harsh.

TIP   Make sure to read all equipment manuals to discover what 0 really means. In an analog or apparently analog meter, does 0 mean 0VU (+4dBu) or 0dBu?

There are multiple points within an audio system where gain can be adjusted. Make sure to set the gain structure of each of these devices:

- Microphone preamplifiers
- Preamplifiers in mixer microphone inputs
- Line-level mixer inputs for external sources

- Audio mixer
- Processing devices (equalizer, compressor, limiter, delays, effects)
- Digital signal processors (DSPs) and loudspeaker processors
- Active/powered loudspeakers
- Amplifiers

Some DSPs are equipped with microphone inputs. Like their analog counterparts, these inputs will include a microphone preamplifier. This is one of the most important settings for extracting the maximum signal-to-noise (S/N) ratio from your system. As a guide, the following input types may require as much gain as listed here:

- **Handheld vocal microphone**   35dB minimum
- **Handheld presentation microphone**   45dB
- **Gooseneck microphone**   45dB
- **Boundary microphone**   55dB

When using line-level inputs, an unbalanced consumer line-level input may need as much as 12dB of gain, while a balanced professional line-level input may need little or no gain. For best results, refer to manufacturers' documentation for an indication of the amount of gain suggested for each piece of equipment or microphone.

Poor Signal-to-Noise Ratio

Low-noise preamps can help you achieve an excellent S/N ratio in your audio system. A wide S/N ratio means your audience will hear much more signal than noise, which increases intelligibility. But first, let us look at what happens when you do not have a wide S/N ratio. Figure 14-5 plots the amount of signal and noise generated by a microphone, a preamplifier, a mixer, DSP boxes, and an amplifier in an audio signal chain. The line along the bottom represents the noise. As you can see, the dynamic microphone at the beginning of this signal chain does not introduce any noise into the system, but each item afterward adds noise.

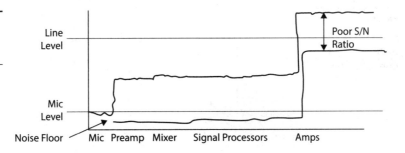

Figure 14-5 Poor signal-to-noise ratio

In this figure, instead of bringing the signal up to the line level (0dBu) in the preamp, the gain was set too low, perhaps around –20dBu. When the signal hits the amplifier, it needs to be amplified significantly to reach the desired sound pressure level (SPL). But do you see what also happens to the noise? It also increases significantly. In systems set like this, you will hear audible hiss from the loudspeakers.

TIP If you are trying to find the source of hiss in your audio system, check to see whether the gain at any stage is turned up to its maximum. This is often a dead giveaway that the hiss is a symptom of poor gain structure.

Good Signal-to-Noise Ratio

Now let's see what happens when the preamp output is brought up to line level (0dBu). As shown in Figure 14-6, you don't need to turn up the amplifier nearly as much to reach the same SPL at the listening position. This means the noise will not be amplified nearly as drastically as it was in the previous example. This gives you a wide S/N ratio, and you probably will not hear hiss in this audio system.

Figure 14-7 compares the two examples. Both systems end with the same SPL, but look at how much less you need to adjust at the amplifier when you set your gain structure at the preamplifier. This prevents too much noise or hiss from getting into your system, which makes the system intelligible and clear.

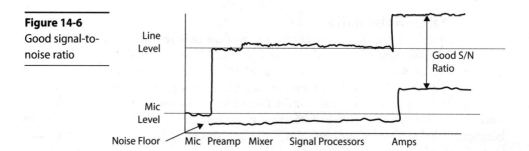

Figure 14-6 Good signal-to-noise ratio

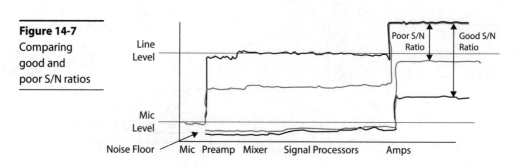

Figure 14-7 Comparing good and poor S/N ratios

Common DSP Settings

Before we dive into DSP settings, let's go over several terms.

- *Threshold* is the level at which a desired function is triggered. Generally speaking, a lower threshold level means it will activate earlier. The recommended starting threshold for many line-level (post-preamp) functions is 0dBu.
- The *attack time* of an audio function determines how quickly the function will be implemented once it is triggered. If the attack time is too slow, then the effect will not be fully active while the system adjusts.
- The *release time* of an audio function is how long the function takes to cease its effect on the audio signal once it is no longer active.
- *Automatic gain control* (AGC) is an electronic feedback circuit that maintains a constant output in a system by varying system gain in response to changing inputs. An AGC function in an audio system can be used to maintain a constant power output in response to variables such as input signal strength or ambient noise level. AGC raises the gain if the signal is too low or attenuates the signal if it is too high. Its primary application is to capture weak signals for recording or transmission. Be careful with this DSP setting, as it can create feedback if used on amplified inputs. When using AGC, start with a threshold set at 0dB. This will help keep the gain centered at line level.
- *Ambient level control* uses a reference microphone to measure a room's noise level. It then automatically adjusts a system to compensate for noisier environments. This function may sometimes be useful in managing music in restaurants. An ambient level controller will ensure that, no matter how loud the patrons are, the background music will remain a specified number of dB above the ambient noise. This may also really annoy patrons and front-of-house staff, who need to continually raise their voices to communicate above the ever-rising levels of music and noise. Use this function sparingly, and only implement it if you have clearly informed consent from the end user.

 TIP The reference microphone used for ambient level control may be a dedicated microphone, used only to take this SPL measurement, or it may be a microphone used elsewhere in the system but also designated to measure the reference signal.

Compressor Settings

Let us begin our look at DSP settings with compressors. A compressor controls the dynamic range of a signal by attenuating the part of the signal that exceeds a user-adjustable threshold. When the input signal exceeds the threshold, the overall signal is attenuated by a user-defined ratio, thus reducing the overall dynamic range.

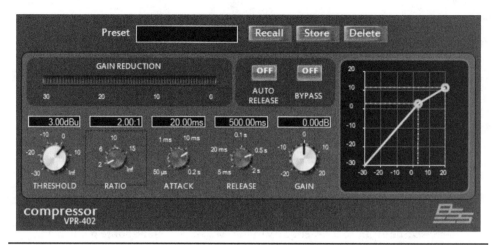

Figure 14-8 Software compressor interface showing sample threshold and ratio settings

Compressors compensate for peaks in a signal level. All signal levels below a specified threshold will pass through the compressor unchanged, and all signals above the threshold will be attenuated. This reduces the variation between the highest and lowest signal levels, resulting in a compressed (reduced) dynamic range. Compressors are useful when reinforcing unpredictable, energetic presenters or vocalists, who may occasionally raise their voice for emphasis or lean in too close to the microphone (see Figure 14-8).

The *compressor threshold* sets the point at which the automatic volume reduction is triggered. When the input goes above the threshold, the audio compressor automatically reduces the output to keep the signal from getting too loud.

The *compressor ratio* is the amount of actual level increase above the threshold that will yield 1dB in gain change after the compressor. For example, a 3:1 ratio would mean that for every 3dB the input signal increases above the threshold, there will be only a 1dB difference in the output. Likewise, if the input level were to jump by 9dB, the output would increase by only 3dB.

You can also set the attack and release times on a compressor. Again, the attack time is how long it takes for the compressor to react after the signal exceeds the threshold. The release time determines when the compressor lets go after the input level settles below the threshold. Both of these functions are measured in milliseconds.

When setting a compressor, try starting with the following settings. You will have to make adjustments depending on the specific needs of your audio system, but these settings are a good place to start. For speech applications, such as conference rooms, boardrooms, or presentation spaces, try these settings:

- **Ratio**   3:1
- **Attack**   10 to 20ms
- **Release**   200 to 500ms
- **Threshold**   0

For these settings, if the initial input gain were set for a 0dB level, it would take a 60dB increase at the microphone to hit the +20dB limit of input, which would result in clipping.

For music or multimedia applications, try these settings:

- **Ratio** 6:1
- **Attack** 10 to 20ms
- **Release** 200 to 500ms
- **Threshold** 0

Extreme compression of high-amplitude signals is called *limiting*.

Limiter Settings

Limiters are similar to compressors in that they are triggered by transient peaks or spikes in the input signal level (see Figure 14-9). Limiters limit the level of all signals above an adjustable threshold. In other words, they prevent high-amplitude signals from getting through. Limiting is used to prevent damage to components such as loudspeakers and to prevent signal clipping in analog-to-digital conversion. It is triggered by peaks or spikes in the input signal (like a dropped microphone, a hit from a drum stick, phantom-powered mics that get unplugged without being muted, and equipment that is not powered up or shut down in the correct sequence), and it reacts quickly to cut them off or reduce them before they exceed a certain point. The amount of limiting above the threshold is determined by a more aggressive ratio than a typical compressor reduction ratio. The reduction limits the variation between the highest and lowest signal levels, resulting in a limited dynamic range. With limiters, signals exceeding the threshold level are reduced at ratios of 10:1 or greater.

Figure 14-9
User interface for a limiter showing sample limiter settings

Always set a limiter's threshold above any compressor's threshold. Otherwise, the compressor will never engage. Here are some suggested settings for your limiter. You will need to adjust these values depending on the specific needs of the audio system.

- **Ratio**   10:1 above the user-adjustable threshold.
- **Threshold**   10dB higher than the compressor's threshold. If the compressor threshold is 0dB, then the limiter's threshold should be 10dB.
- **Attack time**   2ms or more faster than the compressor's attack time.
- **Release time**   200ms or less than the compressor's release time.

Expanders

An expander is an audio processor that comes in two forms: a downward expander or as part of what's known as a *compander*.

Downward expanders increase the dynamic range by reducing—or attenuating—the level below the adjustable threshold setting. They increase gain if the signal is low, such as a presenter with a weak voice. This reduces unwanted background noise and is especially useful in a system using multiple open microphones.

As you can see in Figure 14-10, downward expanders have many of the same settings as compressors and limiters. In this example, the threshold is set to –40dB. When the input signal falls below this threshold, it will reduce at a 4:1 ratio. This allows you to eliminate low-level noise from your audio signal.

Keep in mind that expanders are primarily intended for recording and transmission. When used in a live amplified environment, expanders can drive a system into painfully loud feedback.

Figure 14-10 An example of settings for a downward expander

Gate Settings

A gate is an audio processor that allows signals to pass only after they have exceeded a preset threshold. This can be used to turn off unused microphones automatically. You can control when the gate activates by setting the gate's attack and hold times. Gates are found in some automated mixers and are useful for noise control, such as from a noisy multimedia source, or an open microphone in an environment with high ambient noise (see Figure 14-11). If the gate's threshold is set too high, the signal may be clipped off at the beginning of the wanted sounds: a problem known as "gating" (for pretty obvious reasons).

When setting a gate, try the following settings. As with other settings, you may need to make adjustments based on the situation.

- **Attack**   Start with a relatively fast setting, such as 1ms.
- **Release**   Start at 50ms.
- **Threshold**   Your threshold setting depends on the gate's position in the audio signal chain and the number of mics used. If it's after the input gain and there is one mic, less than 0dB is a good starting place. You may need to go lower than this if you are not getting a reliable start (gating).

NOTE   Compressors, limiters, gates, and expanders are all "dynamic filters" in that they filter audio on the basis of its dynamic range. However, some dynamic filters can also be frequency specific. For example, a compressor with a low-pass filter is good for controlling proximity effects while allowing high frequencies to pass unaffected.

Figure 14-11 User interface for a gate, which mutes the level of signals below an adjustable threshold

Delays

A *delay* is the intentional retardation of the transmission of a signal. In the context of audio processing, it is an adjustment of the time in which an input signal is sent to an output. A delay is often used to compensate for the distance between loudspeakers or for the differential in processing required between multiple signals. In digital processing systems a delay is achieved by storing the input signal in a buffer for the required period of time before outputting it. The *unintentional* signal delay introduced by processing the signal through steps such as compression, equalization, echo cancellation, A-to-D and D-to-A conversions, etc., is usually referred to as *latency*.

Delays are used in sound systems for loudspeaker time-alignment—to align the components within a single loudspeaker enclosure, to align the wavefront generated by a multicabinet array, or to align the multiple cabinets in a distributed-loudspeaker system.

Within a given loudspeaker enclosure, the individual components may be physically offset, causing differences in the arrival time of the wavefronts from those components. This issue can be corrected either physically or by using delays to provide proper alignment.

Electronic delay is often used in sound reinforcement applications. For example, consider an auditorium with an under-balcony area. The audience seated underneath the balcony may not be covered well by the main loudspeakers. In this case, supplemental loudspeakers are installed to cover the portion of the audience seated underneath the balcony.

Although the electronic audio signal arrives at both the main and under-balcony loudspeakers simultaneously, the sound coming from these two separate locations would arrive at the audience underneath the balcony at different times and sound like an echo. This is because sound travels at about 343 meters per second (1,125 feet per second), which is much slower than the speed of the electronic audio signal, which travels at approximately 150,000 kilometers per second (90,000 miles per second).

In this example, an electronic delay would be used on the audio signal going to the under-balcony loudspeakers. The amount of delay would be set so that the sound both from the main loudspeakers and from the under-balcony loudspeakers arrives at the audience at the same time. Similarly, delays can be used to time-align the audio from a video replay to be in synch with the vision throughout a viewing area served by multiple loudspeakers.

Delay can also be introduced to combat the *Haas effect*. The human ear has the ability to locate the origin of a sound with fairly high accuracy, based on where you hear the sound from first. Through the Haas effect—or sound precedence—you can distinguish the original source location even if there are strong echoes or reflections that may otherwise mislead you. A reflection could be 10dB louder and you'd still correctly identify the direction of the original source due to its arrival time.

When setting up a delay, designers can use this effect to their advantage. Rather than timing the delay of the supplemental or "delay" loudspeakers to come out at exactly the same time as the original source, try increasing the delay by an extra 15ms, thereby making the relay speaker appear to be 5m (16ft) behind the original source. This allows the listener to locate correctly the origin of the sound (the lecturer, band, video image, and so on). Without the additional delay, the listener would perceive the source of the sound as the location of the loudspeaker, rather than the location of the original source. Just don't exceed 25ms. Longer might be perceived as echo and compromise intelligibility.

NOTE At 30 degrees Celsius (85 degrees Fahrenheit), sound travels about 350 meters per second (1,145 feet per second). As the standard velocity for sound is calculated at 20 degrees Celsius (68 degrees Fahrenheit), this will have an impact on the delay settings for audio in warmer, outdoor locations, for example. It is also important to take the altitude of a location into account, as figures quoted for the velocity of sound are measured at sea level. At higher elevations the air is less dense, causing soundwaves to propagate more slowly.

Echo Cancelers

In audioconferencing and videoconferencing applications, two distinct types of echo need to be minimized:

- **Electronic echo** This can occur on a line used for bi-directional communication. An electronic echo canceler will attempt to discern which audio came from which direction and cancel it out to avoid echoes, and eventually feedback.
- **Acoustic echo** This refers to the environmental echoes created by the far-site sound bouncing around walls and furniture and returning to the microphones. Acoustic echo cancelers (AECs) are used in conferencing systems. AEC is often one function of a conferencing device or DSP.

The echo-canceling function is rated by the echo's tail length, or the amount of reverberation memory the function has available. Tail lengths may vary from 40ms to as much as 270ms.

Echo cancellation may be implemented in a dedicated echo-canceling device, a DSP, or as a function of a software unified communications system such as a web conferencing application.

TIP Avoid using both hardware and software echo cancellation in the same setup, as the results can be unpredictable and difficult to troubleshoot.

Introduction to Equalization

Equalization is one of the more commonly used functions of the DSP. Equalizers—or EQs—are frequency controls that allow the user to filter and boost (add gain) or cut (attenuate) a specific range of frequencies.

Both the input and output signals of an audio system may be equalized. *Input equalization* is generally for tonality control—adjusting the tonal content so each input sounds as intended. Output equalization is generally used for loudspeaker compensation—adjusting for "quirks" or characteristics of a loudspeaker, or a loudspeaker cluster's, response—and for compensating for the tonal peculiarities of a space ("tuning the room").

TIP   To assist in verifying the uniformity of the frequency response of your design, the AVIXA standard A103.01:202X *Measurement and Classification of Spectral Balance of Sound Systems in Listener Areas* defines the parameters for characterizing the spectral balance of sound in a listening space. The standard, which was in the final stages of development at the time of publication, defines a process to measure, document, and classify a sound system's ability to reproduce a relatively uniform frequency response.

Filters are classified by the rate of attenuation on the signal. This is shown in terms of decibels per octave, where an octave represents a doubling of frequency. A first-order filter attenuates at a rate of 6dB per octave (see Figure 14-12), while a second-order filter attenuates at a rate of 12dB per octave (see Figure 14-13). A third-order filter attenuates at 18dB per octave, a fourth-order filter attenuates at 24dB per octave, and so on. Each order has 6dB more roll-off than the one before it. To show how dramatic this roll-off can become, Figure 14-14 shows an eighth-order filter, which attenuates at 48dB per octave.

NOTE   No amount of equalization can change the acoustics of a space. Acoustic recommendations for AV spaces are covered in Chapter 11.

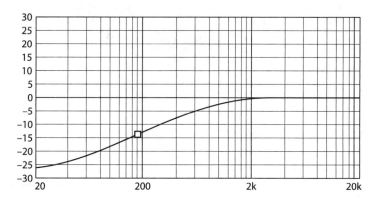

Figure 14-12 First-order filter centered on 180Hz

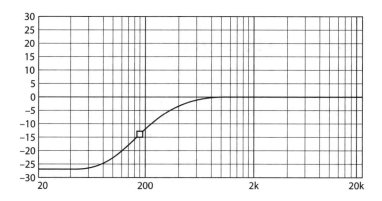

Figure 14-13 Second-order filter centered on 180Hz

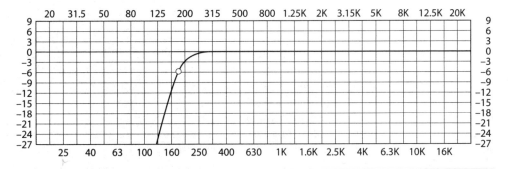

Figure 14-14 Eighth-order filter centered on 180Hz

Parametric Equalizers

A *parametric equalizer*, as shown in Figure 14-15, allows for the selection of a center frequency for boosting or attenuation and the adjustment of the width of the frequency range that will be affected. The filter's width (often called the filter's Q) is the ratio of the height of the peak of the filter against the width of the filter at the 3dB point. A parametric equalizer enables precise frequency manipulation with minimal impact on the adjacent frequencies. It allows users to make many large-scale adjustments to a signal with fewer filters than a *graphic equalizer*, which we will discuss later in this chapter.

Wide Q filters can be used to counteract wide peaks or dips in the frequency response. *High* Q filters, commonly used as notch filters, are used to counteract narrow bands of frequency problems. For example, they are useful for removing primary feedback from fixed microphones or dealing with a specific room resonance.

Figure 14-15
A parametric equalizer screen

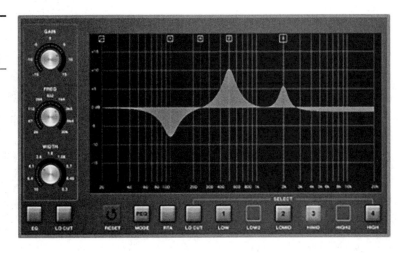

Pass Filters

A *bandpass filter* is a low-Q filter that allows the user to eliminate the highs and lows of a frequency response. While rarely seen in professional audio gear, this type of filter is useful for tone control. Telephone lines also use bandpass filters because most telephone calls do not need to pick up a lot of bass or treble content.

A *low-pass filter* is a circuit that allows signals below a specified frequency to pass unaltered while simultaneously attenuating frequencies above the specified limit. Low-pass filters are useful for eliminating hiss in a system. If you have a source with a lot of hiss, such as a heritage cassette deck or cheap MP3 player, you can apply a low-pass filter to eliminate the high-frequency hiss.

A *high-pass filter* is a circuit that allows signals above a specified frequency to pass unaltered while simultaneously attenuating frequencies below the specified limit. High-pass filters are useful for removing low-frequency noise from a system, such as rumble from a heating, ventilation, and air conditioning (HVAC) system or the proximity effect on a microphone.

A *shelving filter* is similar to a low- or high-pass filter, except instead of removing frequency bands, it simply tapers off and flattens out the sound again. Such filters are useful when the user may want to boost or cut certain frequency bands without eliminating any sound. A *boost shelf*, for example, will let the user boost treble or bass in a car audio system.

Crossover Filters

Crossover filters are used for bi-amplified and tri-amplified systems. By using these filters, you can send low-, mid-, and high-range frequency bands to the appropriate amplifiers. Such filters are extremely useful in large-scale passive (unpowered) concert and event PA systems. Powered (active) concert systems have internal crossover filters in each cabinet.

A crossover separates the audio signal into different frequency groupings and routes the appropriate material to the correct loudspeaker or amplifier to ensure that the individual loudspeaker components receive program signals that are within their optimal frequency range. Crossovers are either *passive* (employing inductor/capacitor/resistor filter networks) or *active* (using electronic signal processing).

Crossover filters often use high-order, roll-off filters, often in the fourth- to eighth-order range. Bass amplifiers will often take the 250Hz or lower range, while the high-pass filter may start at 4kHz.

The crossover point between filters is often 3dB. (In Figure 14-16 the crossover points are represented by the large dots where the filter curves intersect.) In these areas, two different loudspeakers will reproduce that particular frequency band. Because each loudspeaker has a –3dB cut, when they each play that sound, the sound pressure in that area will be doubled, and the audience will perceive that frequency as 0dB. The crossover may not be entirely seamless due to some phase differences between the wavefronts produced by the different loudspeaker drivers.

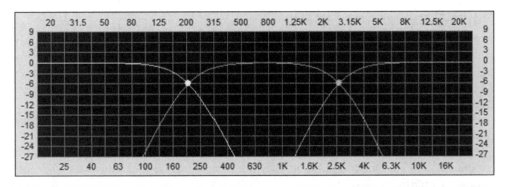

Figure 14-16 Three filters for a smooth crossover

Feedback-Suppression Filters

Feedback-suppression filters are a useful option for unattended speech-only audio systems. They are tight-notch (high-Q) filters that can be used to eliminate specific bands of frequencies. If you've identified a specific frequency band as a source of feedback, you can use a feedback-suppression filter to eliminate the problem.

If you have a fixed microphone in a room, such as a microphone bolted to a witness stand in a courthouse, you may find that the way the microphone interacts with reflections in the room causes feedback along specific frequencies. You can eliminate such problem areas with a feedback-suppression filter. To do so, follow these steps:

1. Turn off all microphones except for the one you're testing.
2. Before you attempt the final system equalization, turn up the gain on the microphone and note the first three frequencies that cause consistent feedback.
3. Construct three tight-notch filters at the input on those frequencies.
4. Once you've created the filters, reset them and equalize the system.

Once the system has been equalized, you are ready to set your feedback-suppression filters. Simply reengage the feedback filters you identified and see whether they improve your operational-level system. Do this for every fixed microphone in your system.

Automated feedback suppression systems track system output, and if the output of any narrow frequency range increases at higher than a preset rate, the suppressor responds by applying a very narrow notch filter to the troublesome frequency. Some automated feedback suppressors reverse the phase of the troublesome frequency to cancel out the acoustic effect.

TIP Like any automated capability (auto volume, auto focus, auto iris, auto white balance, etc.), automated feedback suppression systems may produce a range of unintended consequences, providing a fallback position for occasional, noncritical applications. There is no substitute for a skilled and experienced professional operator where high-quality sound is required.

Noise-Suppression Filters

A noise-suppression (or noise-reduction) filter is based on an algorithm that actively samples the noise floor and removes specific spectral content. In addition to detecting electronic background noise, it listens for regular background acoustic noise from sources such as HVAC systems, exterior traffic rumble, fans, and other gear near microphones (such as laptops), then attenuates that noise.

The canceller depth will depend on the amount of noise present in the room. Quiet conference rooms with little or no noise may not need a noise-suppression filter, while rooms with heavy noise, such as a training room with a large audience, racks of fan-cooled equipment, and loud air conditioning, may need a severe noise-reduction filter. Try these settings as a starting point:

- For eliminating computer and projector fan noise, start at 9dB.
- For eliminating heavy room noise, start at 12dB.

Remember that the purpose of these filters is to remove spectral content. They are not perfect, and they will affect your room response.

Graphic Equalizers

Instead of using parametric equalizers, many audio professionals prefer to use graphic equalizers, particularly across the system output. A common graphic equalizer is the one-third-octave equalizer, which provides 30 or 31 slider adjustments corresponding to specific fixed frequencies with fixed bandwidths, with slight overlaps. The frequencies are centered at every one-third of an octave, starting with the sequence 20Hz, 25Hz, and 31.5Hz and progressing up to 12.5kHz, 16kHz, and 20kHz. The numerous adjustment points allow for shaping the overall frequency response of the system to produce the required effect. The graphic equalizer is so named because the adjustment controls provide a rough visual, or graphic, representation of the frequency adjustments, Active graphic equalizers can provide either boost or cut capability on each filter.

The graphic equalizer in Figure 14-17 has 31 controllable filters. This type of display offers fine control of dozens of specific frequencies, where you can grab any control and add precise amounts of boost or cut to that particular frequency.

In Figure 14-17, the line that runs near each filter control represents the combined interaction between the filters. The bumpy, rippled area around 315Hz is due to phase interference between the filters, which is a downside to using a graphic equalizer. Parametric equalizers will mostly avoid these phase ripples because they use fewer smoother filters. In addition, because they use fewer filters, parametric equalizers in DSPs tend to consume less processing power than a heavily adjusted graphic equalizer.

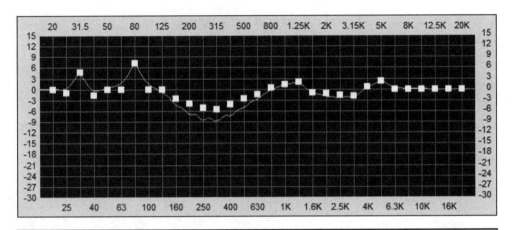

Figure 14-17 A graphic equalizer

Chapter Review

Audio processing is a complex art, and getting it right takes practice. This chapter covered audio sources and destinations, analog versus digital audio, transport methods, DSP architecture types, signal monitoring, DSP functions, equalization, and filters. They're all critical to an audio design that delivers what the client wants from a sound system.

Review Questions

The following questions are based on the content covered in this chapter and are intended to help reinforce the knowledge you have assimilated. These questions are similar to the questions presented on the CTS-D exam. See Appendix E for more information on how to access the free online sample questions.

1. Stereo loudspeaker systems typically use what type of signal?
   A. Analog
   B. RGBH
   C. High gain
   D. Low gain

2. What is the advantage of converting an analog signal to digital?
   A. Digital signals have more signal headroom.
   B. Digital signals can address signal degradation, storage, and recording issues.
   C. Digital signals carry audio and video feeds that analog will not.
   D. Digital signals are more energy efficient to broadcast.

3. When creating a schematic diagram of audio signal flow for a project, what must you include in the diagram?

   A. Microphones, mixers, switchers, routers, and processors

   B. Inputs, outputs, equipment rack locations, ceiling venting

   C. Conduit runs, pull boxes, bends

   D. Construction materials, wall panels, acoustic tiles

4. Two types of meters typically used for signal monitoring are _____ and _____.

   A. VU, PPM

   B. VU, AES

   C. dBU, PPM

   D. EBU, SMPTE

5. For a handheld vocal microphone, you might want set the input gain for at least _____.

   A. 10dB

   B. 20dB

   C. 35dB

   D. 60dB

6. Which of the following is *not* a recommended compressor setting for speech applications?

   A. Ratio: 3:1

   B. Attack: 10 to 20ms

   C. Release: 20 to 50ms

   D. Threshold: 0

7. Eighth-order filters are useful as _____.

   A. Notch filters

   B. Crossover filters

   C. Low-pass filters

   D. Shelving filters

8. When setting a delay filter, it's useful to know that sound travels at how many meters (feet) per second at about 20 degrees Celsius?

   A. 390m (1,280ft)

   B. 305m (1,001ft)

   C. 312m (1,024ft)

   D. 343m (1,125ft))

Answers

1. **A.** Stereo loudspeaker systems typically use analog signals.
2. **B.** An advantage of converting an analog signal to digital is that digital signals can address signal degradation, storage, and recording issues.
3. **A.** When creating a schematic diagram of audio signal flow, you should include microphones, mixers, switchers, routers, and processors.
4. **A.** Two types of meters typically used for signal monitoring are VU (volume unit) and PPM (peak program meter).
5. **C.** For a handheld vocal microphone, you might want set the input gain for at least 35dB.
6. **C.** The recommended release setting for speech applications is actually 200 to 500ms.
7. **B.** Eighth-order filters are useful as crossover filters.
8. **D.** Sound travels at 343 meters (1,125 feet) per second at about 20 degrees Celsius.

CHAPTER 15

Control Requirements

In this chapter, you will learn about
- Modern types of control system and the devices they can control
- Control system configurations
- Control system design and performance verification

Control systems allow people to operate complex AV equipment using simple interfaces. Usually, these interfaces are located in a single convenient location within a room, such as at a lectern or on a wall near a door, or via the user interface of a smart device. Today's users also expect a control system to meet their needs regardless of the interface type.

The best control system designs fulfill users' need to collaborate as well as to control their environments in a familiar and seamless way. When all aspects of a system have been designed correctly, the AV system becomes a powerful business tool for anyone to use.

Because AV systems are highly contextualized, control systems are the "glue" that holds systems together. They make it possible for system designers to create systems out of diverse components from different manufacturers.

Duty Check

This chapter relates directly to the following tasks on the CTS-D Exam Content Outline:

- Duty C, Task 1: Create Draft AV Design
- Duty C, Task 2: Confirm Site Conditions

In addition to skills related to these tasks, this chapter may also relate to other tasks.

Types of Control Systems

What types of devices can users control today? From AV and production equipment to environmental/energy components and security/access control—the number of devices that can be controlled from a central point keeps growing.

What types of control systems can users employ today? The following are just a few:

- **Traditional control system processor (i.e., relays, RS-232, Ethernet, USB, general-purpose input/output [GPIO], and infrared [IR])**   While reliable, this type can also be expensive and require firmware updates and extensive manufacturer training for propriety control software. Examples include Crestron PRO series control processors, Kramer room controllers, AMX NetLinx integrated controllers, and Extron IPCP Pro control processors.

- **IP network-based systems (i.e., all devices, device interfaces, and controllers are nodes on a TCP/IP network, which also carries the AV content)**   While using IP networks, either local, campus area, or wide area, as transport for both the AV content and control signals, this type of control is similar in concept to the traditional control systems with a central controller communicating with individual devices. Like traditional control systems, they require extensive proprietary training for both programming and commissioning. Examples include Q-SYS from QSC, DigitalMedia from Crestron, NAV from Extron, AV over IP from Kramer, and SVSI AVoIP from AMX.

- **AVoIP media server–based systems with interfaces to external I/O devices**   While initially developed as media servers rendering and mapping multistream video and audio for live production, these now include I/O control and interfaces to other devices, including machine control, lighting, effects, and positional-tracking systems. Examples include Disguise Designer, Green Hippo Hippotizer, Dataton Watchout, and AV Stumpfl Pixera.

- **Control built into the system (i.e., performs actions without a secondary interface, and the same manufacturer that provides the system provides the user interface)**   While less expensive and less complex to configure than traditional systems, this type is proprietary and less flexible. Examples include the Aurora range, collaboration systems from Poly, and the MediaLink controller range from Extron.

- **Computer-driven control (i.e., provided and programmed by the end user)**   While highly accessible through the Web and driven by bring your own device (BYOD)/standard mobile devices, this type is dependent on operating system updates and versions and may not be scalable. Examples are Opto 22 and Barco Overture.

- **Cloud-based control**   While capable of controlling large and multiple environments and providing robust backup, failure can be expensive. Examples include Jydo Controls and Surgex Axess Manager.

- **Artificial intelligence–based voice assistant platforms and habitual-learning systems**   While capable of learning user patterns, this type's cloud-based data processing is a continuously developing work in progress, where new end-point device categories spontaneously appear, and may subsequently disappear when the developer has burned through its start-up funding or had a commercial disagreement with the platform owner. Some voice assistants are being used as the primary user interface for AV systems. However, there is a significant risk that such systems may not be expandable or maintainable over the long term. There are also the security, confidentiality, and business continuity risks associated with having all data and processing located in a proprietary cloud. Examples include proprietary cloud-processing voice assistant architectures from Alphabet, Apple, Samsung, and Amazon, each with its own universe of licensed third-party end points.
- **Custom-built control applications**   Perceived as being zero cost, ultimately flexible, infinitely extendable, and free of proprietary lock-ins, custom-built applications, whether built in-house or outsourced to a developer, have never been easier to build than at present, with so many open-source software development tools and packages available. Almost all of the existing commercial control systems grew out of such projects. However, the risks associated with custom-built systems are almost exactly the same as the benefits. The owner has to support, debug, secure, document, maintain, update, and repair the system without the benefit of outside expert help. If the original developer ceases to be part of the organization, the system may become orphaned, frozen in development, and no longer updated with security patches. Changes to the availability or compatibility of the open-source components of the system may require major redevelopment work.

Control System Components

The complete control system typically consists of a central processing unit (CPU), programming, external controls, connected devices, interfaces, software, control points, and the connecting infrastructure. The following sections will discuss the basic subset of those components—the CPU, interfaces, and control points.

Central Processing Unit

In an AV control system, the *central processing unit,* or master control unit, is the "brains" of the control system, typically consisting of a computer processor that communicates with a nominated set of devices over a set of networks. The CPU is rated similarly to a computer system—based on the control protocols, control signal speed, random access and read-only memory (RAM and ROM), number and variety of bus devices, video and audio inputs and outputs, and so on. CPU capabilities determine how many keypads, push-buttons, touch panels, dimmers, media players, computers, audio and video streams, and other elements are possible within the system. Individual control systems can also be linked together to form larger, composite control systems.

Figure 15-1 An AV system CPU master unit (Courtesy of Crestron)

The CPU may be a physical device in the same equipment room or rack as the system(s) it controls; it may be a virtual machine or container hosted on a shared-resource machine within the enterprise data center; or it may be a virtual machine or container hosted somewhere in the cloud.

A CPU can be used to control multiple zones and devices by performing a series of automatic and independent functions. The CPU requires an operating system that allows a custom or fixed AV control program to run the processing operations. The control program defines the choices, variables, calculations, and fixed functions for the processor to choose from.

Processors are a central component of a control system. Some control systems have a single master controller, while others have several. CPU-to-CPU or master-to-master systems refer to systems that enable data processing in more than one location simultaneously, for a combined effect. Small self-contained systems can be linked together to provide better reliability, redundancy, independence, and efficiency. CPU systems provide a hierarchy of control, with one master controller and many subcontrollers and peripheral devices. Automated systems without built-in user interfaces use external interface devices to communicate with the processor.

The master unit (Figure 15-1) provides an example of the range of device interfaces that may be available within a control system. In this case, the master controller provides connections supporting a wide range of devices, including some legacy device interfaces.

Control Interfaces

A *control interface* is a graphical or mechanical interface system that communicates with a controller to initiate command sequences. Let us consider a situation where a user would like to turn on a "Room in Use" sign outside the room by pressing a button on a touch panel. The "button pressed" message from the touch panel is sent to the CPU, which interprets the message as a command to turn on the sign. It then sends out the appropriate signal across a wire/cable/network to the designated control point. At the control point, the appropriate relay is closed to power the sign on. The control system could also be programmed to switch the sign off with a second press of the same touch panel button and/or to vary the appearance of the button (change color or brightness) to indicate that the sign has been activated or deactivated.

Control interfaces can be connected by wired or wireless links. Complex wired control systems include control rooms that operate nuclear plants and systems to control video, lighting, audio, scenery, pyrotechnics, and other elements of live production. Many systems today combine wired and wireless control interfaces with central controllers that have native interfaces for a wide range of input and output modes and protocols. The control interfaces often combine graphical user interfaces, feedback displays, push-buttons, faders, rotary controls, and a wide range of sensors, with visual, haptic (mechanical feedback), and audio indicators. The sensors may include anything from a simple IR beam, IR occupancy detector, smoke density, or positional detector to accelerometers, CO and CO_2 gas detectors, photometers, and noise-level meters.

What type of interface might your users need for their control environment? Control interfaces have different aspects that you'll need to communicate to your client so you can both decide on the best device for the space.

- *Touch-screen devices* (including tablets and touch-screen phones) are a widely adopted style of control (Figure 15-2). They are inexpensive (especially in the case of BYODs) and a familiar operating paradigm. A touch screen offers a great deal of flexibility and style and can be wired or wireless. However, standard touch-screen devices are rarely suitable for operation by the profoundly visually impaired.

- *Smart devices* such as personal tablets, smartphones, smart watches, wireless headsets, and smart loudspeakers are linked via an IP network (wired or unwired) to a server, which parses and interprets gestures and natural-language commands and manages user interaction. Spoken-language user interfaces offer substantial flexibility and interactivity.

- *Control panels* incorporate push-buttons, switches, rotary knobs, display screens, indicators, and sliders. They are the predecessor of touch-screen devices and are often appropriate for use in professional AV environments where critical functions must be controlled.

- *Simple wired panels* are a common user interface. They include button plates, wall switches, indicators, and dedicated buttons.

Figure 15-2
A touch-screen control panel (Courtesy of Crestron)

- *Handheld remotes* and app-enhanced mobile devices may also be used in conjunction with other types of control interfaces. Handheld remotes don't typically include programmable control or sophisticated graphical user interfaces (GUIs). Traditional handheld remotes are preprogrammed and have only push-buttons—devices such as the remote that comes with a TV or Blu-ray Disc player. The signals from the handheld devices can be used to trigger control sequences in the same way as a button on a control panel. Some universal remotes have GUIs, while most incorporate some form of programmable control or macro sequencing capability.

AV design professionals are responsible for creating an easily intelligible, and hopefully intuitive, control system design to ensure the interface is easy for the user to understand and that it is installed in appropriate locations within the space, ensuring easy access within the environment (for example, within reach of a user in a wheelchair and out of direct sunlight that can affect screen legibility).

Smart devices have changed the way people interact with technology. Furthermore, these devices have changed user expectations. Users have been conditioned to expect *discoverability* and *understanding*.

Discoverability allows a user to determine which actions are possible and where or how they can perform them using a device. Understanding gives users a clear sense of how a product or device is supposed to be used. These characteristics are the key elements of *human-centered design* when applied to user interfaces.

Human-Centered Interface Design

Human-centered interface design considers user behaviors, capabilities, and needs before technology and creativity. According to this approach, the design process begins without specifics and focuses on what the user might look for when interacting with devices. The resulting design informs the user if an action is possible, displays the action as it occurs, and notifies the user if something goes wrong.

The following are some of the characteristics of human-centered design:

- **Affordance** This refers to building intuitive interactions between people and their environment (for example, wheels rotate to make adjustments—they do not pull in and out).
- **Signifiers** These signal what actions are possible and how they should be done (for example, doorknobs or handles can signal turning, pushing, or pulling).
- **Constraints** These communicate what the design can or cannot do (for example, push versus pull on a door).
- **Mapping** This refers to laying out the design for clear understanding (for example, turn wheel left to steer left).
- **Feedback** This communicates the results of an action. The feedback must be immediate, informative, and specific (for example, an indicator changes shape, brightness, or color).

Users could get frustrated if they can't figure out how to use the control system to make the lights dim or how to make the screen come down to project their presentations. In some regions, light switches are ON when down; in others, they are ON when up. If users can't figure out the controls, they'll be discouraged from using the interface. A human-centered interface allows the user to utilize a control system without any deep technical knowledge.

For example, users expect some form of feedback confirming a selection they have made in a given space. After pressing the "Screening" button on a touch panel, they expect to see the drapes closed, the projector lowered and powered on, and the button they pressed change in appearance. The actual devices in this scenario do not require two-way communication with the user, but users may still expect some form of positive feedback from the interface.

Consider a few of the elements when designing the user interfaces. You and your client should determine how easily the user can do the following:

- Determine what the system is for
- Tell what device options are possible
- Tell what state the system is in
- Tell if the system is in the desired state
- Determine mapping from system state to interpretation
- Determine mapping from intention to physical movement
- Perform the action with the fewest button presses

Control Points

A control point, such as Figure 15-2, is the vehicle that connects devices to the control system CPU. For example, a touch panel would be a control interface, and the control point could be Wi-Fi, IR, UWB, RS-232, Bluetooth, or Ethernet. When the user touches a button on the touch panel, the control point communicates that information from the CPU to the device(s).

The control point types range from basic (contact closure) to advanced (Internet Protocol) in complexity.

- **Contact closure**   The simplest form of remote control by opening or closing a circuit. Often referred to as general-purpose input/output (GPIO).
- **Analog voltage (a voltage ramp)**   The analog form of variable data control. A voltage is applied to the control point to adjust the level of a specific analog parameter in a given ratio to the control voltage. Analog signals may require one wire per signal (plus a common ground reference/return), or many signals may be multiplexed/sampled over a single wire (plus a common ground reference/return).

- **Wireless optical (IR)**   A mostly one-way optical communication method that commonly uses IR LEDs as transmitters and phototransistors as receivers. In *Li-Fi* data often rides as a modulated signal overlaid on general LED illumination. The device sends a series of coded binary pulses that are recognized by the device's control point. Transmission requires a direct line of sight (or some very good luck).

- **Radio frequency (RF)**   Using radio-frequency wireless transceivers, this medium can carry a wide range of bi-directional control signals across a vast array of frequency bands using many different modulation methods. Formats using RF transmission include Wi-Fi, UWB, AM and FM radio, Bluetooth, Zigbee, DECT, LTE 5G and 6G cellular, and a variety of proprietary techniques over the unlicensed industrial, scientific, and medical (ISM) bands. Coverage is dependent on the operating environment, the frequencies selected, and the chosen modulation method. As many of these wireless signals may use conflicting or adjacent frequencies, it is imperative to include a wireless spectrum plan as part of the system design process.

- **Network**   This works on the basis of transmitting packets of control information. The three forms of such transmission are as follows:

  - **Point-to-point**   Full-duplex or half-duplex serial communication between two devices. Its formats include RS-232, Bluetooth, and USB versions 1 to 3.2.

  - **Broadcast**   A single transmitter of information and multiple receivers. All receivers "listen" to the broadcast but respond only to traffic specifically addressed to that receiving device. Its formats include Ethernet (and Wi-Fi), RS-422 and RS-485, and USB3.2+.

  - **Multidrop**   Multiple transmitters and multiple receivers all interacting on the same wire/cable/fiber/radio channel. This subtype can be complex and requires careful setup for device addressing and line contention. Its formats include RS-422, RS-485, USB3.2+, and Ethernet (and Wi-Fi).

- **Internet Protocol (IP)**   A type of multidrop network that allows for communication between networks via routers and bridges. It provides both control and AV data connections between CPUs, the Internet, control components, applications, and so on.

Control System Design

For AV control systems, the information gathered from the user during the needs analysis should be applied to every step of the design process. AV system designers must determine exactly what kinds of functionality and features the users need. Such an approach helps to avoid and solve many problems early on.

Best Practice for User Experience Design

This approach includes methodologies for incorporating users and their needs in the AV system project process—from the initial identification of a requirement through brief definition, design, prototyping, installation, commissioning, and ultimately, handover, training, and support.

This guides the AV project delivery team in establishing a user-centered design approach for AV systems. The practices can be applied to the design and delivery of any element of the AV user experience, including but not limited to the

- AV products and systems
- Physical and virtual spaces (permanent and temporary) where the products and systems are used
- Content created for use with the AV systems
- Method of user interaction/control
- Professional services that support the AV user experience

The approach defines four phases of user experience design based on the best practices of design-thinking principles:

1. **Understand** Research and gather as much information as possible about the users and stakeholders to understand the project's success factors and specific tasks that end users need or want to accomplish.
2. **Plan/Specify** Using information gathered during the Understand step, identify the priorities and requirements of users to enable design and prototype production.
3. **Produce** Generate a wide range of prototype ideas to solve the identified design challenge. Develop and test the design concepts iteratively.
4. **Evaluate** Formally capture and measure the design/prototype's ability to meet the documented user requirements.

Each phase should be completed at least once during any project. Ideally, these four phases would be repeated at more than one point during the project.

At the time of publication, an AVIXA volunteer task group of subject matter experts was working toward developing this best practice into an AVIXA standard. Watch for its future appearance on the AVIXA website at avixa.org/standards.

Needs Analysis

These are some general concepts to ask about or listen for when interviewing the client:

- **Simplicity of operation** Who are the users of the system? What is their level of technical experience?
- **Device automation** What do they want the system to do autonomously?
- **Multiple device integration** How many devices do they want to control?

- **Device operation from multiple locations**   Where should interfaces be located?
- **Device operation at a distance**   How do they want to control devices? Consider any preexisting systems or future expansion issues.
- **Security of access**   Do any control interfaces require secured access?
- **Cost**   What is the project budget?

For each device to be controlled by the AV system, choose an appropriate control interface by considering the following issues:

- Can the CPU accommodate all of the system's needs?
- What control interfaces are available and required for this device?
- How does the selection affect other subsystems?

CPU Configurations

Devices may need to be controlled in multiple rooms, across a college campus, or even in another part of the world. In these advanced configurations, multiple interconnected CPUs may be required. These configurations are often referred to as *centralized, client-server,* and *distributed*. Control system professionals use these terms for one or many processors.

A centralized system consists of a single CPU and many devices and is the most common configuration in the field.

Figure 15-3 shows a configuration known as *client-server*. If the control CPU fails, the client CPU also eventually fails because it cannot receive further instructions from the primary CPU. The client CPU can act like a pass-through or limited instruction buffer. In other words, the client CPU is receiving control from the actions of the user interface (UI) connected to the control CPU. In Figure 15-3, the client has no associated control surface or UI (touch panel, button panel, control panel), so it only relays the commands from the control CPU.

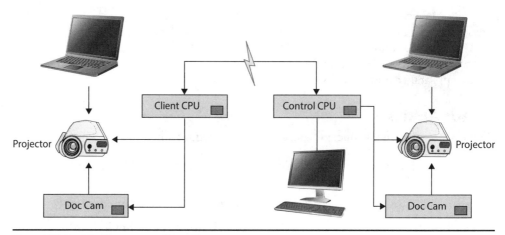

Figure 15-3   Client-server configuration

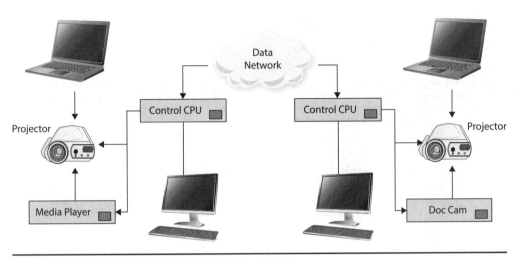

Figure 15-4 Distributed processing

The configuration shown in Figure 15-4 is known as *distributed processing*. It has multiple independent systems linked via a data network. If the network link fails, each system can still function independently, but task coordination and synchronization between the systems are lost.

 NOTE The figures display CPU configurations, not signal flows or wiring schemes.

Programming for Control

After they understand which devices the client needs to control, AV system designers should draw or write out how these devices can be controlled with control points. This step includes transferring the list of devices and device actions to a *control functions script*.

As shown in Table 15-1, a control functions script is an itemized list of devices to be controlled that helps to organize what needs to be done by everyone on the project, such as the client, programmers, designers, and facility engineers. It contains the required actions for each device and interconnection information, such as the control point and connector type. Some of the items on the list may not be directly AV related or implemented by AV installers (for example, drapes, lights, rigging, atmospheric effects, and temperature sensors).

A control functions script also assists the designer in creating a control system that meets the required client needs. System designers must discuss the control functions script and sketches with the client to receive their approval of the overall system functionality.

Control Functions Script

For Upgraded Conference Room A719 AV System

Purpose

This document complements the preliminary design documentation for the new AV system in conference room A719 at the Tyrell Corporation headquarters. It will be updated with more detail as the system design is finalized and approved by stakeholders and Tyrell Corp IT. The new AV system will support team meetings and general building and/or corporate communications—see AV program report for further detail.

AV Devices

Device ID	Description	Make	Model	Control Port	Summary Actions
AV1	Local AV control interface (new)	Hyperdyne	Int123	LAN "AV12," RS232, digital I/O, USB 3.0	Interface between cloud control processor and room-based devices
AV2	Cloud remote processor and software (existing)	Hyperdyne	CC2.0	LAN – private subnet "AV12"	Monitoring and management of AV system(s)
AV3	Control panel (new)	Hyperdyne	TPA8	LAN "AV12"	End-user UI
AV4	75-inch monitor (new)	TBD	TBD	RS232	On, Off, Input select
AV5	Wireless presentation and conferencing hub (new)	Fabulous	Tech 1A	LAN "AV12"	API integration of all core meeting functions
AV6	Integrated conference camera and soundbar (new)	TBD	TBD	USB 3.0	Camera Mute, Auto Tracking, Sound Up, Sound Down, Mic On, Mic Off
AV7	Digital signage player (new)	PCPlay	Sgn4	LAN "DS2"	API for setting streaming mode and remote management

Non-AV Devices

Device ID	Description	Make	Model	Control Port	Summary Actions
FM1	Room occupancy sensor (existing)	Towertech	BA1	Contact closure	Occupied, Non-Occupied (auto presets)
FM2	DALI lighting interface (existing)	Towertech	LED7	Digital I/O	On, Off, 50%, 75%

Control actions narrative:

Unoccupied room (FM1) *with no motion activity or user interface button pushes* (AV3) for >7 minutes turns off room AV equipment via AV1. Note: Unless activity by user on AV3, the room lighting is controlled by non-AV third-party building management system.

Occupied room (FM1) *with no activity on user interface* (AV3) turns on room monitor (AV4) via AV control interface (AV1) and switches input to display building digital signage (AV7).

Occupied room (FM1) *with activity on user interface* (AV3) turns on room monitor (AV4), if not already on, via AV1 and switches input to display UI selection.

Occupied room (FM1) *with activity on user interface* (AV3) with user selection of "Conferencing" executes AV3 UI page flip to conferencing options for presentation hub (AV5) and conference camera/soundbar (AV6).

Table 15-1 A Control Functions Script

Establishing Control Points

In some cases, the design is complex and requires a careful selection of control points, such as in a system that has a large number of conference microphones, simultaneous interpretation systems, or voting systems, or is an immersive experience, a live production, or a complex museum exhibit. Regardless of the project complexity, system designers should refer to the approved control functions script and choose control points that most efficiently meet the design need.

Once the control points are selected, the control functions script should be updated to reflect the choice. The next step in the process is to verify that the design can be implemented.

Verifying System Performance

During the system's performance verification, the AV team ensures that devices can be controlled correctly and that the system matches up with the client's narrative. It is important to stay informed about product changes. For example, a user might not be able to control a device using a previous version of a control point three months after the implementation.

Communicate the verification results to the client, even if the system passes the verification test. Refer to ANSI/AVIXA 10:2013, *Audiovisual Systems Performance and Verification,* and the related guide for more information on how you can test your system's performance.

Finally, start formally designing your control systems around what you've discovered from the previous steps. This way, the problems will be taken care of and you won't have to redesign the system.

Chapter Review

In this chapter, you learned that the success of a control system project largely depends on the choices you as an AV professional make. If your control system focuses on human behavior first and technology second, then users will be even more empowered to take advantage of all they have to offer.

You also learned that when designing a control system, things can get pretty complex. It's especially important to establish a good rapport with clients and the users and communicate the project progress and changes on a regular basis. The next chapter will cover network design for AV systems, which you will find important for control system design.

Review Questions

The following questions are based on the content covered in this chapter and are intended to help reinforce the knowledge you have assimilated. These questions are similar to the questions presented on the CTS-D exam. See Appendix E for more information on how to access the free online sample questions.

1. Which of the following is a type of control system?
   A. Computer-driven
   B. Cloud-based
   C. Habitual learning
   D. All of the above
2. Which of the following control system components is used to control multiple zones and devices by performing a series of automatic and independent functions?
   A. Control interface
   B. Control point
   C. Central processing unit
   D. Control system software
3. What is the purpose of mapping in human-centered interface design?
   A. Communicate the results of an action
   B. Communicate what the design can or cannot do
   C. Lay out the design for clear understanding
   D. Signal what actions are possible and how they should be done
4. Which CPU configuration has multiple independent systems linked together by a data network?
   A. Client-server
   B. Centralized
   C. Scattered
   D. Distributed
5. What list helps to organize what needs to be done by everyone on the project?
   A. System component list
   B. Control functions script
   C. Materials list
   D. Control object list

Answers

1. **D.** Control systems can be computer-driven, cloud-based, or habitual-learning systems.
2. **C.** A central processing unit is the part of a control system used to control multiple zones and devices by performing a series of automatic and independent functions.

3. C. The purpose of mapping during human-centered interface design is to lay out the design so it's easier to understand.

4. D. A distributed CPU configuration includes multiple independent systems linked together by a data network.

5. B. A control functions script lists the devices to be controlled and helps organize what needs to be done by the client, programmers, designers, and facility engineers.

CHAPTER 16

Networking for AV

In this chapter, you will learn about
- Common networking components
- The layers of the Open Systems Interconnection model and the functions of each layer
- Common physical network connections and their capabilities
- The function and capabilities of Ethernet technologies
- The function and capabilities of Internet Protocol technologies
- The pros and cons of various Internet Protocol address assignment methods
- The differences between Transmission Control Protocol and Universal Datagram Protocol
- The characteristics of AV over Internet Protocol networks

Modern AV systems use IT networks to support their systems in all kinds of ways. Technicians can monitor, control, and troubleshoot devices over the network. AV content is stored on the network. Audio and video signals can be transported over the network infrastructure in real time.

Realizing the potential of networked AV systems requires you to work closely with IT professionals. You need to be able to understand and use the correct technical terminology and jargon to explain your needs and understand their concerns. In this chapter, you'll get an overview of networking technologies and gain an understanding of your networking needs so you can communicate them better with your IT counterparts.

> **Duty Check**
>
> This chapter relates directly to the following tasks on the CTS-D Exam Content Outline:
>
> - Duty A, Task 3: Educate AV Clients
> - Duty A, Task 4: Review Client Technology Master Plan
> - Duty A, Task 5: Identify Client Expectations
> - Duty B, Task 7: Coordinate with IT and Network Security Professionals
> - Duty B, Task 8: Coordinate with Acoustical Professionals
> - Duty C, Task 1: Create Draft AV Design
> - Duty C, Task 2: Confirm Site Conditions
>
> In addition to skills related to these tasks, this chapter may also relate to other tasks.

What Is a Network?

A network can consist of anything that is interconnected in a netlike manner or that is linked together to pass information. In the IT world, *network* is generally short for "computer network" or "data network."

All networks consist of two basic parts: nodes and connections. Nodes are the devices that send and receive data. In the early days of networking, a node was basically a computer. Today, a node could be a computer, a mobile phone, a media server, an Internet of Things (IoT) device, a mixing console, a video projector, or any other content-carrying device. On Internet Protocol (IP) networks, a node is a device that has an IP address. Active network components such as routers, wireless access points, and switches are also nodes.

Connections are the means by which data travels from one node to another. The base-level connection could be any physical signal transmission medium such as radio frequency (RF) waves, copper cable, or fiber-optic cable. Passive devices such as patch panels also fall into the category of physical connections or links.

Network Components

Beyond nodes and connections, network components (Figure 16-1) can be broken down into the following:

- Clients such as
  - Amplifiers
  - Audio mixers
  - Cameras

- Desktop computers
- IPTV receivers
- Laptop computers
- Lighting consoles
- Loudspeakers
- Luminaires
- Media recording and replay systems
- Mobile telephones
- Printers and plotters
- Projectors
- Tablets/touch screens
- Talkback systems
- Video displays
- Videoconference/teleconference systems
- Vision mixers/switchers
- VoIP telephony/unified communications (UC) systems
- Winch controllers
- Network interface cards (NICs)
- Servers such as
  - Database servers
  - File servers
  - Media servers
  - Web servers
- Switches and routers such as
  - Bridges
  - Ethernet switches
  - Gateways
  - Modems
  - Routers
  - Wireless access points
- Links such as
  - Fiber
  - Wired
  - Wireless

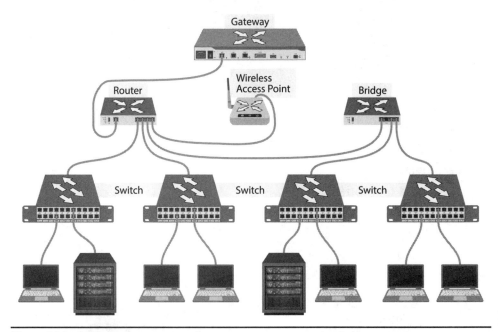

Figure 16-1 Network components: clients, servers, routers, and switches

- Protocols such as
  - Ethernet
  - Internet Protocol (IP)
  - Transfer Control Protocol (TCP)
- Applications such as
  - AV over IP
  - Dante
  - File Transfer Protocol (FTP)
  - Hypertext Transfer Protocol (HTTP)

Of these components, clients, servers, NICs, switches, routers, gateways, and links are considered hardware, while protocols and applications are considered software parts of a network. In this section, we will cover hardware components and focus on protocols later in the chapter.

Clients and Servers

Servers are used to share resources among connected nodes. They may be a stand-alone device devoted to providing a single service such as database storage, media storage, or video rendering, or they may be a powerful device providing a range of services to the network clients, including such things as website services, file storage and transfers, a network time reference, and printing services. If a hardware device does nothing but provide services, it is called a *server*. However, the term *server* can also refer to the software that provides a service alongside other programs on a hardware device. In a client-server network architecture, a server provides services to client nodes.

Servers on larger networks are usually hosted on many separate computers. Organizations may host several servers on each computer, or they can have a dedicated computer for each service. A computer whose only task is to perform a particular service is known as a *thin server*. Typically, a thin server resides on a dedicated computer, virtual machine, or container, configured with only the functionality required to perform the service. This increases the resources available for the server's dedicated task. A thin server is not necessarily a low-powered device. A video-rendering server, for example, may be equipped with many parallel-video processors and vast amounts of memory to perform its intensive task, but as it only performs a single task, it is classified as a thin server.

NOTE It is standard security practice to disable all unrequired features on a system to avoid providing unintended access via unused ports, services, or functions.

Network Interface Cards

Every device that connects to a network must have a *network interface card* and the associated *Media Access Control* (MAC) address.

A NIC is an interface that allows you to connect a device to a network. At one time, most devices had a separate card or adapter. Today, though, NICs are typically integrated into the device's main circuitry.

A MAC address is the actual hardware address, or number of the NIC device. Every NIC has a globally unique MAC address that identifies its connection on the network. MAC addresses use a 48-bit number, expressed as six groups of two hexadecimal numbers, separated by a hyphen or colon. For example, *24:4b:fe:8d:87:97* is the MAC address of the Ethernet port connecting the author's computer to its local network. The first part of the address (*24:4b:fe*) indicates the interface controller manufacturer (in this case, Asustek Computer, Inc., in Taiwan), and the second part of the address is a serial number for the product or circuit component. In devices with multiple network connections, each NIC will have its own associated MAC address.

TIP As it is possible to change the MAC address on some devices, it is not entirely unknown for two NICs on a same network to have identical MAC addresses. In cases of inexplicably strange communications problems when installing devices on a network, it is worth checking for duplicate MAC addresses.

Network Devices

As shown in Figure 16-1, devices that are used in data networks include switches, routers, gateways, and wireless access points. Blended devices combine the functionality of several devices. We shall take a look at each device's functionality:

- **Blended devices**   It is important to note that while the functionality of gateways, routers, and switches is discussed separately here, they do not have to be separate physical devices. Much like an audio digital signal processor (DSP) may combine the separate functions of a mixer, equalizer, delay, and compressor/limiter, networking devices often include several functions in a single box. For example, routers often include the functionality of a switch and a Wi-Fi controller. On a small network, there will probably not be a separate gateway, router, Wi-Fi controller, and switch. In fact, there may be only one router, which acts as a gateway to the Internet, a wireless access point for local Wi-Fi connections, and a switch directing traffic among the devices on the network.

- **Switch**   A network switch provides a direct physical connection between multiple devices through storing and forwarding data. As each device is initially connected, the switch collects and stores its MAC address. This allows devices connected to the same switch to communicate directly with each other via Ethernet, using only MAC addresses to locate each other. A switch acts as a multiport bridge between the network segments connected to its ports. There are two types of switches: unmanaged and managed.

    - Unmanaged switches have no configuration options. You just plug in a device, and it connects.

    - Managed switches allow the network technician to perform functions such as adjust port speeds, set up virtual local area networks (VLANs), configure quality of service (QoS), and monitor data traffic. Managed switches are commonly used in corporate and campus environments and are a requirement for some networked AV applications.

- **Router**   A router forwards data between devices that are not directly physically connected. Routers mark the border between a local area network and a wide area network. After data leaves a local area network (that is, travels beyond the switch), routers direct it until it reaches its final destination. When data arrives at a router, it examines the data packet's logical address (the IP address) to determine its final destination and/or next stop.

- **Gateway**   A gateway connects a private network to outside networks. Similar to a router, a gateway examines a data packet's IP address to determine its next destination. A gateway can also forward data between devices that are not directly physically connected. All data that travels to the Internet must pass through a gateway. Routers below the gateway will forward packets destined for any device that cannot be found on the private network to the gateway. When traffic arrives from outside the private network, the gateway forwards it to the appropriate router below. Gateways also translate data from one protocol to another. For example, when data leaves a private network to travel across the Internet, a gateway may translate it from a baseband to a broadband protocol.

- **Bridge**   A bridge forwards data between multiple networks or network segments, creating a single aggregated network. The bridge reads the destination on an incoming packet, stores the data, then forwards it to whichever port connects to the destination device. In a local area network (LAN), the devices are identified only by their MAC address. Most network switches are multiport bridges.
- **Wireless access point (WAP)**   A WAP is a common blended device, combining a Wi-Fi controller and an Ethernet bridge. It forwards packets between connected Wi-Fi devices and a LAN, based on the MAC address of the connected devices. In an access point, the wireless devices are in the same subnet as the Ethernet port.

Links

There are a lot of different physical technologies used to transport data over long distances for wide area networks: optical fiber, coaxial cable, satellite, twisted-pair, DSL, fixed wireless, and so on. The network connections that AV professionals have to deal with directly, though, are primarily those used within local area networks—the physical connections within a system or building. For local area network connections, the three most common physical transmission methods are as follows:

- Over copper wire, as a current
- Over glass or plastic fiber, as light
- Over the air, as radio frequency wireless

Of course, it's not as simple as it sounds. Within each transmission medium, there are several options. One copper wire network cable may have very different capabilities, limitations, and internal design than another. The same is true for fiber and wireless networks. As an AV professional, you should know which medium is best for your application and why.

Twisted-Pair Network Cables (Cat x)

Since the adoption of four-pair, 100Ω, UTP Category 3, 16MHz-rated cable for telephony, and later for 10Mbps Token Ring and Ethernet networking, the noise immunity and data speed ratings of four-pair 100Ω twisted-pair data cables have progressively improved. Each new generation of cable is tested and assigned a data category rating by the Telecommunications Industry Association (TIA) and the Electronics Industries Association (EIA). The most widely used of these are

- **Category 5 (Cat 5)**   Bandwidth 100MHz for data speeds up to 100Mbps over distances up to 100m (328ft) (see Figure 16-2).
- **Category 5e (Cat 5e)**   An enhanced version of the 100MHz Cat 5 cable standard that adds specifications to reduce far-end crosstalk (FEXT) for data transmission up to 1Gbps over distances up to 100m (328ft).
- **Category 6 (Cat 6)**   Cat 6 has even more stringent specifications for crosstalk and system noise than Cat 5e, with a bandwidth of 250MHz and data speeds up to 10Gbps over distances up to 55m (180ft). Sometimes used with an overall shield around the four pairs (S/FTP).

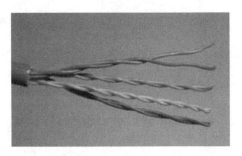

Figure 16-2
Category 5 cable with four unshielded twisted-pairs

- **Category 6A (Cat 6A)**   Constructed using a mechanism to physically separate the twisted pairs, with more stringent specifications for alien and near-end crosstalk (NEXT), with a bandwidth of 500MHz and data speeds up to 10Gbps at distances up to 100m (328ft). Sometimes used with an overall shield around the four pairs (S/FTP).
- **Category 7 (Cat 7)**   To reduce crosstalk, each twisted-pair is shielded and the entire cable is shielded, for a bandwidth of 600Mhz and data speeds up to 10Gbps at distances up to 100m (328ft). Cat 7 is not recognized as a standard by the TIA or EIA and is not terminated with an RJ45 (8P8C) connector.
- **Category 7A (Cat 7A)**   To reduce crosstalk, each twisted-pair is wrapped in a foil shield and the entire cable is covered with a braided shield, for a bandwidth of 1GHz and data speeds up to 10Gbps at distances up to 100m (328ft). Cat 7A is not recognized as a standard by the TIA or EIA and is not terminated with an RJ45 (8P8C) connector.
- **Category 8 (Cat 8)**   Constructed with a shield around each twisted-pair and a further shield enclosing all twisted-pairs, with a bandwidth of 2GHz and data speeds up to 40Gbps for runs up to 30m (100ft). Cat 8.1 is terminated with an RJ45 (8P8C)–compatible connecter; Cat 8.2 uses a lower-noise connecter that is not RJ45 compatible.

Shielded Twisted-Pair Network Cables

Shielding isolates and protects a signal from sources of electromagnetic (EM) and RF interference. Shields can be implemented in a variety of ways. They can be used to provide overall coverage around a single insulated conductor, or around individual insulated conductors, or twisted-pairs of conductors in a multiconductor cable. Shielding is also used to provide all types of data with protection from electromagnetic interference (EMI) and radio frequency interference (RFI) on longer cable runs or within severe RF fields.

Overall shielding is often added to twisted-pair network data cables both to exclude external EM and RF noise from affecting the data on the twisted-pairs inside and to prevent the emission of RF from the signals on the twisted-pairs into the outside world.

Shielding may also be placed around the individual twisted-pairs to further isolate them from exterior EM and RF interference and, more importantly, to isolate each pair from the emissions of the other pairs in the cable (crosstalk). Cat 5, Cat 5e, Cat 6, and Cat 6A are available as either unshielded or shielded cables, but Cat 7, Cat 7A, and Cat 8 must be shielded to meet their specifications.

Chapter 16: Networking for AV

ISO Format	Common Industry Names	Overall Shielding Type	Twisted-Pair Shielding Type
U/UTP	UTP	None	None
F/UTP	FTP, STP, ScTP (screened twisted-pair)	Foil	None
S/UTP	STP, ScTP	Braiding	None
SF/UTP	SFTP, S-FTP, STP	Braiding and foil	None
U/FTP	STP, ScTP, PiMF (pairs in metal foil)	None	Foil
F/FTP	FFTP	Foil	Foil
S/FTP	SSTP, SFTP, STP, PiMF	Braiding	Foil
SF/FTP	SSTP, SFTP	Braiding and foil	Foil

Table 16-1 Some Twisted-Pair Cable Shielding Types and Their Common Industry Names

The shielding may be constructed of either foil tape (type F) or braided wire (type S), or sometimes both (type SF). Unshielded cable is identified as type U. The variations of shielding on twisted-pair data cables are identified by the ISO using the following format:

X/YTP
Where X is the overall cable shield type
Y is the shield type on the individual twisted-pairs
TP simply indicates that the cable contains twisted-pairs

As you can see from Table 16-1, there are often several different industry names for the same cable, and sometimes the same industry name for different cables. It is essential that an AV systems design specifies the correct network cabling for the project and is able to communicate with electricians and system installers to ensure the correct cable has been supplied.

Fiber-Optic Cable

As shown in Figure 16-3, fiber-optic cable uses fibers made from transparent glass or plastic. These cables are commonly used for both long-distance and high-bandwidth data transmission. They are also used to provide galvanic isolation between devices where high voltages such as lightning strikes are possible, for RFI immunity where high levels of RF are present in locations near high-power transmitters, and where EMI is present near high-voltage electrical transmission lines. Fiber-optic cable offers many benefits for the transport of audiovisual content and other high-bandwidth data.

Figure 16-3 A multicore fiber-optic cable with an outer sheath

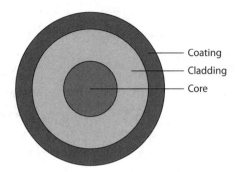

Figure 16-4 A cross-section of optical fiber

Fiber-optic cable has a transparent glass or plastic core used for carrying modulated light, as shown in Figure 16-4. The cladding is the next layer of the cable, which is used to reflect the light back onto the core. This keeps the signal moving down the fiber. The next layer is the coating, which protects the fiber from damage. A sheath is usually added to provide mechanical strength and additional protection to the cable in harsh environments.

Two modes or paths of light transmission are used in fiber-optic cable, known as single-mode and multimode. As shown in Figure 16-5, single-mode fiber is narrow in diameter, which makes the signal path mostly straight, with little of the light reflecting off the walls of the glass.

In multimode transmission some of the signal goes straight down the fiber, while the rest repeatedly reflects off the interface with the cladding. As a result of these multiple reflection paths, the signals take slightly different times to reach the end of the fiber, making it more difficult to detect the edge of each data pulse being received. As signal dispersion increases with the length of a multimode fiber, signals can typically travel more reliably over extended distances using single-mode fiber than with multimode. However, the wider multimode fiber is more easily able to carry multiple signals at different wavelengths (colors), a technique known as wavelength division multiplexing (WDM), which enables multiple channels of data to be carried.

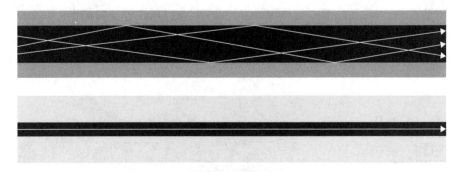

Figure 16-5 Longitudinal sections of multimode (top) and single-mode fiber-optic cable showing light paths

Type	Typical Jacket Color	Max Range at 100Mbps	Max Range at 1Gbps	Max Range at 10Gbps	Max Range at 40Gbps	Max Range at 100Gbps
OM1 multimode	Orange	2km (1.24mi)	275m (900ft)	22m (72ft)	Not specified	Not specified
OM2 multimode	Orange	2km (1.24mi)	550m (1,800ft)	82m (270ft)	Not specified	Not specified
OM3 multimode	Aqua	2km (1.24mi)	550m (1,800ft)	300m (980ft)	100m (330ft)	100m (330ft)
OM4 multimode	Aqua or violet	2km (1.24mi)	1km (3280ft)	400m (1,300ft)	150m (490ft)	150m (490ft)
OM5 multimode	Lime green	2km (1.24mi)	1km (3280ft)	400m (1,300ft)	150m (490ft)	150m (490ft)
OS1 single mode	Yellow	10km (6.2mi)	10km (6.2mi)	10km (6.2mi)	10km (6.2mi)	10km (6.2mi)
OS2 single mode	Yellow	200km (124mi)	200km (124mi)	200km (124mi)	200km (124mi)	200km (124mi)

Table 16-2 Optical Fiber Types and Data Ranges

Distance Limitations of Fiber

There are distance limitations with fiber-optic cables based on the type or mode of the cable, as shown in Table 16-2. Single-mode fiber is loss limited, meaning you will run out of light before the quality is compromised. Multimode fiber is bandwidth limited because although there is a lot of light, there is high dispersion, resulting in the gaps between pulses becoming more difficult to detect.

Fiber-optic cabling can be used in many different environments: indoor, outdoor, hybrid, or composite. Indoor fiber cabling is similar to copper cabling, with plenum, riser-rated, and *low smoke zero halogen* (LSZH) cable options. Riser-rated cable can support its own weight as it rises up the building floor by floor.

Outdoor fiber cabling is mainly used for aerial or in-ground installations, but is also used for rental and staging applications.

Hybrid cabling combines both fiber and copper. For example, you could use hybrid cabling for a cable run to a high-definition camera, where fiber will carry the video signal and copper will power the camera electronics.

Composite combines both single-mode and multimode fiber. For instance, you might need to send 12 fibers down a cable. Six would be for single-mode, and six would be for multimode. Typically, you would use only one of these modes, but if two companies are sharing a space, the composite option would be more appropriate. Your selection should be based on the application.

 NOTE It is important to check that the fire-safety rating of the cables you are specifying matches or exceeds the requirements of the AHJ over an installation.

Fiber Applications

Optical fiber cables are constructed for specific applications and to meet specific installation requirements.

OFC	Optical fiber, conductive
OFN	Optical fiber, nonconductive (can replace OFC)
OFCG	Optical fiber, conductive, general use
OFNG	Optical fiber, nonconductive, general use (can replace OFCG)
OFCP	Optical fiber, conductive, plenum
OFNP	Optical fiber, nonconductive, plenum (can replace OFCP)
OFCR	Optical fiber, conductive, riser
OFNR	Optical fiber, nonconductive, riser (can replace OFCR)
OPGW	Optical fiber, composite overhead ground wire
ADSS	All-dielectric self-supporting
OSP	Optical fiber, outside plant
MDU	Optical fiber, multiple dwelling units

Table 16-3 Some Common Abbreviations for Fiber Cable Types

Risers To allow fiber-optic cables to be safely suspended in such places as vertical shafts in a building core or in runs between different floors of a building (risers), some cables have strengthening elements (often flexible metal-wire or heavy nonoptical glass fiber) included to carry the weight load of the cable without stressing the optical communication fibers. If the cable is also fire-resistant, it is considered to be riser-rated.

Whether or not there are current-carrying conductors in a riser cable, those using conductive metal wire as the load-carrying element may be rated as "conductive" cables by fire-safety authorities.

Plenums Every cable, electrical or optical, that is to be used in ceiling spaces, air ducts, and spaces that carry air for HVAC systems (plenums) must be constructed of materials that are fire-resistant and that produce no toxic gases on combustion, e.g., LSZH cables. This construction reduces the risk of toxic fumes entering a building's ventilation systems during a fire outbreak. These are classified as plenum-rated cables.

Fiber cable types may be identified in a number of ways, as listed in Table 16-3, with the ratings in the U.S. National Electric Code being one of the more common naming schemes.

Common Fiber Connectors

A variety of fiber connectors are used in AV applications, some of which can be seen in Figure 16-6. Some are suitable for permanent installations, while others are more suited for flexible, and more hostile, operating environments. Take a look at each in the following sections.

ST Connector The ST, or straight-tip, connector can be found on transmitter-receiver equipment and is similar to the BNC connector used in video and RF applications. It is a bayonet connector, meaning that all you have to do is "stab and twist" to lock it into

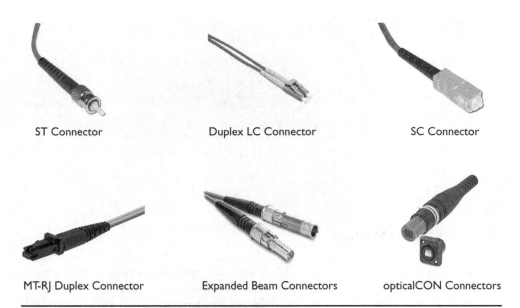

Figure 16-6 Some commonly used optical connectors

place, which keeps the fiber and ferrule from rotating during connection. This connector can be used on both multimode and single-mode fiber-optic cable.

LC Connector The LC, or Lucent, connector (available in single and duplex formats), is much smaller in diameter than the ST and is used for basic wiring applications and AV systems from many manufacturers. It has great low-loss qualities and is known as a "push-pull" connector.

SC Connector The SC, or subscriber, connector is larger in diameter than the LC, as shown in Figure 16-6. It is a "stab and click" connector, which means that when the connector is pushed in or pulled out, there is an audible click because of the attachment lock. This is a great connector to get in and out of tight spaces.

MT-RJ Connector The Mechanical Transfer Registered Jack (MT-RJ) connector shown in Figure 16-6 is a compact, small form-factor, dual-fiber (duplex) connector, similar in style to the RJ45 (8P8C). It was designed as a lower-cost, more compact replacement for the SC connector in data network applications.

Expanded Beam Connector The expanded beam connector is easy to use in an outside environment because it is much more tolerant to dust and dirt and easier to clean than the other connectors.

opticalCON Connector The opticalCON is a rugged, easy-to-use connecter designed for repeated connection and disconnection in hostile environments such as live production, roadshows, outside broadcasts, and exhibitions. The panel connector is actually a pass-through connector that accepts one (duo variant) or two (quad variant) duplex LC connectors on its back end.

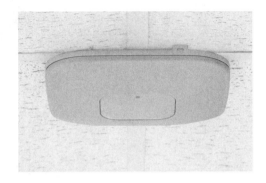

Figure 16-7 A typical wireless access point (Image: thongseedary / Getty Images)

Wireless Connections

In addition to wired connections, you are likely to encounter the need for wireless connection from clients who want to control devices, connect to portable devices, or send content wirelessly over the network. Figure 16-7 shows a typical wireless access point.

The wireless connection known as Wi-Fi is defined by the IEEE 802.11 standard. This standard is a set of MAC and Physical layer specifications for implementing wireless LANs. It has been amended several times to provide the basis for the latest wireless network products. The speed of a Wi-Fi connection depends on the RF signal strength and the revision of 802.11 with which you connect. As signal strength weakens, the speed of the connection slows. The number of users accessing the wireless devices also affects connection speed. On the other hand, the 802.11ah standard, known as Wi-Fi HaLow, is intended for use over distances of up to 1km (3,280ft) at lower data rates for IoT applications. It is quite suitable for AV control (not AV over IP) applications.

The most commonly used versions of Wi-Fi today are 802.11n (Wi-Fi 4), 802.11ac (Wi-Fi 5), and 802.11ax (Wi-Fi 6/6E). To reduce the need to explain the differences between various revisions of the 802.11 standard to end users, the Wi-Fi Alliance has taken to naming these revisions "*Wi-Fi N*," with the implication being that the larger the value of "*N*," the faster the connection. Table 16-4 shows the frequency band, typical throughput, and maximum throughput for each standard.

Wi-Fi is extremely popular, both among users and manufacturers. It is difficult to find laptop and desktop computers today that do not have factory-integrated wireless technology. Computer manufacturers assume users would rather connect wirelessly than via a network cable. This might be true for the home user. For the enterprise network, however, Wi-Fi can create more problems than it solves. Before choosing Wi-Fi as a physical connection medium, you should carefully weigh its advantages and disadvantages.

Wi-Fi Advantages

Wi-Fi allows users great flexibility. It is a lot cheaper than cable. It is also the default customer expectation. Users expect to be able to walk into a meeting room or presentation space, connect to the Internet, and launch a presentation from their laptops, tablets, or mobile phones. You could try to address this requirement by providing a connection for every device imaginable. Or, you could let users connect to the presentation system using

Revision	Release Date	Frequency Band	Typical Throughput	Maximum Throughput
802.11a Wi-Fi 2	October 1999	5GHz	27Mbps	54Mbps
802.11b Wi-Fi 1	October 1999	2.4GHz	~5Mbps	11Mbps
802.11g Wi-Fi 3	June 2003	2.4GHz	~22Mbps	54Mbps
802.11n Wi-Fi 4	September 2009	5GHz and/or 2.4GHz	~144Mbps	600Mbps
802.11ac Wi-Fi 5	December 2013	5GHz	~433Mbps	3.5Gbps
802.11ad	December 2012	60GHz	~1.5Gbps	6.7Gbps
802.11ah Wi-Fi HaLow	December 2016	~900MHz (ISM/unlicensed)	150kbps over 1km	78Mbps
802.11ax Wi-Fi 6/6E	December 2018	5/6GHz and/or 2.4GHz	~4.8Gbps	11Gbps
802.11ay	November 2019	60GHz	~6Gbps	20Gbps
802.11be Wi-Fi 7	Early 2024*	5/6GHz and/or 2.4GHz	~9.6Gbps*	40Gbps

*Based on information available at the time of publication.

Table 16-4 802.11 Wi-Fi Standard Revisions

an ad hoc Wi-Fi network. Wi-Fi just seems easier for several reasons. The following are the advantages of Wi-Fi:

- Wireless hotspots are everywhere. If the user's device is Wi-Fi capable, they can nearly always find a way to connect to the network.
- The convenience of Wi-Fi encourages mobility and therefore productivity. Workers can reasonably expect to walk to a park, sit in a cafeteria, or ride in an airplane and still conduct business. Their work could feasibly be accomplished from anywhere.
- Wi-Fi requires little infrastructure. Installing a small-scale wireless connection is almost as simple as pulling an access point out of a box and plugging it into the wall. Most wireless access points are sold with a functional default installation for small networks.
- Wireless networks are reasonably scalable. If you need to add more client nodes to a wireless network, all you have to do is add more access points and the supporting infrastructure.
- Wireless networks are cheaper to install than copper or fiber networks, which require the installation of cables, connectors, wall sockets, switches, and patch panels for each connection point. With Wi-Fi, all you need is an access point.

Wi-Fi Disadvantages

Although Wi-Fi offers numerous benefits, the disadvantages can be potentially catastrophic. For some applications, you simply cannot use Wi-Fi. Here are some of the disadvantages:

- Wi-Fi has a limited range. Restrictions on the range of Wi-Fi devices are established by the 802.11 equipment standards and the International Telecommunications Union Radiocommunication Sector (ITU-R).
- Wi-Fi devices are susceptible to radio frequency interference, intermodulation, and jamming, particularly in the 2.4GHz band.
- Equipment selection and placement can be difficult. Proper placement, antenna selection, and signal strength are key. Building construction materials affect RF propagation. Some construction materials will dampen, or even block, RF signals. Others act as reflectors, enhancing signal quality inside the space.
- Wi-Fi devices can be expensive. Although a Wi-Fi network may seem cheap at first, the purchase of additional access points, repeaters, or highly directional antennas to expand the network's range will increase costs rapidly. The cost and complexity of building an extended wireless network can grow quickly.
- Even as Wi-Fi speeds increase, the fastest Wi-Fi is no match for the dependability and throughput of wire or fiber networks. If you need to stream live, high-quality UHD video, Wi-Fi is often not a dependable option.
- Wi-Fi networks have finite limits on the numbers of connections that can be supported by each access point. If public-access Wi-Fi is to be provided, resource planning and access-management policies are required to ensure that satisfactory services can be delivered.
- Wi-Fi networks are not secure. They are far more susceptible to malicious attacks than wired networks because it is easy for devices to connect with each other. For this reason, they are often severely restricted or completely prohibited in certain business, financial, and government/military facilities.
- Like any other means of RF communication, Wi-Fi is susceptible to jamming—whether deliberate or accidental—from other RF sources. An old laptop in the hands of a savvy and malicious individual can black out Wi-Fi coverage for an entire conference center.

If your installation requires wireless connectivity, you may simply need to coordinate access and bandwidth requirements with network management. Be prepared for potential objections. You may even find yourself in the position of explaining why Wi-Fi may not be the best option.

Spectrum Management

With the near-universal use of Wi-Fi and Bluetooth on portable computing and personal communication devices, it is almost impossible to guarantee that the required wireless communications channels will be available and free from interference to use in any given space.

The coordination of frequencies allocated to the wireless technologies involved in an audiovisual installation is critical. A full-spectrum wireless frequency allocation plan is an important component of the design of any AV installation.

The OSI Model

The Open Systems Interconnection (OSI) Model defines seven layers of networking functions and provides a common reference for describing how data is transported across a network. Figure 16-8 shows the layers in the OSI Model.

The OSI Model can be used to describe the functions of any networking hardware or software, regardless of equipment, vendor, or application. However, not all networking technologies or AV devices fit into the strict categories of the OSI Model. Many operate at several different layers. For example, the TCP/IP model, the basis for most AV data communications, compacts all the functions into just four layers, as shown in Table 16-5. Still, the OSI Model provides a useful shorthand for discussing networking software and devices. You will often hear networking professionals and manufacturers talk about the "layer" at which a technology operates or the layer at which a problem is occurring.

Figure 16-8 Layers in the OSI Model

OSI Model		TCP/IP Model
7	Application layer	Application layer
6	Presentation layer	
5	Session layer	
4	Transport layer	Transport layer
3	Network layer	Internet layer
2	Data Link layer	Network Access layer
1	Physical layer	

Table 16-5 The Seven Layers of the OSI Model Compared to the TCP/IP Model

Layers of the OSI Model

Knowing the OSI layer (or layers) at which a technology operates can be useful in several ways. A layer in the OSI Model can reveal what a technology does and when those events occur in the data transfer process. For instance, Application-layer error checking occurs at the host and may be aware of the kinds of errors that really matter to the software application. Transport-layer error checking has no awareness of the application; it just looks for any missing packets.

The OSI Model provides a road map for troubleshooting data transfer errors. It describes the signal flow of networked data. Just as you would use a signal flow diagram to troubleshoot a display system in a conference room, checking at each point in the path, you can troubleshoot a network by observing the data transfer process one layer at a time.

The OSI Model can also indicate which service providers are responsible for each stage of data transfer. Layers of the OSI Model often represent a service provider handoff. For instance, an AV technology manager may be responsible for layer 1 and 2 devices and layer 5 to 7 software, while the network manager controls layer 3 and 4 technology.

The Layers

The OSI Model uses a stack of seven layers to communicate or transmit a file from one computer to the next. Essentially, layers 1 to 3 get data from point A to point B, layers 5 to 7 define what the data does when it gets there, and layer 4 does a little bit of both.

- **Layer 1 – Physical layer**   The Physical layer can be copper, fiber, or even a wireless link; the devices need to be plugged into the network. The Physical layer sends and receives electrons, light, or electromagnetic flux.
- **Layer 2 – Data Link layer**   This layer is the interface to the Physical layer; it uses frames of information to talk back and forth. The addressing scheme uses MAC addresses. One MAC address talks to another MAC address. Switches use layer 2 to send and receive frames.
- **Layer 3 – Network layer**   This layer addresses data packets and routes them to addresses on the network. Packets "ride" inside layer 2 frames. Layer 3 adds IP addressing. Routers can send, receive, and route IP addresses. Note that many routers include an integrated switch.
- **Layer 4 – Transport layer**   This layer works with ports to identify the destination of the data within the host. These virtual ports can be found at the end of an IP address such as 192.168.1.35:80. The :80 is the port number. Port 80 is generally associated with HTTP, used by the World Wide Web. Routers can route by ports.
- **Layer 5 – Session layer**   This layer controls (starts, stops, monitors, keeps track of) layer 4 and layer 3. For example, Real-Time Streaming Protocol (RTSP) keeps track of the UDP layer 3 traffic. It keeps track of the data as it makes it to the far-end address. UDP cannot do it on its own; it requires a session protocol to do the work.

Layer	Name	Carries	Sent to/From	Device/Function
L7	Application	Data	Application protocols	Programs (HTTP, FTP, e-mail)
L6	Presentation	Data	Data translation protocols	Encryption
L5	Session	Data	Session protocols	Session
L4	Transport	Segments	Ports	Firewalls
L3	Network	IP packets	IP addresses	Switches, routers
L2	Data Link	Ethernet frames	MAC addresses	Switches
L1	Physical	Pulses	NIC, wire, Wi-Fi controller	Copper/fiber/wireless

Table 16-6 OSI Model Layers with Type of Signal, Protocol, and Device or Function Involved in Transport

- **Layer 6 – Presentation layer** This layer transcodes or translates data between the Session layer and the Application layer. It is a kind of go-between. Data encryption and decryption are done at this layer. An application cannot talk to data directly; it needs an interpreter, and this is it.

- **Layer 7 – Application layer** This layer defines human interaction with data on the network. Using the streaming example, the Application layer is occupied by the VLC Media Player. The application has the controls to manipulate the output of audio and video. It takes *all* the layers for a player application to retrieve the stream and play it on a desktop. No layer is skipped or bypassed. When troubleshooting, all layers need to be considered.

Table 16-6 shows the seven layers of the OSI Model, the type of signal each carries, the type of protocol involved in transport with each layer, and the device or function engaged in each layer.

Data Transmission and OSI

Layers 1 to 3, known as the media layers, define hardware-oriented functions such as routing, switching, and cable specifications. These are the areas that most concern AV professionals.

Layers 4 to 7, the host layers, define the software that implements network services. Each layer contains a broad set of protocols and standards that regulate a certain portion of the data transfer process. A data transfer on any given network likely uses several different protocols at each layer to communicate. Layer 4, the Transport layer, is also important to AV professionals because it is where the transition between gear and software occurs. This layer tells the media layers which applications are sending the data. It also divides and monitors host-layer data for transport. Data is sent across a network by applications. That means when a computer sends a message, that message starts out at layer 7, the Application layer, and moves down through the layers until it leaves the sending device on layer 1, the Physical layer. The data travels to the receiving device on layer 1 and then moves up through the layers until the receiving device at layer 7, the Application layer, can interpret it.

TIP This simple mnemonic may (or may not) help you remember the names of all the layers of the OSI model, from one to seven: Please Do Not Throw Sausage Pizza Away. This stands for the following:
Physical
Data Link
Network
Transport
Session
Presentation
Application

Ethernet

When data is transmitted within an enterprise, it is typically sent via Ethernet across a local area network. Here, you will explore typical LAN topologies and the capabilities and limitations of Ethernet data transmission.

Local Area Networks

Data sent across a network must be sent to a destination address. That address will be either physical or logical. Local area networks use physical addresses to communicate. The physical address is the MAC address—the unique address that is hard-coded into each node and never changes. LANs require devices to be directly physically connected by wire, fiber, or wireless, which effectively limits their geographical size.

Stated as simply as possible, data travels across a LAN in this manner:

- Data sent across a LAN is addressed to the MAC address of one of the devices on the LAN.
- A switch or bridge receives the packet and examines the MAC address to which it is addressed.
- The switch/bridge forwards the packet to the appropriate device.

LANs are usually privately owned and operated. They are fast and high capacity.

Most real-time AV network protocols, including AVB, SDVoE, SMPTE ST 2110, Dante, and RAVENNA, are designed for LAN speeds and capacity.

Topology

One of the most important characteristics of any given network is its topology. Network topology is a determining factor in how far data must travel to reach its destination. A network's topology will show which nodes and connections data must pass through to get from its source to its destination and how many stops there will be along the way. It will also reveal which network devices and connections will have to carry the most data. This helps network engineers calculate which parts of their network need the most capacity. Both of these factors are crucial in determining whether and how to send AV signals over a network.

Figure 16-9 Physical topology maps the physical placement of each device

All networks, local and wide area alike, have a layout, or *topology*.

Physical topology maps the physical placement of each network device and the physical path (see Figure 16-9). Where will the devices and cables actually be located? Physical topology is constrained by the actual space the network equipment occupies.

Logical topology maps the flow of data within a network—which network segments and devices must data pass through to get from its source to its destination (see Figure 16-10). Logical topology is not defined or constrained by physical topology. Two networks with the same physical topologies could have completely different logical topologies, and vice versa.

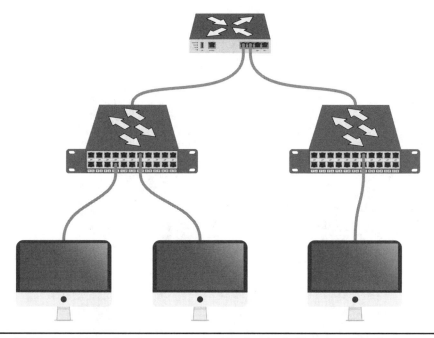

Figure 16-10 Logical topology is not constrained by the physical location of the device.

The topology can reveal the location of a network's potential weak spots. When you look at a local area network topology, you should always look for single points of failure. A single point of failure is any one device or component whose failure will cause the entire system to fail. A single point of failure could be any device that a number of other devices depend on. For example, it could be a network switch that many other devices are connected to, or it could be an audio DSP that handles inputs and outputs from many other devices. Whenever possible, you want to create network redundancy at every potential single point of failure, so if one device or signal path fails, another device or signal path is available and waiting to take its place. As a best practice, no more than 20 devices should be affected by any one single point of failure.

LAN Topologies

Devices can be connected on a local area network in several arrangements, and the layout of connected devices is identified as the network's topology. The basic types are bus, ring, star, extended star (sometimes referred to as *tree topology*), and mesh. Figure 16-11 shows some common local area network topologies you may encounter.

- **Star topology** All nodes connect to a central point; the central point could be a router or switch. Star networks are hierarchical. Each node has access to the other nodes on the star through the central point. If any node fails, information still flows. The central device is a single point of failure; if it fails, communication stops.

Figure 16-11 Common LAN topologies

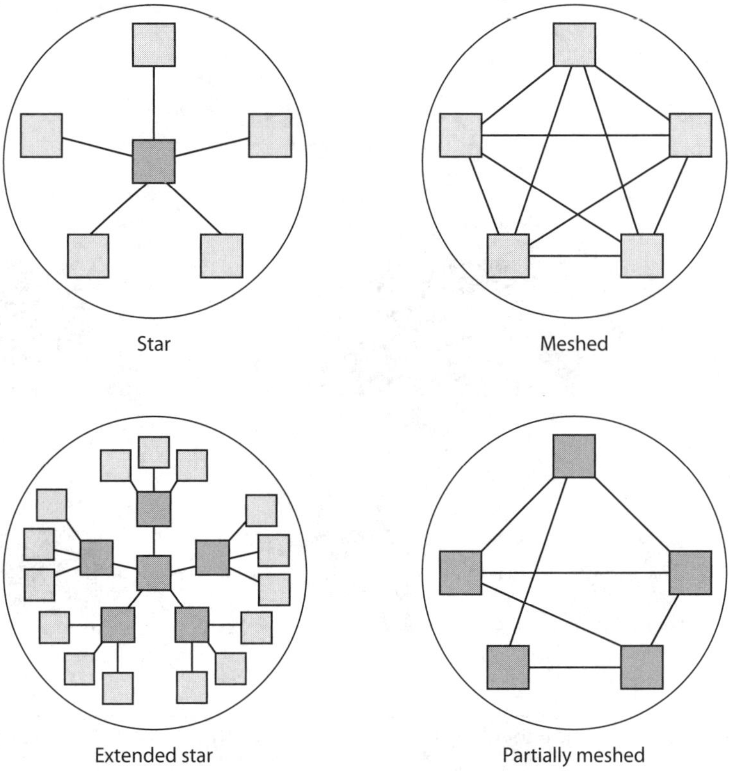

- **Extended star topology** When there is more than one level of star hierarchy, it is known as an extended star topology. If any device fails, access to the devices below it is cut off, but the rest of the network continues to work. The central device remains a single point of failure; if it fails, communication stops.

- **Meshed topology** Each node connects to every other node. Meshed topologies provide fast communication and excellent redundancy, ensuring that the failure of no one device can bring down the whole network. Providing physical connections between every device is really expensive, though. Fully meshed networks are rare.

- **Partially meshed topology** Each node connects to several other nodes, but not all. Partially meshed topologies provide good redundancy, ensuring that several devices must fail at the same time (see Murphy's Law) before communications cease.

What Is Ethernet?

Ethernet is a standard for how data is sent across LANs from one physically connected device to another. It has become the de facto standard for LANs. It is defined in the IEEE 802.3 suite of standards, which define a data frame format, network design requirements, and physical transport requirements for Ethernet networks.

When IP data is sent across a LAN, it is encapsulated inside an Ethernet frame, as shown in Figure 16-12. The Ethernet frame is generated by the device's NIC. The frame has a header and footer enveloping the packet, making sure it gets to its destination and arrives intact.

Ethernet Speed

The speed at which a device sends data over Ethernet depends on the capability of its NIC and the bandwidth of the connecting cable. Not every device can handle sending and receiving data at a rate of 1Gbps or more. Bear in mind that the overall speed of an Ethernet connection is no faster than the slowest link in its path.

The following types of Ethernet are in use today:

- 100Mbps Ethernet
- 1Gbps Ethernet
- 10Gbps Ethernet
- 40/100Gbps Ethernet
- 200/400Gbps Ethernet

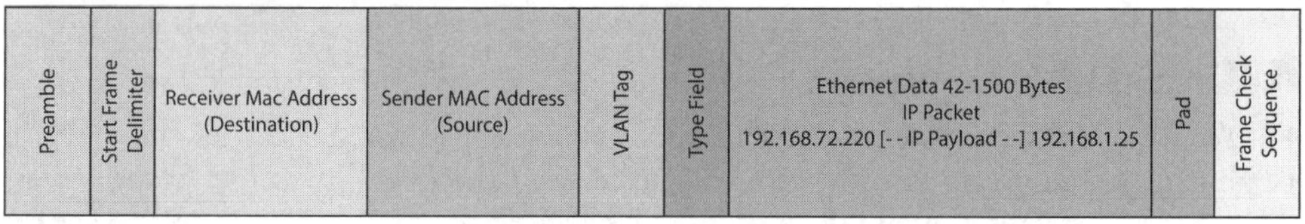

Preamble	Start Frame Delimiter	Receiver Mac Address (Destination)	Sender MAC Address (Source)	VLAN Tag	Type Field	Ethernet Data 42-1500 Bytes IP Packet 192.168.72.220 [- - IP Payload - -] 192.168.1.25	Pad	Frame Check Sequence

Figure 16-12 Ethernet frame with VLAN tag encapsulating the IP data packet

Each of these types of Ethernet has its own capabilities, intended applications, and physical requirements. You will likely encounter only legacy systems that are 100Mbps (or slower) Ethernet. The common 1Gbps Ethernet switches are generally compatible with the legacy 10Mbps and 100Mbps formats with either T568A or T568B pin-outs and are referred to as 10/100/1000Mbps devices. Gigabit Ethernet (1Gbps and 10Gbps) is used for delivery of audiovisual data over a network.

The higher-bandwidth 40/100Gbps Ethernet, although originally only intended for use in the network backbone to connect devices such as routers and switches, is now frequently used to transport uncompressed UHD (4K/8K) video. The more recent 200/400Gbps Ethernet and the forthcoming Terabit Ethernet (800Gbps/1.6Tbps) are primarily intended for use inside data centers, but AV will always find a way to use more bandwidth—if it is even vaguely affordable.

All of these Ethernet technologies are interoperable. A frame can originate on a 100Mbps Ethernet connection and later travel over a 1Gbps or 10Gbps Ethernet backbone. As a general rule, you need a faster technology to aggregate multiple slower links. If your end nodes communicate with their switches by using 1Gbps Ethernet, the next layer of the network hierarchy may need to be 10Gbps Ethernet, and so forth. As devices become capable of faster network speeds, even faster backbones are required.

Ethernet Types and Their Meanings

In the field, you will hear the type of Ethernet referred to by common abbreviations, such as 1000Base-T. To decipher the meaning, it helps to know the following:

- The number indicates the nominal transmission speed in megabits or gigabits per second. For example, 100Base-T transmits at 100Mbps, 1000Base-T transmits at 1,000Mbps or 1Gbps, and 10GBase-T transmits at 10Gbps.

- The word in the middle indicates whether the connection is baseband or broadband. Almost all Ethernet connections are baseband, so you will usually see *Base* associated with Ethernet speeds.

- The characters at the end give you information about the Physical-layer technology used. This varies from standard to standard, but in general, *T* stands for twisted-pair cable, *S* stands for short wavelength (multimode) fiber, *L* stands for long wavelength (single-mode) fiber, and most other character combinations refer to other types of fiber media.

There are many IEEE standards for Gigabit Ethernet, and different characters are used to indicate the type of cable. For example, the IEEE 802.3z standard includes 1000BASE-SX for transmission over multimode fiber and uses 1000BASE-LX for transmission over single-mode fiber.

Isolating LAN Devices

For security or other reasons, your client may not want all the devices on the LAN to be able to communicate directly with each other via Ethernet all the time. Some servers and devices host confidential information that other devices should not be able to access.

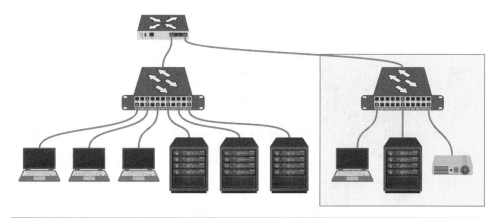

Figure 16-13 Isolation of AV devices (right) on the network with a dedicated switch

You may need to ensure that certain devices have access only to one another and that outside devices cannot easily talk to them. You may also want to isolate certain groups of devices to limit the traffic they receive.

All networked devices send out occasional administrative messages, called *broadcast* messages, to every device on the LAN. Sometimes you want to limit the amount of broadcast traffic a group of devices receives. The need to limit traffic is particularly relevant to networked audio and video. For some protocols, broadcast traffic from other devices on the LAN can cause serious problems to the quality of the audio or video. You need a way to isolate certain AV devices in a cone of silence, shutting out traffic from outside the AV system. One way to handle the issue is to isolate AV devices on a separate LAN by using dedicated switches behind an isolated router port. Figure 16-13 shows a simple networked AV system with a dedicated switch with a direct connection to the network. This simple solution may sometimes be a viable option.

Virtual Local Area Networks

A VLAN is a way to isolate certain devices from others connected to the same network. The devices in a VLAN can be configured on a managed switch, which forwards the data from a device in a VLAN by adding the VLAN identifier to the Ethernet frame, as shown earlier in Figure 16-12.

Devices in a VLAN can be connected to different switches in completely different physical locations. As long as those switches are in the same LAN (that is, they do not have to go through a router to contact each other), the devices can be placed in a VLAN together. Because VLANs are virtual, you can even place a device on more than one VLAN on the same physical LAN. Devices in a VLAN can send each other Ethernet traffic directly. They also receive each other's broadcast traffic. They do not receive any broadcast traffic from the other devices on the LAN, and they cannot communicate with devices outside the VLAN via Ethernet. Traffic in and out of the VLAN has to go through a router.

Applications of VLANs Communication among devices on a VLAN is switched rather than routed. This makes communication among VLANs very efficient. Because devices in a VLAN may be connected to different switches, a VLAN allows you to directly connect devices in different physical locations. For example, in a campus-wide deployment of digital signage, a centrally located media server could send content to display screens at various locations. In a simple signage application with only store-and-forward content to display screens, the devices probably belong on a VLAN. Isolation of the signage devices will enable the transport of video content without issues, and you are unlikely to need to access data on other network devices.

Some locations have a single high-bandwidth LAN covering an entire campus with a separate VLAN for each application group, such as accounting, ticketing, IT, VoIP, operations, streaming video, AV, paging, outside broadcast, audio recording, lighting, stage machinery, system control, BIM, HVAC, communications, hearing assistance, production management, CCTV, security, etc.

Requesting VLANs If you determine that a set of devices should be isolated on a VLAN for the good of either AV signal distribution or the devices on the network, you will need to provide the network manager with the following information:

- Why a VLAN is necessary and what devices will be in it
- Whether any routing between the VLAN and other network locations is required or permitted

This information should all be documented as part of your system device inventory.

Wide Area Networks

When traffic needs to be transported from one LAN to another, it travels over a wide area network (WAN) like the Internet. The difference between LANs and WANs can seem hazy. Some "local" area networks belonging to universities or large enterprises are vast and are often known as campus area networks (CANs). Larger networks that cover a geographic area such as a suburb or a city are identified as metropolitan area networks (MANs). The true distinction between a LAN and WAN lies in how the data is addressed and transported: WAN data travels in IP packets rather than Ethernet frames. A wide area network is a network that connects one or more LANs, as shown in Figure 16-14.

The nodes on a WAN are routers. If a LAN is connected to outside networks via a WAN, a router sits at the top of its network hierarchy. Any data that needs to travel to a device outside the LAN gets forwarded to the router. The router strips the packet of identifying LAN information, like the MAC address of the originating device, before forwarding the packet to its intended IP address.

A WAN can be any size. It may connect two LANs within the same building, or it may span the entire globe, like the world's largest WAN, the Internet. Unlike LAN connections, long-distance WAN connections are rarely privately owned. Usually, WAN connections are leased from Internet service providers (ISPs). The speed of a WAN connection is often directly related to its cost.

Figure 16-14
A simple partially meshed WAN

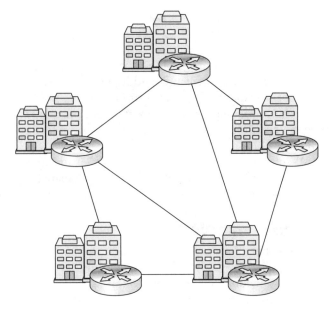

WAN Topologies

Wide area network topologies can be placed into three common categories, as shown in Figure 16-15:

- **Star topology**   In a star WAN, each LAN connects to a central location. For example, several branch offices may connect to a corporate headquarters. Like a star LAN, a star WAN may have several nested layers of hierarchy, with several hubs connecting to several spokes.
- **Common-carrier topology**   In a common-carrier WAN, each LAN site connects to an ISP backbone. The backbone itself is likely to be part of a meshed WAN.
- **Meshed WAN topology**   In a fully meshed WAN, every LAN connects to every other LAN. This provides excellent redundancy. Most meshed WANs are only partially meshed, although this also provides a degree of redundancy.

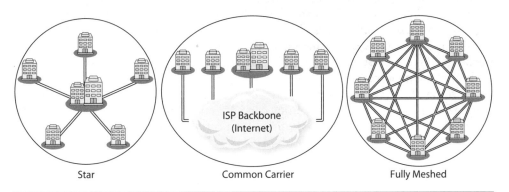

Figure 16-15   Three common WAN topologies

One of the major advantages of networking is the ability to share resources. When examining a WAN topology, try to identify the most effective place to locate shared resources. With a star topology, shared resources should be located at the most-central hub of the nested stars. In a common-carrier topology, the enterprise may choose to lease space at the ISP or a data center and host shared resources there, or it may choose to pay for a lot of bandwidth to and from one of its sites and locate resources there. In a meshed WAN topology, resource location is generally accomplished by building a network map, including the data throughput of all connections, and locating resources "in the middle."

Shared AV resources, such as streaming servers or multipoint control units (MCUs), will ideally be stored in the same physical location as shared IT resources. In some cases, however, you may not have access to IT server spaces for security reasons.

Private and Public WANs

A network could be private or public, but the access or connection to a WAN is usually owned and provided by an ISP. An ISP may lease secure, dedicated communication connections to individual organizations. Some companies and organizations that need to ensure secure connections may own the access and connections to their long-distance communication networks; these are often referred to as *enterprise networks*.

Virtual Private Networks

A virtual private network (VPN) uses the Internet to create a virtual (software) tunnel between points on two or more LANs by adding a layer of secure encryption to the data packets sent between the end points. A VPN is used to create a virtual wide area network for managed communication and for remote monitoring, troubleshooting, and control. VPNs are typically controlled and configured by the enterprise network administrator. Each host or user requires the proper software, access rights, and password to log in to the client network.

Virtual private networks provide support for services that are particularly necessary for AV applications, such as QoS (discussed in a later section), low-latency streaming, managed routing, and multicast transmission. VPN end points may be built into routers, network servers, or security appliances such as firewalls. Organizations using VPNs on a large scale may require dedicated VPN devices.

Using a VPN increases the required bandwidth because an encryption and tunneling wrapper must be added to each data packet. This additional overhead may not be significant in its bandwidth requirement, but it can increase the Ethernet frame size to the point where packets must be fragmented before they can be sent across the network. Packet fragmentation can be disastrous for the quality of streamed video or videoconferences. Always be sure the frame size is set low enough to account for VPN overhead. If you are having difficulties with streaming content over a VPN, this may be the cause, and the problem should be brought to the attention of the network manager.

Network-Layer Protocols

Once AV data joins other data on a routed WAN, Network-layer protocols take care of logical addressing, routing, and, in some cases, prioritizing data as it moves from device to device. The rise of the Internet has made the Internet Protocol the de facto Network-layer protocol for enterprise applications and networked AV systems.

Internet Protocol

Internet Protocol is the communications protocol for relaying data across an IP network. It establishes how data packets are delivered from the source host to the destination host, based solely on the IP addresses in the packet headers. It defines the rules for addressing, packaging, fragmenting, and routing data sent across an IP network.

This protocol is on the Network layer (layer 3) of the OSI Model and addresses these crucial functions that make wide area networking possible:

- **Addressing** Rules for how each host is identified, what the addresses look like, and who is allowed to use which addresses
- **Packaging** What information must be included with each data packet
- **Fragmenting** What size each packet may be and how overly large packets will be divided
- **Routing** What path packets will take from their source to their destination

Internet Protocol is basically the postal service of the Internet. A postal service sets rules for how to package and address mail. If you do not include the right information in the address, your package will not be delivered. If you do not package it correctly, it may get damaged in transit. Just like a postal service, IP assumes responsibility for making sure your data arrives at its destination, although, just like with a postal service, some messages do get "lost in the mail." Addressing, packaging, fragmenting, and routing all help ensure that as many messages as possible reach their intended destinations.

IP Addressing

An IP address is the logical address that allows devices to locate each other anywhere in the world, no matter where they are physically located. The numerical address defines the exact device and its exact location on the network. Even if AV devices such as videoconferencing systems are isolated on a LAN, they are assigned an IP address.

An IP address has three components:

- **Network identifier bits** These bits identify the network. They help the IP packet find its destination LAN. The network bits are always the first digits or prefix in an IP address.

- **Host identifier bits**   These bits identify a specific network node. They help the IP packet find its actual destination device. The host bits are always the last bits or least significant bits in a network address.
- **Subnet mask**   These bits tell you which bits in the IP address are the network bits and which bits are the host bits. The subnet mask also reveals the size of the network. The subnet mask is a separate address that must be included with the IP address.

IPv4 Addressing

An IP address can look very different depending on which version of Internet Protocol addressing is used to create them. Two versions of the protocol are currently in use: version 4 (IPv4) and version 6 (IPv6). Originally defined in the IETF standard RFC 760 in 1980, Internet Protocol version 4 addressing is slowly and cautiously being phased out in favor of IPv6. IPv4 remains the most prevalent addressing scheme by far, and with many ancient IPv4-only devices liberally scattered throughout the Internet, rumors of its death have been greatly exaggerated. During the transition to IPv6, you will need to know the IPv4 addressing scheme and structure.

An IPv4 address consists of four 8-bit groups, known as *octets,* or bytes. These four octets are usually expressed as decimal numbers (from 0 to 255) separated by dots: a format known as *dotted-decimal* or *dotted-quad* notation. Hence, an IP address looks like this:

192.168.1.25

An IPv4 address given to a device on the network has a network prefix (number on the left) and a network host number on the right. Note that each decimal number actually represents eight bits. That same address, written in binary, looks like this:

11000000 10101000 00000001 00011001

The entire range of IPv4 addresses includes every possibility, from all 0 bits to all 1s. In dotted-decimal notation, that range is expressed as follows:

0.0.0.0 to 255.255.255.255

In total, there are almost 4.3 billion possible IPv4 addresses. Although a few addresses are reserved for specific purposes, there are still a lot of possible addresses. However, 4.3 billion addresses will not be enough for the ever-growing IoT and other networks of the future. That is why IPv4 is being phased out, although the widespread adoption of network address translation (NAT) has slowed the adoption process.

IPv4 Subnet Masks

Looking at an IPv4 address on its own, you cannot know which groups of numbers are the network ID and which ones are the host ID numbers. To interpret any IPv4 address, you need a separate 32-bit number called a *subnet mask.*

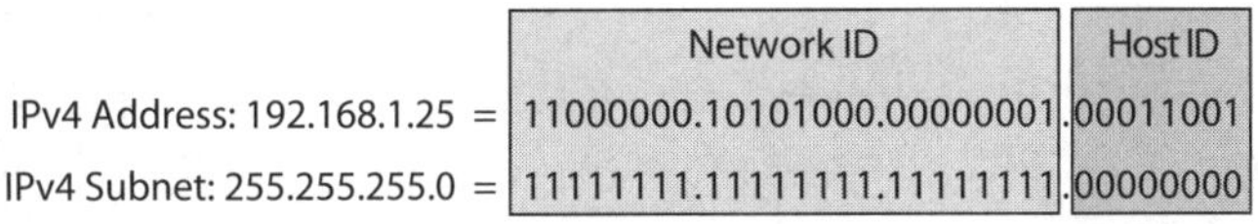

Figure 16-16
IPv4 address and subnet mask

		Network ID	Host ID
IPv4 Address: 192.168.1.25 =		11000000.10101000.00000001.	00011001
IPv4 Subnet: 255.255.255.0 =		11111111.11111111.11111111.	00000000

A subnet mask is a binary number whose bits correspond to IP addresses on a network. As shown in Figure 16-16, the structure of an IPv4 subnet mask looks a lot like an IPv4 address. It consists of four octets, expressed in dot-decimal notation. Bits equal to 1 in a subnet mask indicate that the corresponding bits in the IP address identify the network. Bits equal to 0 in a subnet mask indicate that the corresponding bits in the IP address identify the host.

For example, subnet mask 255.255.255.0 shown in Figure 16-16 would indicate that the first three octets of any corresponding IP addresses are the network address and the last octet is the host address.

Also notice the difference in the binary IPv4 network ID and IPv4 host ID in Figure 16-16. The IP address assigns the numbers 1 and 0 in combination, but in the subnet mask, the first part of the subnet is all 1s and the second part (right) is all 0s. All the devices on the same network have the same network identifier bits in their IP addresses. Only the host bits will differ.

In IPv4, there are two ways to express a subnet mask:

- You can write it out as its own full dotted-decimal number.
- You can attach it to the end of an IP address using the shorthand format: Classless Inter-Domain Routing (CIDR) notation. CIDR notation is simply a slash, followed by a number that indicates how many of the address bits are network bits, with the remaining bits being host bits.

Here is an example of the two ways the same subnet mask can be written:

- Dot-decimal number: 255.255.192.0
- CIDR notation: /18

In binary, both are equal to 11111111 11111111 11000000 00000000.

Networks are both physical and logical. Two devices may be physically attached to the same switch, but that does not mean they are logically on the same network. They must also have the same network identifier bits and subnet mask. When you are working with network management to obtain IP addresses for your network-connected devices, make sure everyone knows which devices need to be in the same subnet.

Types of IP Addresses

The Internet Assigned Numbers Authority (IANA) is in charge of issuing IP addresses or reserving them for specific purposes. The IANA maintains three categories of addresses: local, private or reserved, and global.

Network Address	Address Range	Number of Addresses	Application	Purpose
0.0.0.0/8	0.0.0.1–0.255.255.255	16,777,216	Software	Used as a placeholder to represent an unknown IP address.
10.0.0.0/8	10.0.0.0–10.255.255.255	16,777,216	Private network	Private (unrouted) network address range for very large networks.
127.0.0.0/8	127.0.0.0–127.255.255.255	16,777,216	Local host	Used as a loopback address for the current host.
169.254.0.0/16	169.254.0.0–169.254.255.255	65,536	Private network	Addresses used for local (unrouted) Automatic Private IP Addressing (APIPA). Used in the absence of a static IP addressing or DHCP address assignment.
172.16.0.0/12	172.16.0.0–172.31.255.255	1,048,576	Private network	Private (unrouted) network address range for large to medium networks.
192.168.0.0/16	192.168.0.0–192.168.255.255	65,536	Private network	Private (unrouted) network address range for medium to small networks.
224.0.0.0/4	224.0.0.0–239.255.255.255	268,435,456	Internet	Used for multicast communications.
240.0.0.0/4	240.0.0.1–255.255.255.254	268,435,456	Internet	Reserved for experimental purposes/future use.
255.255.255.255/32	255.255.255.255	1	Local network	Broadcast address used to talk to every host on a local network.

Table 16-7 Examples of Some IANA-Reserved IPv4 Addresses

Table 16-7 shows the more important reserved IPv4 addresses, their ranges, sizes, and purposes.

Local Addresses

Not all devices need to access the Internet directly. Many devices need to communicate with other devices only on their local area network. The Internet Assigned Numbers Authority reserves three IPv4 address ranges and one IPv6 address range for local networking. The addresses in these ranges are private and are not routed outside the network by any routers. Devices with private IP addresses cannot access the Internet or communicate with devices on other networks directly.

IPv4 Private Address Ranges

Private addresses are not routed on the public Internet. As can be seen in Table 16-7, there are three IPv4 private address ranges: 10.0.0.0/8, 172.16.0.0/12, and 192.168.0.0/16. The range your client will use depends on the size of their network.

The principal advantage of private network addresses is that they are reusable. Global addresses have to be unique so no two hosts can use the same IP address to access the Internet. Otherwise, when data is sent to or from an address, there would be no way to know which host was the intended recipient. Since private addresses are not exposed to the Internet, different organizations can use the same private address range. Devices on different networks can have the same private IP address because those devices will never attempt to communicate with each other. As long as no devices on the same network have the same IP address, there is no confusion.

Subnetting Local Networks

Subnetting is the act of logically dividing a network into smaller networks. Each smaller network is called a *subnet*. The networks to which you are connecting AV devices may be divided into several subnets. You will need to work with network management to make sure that any devices that need to communicate via Ethernet are in the same subnet.

Subnets are created when the subnet mask of an IP address is extended. If the subnet mask is extended by one bit, you end up with two subnets that are each approximately half the size of the original network. Similarly, if the subnet mask is extended by two bits, you end up with four similarly sized subnets, each one-quarter the size of the original network. For devices to communicate directly by Ethernet or to belong to the same VLAN, they must be in the same subnet.

The main reason to create subnets is to increase network efficiency. A subnet has fewer addresses than a full class network, so address resolution is faster. Fewer devices also mean less broadcast traffic, as devices on a subnet receive broadcast messages only from other devices on the same subnet, not from the entire undivided network.

Global Addresses

Most network devices will need to connect to the Internet at some point. To access the Internet, a device needs a global IP address, one that any Internet-connected device can locate.

Global addresses go by many names in the networking community, including *globally routable addresses, public addresses,* or *publicly routed addresses*. Any IP address that is not in one of the local or reserved address ranges can be a global address. If an IP address is not a local or reserved address, it is a global address.

Network Address Translation

The obvious disadvantage of private IP addresses is that they can communicate only with devices on the same network because their addresses cannot be routed to the Internet. Initially, this made the networking community reluctant to use them. However, that problem has been solved through the use of NAT.

NAT is a method of altering IP address information in IP packet headers as the packet traverses a routing device. NAT is used to allow devices with private, unregistered IP addresses to access the Internet through a device with a single registered IP address. NAT conserves address space, which is a concern in IPv4 implementations (though it is also used for IPv6 networks). NAT hides the original source of the data. From outside the network, all data appears to originate from the NAT server. Any data that arrives at the NAT server without being requested by a client has nowhere to go; it has the address of the building but not the apartment number. Using NAT, all unrequested data is blocked by the firewall or router, and a malicious intruder can't trace the data's path beyond the edge of the network.

Broadcast Addresses

Broadcast addresses are used as the destination IP address when one node wants to send data to all network devices. Broadcast messages are one-way; there is no mechanism for the other nodes to reply. An IPv4 broadcast address is any IPv4 address with all 1s in the host bits. When data is sent to that address, it goes to every device with the same network bits.

For example, if your network address is 192.168.0.0 and your subnet mask is 255.255.0.0, your broadcast address is 192.168.255.255. Any data sent to that address will go to every device in the address range from 192.168.0.1 to 192.168.255.254. If instead the subnet mask was 255.255.255.0, your broadcast address is 192.168.0.255, and data sent to that address will go to every device in the address range 192.168.0.1 to 192.168.0.254.

Loopback Addresses

Data addressed to a loopback address is returned to the sending device. The loopback address is also known as the *localhost*, or simply *home*.

The loopback address is used for diagnostics and testing. It allows a technician to verify that the device is receiving local network data. Essentially, it allows you to ping yourself.

Any IP address in the range 127.0.0.0 to 127.255.255.255 can be used for loopback, but most network devices automatically use 127.0.0.1.

IPv6 Addresses

IPv6 addresses look very different from IPv4 addresses. An IPv6 address consists of 16 bytes (128 bits), four times as long as an IPv4 address. Because IPv6 addresses are so long, they are usually written in eight lowercase, four-character hexadecimal groups (hextets), separated by a colon, as shown in Figure 16-17.

To compress the representation of an IPv6 address, the leading zeros may be omitted from each hexadecimal group. For example:

> The address 2006:0fe8:85a3:0000:0002:8a2e:0a77:c082
> may be written as 2006:fe8:85a3:0:2:8a2e:a77:c082

The representation may be further compressed by replacing a single sequence of consecutive zero-value hexadecimal groups by a double colon (::) symbol. For example:

> The address 2006:0fe8:0000:0000:0000:8a2e:0000:00c8
> may be written as 2006:fe8::8a2e:0:c8

Note that to avoid any ambiguity, only one run of consecutive zero-value hexadecimal groups may be replaced by a :: symbol.

Figure 16-17 An IPv6 address is written in eight 4-character hexadecimal groups.

IPv6 Address: fec8 : ba98 : 7654 : 0080 : fdec : ba98 : 7654 : 3201
(Network ID: fec8 : ba98 : 7654; Subnet ID: 0080; Host ID: fdec : ba98 : 7654 : 3201)

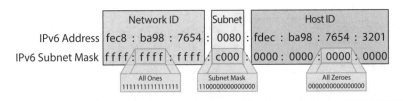

Figure 16-18 An IPv6 subnet mask "masks" only the fourth hexadecimal word of the IPv6 address.

Since each hexadecimal character represents 4 bits, each group represents 16 bits. You can interpret an IPv6 address as follows:

- The first three hexadecimal groups are the network identifier bits (48 bits ~281 trillion networks).
- The next hexadecimal group identifies the subnet within the network (16 bits = 65,536 subnets).
- The last four hexadecimal groups are the host identifier bits within the network (64 bits ~18,446,744 trillion hosts).

The host identifier portion of an IPv6 address is long enough to include a 48-bit MAC address, so IPv6 can actually use a device's MAC address as part of the host identifier. Some IPv6 implementations even do this automatically. Because a MAC address uniquely identifies a device, using the MAC address in the host identifier should ensure that no two devices ever have the same IPv6 address.

IPv6 Subnet Masks

An IPv6 subnet mask can be written out in eight full hexadecimal words, but the first three words of the netmask will always be all 1s and the last four will always be all 0s. As a result, many implementations of IPv6 allow you to enter the subnet mask as a single four-character hexadecimal word. The subnet mask could be written as simply c000, as shown in Figure 16-18. You can also express an IPv6 subnet mask using CIDR notation. The CIDR suffix for an IPv6 subnet mask will almost always be between /48 and /64.

Address Assignment

Every device connected to the IP network must have an IP address. That address may be obtained automatically or configured manually. Either way, you will need to work with network management to make sure there are enough addresses available for your devices. You will also need to make sure your devices have the right kind of address (local or global). Finally, you need to make sure that any device that needs a permanent address gets one.

Static and Dynamic IP Addresses

IP addresses can be either static or dynamic. A static IP address is assigned manually to a device, and the address will not change. It is necessary to have a static address for certain devices such as those used in videoconferencing systems or IP-controlled AV equipment

Manufacturer	Model #	Software Version	Firmware Version	MAC Address	IP Address	Subnet Mask	Gateway WAN IP Address
ProjectTech	4000ZT	8.0	11.4.5	78:ab:0f:23:32:89:0c:7a	192.168.38.4	255.255.255.224	202.38.192.1

Table 16-8 Log of Static IP Addresses and Related Data

so that they can be easily found on the network. As shown in Table 16-8, the network manager has to document and manually keep track of which devices have static addresses and what addresses they have been assigned.

If any two devices have the same addresses, most likely neither will work because of the conflict or perhaps either device may work intermittently. As a result, keeping track of static addresses can be a pain point for network managers.

A dynamic IP address is a temporary IP address that is automatically assigned to a device when it connects to the network. Dynamic addresses can change and therefore are not practical for applications that require locating a device through its IP address. If you need a control system to be able to locate a device by its IP address, you should assign the device a static address.

Dynamic Host Configuration Protocol

Dynamic Host Configuration Protocol (DHCP) is an IP addressing scheme that allows network administrators to automate address assignment. When a device connects to the network and the device has the "obtain IP address automatically" option activated, the network DHCP service will read the MAC address of the device and assign it an IP address. The pool of available IP addresses is based on the subnet size and the number of addresses that already have been allocated.

A DHCP server will allow a device to hold the IP address for only so long; the amount of time is called the *lease time*, which is set by the network administrator. After the lease time has expired, the lease will usually be renewed automatically if the device is still connected to the network; otherwise, another device connecting to the network can reuse that same address. There is no guarantee that a device will be allocated the same address when it next connects to the network.

A single DHCP server can assign addresses to devices on multiple subnets. The server keeps track of the following:

- The range of available network addresses
- Which addresses are available for DHCP assignment
- Which addresses are currently in use by which devices
- The MAC addresses of the connected devices
- The remaining lease time on each allocated address

DHCP Advantages

DHCP is simple to manage. It takes care of making sure no two devices get the same address, relieving potential conflicts. It allows for more people to connect to the network, as the pool of addresses is continually updated and allocated. For example, a conference center may have thousands of visitors using their Wi-Fi in the course of a week. With DHCP, you only need enough host addresses to cover the devices that are using the network at any given time.

A disadvantage of DHCP is that you never know what your IP address will be from connection to connection. If you need to reach a certain device by IP address, you must have a high level of confidence that the device will be there all the time, and DHCP may not give you that confidence.

Reserve DHCP

Reserve DHCP is a hybrid approach to DHCP that reserves a block of addresses for static addresses and dynamic addresses. The pool of addresses for DHCP is reduced by the number of addresses reserved for static devices. To make this happen, an IT manager will need the MAC address of each device that must be statically set. The static (manually assigned) IP address and MAC address are entered into a table.

When a device configured for automatic addressing connects to the network and reveals its MAC address, the DHCP server will see that an IP address is reserved for the device and will assign it. The IP address cannot be given to any other device or MAC address. This eliminates the possibility of devices being manually assigned static addresses that have also been allocated to the DHCP pool.

Note that if a device with a reserved IP address is replaced, the MAC address of the replacement device needs to be reported and reconfigured on the DHCP server so that the new device can assume the old IP address.

> ### Best Practice for Reserve DHCP
> If you have a DHCP server, it is best to use reserve DHCP rather than manually assigning an IP address to each AV device.

Automatic Private IP Addressing

A device on a DHCP-enabled network can fail to get an address from a DHCP server under the following conditions:

- The DHCP server is down.
- DHCP is not configured properly on the client.
- The DHCP server has exhausted the IP address pool.
- The DHCP server is improperly configured.

In these circumstances the device will not be able to access other network segments or the Internet, but it may still be able to communicate with other devices on its own network segment using Automatic Private IP Addressing (APIPA), if it has been enabled.

Intended for use with small networks with fewer than 25 clients, APIPA enables plug and play networking by assigning unique IP addresses to computers on private local area networks. APIPA uses a reserved range of IP addresses (169.254.0.1 to 169.254.255.254) and an algorithm to guarantee that each address used is unique to a single computer on the private network.

It works seamlessly with the DHCP service, yielding to the DHCP service when DHCP is deployed on a network. A DHCP server can be added to the network without requiring any APIPA-based configuration. APIPA regularly checks for the presence of a DHCP server, and upon detecting one replaces the private networking addresses with the IP addresses dynamically assigned by the DHCP server. A device with an APIPA address may not be able to communicate outside the local network, but it can communicate with other devices (if any) on the same subnet.

When a device has an IP address but cannot connect to the Internet, network troubleshooters often look for link-local (local network only) addresses such as APIPA addresses.

Domain Name System

Devices on a network must have unique identifiers. At the Data Link layer (layer 2), devices are uniquely identified by their MAC addresses. At the Network layer (layer 3), devices are uniquely identified by their IP addresses. Naming services allow people to identify network resources by a name instead of a number. From the human perspective, names are a lot easier to memorize than numeric or alphanumeric addresses. For example, it is quite easy to remember *store.avixa.org* but not so easy to remember 101.53.188.4.

DNS is the most widely used system for name-to-address resolution. It is a hierarchical, distributed database that maps names to data such as IP addresses. The web addresses you type into a browser are not actually addresses; they are DNS names. Every system that connects to the Internet must support DNS resolution.

The goal of DNS is to translate, or resolve, a name into a specific IP address. DNS relies on universal resolvability to work: Every name in a DNS must be unique so that information sent to a domain name arrives only at its intended destination.

A DNS uses domain name servers to resolve names to addresses. The server contains a database of names and associated IP addresses. These servers are arranged in a hierarchy. Each server knows the names of the resources beneath them in the hierarchy and the name of the server directly above them in the hierarchy. No one device has to keep track of all the names and IP addresses on the Internet. That information is distributed across all the DNS servers on the network.

As shown in Figure 16-19, a domain name or alias has three main parts:

- A computer name or alias
- The domain itself
- The top-level domain (TLD)

Figure 16-19 Three main parts of a domain name

A domain may be further divided into subdomains. This system helps prevent any two devices from being assigned the same name. Many World Wide Web servers are assigned the subdomain "www," many e-mail servers are assigned the subdomain "mail," and many e-commerce servers are assigned the subdomain "shop." Any number of computers may be called www, mail, or shop, provided they belong to different domains.

Dynamic DNS

Using a service called dynamic DNS (DDNS), DNS can work hand in hand with DHCP. The service links and synchronizes the DHCP and DNS servers. Whenever a device's address changes, DDNS automatically updates the DNS server with the new address. When DHCP servers and DNS are working together, you may never need to know the IP address of a device, only its name. This makes managing the network simpler, as the IP address does not need to be static. The entire addressing scheme could change without affecting the communication between devices.

Internal Organizational DNS

You do not need your own DNS server to resolve the names of web addresses on the Internet. You can do that through your Internet service provider's DNS server. However, many organizations use DNS internally to manage the names and addresses of devices on their private networks. In this case, you will need your own DNS server. Usually, you will have a master DNS server and at least one secondary DNS server that runs a copy of the database stored on the master. This provides a backup in case the master ever fails. If an organization is really dispersed, you may want to locate a secondary DNS server at each physical site. This keeps DNS traffic off the wide area network.

Internal DNS Adoption

The advantages of using internal DNS address assignment for a networked system seem obvious:

- Devices can be identified by easily remembered names.
- Dynamic DHCP makes control and remote monitoring systems much easier to maintain.

- Device replacement and reallocation are simplified.
- DNS includes load balancing functionality.

Many AV devices still do not natively or fully support DNS address assignment. Connecting a device without native DNS support to a network with a DNS addressing scheme may require the network manager to perform many, and ongoing, manual DNS server updates. If your device supports DHCP but not DNS, you are usually better off reserving a pool of DHCP addresses for your devices by MAC address rather than manually configuring the DNS servers.

As IPv6 is adopted alongside and eventually replaces IPv4, DNS should become more commonplace on AV devices. Since IPv6 addresses are long and susceptible to errors when entered manually, the transition to IPv6 will likely encourage the widespread adoption of DNS.

Transport Protocols

Layer 3, the Network layer, handles assigning IP addresses to network devices and identifying paths from one network to another. The actual end-to-end transportation of data, however, is handled by layer 4, the Transport layer. Transport-layer protocols fragment IP packets into smaller chunks that fall within the maximum transmission unit (MTU) size of the network connection. This process is known as *segmentation*. The transport protocol is responsible for segmenting data for transmission and reassembling it at its destination.

A transport protocol may be connection-oriented or connectionless. Connection-oriented transport protocols are bi-directional. The source device waits for acknowledgment from the destination before sending data. It checks to see whether data has arrived before sending more. Connection-oriented transport includes error checking and flow control.

Connectionless communication is one-way. The source device sends. The destination device may or may not receive. Connectionless protocols are less reliable than connection-oriented protocols, but they are also faster because there are no pauses for replies. Many media-oriented applications, including practically all real-time protocols, use connectionless transportation protocols.

In IP networks, the commonly used connection-oriented transport protocol is TCP and the commonly used connectionless protocol is UDP.

TCP Transport

TCP transport uses two-way communication to provide guaranteed delivery of information to a remote host. It is connection oriented, meaning it creates and verifies a connection with the remote host before sending it any data. It is reliable because it tracks each packet and ensures that it arrives intact. TCP is the most common transport protocol for sending data across the Internet.

TCP data transfer involves the following steps:

- TCP communication starts with a *handshake* that establishes that the remote host is there and negotiates the terms of the connection, including the sliding window size (how many packets can be sent at once before verification is required).

- The origin device sends one window at a time to the destination device.
- The destination device acknowledges receipt of each window, prompting the origin device to send the next one. The sliding window cannot move past a packet that has not been received and acknowledged. If any packets are damaged or lost in transmission, they will be resent before any new packets are sent.

Because TCP is reliable and connection oriented, it is used for most Internet services, including HTTP, FTP, and Simple Mail Transfer Protocol (SMTP).

UDP Transport

UDP is a connectionless, unacknowledged protocol. It begins sending data without attempting to verify the origin device's connection to the destination device and continues sending data packets without waiting for any acknowledgment of receipt.

In UDP data transfer, the following happens:

- The origin computer does not attempt a "handshake" with the destination computer. It simply starts sending information.
- Packets are not tracked, and their delivery is not guaranteed. There is no sliding window.

UDP lacks TCP's inherent reliability. That does not mean all data transmitted using UDP is unreliable. Systems using UDP may manage reliability at a higher level of the OSI Model, such as the Application layer.

UDP is used for streaming audio and video. When packets are lost in transport, UDP transport just skips over missing bits, inserting a split second of silence or a repeated image instead of coming to a full stop and waiting for the packets to be resent.

UDP may also be used to exchange very small pieces of information. In some cases, such as retrieving a DNS name, a TCP "handshake" takes more bits than the actual exchange of data. In such instances, it is more efficient to use the "connectionless" UDP transport.

TCP vs. UDP

TCP transport is used when the guaranteed delivery and accuracy or quality of the data being sent is most important, for example, when sending AV control signals. UDP transport is used when speed and continuity are most important, for example, during any real-time AV communications. However, many enterprises have policies against UDP transport because of security issues. UDP streams can be used in malicious attacks such as denial-of-service attacks, which swamp network equipment with useless requests, or self-replicating Trojan horse viruses. If you recommend the use of UDP transport for streaming media, be prepared to defend its necessity.

Ports

After data arrives at a device, how does the device know what to do with the data? The Transport-layer protocol, either TCP or UDP, will include a port number. Transport-layer ports are not physical ports: They are logical ports, telling the network data what application it should "connect" to on the network device.

Essentially, the port number indicates to the server what you want the data to do. The Internet Assigned Numbers Authority permanently assigns many port numbers to standardized, well-known services. Every IP network has a Services file that contains a list of permanently assigned ports and their associated services:

- System ports, 0 to 1023, are assigned to standard protocols. These are also known as *well-known* ports.
- User ports, 1024 to 49151, are assigned by IANA upon request from an application developer.
- Dynamic ports, 49152 to 65535, cannot be assigned or reserved. Applications may use any dynamic port that is available on the local host. However, the application cannot assume that any specific dynamic port will always be available. Dynamic ports are also known as *ephemeral* ports.

Although you can often identify the originating application by the data packet's port number, this is not always possible. Sometimes, a service with a permanently assigned port has to open one or more dynamic ports to run several instances of the service on the same host. Many applications will choose a port from the dynamic port range at random.

Table 16-9 shows some well-known ports that are commonly used by AV traffic.

Ports are one of the most important points of coordination between AV and IT. You will need to consult the manufacturer's specifications of each piece of networked gear to discover what ports it uses. Your installer and the network manager may also want to test the gear to ensure that port documentation is complete. This information will need to be documented in the network device inventory. The network manager will use this information to make sure the right ports are opened on the right devices. That way, your AV traffic will not be blocked by a firewall or router because it uses an unrecognized port.

A port number may be specified in a URL by appending it to the domain name after a colon. For example, *http://www.domain.com:8080* would direct the web browser to connect to port 8080 on the domain.com web server.

Protocol	TCP Port(s)	UDP Ports
HTTP	80	
Secure HTTP (HTTPS)	443	
File Transfer Protocol (FTP)	20 (data), 21 (control)	
Secure Shell (SSH)	22	22
Network Time Protocol (NTP)	123	
Simple Network Management Protocol (SNMP)	161	161
Domain Name System (DNS)	53	53

Table 16-9 Well-Known Ports Commonly Used in AV

The Host Layers

In the OSI Model, layers 5 to 7 are host layers (layers 1 to 3 are media layers, and layer 4 is the transport layer). The media layers are where most of the AV–IT coordination needs to take place. IP and Ethernet networks do not care what kinds of applications they are carrying, as long as those applications are sending out data in the right format. Most network troubleshooting takes place at the lower levels as well. Still, you need to be familiar with the terminology and functions of the host layer.

The Session Layer

Layer 5, the Session layer, manages sustained connections between devices. TCP includes some Session-layer functionality in that it verifies that the receiving device is listening and negotiates how much data it can send before transmitting packets. True Session-layer protocols negotiate even more parameters. A Session-layer protocol formally begins and ends sustained communication among devices. It regulates which devices transmit and which receive. It also regulates what kind of data each device can send and receive and at what bandwidth. Session-layer protocols are important in streaming media and conferencing applications. They negotiate to make sure each device sends and/or receives the best quality of which it is capable. In conferencing applications, they also manage which device talks and which devices listen at any given moment.

The Presentation Layer

Layer 6, the Presentation layer, is responsible for making data look the same to the lower-level protocols. The Presentation layer is also responsible for encoding and compressing data to reduce its required bandwidth. Codecs are a Presentation-layer technology. The Presentation layer is also sometimes responsible for encrypting and decrypting data for security purposes, although this can also take place at the Application layer.

Codecs

The term *codec* is short for enCOder/DECoder. A codec is an electronic device or a software process that encodes or decodes a data stream for transmission and reception over a communications medium.

Codecs may be one-way or two-way, encrypted or unencrypted, symmetrical or asymmetrical, and compressed or uncompressed. They may also include analog-to-digital or digital-to-analog conversion. The decision as to what codecs to use for a client's streaming services will be determined by a number of factors, including the following:

- IT policies. The software that users currently have or may be allowed to have will determine what codecs are available on the end-user playback device, which in turn determines what you can encode to.
- Licensing fees associated with the codec.
- Resolution and frame rate of the source material.
- The available processing power of the encoding and decoding devices.

- Desired resolution and frame rate of the stream.
- Latency introduced by the codec processes.
- Bandwidth required for the desired quality. You may not be able to find specification for bandwidth; therefore, some network testing will be necessary.

The Application Layer

The Session layer hides the differences between data. Layer 7, the Application layer, unpacks them. The Application layer is responsible for presenting data to the right software in a way that the software can understand. It turns the data it receives from the Presentation layer into e-mails, web pages, FTP files, databases, media streams, and so on, depending on the port number identified by the Transport layer. This is the layer that turns network data into data the user can actually interact with. There are as many Application-layer protocols as there are software applications; there are too many to count, and more every day.

Bandwidth

Bandwidth is a critical networking concern that spans multiple layers. In fact, as an AV professional, the network's bandwidth is one of the attributes you care most about. If the network does not have sufficient bandwidth, the AV signal quality will plummet or the signal simply will not arrive at its destination. In analog signals, system bandwidth is measured in hertz (Hz); however, in most complex digital encoding systems, a single signal cycle may encode more than one bit of data, so *data throughput* is measured in bits per second (bps).

The term bandwidth refers to the following:

- The capacity of the network connections, such as "This switch has a bandwidth of 100Gbps."
- The throughput requirements of the data or devices, such as "This videoconference system requires 4Mbps of bandwidth per end point."

When using an IT network to transport AV data, your concern should be about bandwidth availability, not about bandwidth capacity. You need to make sure the network has enough free unused bandwidth to handle AV signals.

In reality, no more than 50 percent of a network's ready capacity should be allocated for routine use. Your client's AV devices may be on a separate LAN, in which case you do not have to worry about other data traffic crowding the client's network. This allows you to comfortably plan to use a substantial percentage of the available bandwidth. Work with your client's network engineer or IT manager to find out how much bandwidth you can actually plan to use for AV.

Quality of Service

If you run into bandwidth limitations, there are strategies you can use to make sure your AV traffic gets through. *Quality of service* is a term used to refer to any method of managing data traffic to preserve system usefulness and provide the best possible user experience. Typically, QoS refers to some combination of bandwidth allocation and data prioritization.

Many different network components have built-in QoS features. For example, video-conferencing codecs sometimes have built-in QoS features that allow various devices on the call to negotiate the bandwidth of the call.

Network managers may also use software to set QoS rules for particular users or domain names. During the network design stage or with AV device installation, you should be concerned with QoS policies that need to be configured directly on network switches and routers. You will need to work with network management to make sure they are aware of any networked device QoS requirements and that those settings have been configured on the relevant network devices.

DiffServ

The underlying strategy of network-based QoS is to prioritize time-sensitive traffic over other traffic. One way to accomplish this is to assign each type of traffic on the network to a particular QoS differentiated service (DiffServ) class. Each class is handled differently by the managed network switches and routers, which is why it is called differentiated service. Some classes are designed as *low loss* to preserve data without losing any packets. Some classes are designed as *low latency* to transport data as quickly as possible. Some classes prioritize data arriving in the exact order in which it was sent, that is, *low jitter*. The lowest-priority class is *best effort*, where data in this class will arrive when and how it arrives, with no guarantees of integrity or timeliness.

Each application your customer will use is assigned a DiffServ class on the network routers and switches. When traffic enters the network, these devices automatically detect which application it comes from and tags it with a DiffServ class. The DiffServ class then defines how the network devices prioritize the traffic.

Signaling Service The *signaling service* class is for traffic that controls applications or user end points. For example, signals that set up and terminate a connection between conference call end points would belong in this class.

Telephony The *telephony* class is intended for Voice over IP (VoIP) traffic, but it can be used for any traffic, such as streaming video and audio, that transmits at a constant rate and requires low latency.

Real-Time Interactive The *real-time interactive* class is for interactive applications that transmit at a variable rate and require low jitter and loss and very low delay. Examples include interactive gaming and some types of videoconferencing.

Multimedia Conferencing The *multimedia conferencing* service class is for conferencing solutions that can dynamically reduce their transmission rates if they detect congestion. If a conferencing class cannot detect and adapt to network congestion, the real-time interactive class should be used instead.

Broadcast Video The *broadcast video* service class is for inelastic, noninteractive media, that is, media streams that cannot change their transmission rate based on network congestion. This class is used for live events, AV streaming, and broadcast video.

Multimedia Streaming   The *multimedia streaming* service class is for noninteractive streaming media that can detect network congestion and/or packet loss and respond by reducing its transmission rate. This class is used for video-on-demand (VOD) services, video that is stored before it is sent and buffered when it is received to compensate for any variation in transmission rate.

Low-Latency   The *low-latency* data class is for applications where data arrives in big, short-lived bursts.

High-Throughput   The *high-throughput* data class is for longer, high-volume traffic flows. It is used by applications that store data and then forward it, like FTP service and e-mail.

Low-Priority   The *low-priority* data class is used for any applications that can tolerate long interruptions.

Standard   The *standard* class provides best-effort delivery. Any applications that are not specifically assigned to another class will fall into the standard class.

Security Technologies

Networks make resources easily accessible worldwide. You no longer have to be physically present at a device to use it, configure it, or troubleshoot it. In theory, you can do so from anywhere. The problem is, unless your client's network is properly secured, anyone can gain access to it. If the wrong people gain access to your client's network resources, they can do a lot of damage.

Security is often cited as the number-one concern regarding attaching AV devices to an enterprise network. What are IT professionals so worried about? In this section, you will learn about the common security risks network professionals must face. You will also learn what can be done about those risks. Security for networked AV applications is the subject of Chapter 18.

Network Access Control

User authentication and authorization are key aspects of security on an enterprise network. All the network's known users, including administrators, have a user profile. This profile identifies the user's e-mail account, access privileges, group memberships, and other relevant information. When users log on to the network, they have to prove their identity to gain access to their user profiles. This is usually accomplished with a username and password combination (something you know); increasingly, network administrators are demanding additional multifactor requirements, such as physical access cards, digital token devices, code generators, mobile phone applications, or biometric scans (something you have).

The guiding philosophy of user access is *least privilege*. That is, users should have the least level of privileges they can get by with and still do their jobs. From time to time, administrators will scan the network and systems to verify user access. If a user needs additional access, they will go through a formal approval process. Then the administrator will escalate the privileges but will still enforce least privilege.

Permissions to enter a network and what rights you have once there are governed by a group of technologies and policies known collectively as Network Access Control (NAC). NAC is based on an idea that is simple to understand but challenging to implement: When you log on to a network, who you are should determine what you can do.

In a robust NAC environment, "who you are" is determined by more than your username. Identification and authentication are part of NAC, and the right to access certain VLANs, files, or programs may be directly associated with the user's login. However, NAC may look at other factors to determine what rights a user should have. For instance, NAC may examine the end point you are using and limit your rights if it is not sufficiently secure. For example, is the antivirus software on the computer up to date? If not, you may be denied access to sensitive areas of the network. NAC may also examine what type of connection you are using. Are you connected via Wi-Fi, remote VPN, or a cable connected to an onsite wall port? You may have access to more parts of the network via a physical, onsite connection than a remote or wireless one. NAC may examine some or all of these factors when deciding what rights to grant a user.

Access Control List

After a user is granted access to a network in general, their specific rights within the network may be governed by an access control list (ACL). The ACL is typically configured on the network router or on the device being accessed. It controls what is permitted to travel through the router based on type of traffic, source, and/or destination. The ACL may also contain network privileges regarding who can access what parts of the network. If your client's AV system will require special access rights, be sure that network security personnel create an ACL for the system and add the appropriate end users.

Firewall

A firewall is any technology—hardware or software—that protects a network or device by preventing intrusion by unauthorized users and/or regulating traffic permitted to enter or exit the network. Firewalls may control access across any network boundaries, including between an enterprise network and the Internet, between LANs within an enterprise, and between a host and its local network.

An enterprise network is usually protected by a dedicated hardware firewall appliance, while LANs are often protected by firewall applications on the gateway router, and individual hosts are usually protected by the firewall service built into their operating system.

A firewall is really a set of policies implemented across a range of devices. In essence, a firewall policy within an enterprise can be distilled to one of two approaches:

- All network traffic will be allowed unless it is specifically forbidden; the default is "allow."

- All network traffic will be forbidden unless it is specifically allowed; the default is "deny."

The former emphasizes ease of use but forces the network administrators to try to predict how the network may be attacked. The latter is more secure but makes new systems more difficult to configure. In either case, the responsibility of the AV designer is the same: Document system ports and protocols and coordinate those needs with the network manager.

CAUTION A firewall cannot protect users from traffic they invite onto the network. For example, it will not stop a virus or malware that they download. This is why user awareness training will always be the front line of network defense.

Types of Firewalls

The number of firewalls an enterprise network will need is a critical network design decision. There will be firewalls protecting gateways to the Internet. Firewalls may also be deployed within an organization's private network to protect certain areas from internal intruders. Firewalls use several different strategies or a combination of strategies to protect the network, including the following:

- Packet filtering rules determine whether a data packet will be allowed to pass through a firewall. Rules are configured by the network administrator and implemented based on the protocol header of each packet.
- Packet inspection tracks the state of ports and protocols in network connections and determines whether the data in each packet is part of a permitted connection initiated by devices behind the firewall. Deep packet inspection may work at the Application layer (layer 7) to verify that the data in the packets matches with the packet headers and is in the permitted format for the intended application.
- Port address translation (PAT) is a method of network address translation whereby devices with private, unregistered IP addresses can access the Internet through a device with a registered IP address. Unregistered clients send datagrams to a NAT server with a globally routable address (typically a firewall). The NAT server forwards the data to its destination and relays responses to the original client.

PAT is also known as *one-to-many NAT, network and port translation* (NAPT), or *IP masquerading*.

By enabling multiple devices to access the Internet without globally routable addresses, PAT conserves address space, which is a concern in IPv4 implementations. Even though the number of available global addresses in IPv6 is effectively unlimited, PAT is used with IPv6 networks because it hides the original source of the data. From outside the network, all data appears to originate from the NAT server. Any data that arrives at the NAT server without a client's request has nowhere else to go (analogous to having the address of a building without an apartment number, resulting in entry blocked by the doorman). PAT blocks all unrequested data with a firewall. A malicious intruder cannot trace the data's path beyond the edge of the network.

Port forwarding combines PAT and packet filtering. The firewall inspects the packet based on packet filtering rules. It is also configured to translate certain ports to private addresses on the network.

By combining packet filtering and PAT, the network administrator can allow incoming, unrequested traffic under controlled conditions. For example, the computer with a specific IP address may be allowed to send Telnet commands over port 23 to AV devices, but port 23 packets from any other address will be rejected. The firewall detects the IP address and the port and translates that port to an address, automatically forwarding the Telnet command to the designated device.

AV over Networks

Streaming audio and video in real time requires bandwidth allocation to deliver all the data. If insufficient bandwidth is available, the data gets lost or delayed, resulting in unintelligible, clipped audio and blocky video. Some amount of latency may be permissible in certain applications. In general, the more interactive the AV application, the less latency is acceptable in the displayed content. For example, in a lecture that is being streamed live to a remote location, several seconds of latency would be tolerable. However, in a two-way live conversation, even a couple of seconds delay would be unacceptable. Similarly, a networked audio system in a stadium cannot tolerate much latency because the announcers' commentaries should not lag behind the action on the field.

Many protocols are designed to deliver real-time audio and/or video over LANs. Some are proprietary, which means a private company developed the protocol and owns the rights to its use and then licenses the technology to manufacturers to use in their products. This enables the licensed manufacturers to produce interoperable products. Other protocols are either free, open-standards or proprietary protocols that have been released into the public domain for general use (which is not exactly the same thing). Open protocols may be developed, improved, extended, and released by anyone who has the inclination. Proprietary protocols, whether strictly licensed or available for public use, are maintained, updated, and extended by the original developer and released at that developer's discretion, which reduces the possibility of slightly incompatible variants circulating in the wild.

AV over Ethernet

Early on in the development of networked audio protocols, a number of proprietary protocols were created that squeezed digital audio over the Physical layer (layer 1) and the Data Link layer (layer 2) of the 10Mbps, and later, the 100Mbps networks then coming into use. The most widely adopted of these were the CobraNet protocol from Peak Audio and Ethersound from Digigram, both of which could handle up to 64 channels of 20- to 24-bit audio over dedicated audio networks. These layer 2 protocols may still occasionally be found in existing installations.

AVB

In 2011 the IEEE released 802.1-AVB into the public domain. This Audio Video Bridging (AVB) suite of Ethernet (layer 2) standards enables the transport of low-latency AV on 100Mbps and faster Ethernet networks. Unlike its predecessors on layer 2, AVB can coexist on an Ethernet network with other traffic. To achieve this feat, the AVB standards include prioritization and traffic-shaping functions to ensure that AV data is not unduly delayed by other traffic. As standard Ethernet switches and bridges do not usually include the required quality of service prioritization capabilities, only AVB-qualified bridges and switches can be used in AVB networks. Over 1Gbps Ethernet, AVB can carry uncompressed video and up to 200 channels of 48kHz, 24-bit audio, plus embedded control and monitoring, in real time. AVB has also added ASE67 interoperability to its audio protocols. As a level 2 protocol, AVB data cannot be routed outside its LAN.

AV over IP Networks

As network bandwidth became much less expensive, the early proprietary Data Link–layer protocols were replaced by a wide range of IP-based Network-layer (layer 3) protocol suites that can be routed across multiple networks, including the Internet, if sufficient bandwidth can be made available. Some IP-based protocols are widely licensed to many manufacturers and have significant ecosystems that provide a huge range of solutions for control and system integration, while some others are tightly limited to a particular manufacturer's range of products. As a system designer you have a wide range of AV over IP network architectures to choose from. Each protocol and its accompanying technologies use a different approach to installation, configuration, and maintenance and will therefore require research and reading before you can specify it with confidence.

AES67 Interoperability

Most of the early networked AV protocols were incompatible, leaving isolated islands of one or another technology scattered throughout the AV world. To alleviate this chaos, in 2013 the Audio Engineering Society (AES) released AES67, the AES Standard for Audio Applications of Networks – High-Performance Streaming Audio-Over-IP Interoperability, which has now been incorporated into the major audio-over-IP protocols from all developers. This allows substantially different protocol suites to include gateways and bridges for the seamless movement of audio between varying networks and AV architectures.

Each AES67 link carries up to 120 channels of 16- or 24-bit audio at sample rates of 44.1, 48, or 96kHz.

Dante

Dante is a proprietary IP-based, Network-layer (layer 3) bi-directional protocol developed by Audinate and used in thousands of products from hundreds of manufacturers. Dante is a combination of hardware, control software, and the transport protocol itself.

Dante requires a switched Ethernet network of 100Mbps or better, with at least a 1Gbps Ethernet backbone. Dante does not require dedicated bandwidth, as it uses the QoS DiffServ VoIP (telephony) category to prioritize AV and control data over other traffic. Over a 100Mbps Ethernet network, a single Dante connection can carry 96 channels of 24-bit, 48kHz audio or 48 channels of 24-bit, 96kHz audio. Over Gigabit Ethernet, a single Dante connection can carry 1,024 channels of 24-bit, 48kHz audio or 32-bit video with eight channels of 16-, 24-, or 32-bit audio at sample rates of 44.1, 48, 88.2, or 96kHz.

SMPTE ST2110

Developed by the Society of Motion Picture and Television Engineers (SMPTE), SMPTE ST 2110 Professional Media Over Managed IP Networks is a suite of industry standards aimed at a single, common, IP-based delivery mechanism for the professional media industries. Its target applications include film, Internet streaming and broadcast production and post-production, live event production, museums, digital video distribution, and theme parks. It is intended by SMPTE to replace its traditional SDI transport protocols for high-quality video and audio distribution over dedicated coax and fiber networks. SMPTE ST2110 is a publicly available suite of standards that can be incorporated into products by any manufacturer.

The suite of protocols specified include a transport mechanism for uncompressed video and audio streams in any resolution and format, control signals and metadata, traffic shaping and delivery timing for the data, and frame and clock synchronization across the network. It also allows the synchronized splitting, combining, embedding, and de-embedding of clock, control, and audio from the video stream. The audio components of the suite are AES67 compliant.

SMPTE ST2110 is a publicly available suite of standards that can be incorporated into products by any manufacturer.

RAVENNA

Realtime Audio Video Enhanced Next generation Network Architecture (RAVENNA) is an IP-based audio protocol suite developed by ALC NetworX, who continue to contribute to its development, even though it is now an open technology. Initially developed for the broadcast market, RAVENNA is based around a range of existing IP standards for clock synchronization, QoS, and interchange via AES67. It includes direct compatibility with Q-SYS, Dante, and the audio sections (30 and 31) of SMPTE ST2110 protocols.

Each RAVENNA device may have two independent network outputs, which can be connected to independent physical networks, providing for high levels of redundancy. Built-in functionality provides seamless receiver failover between independent streams.

RAVENNA is implemented by many manufacturers in a broad range of products across the professional audio spectrum. It is used in production facilities, live events, television outside broadcasts, recording facilities, and interstudio links across WANs.

RAVENNA requires Gigabit Ethernet (1Gbps, 10Gbps, 40Gbps) to handle up to 768 channels of uncompressed audio data at bit depths up to 24 bits and at sample rates up to 384kHz.

Crestron DM NVX

AV control and automation system supplier Crestron includes the proprietary DM NVX video-over-IP technology as part of its all-encompassing AV ecosystem. The 1Gbps Ethernet-hosted protocol suite includes low-loss, low-latency video compression for a range of video formats, together with HDCP, Dante/AES67-compliant audio, USB 2.0 signaling, and the Crestron family of control protocols.

DM NVX AV over IP is a component of Crestron's integrated environment and is usually installed, configured, and maintained by Crestron-trained and -certified technology specialists.

AMX SVSI

AV control and automation system supplier AMX includes the proprietary SVSI video over IP technology as part of its all-encompassing AV ecosystem. The 1Gbps Ethernet-hosted protocol suite includes low-loss, low-latency video compression for a range of video formats, together with AES67-compliant audio, EDID and HDCP management, complex matrix switching, and the AMX family of control protocols.

SVSI AV over IP is a component of AMX's integrated environment and is usually installed, configured, and maintained by AMX-trained and -certified technology specialists.

Extron NAV

AV control and automation system supplier Extron includes the proprietary NAV Pro video over IP technology as part of its all-encompassing AV ecosystem. The 1Gbps or 10Gbps Ethernet-hosted protocol suite includes its PURE3 codec for low-loss, low-latency video compression for a range of video formats, together with an AES67-compliant audio, USB 2.0, RS-232 and IR control, EDID and HDCP management, complex matrix switching, and the Extron family of control protocols.

NAV Pro's AV over IP is a component of Extron's integrated environment and is usually installed, configured, and maintained by Extron-trained and -certified technology specialists.

BlueRiver and SDVoE

The BlueRiver Field-Programmable Gate Array (FPGA)-based AV over IP transmitter and receiver chipset was originally developed by Aptovision for the transport of uncompressed 4K HDMI 2.0a video, 32 channels of HDMI audio, USB 2.0, RS-232, and infrared signaling via IP over Gigabit Ethernet networks with a 10Gbps backbone.

Semtech took BlueRiver and developed it into a more energy-efficient ASIC chipset, together with a set of application programming interfaces (APIs) that operate in all seven layers of the network. This more advanced AVP version of BlueRiver technology has been incorporated into products from many different original equipment manufacturers (OEMs) and has become the core technology for the SDVoE (Software Defined Video over Ethernet) alliance. BlueRiver is targeted at a broad range of networked AV applications, including live events, e-sports, medical imaging, high-end residential, and command and control.

SDVoE

SDVoE is a combination of 10Gbps Ethernet and BlueRiver technology to produce a standard method for the distribution of video (and its audio) over Ethernet networks. An SDVoE network includes a control server providing device discovery, video scaling and cropping, matrix switching, signal routing, and EDID processing. All SDVoE sources and display devices are equipped with BlueRiver chipsets to handle signal encoding and decoding. At the time of publication, the SDVoE Alliance included dozens of high-profile AV technology companies.

NewTek NDI

Network Device Interface (NDI) is a royalty-free proprietary platform, originally developed by NewTek for use in its TriCaster video production networks, and retains an emphasis on video production and distribution. The technology delivers multiple compressed video streams (up to 4K and beyond) and at least 16 audio streams over a minimum 1Gbps Ethernet network. It allows video devices to discover, identify, and communicate over IP in real time and to encode, transmit, and receive high-quality, low-latency, frame-accurate video and its accompanying audio. A 3840 60p NDI stream requires approximately 250Mbps of bandwidth. The highly integrated NDI ecosystem includes dozens of hardware and software suppliers.

NDI systems are usually installed, configured, and maintained by NDI-trained and -certified technology specialists.

Q-SYS

Audio systems company QSC has extended its original proprietary Q-SYS networked audio platform to become a fully fledged, IP-based network audio, video, and control ecosystem. The 1Gbps Ethernet-hosted protocol suite includes low-loss, low-latency video compression for a range of video formats, together with Dante-based AES67-compliant audio, USB integration, AVB compatibility, acoustic echo cancellation, a VoIP/SIP conferencing interface, complex matrix switching, and a complete suite of control protocols.

Q-SYS is at the core of QSC's integrated environment and is usually installed, configured, and maintained by QSC-trained and -certified technology specialists.

Chapter Review

Before you can send any signals across a network, especially AV signals, you have to understand how data travels through the network. In this chapter, you learned about network components, types, and topologies, as well as how different protocols play out in the design and operation of AV/IT networks. You're now ready to take on streaming design, which is the focus of Chapter 17.

Review Questions

The following questions are based on the content covered in this chapter and are intended to help reinforce the knowledge you have assimilated. These questions are similar to the questions presented on the CTS-D exam. See Appendix E for more information on how to access the free online sample questions.

1. A _____ uses unique, hard-coded physical addresses, known as Media Access Control (MAC) addresses, to send data between nodes.
    A. Metropolitan area network (MAN)
    B. Local area network (LAN)
    C. Global area network (GAN)
    D. Wide area network (WAN)

2. What standard best identifies the purpose and use of a wireless connection?
    A. SCP X.25
    B. IEEE 802.11
    C. RFC 761
    D. Signaling system 7

3. Twisted-pair cables offer protection from electromagnetic interference by _____.
    A. Shielding each pair inside a foil or braided shield
    B. Physically separating conductors from higher-voltage signals
    C. Exposing each wire to the same outside interference, allowing it to be canceled at the input circuit
    D. Comparing the signal that arrives at the destination device against a checksum

4. An organization needs its own DNS server if it _____.
    A. Needs to statically associate certain names and IP addresses
    B. Uses DNS internally to manage the names and addresses of devices on the private network
    C. Has more than one LAN
    D. Needs to access the Internet

5. Reserve DHCP allows you to _____.
    A. Limit the use of a pool of addresses to a particular VLAN or subnet
    B. Establish a pool of additional addresses in case the primary pool of addresses runs out
    C. Assign static IP addresses to devices using a DHCP server
    D. Set an unlimited lease time for all devices using a DHCP server

6. TCP transport should be used instead of UDP transport when _____.
   A. Speed and continuity of transmission are more important than guaranteed delivery or accuracy of data
   B. Data accessibility is more important than security of the data transmission
   C. Security of the data transmission is more important than data accessibility
   D. Guaranteed delivery and quality of the data are more important than the speed or continuity of the transmission

7. The term *Gigabit Ethernet* refers to _____.
   A. 10Mbps Ethernet
   B. 100Mbps Ethernet
   C. 800Gbps Ethernet
   D. 10Gbps Ethernet

8. The speed of Ethernet your device can send and receive depends on the _____.
   A. Type of WAN physical medium the organization uses to access the Internet
   B. Capability of its network interface card (NIC)
   C. Type of cable used to connect the device to the network
   D. Speed of the switch to which it is directly attached

9. To transport audio using the AVB protocol, you must have at least _____ Ethernet.
   A. 10Mbps
   B. 100Mbps
   C. 1Gbps
   D. 10Gbps

Answers

1. **B.** A local area network uses unique, hard-coded physical addresses, known as Media Access Control (MAC) addresses, to send data between nodes.
2. **B.** The IEEE 802.11 standard defines many aspects of wireless connectivity.
3. **C.** Twisted-pair cables offer protection from electromagnetic interference by exposing each wire to the same outside interference, allowing it to be canceled at the input circuit.
4. **B.** An organization needs its own DNS server if it uses DNS internally to manage the names and addresses of devices on the private network.
5. **C.** Reserve DHCP allows you to assign static IP addresses to devices using a DHCP server.

6. D. TCP transport should be used instead of UDP transport when guaranteed delivery and quality of the data are more important than the speed or continuity of the transmission.

7. D. The term *Gigabit Ethernet* refers to 1 to 10Gbps Ethernet, while 800Gbps Ethernet is usually classified as *Terabit Ethernet*.

8. B. The speed of Ethernet your device can send and receive depends on the capability of its network interface card (NIC).

9. B. To transport audio using the AVB protocol, you must have at least 100Mbps Ethernet.

CHAPTER 17

Streaming Design

In this chapter, you will learn about
- Conducting a needs analysis for streaming AV systems
- Conducting a network analysis for streaming AV systems
- Designing an AV streaming system that takes into account bandwidth, latency, and other requirements
- Quality of service (QoS) and its effect on streaming AV systems
- The network protocols used for streaming
- The difference between unicast and multicast and the basics of implementing multicast distribution on an enterprise network

The network environment is as important to a streaming application as a physical environment is to a traditional AV system. When it comes to designing for a streaming AV system, you need to analyze the network as carefully as you would a physical room by exploring its potential, discovering its limitations, and recommending changes to improve system performance.

Streaming media comprises live video and audio streaming, video on demand, and Internet Protocol television (IPTV). It is also the foundation for other networked AV systems, such as digital signage and conferencing.

To design a successful streaming application, you must think carefully about the client's needs and how they impact service targets. For example, how much bandwidth will your AV streams require? How much latency and packet loss can users tolerate? Some answers depend on the network itself; others depend on how content is encoded to travel over that network. As you begin to discover the number and quality of streams your client needs, you may set off some bandwidth alarm bells. Video and audio streams require significantly more bandwidth than other AV signals, such as control. If you understand the concepts underlying the compression, encoding, and distribution of digital media, you're on your way to designing a variety of streaming solutions.

> **Duty Check**
>
> This chapter relates directly to the following tasks on the CTS-D Exam Content Outline:
>
> - Duty B, Task 7: Coordinate with IT and Network Security Professionals
> - Duty C, Task 1: Create Draft AV Design
> - Duty C, Task 2: Confirm Site Conditions
>
> In addition to skills related to these tasks, this chapter may also relate to other tasks.

Streaming Needs Analysis

When conducting a needs analysis for a streaming solution, you need to discover not only what the client wants but also what the current networking environment can deliver. Typical questions related to traditional audiovisual applications—regarding room size, viewer positioning, and so on—may be impossible to answer because the client may not know with any certainty where end users will be when they access streaming content. Instead of asking about the room where an AV system would reside, you need to delve into issues relating to streaming quality, bandwidth, latency, and delivery requirements. You need to think in terms of tasks, audience, end-point devices, and content. And you need to explore these issues in the context of an enterprise network, not a particular venue.

Ultimately, what you learn in the needs analysis will help inform the service-level agreement (SLA) for your streaming AV system—and possibly other, more encompassing SLAs. Not to mention, as you're collecting information about the client's need, you should always ask yourself, "How is this going to impact network design?" because, eventually, you may have an IT department to answer to.

Streaming Tasks

In the design of any AV system (networked or non-networked), form follows function. The needs analysis begins with discovering the purpose of the system. Assume you've already established your client needs to stream AV. Why? What tasks will streaming AV be used for?

Decisions regarding bandwidth allocation, latency targets, and port and protocol requirements should be driven by a business case that is established in the needs analysis. How do the tasks that the prospective streaming system will perform contribute to the organization's profitability, productivity, or competitiveness? If you can make a valid business case for why your streaming system requires a high class of service, opening User

Datagram Protocol (UDP) ports, or 1Gbps of reserved bandwidth, you should get it. But you need to understand the following:

- "The picture will look bad if we don't reserve the bandwidth" is not a business case.
- "The picture will look bad if we don't reserve the bandwidth, and if the picture looks bad, the intelligence analysts won't be able to identify the bunker locations" is an excellent business case.

If there's no business case for high-priority, low-latency, high-resolution streaming video, for example, then the user doesn't need it, and it probably shouldn't be approved. In that case, be prepared to relinquish synchronous UHD streaming video if the client doesn't really require it. Just be sure you document the lower expectations as your system's service target.

Streaming Task Questions

What tasks will the streaming system be used for? This is the most basic question of a needs analysis. Answering it in detail should reveal the following:

- The importance of the system to the organization's mission
- The scope of use in terms of number and variety of users
- The frequency of use

If the task is a high priority, the streamed content will require priority delivery on the network, which will affect the bandwidth allocated and the class of service assigned to it. The nature of the task will also impact latency requirements. As you delve further into the tasks that the streaming system must support, you may ask questions such as, "Do users need to respond or react immediately to the streamed content?" If so, latency requirements should be very low—such as 30 milliseconds.

Furthermore, find out the answer to this question: "Will any delay in content delivery undermine its usefulness?" This question is also aimed at determining how much latency is acceptable in the system, outside of immediate-response situations. And the answers will help address follow-on questions: "Can we use TCP transport, or is UDP necessary?" and "How should the data be prioritized in a QoS architecture?"

Audience

The user is the star of the streaming needs analysis (or any needs analysis, for that matter). In fact, *who* the audience is specifically may be a major factor in a business case for a high-resolution, low-latency streaming system. We're talking about the organization's hierarchy here. If the main users are fairly high in the hierarchy—management, executives—they may demand an unnecessarily high-quality system to match their status.

If, on the other hand, the system will be used broadly throughout the organization, you may need to consider the bandwidth implications of such widespread use. Would a multicast work? (We will cover multicasting in detail later in this chapter.) Should different groups of users be assigned different service classes? You'll understand better when you identify the audience.

Where the audience is located, however, may be your primary challenge in offering a streaming system that meets the client's need. It's actually a far more complex issue when it comes to streaming applications than traditional audiovisual systems. Asking about location should reveal whether all users (the audience) are on the same local area network (LAN) or wide area network (WAN) and whether some users will access streaming content over the open Internet. How does that impact design? There are various reasons, including the following:

- Bandwidth over a WAN is limited.
- QoS and multicast are impossible over the open Internet.
- LAN-only solutions have a far greater array of data link and quality options.

End Points

Related to audience are the end points people will use to access streaming content. Your client may need to stream data to a handful of overflow rooms, hundreds of desktop PCs, or thousands of smartphones and tablets. The number and type of end points have a direct impact on service requirements.

How many different end points does the client need to stream to? With unicast transmissions, the bandwidth cost will rise with each additional user. If the user needs to stream to a large or unknown number of users within the enterprise network, you need to consider using multicast transmission, reflecting servers, or even a content distribution network (CDN) to reduce bandwidth. What *kind* of end points will people be using? Different end points support different resolutions, for example. If end points will be especially heterogeneous, you may want to implement a variable bit-rate intelligent streaming solution. If most users will be using mobile devices, content must be optimized for delivery over Wi-Fi or 4G/5G/6G cellular networks.

Content Sources

The type of content the client will stream has a direct impact on the network, which in turn affects the streaming system design. Like other parts of the needs analysis, the content will factor in its bandwidth usage, latency targets, and class of service. The following are some basic questions to ask:

- Do you need to stream audio, video, or both? Video requires far more bandwidth than audio. Codec options will also differ based on type of media.
- How will content be generated? Video content generated by a computer graphics card or an interactive media server may require more bandwidth than content captured by a camera. You'll also have different codec options for different sources.
- Will streaming images be full-motion or static? Still images, or relatively static images, require a lower frame rate than full-motion video.
- Will the client be streaming copyrighted material or material that's protected by digital rights management (DRM)?

You will also need to consider how and where content will enter the network. This is often referred to as the place where content is *ingested* (or uploaded) and can be a computer, a server, a storage system, or a purpose-built network streaming appliance. You should determine how many ingestion points there will be and ensure each has the bandwidth required for expedient uploading. In addition, how many concurrent streams, or "channels," must be uploaded? The answer should help establish how many video servers or server input interfaces will be needed.

Moreover, can you exercise any control over the format, bandwidth, and so on of the content that will be ingested for streaming? If your clients are ingesting several different formats, they may need a transcoder, which translates streams into formats that are compatible with the desired end points or are bandwidth-friendly toward certain connections.

And when it comes to the streaming content itself, you may want to make recommendations on bit depth, frame rates, resolutions, and other streaming media characteristics based on the network design you anticipate. For video, how much motion fidelity do users need to accomplish their tasks? Lower acceptable fidelity will allow you to use a lower frame rate and/or more aggressive compression. Higher fidelity will result in increased latency or higher bandwidth.

What level of image quality or resolution is required for the task? Again, the focus is on *requirement*, not the desire. Use the answer to these questions to help establish a business case: At what point will limiting the video resolution or increasing the compression detract from the system's ability to increase the client's productivity, profitability, or competitiveness? How low can the video resolution be and how lossy can the compression get before the image is no longer usable?

Using Copyrighted Content

Your clients may want to stream content that they didn't create, such as off-air radio and television or video or music from a media library or streaming service. Unfortunately, such digital distribution can violate copyright laws.

Make sure clients are aware of the potential licensing issues related to the content they want to stream. You may need to negotiate a bulk commercial license with a content provider, such as a broadcaster or a streaming video or music service.

If you fail to obtain the proper licenses to stream content, you aren't just risking legal repercussions; you're risking the system's ability to function at all. Publishers and copyright owners use DRM technology to control access to and usage of digital data or hardware. DRM protocols within devices determine what content is even allowed to enter a piece of equipment. Copy protection systems such as the Advanced Access Content System (AACS) and BD+ used on Blu-ray content and the Content Scrambling System (CSS) used in DVD players are each a subset of DRM. Actual legal enforcement of DRM policies varies by jurisdiction.

High-Bandwidth Digital Content Protection (HDCP) is a form of DRM developed by Intel to control digital audio and video content as it travels across DVI or HDMI connections. It prevents the transmission or interception of nonencrypted HD content.

HDCP support is essential for the playback of protected high-definition (HD and UHD) content. Without the proper HDCP license, material will not play.

It can be difficult—though possible—to stream to multiple DVI or HDMI outputs. All the equipment used to distribute the content must be licensed. When in doubt, always ask whether a device is HDCP-compliant.

NOTE IPTV and streaming are similar, given that both utilize much of the same technology, so for the purposes of this book, the two will be treated as one topic. IPTV is a system that delivers television services over a packet-switched network, such as a LAN or the Internet. Streaming is traditionally the transfer of audio and video files that are played at the same time they're temporarily downloaded to a user's computer or other device.

Streaming Needs Analysis Questions

Use the following questions to gather information from clients about streaming applications. Consider how the users' needs with respect to each item could impact the system's design, cost, or network.

Tasks

- What tasks will the system be used to perform?
- Do users need to respond or react immediately to the streamed content?
- Will any delay in content delivery undermine the usefulness of the content?

Audience

- Who is the intended audience?
- Where is the intended audience (onsite, within the LAN; offsite, within the company WAN; offsite, outside the company WAN; and so on)?
- What are the audience access control requirements?

End Points

- How many different end points (devices) do you need to stream to?
- What kind of end points will your end users be using to view content (desktops, mobile devices, large displays, videowalls, projectors, etc.)?

Content

- What kind of content do you need to stream (for example, full-motion video and audio, full-motion video only, video and audio with captioning and/or audio description, audio only, still images only, still images and audio, etc.)?
- How will content be generated?
- Will you be streaming Voice over IP (VoIP)?
- How will content be ingested into the network?

- For motion video, how much motion fidelity is required?
- What level of quality and/or image resolution do you require (standard definition, high definition, ultra-high definition, high dynamic range, high frame rate, best possible, adequate, and so on)?
- How many concurrent upload streams, or "channels," do you require?
- How many concurrent download streams, or "channels," do you require?
- What are the accessibility requirements, if any?

Storage
- Will content be available on demand?
- How long will content need to be stored?
- How quickly does content need to be propagated from storage?
- What are the backup or redundancy requirements?

Streaming Design and the Network Environment

You've talked to users to find out what they need from a streaming media solution, so now it's time to speak with the client's IT department. The network environment is as important to a streaming application as the physical environment is to a traditional AV system. You need to analyze the network as carefully as you would a physical room by exploring its potential, discovering its limitations, and recommending changes to improve system performance. Let's get started.

Topology

In the needs analysis stage, you determined whether streaming users would be accessing content inside or outside a LAN. Remember that a LAN in this case is a single, openly routable location. Until you've had to traverse a router, you're still on a LAN.

If your streaming system will remain within a single LAN, you're lucky. The system's routing will be considerably less complex, multicasting will be far easier to implement, and bandwidth availability is unlikely to be a problem. If you're streaming content over a WAN, however, you have several additional factors to consider, including the following:

- The physical location of streaming servers
- Bandwidth availability on every network segment
- The possible need for hybrid unicast/multicast implementation
- The addressing scheme of the organization

The addressing scheme will determine what information you need to gather about your ports and protocols. It will also impact how access control and system control will be managed. Can the system identify authorized users via the DNS? Do you need to reserve a set of IP addresses on the DHCP server for the streaming servers?

Bandwidth: Matching Content to the Network

Bandwidth availability represents the single largest constraint on most networked AV systems. It will drive your choice of transmission method and codec for streaming applications. We will cover these considerations later in this chapter.

Remember that the network is only as fast as its slowest link. The rated capacity of the network should be based on the bottlenecks—the lowest-throughput link that data will have to traverse. Whatever the rated capacity of the network is, you can't use it all. That would be like mounting a display with bolts that can hold exactly the display's weight; the first time someone bumps into the screen, it will tear right off the wall. How much of the network can you use?

Realistically, only about 70 percent of rated network capacity is available. The remaining 30 percent should be reserved to accommodate peak data usage times and avoid packet collision. Some industry experts recommend reserving as much as 50 percent of the network for this purpose. Ask the network manager what network capacity is available for each client.

In a converged network, of the 70 percent of bandwidth considered "available," only a portion should be used for streaming media—industry practice is to allow about 30 percent of the available 70 percent. Otherwise, you may not have enough bandwidth left for other network applications.

Depending on the importance of the streaming application, you may want to reserve this bandwidth with Resource Reservation Protocol (RSVP). RSVP is a Transport-layer protocol used to reserve (set aside) network resources for specific applications. The reservation is initiated by the host receiving the data and must be renewed periodically. RSVP is used in combination with DiffServ. At the least, DiffServ QoS will be required for WAN streaming.

Your goal, then, is to design a streaming application that consumes no more than 30 percent of the available network capacity. You may also need to implement some form of bandwidth throttling to prevent the streaming application from overwhelming the network during peak usage, setting the limit at that 30 percent mark. Traffic shaping will introduce latency into the stream, causing it to buffer but preserving its quality. Traffic policing drops packets that are over the limit, which reduces video quality but avoids additional latency. What's more important to your client: timeliness or quality?

NOTE Based on estimates that only about 70 percent of rated network capacity is considered available and that only about 30 percent of that capacity is available for streaming, you should work on the basis that only approximately 20 percent (30 percent of 70 percent) of a network's rated capacity should be considered available for streaming.

Image Quality vs. Available Bandwidth

Working with the client and the client's IT group, you must determine whether the amount of bandwidth allocated for streaming will be driven by current network availability or the end user's quality expectations. It may be anathema to AV professionals, who pride themselves on high-quality experiences, but the client may be willing to sacrifice quality in favor of network efficiency. In that case, it's your job to ensure that the client knows what they are giving up and that the SLA reflects an acceptable balance between efficiency and quality.

In general, streaming bandwidth will be driven by network availability when the client is extremely cost-conscious or the streaming service is being added to an existing network that's difficult to change. Streaming bandwidth will be driven by the user's need for quality when high image resolution and frame rate are required to perform the tasks for which the streaming service will be used or when the tasks for which the streaming service will be used are mission critical. The need for quality also seems to get a boost when your client audience is from the C-level suite.

What can you do when you discover that streaming requirements exceed available network resources? If quality takes precedence, either add bandwidth, optimize existing bandwidth, assign streaming a high QoS class, or reserve bandwidth using RSVP. We will cover QoS and RSVP in more depth later in this chapter.

If content availability takes precedence, you may need to reduce image resolution, frame rate, or bit rate; cut the number of channels by implementing multicast streaming or cache servers; or implement intelligent streaming.

TIP When discussing bandwidth capacity with a network administrator, make sure you're on the same page. Ask yourself, "Is the admin telling you how much bandwidth the whole network is rated for?" If so, you can use only about 20 percent of that for streaming. Is the admin telling you the average bandwidth availability? If so, you can use about 30 percent of that for streaming. Is the admin telling you the availability during peak usage? If so, it will help you determine QoS requirements.

Streaming and Quality of Service

Quality of service (QoS) is a term used to refer to any method of managing data traffic to provide the best possible user experience. Typically, QoS refers to some combination of bandwidth allocation and data prioritization.

Many different network components have built-in QoS features. For example, videoconferencing codecs sometimes have built-in QoS features that allow various devices on a conference to negotiate the bandwidth requirements. Network managers may also use software to set QoS rules for particular users or domain names. AV designers need to be more concerned with QoS policies that are configured directly into network switches and routers.

The bandwidth required for streaming video makes QoS a virtual requirement for streaming across a WAN. If you're adding a streaming application to an existing network, you need to find out the following:

- **Whether QoS has been implemented across the entire WAN** To provide any benefit, QoS must be implemented on every segment of the network across which the stream will travel.

- **What differentiated-service classes have already been defined for network-based QoS (NQoS)** Every organization prioritizes data differently. You may think that AV data should always be assigned a high service class because it requires so much bandwidth and so little latency. However, a financial institution may value the timely delivery of transaction data or stock quotes above streaming video. Where will the streaming application fit within the enterprise's overall data priorities?
- **Whether RSVP can be implemented to reserve bandwidth for the streaming application** RSVP must be implemented on every network segment to function, which is a labor-intensive process. Start by finding out whether RSVP is currently implemented on the network—whether or not you should use it is a separate question. Does the importance of the streaming application really merit permanently reserving 20 percent of the network capacity for its traffic?
- **What policy-based QoS rules are already in place, if any** Assigning QoS policies to particular user groups can be helpful for networked AV applications. You may want to place streaming traffic to and from the content ingest points in a higher class of service than on the rest of the network, for instance. If remote users are accessing the system via an Internet Protocol Security (IPSec)–secured link, you may have to use policy-based QoS. The ports, protocols, and applications of IPSec traffic are encrypted and hence hidden from the router. (IPSec is discussed in more detail in Chapter 18.)
- **Whether traffic shaping can be used to manage bandwidth** Again, traffic shaping policies will have to be implemented on every router on the network. However, if your client doesn't mind the additional latency, traffic shaping can be an effective way to prevent the network from being overwhelmed by streaming traffic. Of course, if traffic policing has been implemented, you need to know this too. Find out where bandwidth thresholds have or will be set for streaming traffic and set expectations for dropped or delayed packets accordingly.

Latency

If your application will include live streaming, the amount of latency inherent to the network can be nearly as big a concern as bandwidth availability. What latency is present, and what causes it?

Find out whether your client has an internal speed-test server. If it does, work with the network manager to determine the inherent latency of the network. You can use a WAN speed test to verify network upload and download bit rates, as well as latency inherent between IP addresses.

If your client does not have an internal speed-test server, free WAN speed-test tools are available from many websites; try the search term "internet speed test" in your web search engine.

Varying degrees of latency are acceptable, depending on your client's needs. Here are some examples:

- **Videoconferencing**   200 milliseconds
- **High-fidelity audio**   50 microseconds
- **Streaming desktop video**   1 second

Eliminating latency entirely from streaming applications is impossible. Your SLA should document how much latency your client can tolerate for the application. Keep in mind, if your client requires secure data transport, it will induce more network latency, as encryption and decryption take time. Some codecs that use complex software algorithms to compress and/or decompress the video stream may also add to overall streaming latency.

Network Policies and Restrictions

Remember, what the network is *capable* of doing and what the network is *permitted* to do are two separate issues. You must determine not only whether bandwidth optimization and latency mitigation strategies, such as differentiated service, policy-based QoS, multicasting, and UDP transport, are possible but also whether they are even permitted. Many of these strategies are extremely labor-intensive to implement, and some, such as multicast and UDP transport, may represent significant security risks. For a review of UDP, see Chapter 16.

You should always investigate what software and hardware are already in use in the enterprise. You can bet that staying within an organization's existing device families will ease the process of getting approval to connect devices to the network. For example, the software that users currently have (or are allowed to have) will determine what codecs are available on the end-user playback device, which in turn determines the encoder your system should use.

Cheat Sheet: Streaming Network Analysis Questions

Use the following questions to gather information from your client's IT department about supporting streaming media applications. Consider how network policies with respect to each item could affect the streaming system's design, cost, and network impact. Some of the technical issues in these questions are covered in greater depth later in this chapter.

Topology

- Will content be streamed within a LAN, within a WAN, or over the open Internet?
- If content will be streamed over a WAN, what is the network topology?
- What is the addressing scheme (DNS, DHCP, static)?

Bandwidth Availability

- What is the network's total available end-to-end upload bandwidth?
- What is the network's total available end-to-end download bandwidth?
- What is the network's typical available worst-case upload bandwidth?
- What is the network's typical available worst-case download bandwidth?
- If content will be streamed to the Internet, what upload and download bandwidth is provided by the Internet service provider (ISP)?
- If traffic will be streamed over a WAN, has QoS been implemented? If so, what level will AV traffic occupy?
- If traffic will be streamed over a WAN, has traffic shaping been implemented?
- If traffic will be streamed within a LAN, has Independent Group Management Protocol (IGMP) been implemented? If so, what version?
- Is a hybrid solution required to preserve bandwidth? If so, is Protocol-Independent Multicast (PIM) available to transport?
- Do you need to relay multicast streams across domains?

Latency

- How much latency is inherent to the network?
- Will there be transport-level security requirements?

Network Policies

- Are multicast and UDP transport permitted?
- Are there restrictions on protocols allowed on the network?
- Are there restrictions on what software may be installed on hosts?
- How is access control currently managed on the network?
- What video player software is currently installed on the hosts? What codecs does it support?

Designing the Streaming System

Most live video with low latency requirements will be delivered via UDP transport. That's because if a stream is time sensitive, it's generally preferable to drop a few packets than wait for the source to confirm delivery and have to resend dropped packets.

Multicast transmissions are always delivered via UDP. UDP packets include little data, however, because they're designed to deliver data as efficiently as possible.

Real-Time Transport Protocol

Real-Time Transport Protocol (RTP) is a Transport-layer protocol commonly used with UDP to provide a better AV streaming experience. (RTP also works with TCP, but the latter is more concerned with reliability than with speed of transmission.) In addition to a data payload, RTP packets include other information, such as sequence and timing. RTP helps prevent jitter and detects when video packets arrive out of sequence. It also supports one-to-many (multicast) video delivery.

RTP is often deployed in conjunction with the Real-Time Transport Control Protocol (RTCP), a Session-layer protocol for monitoring quality of service for AV streams. RTCP periodically reports on packet loss, latency, and other delivery statistics so that a streaming application can improve its performance, perhaps by lowering the bit rate of the stream or using a different codec. RTCP does not carry any multimedia data or provide any encryption or authorization methods. In most cases, RTP data is sent over an even-numbered UDP port, while RTCP is sent over the next higher odd-numbered port.

Other Streaming Protocols

When setting up a streaming system, in addition to transport protocols, you should also consider streaming protocols. Streaming-specific protocols serve different functions and are based largely on the types of end points clients will use to view content.

Real-Time Streaming Protocol

Real-Time Streaming Protocol (RTSP) comes in handy if all end points are desktop computer clients. This is because RTSP supports RTCP, while MPEG Transport Stream (MPEG-TS), for example, does not. RTSP is a control protocol that communicates with a streaming server to allow users to play, pause, or otherwise control a stream. For its part, RTCP sends back user data, allowing the streaming device to dynamically adjust the stream to improve performance.

MPEG Transport Stream

If any of the end points are set-top boxes or similar devices (maybe a streaming box behind a digital signage screen in a restaurant), you'll probably use MPEG-TS. The frequency of some displays used with set-top-style boxes can sometimes cause a perceived lag between audio and video. MPEG delivery prevents this from happening by combining audio and video into a single stream. MPEG-TS is defined as part of MPEG-2, but it is the transport stream used to deliver MPEG-4 audio and video as well.

Session Description Protocol

Session Description Protocol (SDP) is a standardized method of describing media streamed over the Internet. The information carried by SDP generally includes session name, purpose, timing, information about the media being transported (though not the media itself), and contact information for session attendees. In short, SDP is used to kick off a streaming session—ensuring all invited devices are in contact and understand what's coming next. The information contained in SDP can also be used by other protocols, such as RTP and RTSP, to initiate and maintain a streaming session.

High-Quality Streaming Video

As you probably know, the Motion Picture Engineering Group defines many of the compression formats commonly used for streaming high-quality video: MPEG-1, MPEG-2, MPEG-4, and MPEG-5. There used to be an MPEG-3 format, but it's no longer used and shouldn't be confused with the MP3 audio compression format.

All major MPEG standards stream over UDP. MPEG-2 and MPEG-4 are the most prominent for networked AV systems, though MPEG-1 is also ubiquitous—MP3 audio is part of the MPEG-1 standard. We'll explore further the flavors of MPEG you're likely to use for a streaming design.

MPEG-2/H.222/H.262

MPEG-2, also known as H.222/H.262, is the most common digital AV compression format. It's an international standard, defined in ISO/IEC 3818. MPEG-2 is used to encode AV data for everything from DVD players and digital cable to satellite TV and more. Notably, MPEG-2 allows text and other data, such as program guides for TV viewers, to be added to the video stream.

There are various ways to achieve different quality levels and file sizes using MPEG-2. MPEG-2 streams have a minimum total bit rate of 300kbps. Depending on the frame rate and aspect ratio of the video, as well as the bit rate of the accompanying audio, the total bit rate of an MPEG-2 stream can exceed 10Mbps.

MPEG-4

MPEG-4 is designed to be a flexible, scalable compression format. It is defined in the standard ISO/IEC 14496. Unlike MPEG-2, MPEG-4 compresses audio, video, and other data as separate streams. For applications where audio detail is important, such as videoconferencing, this is a major advantage. MPEG-4 is capable of lower data rates and smaller file sizes than MPEG-2, while also supporting high-quality transmission. It's commonly used for streaming video, especially over the Internet.

The MPEG-4 standard is still developing. It is broken down into parts, which solution providers implement or not, depending on their products. It's safe to say there are a few complete implementations of MPEG-4 on the market, but it isn't always clear what parts of MPEG-4 an MPEG-4 solution includes. Therefore, it's important to understand the major components of MPEG-4.

MPEG-4 Levels and Profiles

Within the MPEG-4 specification are levels and profiles. These let manufacturers concentrate on applications without getting bogged down in every aspect of the format. Profiles are quality groupings within a compression scheme. Levels are specific image sizes and frame rates of a profile. This breakdown allows manufacturers to use only the part of the MPEG-4 standard they need while still being in compliance. Any two devices implementing the same MPEG-4 profiles and levels should be able to interoperate.

MPEG-4 Part 10–Advanced Video Coding/H.264 MPEG-4 Part 10 was once the most commonly implemented part of the MPEG-4 standard for recording, streaming, or compressing high-definition audio and video. It is also known as H.264 Advanced Video

Coding (AVC). AVC can transport the same quality (resolution, frame rate, bit depth, and so on) as MPEG-2 at far lower bit rates—typically half as low as MPEG-2.

AVC profiles include but are not limited to the following:

- Baseline profile (BP), used for videoconferencing and mobile applications
- Main profile (MP), used for standard-definition digital television
- Extended profile (XP), used for streaming video with high compression capability
- High profile (HiP), used for high-definition broadcast and recording, such as digital television and Blu-ray Disc recording

AVC also includes intraframe compression profiles for files that might need to be edited—High 10 Intra Profile (Hi10P), High 4:2:2 Intra Profile (Hi422P), and High 4:4:4 Intra Profile (Hi444P).

MPEG-H–High Efficiency Video Coding/H.265 MPEG-H includes H.265, also known as High Efficiency Video Coding (HEVC), which is the successor to H.264. With H.265, AV designers can double the data compression ratio of a stream compared to H.264/MPEG-4 AVC without sacrificing video quality. Conversely, they can offer much better video quality at the same bit rate. H.265 can support 8K ultra-high-resolution video up to 8192×4320.

Because networked AV applications can have voracious bandwidth appetites, H.265 was considered a significant advance when it was approved as an ITU-T standard in 2015 and consequently it too has been superseded as the fastest codec on the block.

MPEG-I Part 3–Versatile Video Coding/H.266 The successor to H.265, the compression in Versatile Video Coding (VVC)/H.266, once again reduces the bandwidth and encoding overhead required for video streams by between 30 and 50 percent of that required by H.265 for the same quality of images. Rather than replacing H.265 for general use, it is expected to be used initially for high dynamic range (HDR) and high frame rate (HFR), 8K, 10K, 16K, and 360-degree video streams.

MPEG-5 Part 1–Essential Video Coding The royalty-free Essential Video Coding (EVC) codec had only been implemented as an open-source project at the time of publication. In this format it uses half the processing resources of H.265 and produces superior video quality at comparable bit rates. As this project was proposed by Samsung, Huawei, and Qualcomm, it is likely to be widely available.

MPEG-5 Part 2–Low Complexity Enhancement Video Coding Low Complexity Enhancement Video Coding (LCEVC) is a recent video standard from MPEG. It specifies an enhancement layer which, when combined with a base video encoded with a separate codec (e.g., AVC, HEVC, VP9, AV1, EVC or VVC), produces an enhanced video stream. It is suitable for software processing implementation with sustainable power consumption. The enhancement stream provides new features such as extending the compression capability of the base codec and lowering encoding and decoding complexity.

The key attributes of LCEVC are

- The video stream can be decoded without specific firmware or OS support by all devices capable of decoding the base codec.
- All web browsers are able to decode high-resolution video without plug-ins and/or a browser upgrade.
- The additional data stream is compatible with the existing ecosystem (e.g., ad insertion, metadata management, CDNs, DRM/CA).
- The overall processing power requirement to encode a video stream is comparable with that of the base codec when used alone at full resolution.

Unicast and Multicast

Two major architectures are used for streaming AV data across networks: unicast and multicast.

Unicast

Unicast streaming establishes one-to-one connections between the streaming server that sends the AV data and client devices that receive it. Each client has a direct relationship with the server. The client sends a request to the server, and the server sends the client a stream in response. Because the server is sending out a separate stream to each client, each additional client consumes more of the available bandwidth. Streaming media to three clients at 100kbps actually uses 300kbps of bandwidth. Unicast streams may use either UDP or TCP transport, although with TCP transport, you can assume there will always be some *buffering,* or client devices waiting for parts of the stream to arrive before playing it.

An encoder typically produces only five or six unicast streams, depending on the streaming device's available resources. If your client needs more than a handful of unicast streams, you'll need a streaming or caching server to replicate and manage them.

That said, unicast is easier to implement than multicast, and it's cheaper for small applications. Consider using unicast for point-to-point streaming, and even point-to-point plus recording. If you're unicasting to thousands of users, however, you'll have to invest in a facility like a CDN at the network's edge, which can be expensive.

Multicast

Multicast streaming is a one-to-many transmission model. One server sends out a single stream that multiple clients can access. Multicast streams require UDP transport. They can be sent only over LANs or private networks, not over the open Internet. A group of IPv4 IP addresses 224.0.0.0/4 (224.0.0.0 to 239.255.255.255) (see Table 17-1 later in this chapter) and a group of IPv6 IP addresses using the prefix ff00::/8 (see table 17-2 later in this chapter) are set aside for multicast transmissions.

In multicast streaming, the following happens:

1. A server sends the stream to a designated reserved IP address, called the *host address*.
2. Clients subscribe to the host address.
3. Routers send the stream to all clients subscribing to the host address.

Subscribing to an active multicast host address is like tuning into a radio station: All the users receive the same transmission, and none of them can control the playback. There is no direct connection between the server and the clients. Because the server is sending only one stream, the transmission should theoretically take up the same amount of bandwidth no matter how many clients subscribe. Sounds efficient, right?

However, not all networks are capable of passing multicast streams. For multicast to work, every router in the network must be configured to understand multicast protocols and the reserved address ranges. If the router doesn't recognize these IP address as a multicast host address, the clients will have no way to access the stream. As a result, multicasting can be very labor-intensive to implement. Only small portions of the Internet are multicast-enabled, so it's usually not possible to multicast over the open Internet. If you want the client's IT department to allow multicast, you will need to make a convincing case. Figure 17-1 illustrates the main differences between unicast and multicast streaming.

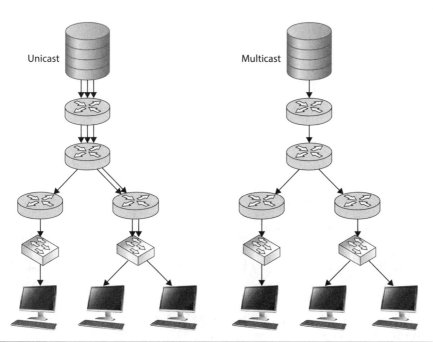

Figure 17-1 Differences between unicast and multicast

Unicast vs. Multicast

How do you choose whether to stream using unicast or multicast? The decision should be based on your streaming needs and network capabilities. In many cases, you won't have the option to use multicast streaming, particularly if the network you're working with isn't multicast-enabled. Managed switches must be capable of multicasting. IGMP should be implemented on the router, and if you want to send multicast streams over a wide area network, you have to set up a PIM relay, which forwards only the unicast streams that are in use. (We'll cover IGMP, PIM, and more later in this chapter.) All the routers in the network have to support this functionality.

If the network isn't ready for multicasting, use unicast streaming. Moreover, as long as the projected number of clients and the bit rate of the streams won't exceed the network's bandwidth capacity, stick with unicast because it's much easier to implement. And if you're implementing video on demand, you must use unicast to allow users to control playback.

On the other hand, when you need to stream to a large number of client devices on an enterprise network, multicast can help limit your bandwidth usage. For example, streaming a live lecture to all the student computers on a campus is likely to benefit from multicasting. Multicasting is also typically used for IPTV systems, with each channel sent to multiple hosts.

Compromise between unicast and multicast is possible. For example, using a caching or reflecting server, you can convert a multicast stream to a unicast stream to pass it over a network segment that cannot accept multicast transmissions. This hybrid approach is the most common approach for WAN implementations.

You Decide: Unicast or Multicast?

Q: You want your server to be able to send out the same stream at multiple bit rates so that each client can receive the highest-quality stream of which it is capable. Should you use unicast or multicast?

A: To take advantage of multiple bit-rate encoding and intelligent streaming, you must use unicast transmission. Multicast transmission sends out a single stream at a single bit rate. Unicast servers send out a different stream for each client and may vary the bit rate according to information in the client handshake.

Q: You're designing a streaming application for a high school that will allow video announcements to be streamed live to 100 classrooms at 300kbps every morning. Should you use unicast or multicast?

A: It depends on the size of the LAN. If the school has Gigabit Ethernet, you should probably use unicast because it's easier to manage. If it has 100Mbps Ethernet, however, use multicast.

Q: You want to be able to retrieve a detailed record of which clients have accessed the stream. Should you use unicast or multicast?

A: If you want to be able to retrieve a detailed client list, use unicast. Unicast clients have a direct connection to the server, while multicast transmissions are connectionless. No connection, no record.

Implementing Multicast

Multicast is great at persevering bandwidth, but it takes a substantial amount of configuration to implement. For starters, every router on the network must be configured to understand what are called *group-management protocol requests*. These requests allow a host to inform its neighboring routers that it wants to start or stop receiving multicast transmissions. Without group-management protocols (which differ between IPv4 and IPv6), multicast traffic is broadcast to every client device on a network segment, impeding other network traffic and overtaxing devices.

IGMP is the IPv4 group-management protocol. It's gone through a few revisions. IGMPv1 allowed individual clients to subscribe to a multicast channel. IGMPv2 and IGMPv3 added the ability to unsubscribe from a multicast channel. IGMP is a communications protocol used by hosts and their adjacent routers to allow hosts to inform the router of their desire to receive, continue receiving, or stop receiving a multicast.

Multicast Listener Discovery (MLD) is the IPv6 group-management protocol. IPv6 natively supports multicasting, which means any IPv6 router will support MLD. MLDv1 performs roughly the same functions as IGMPv2, and MLDv2 supports roughly the same functions as IGMPv3.

IGMPv3 and MLDv2 also support source-specific multicast (SSM). SSM allows clients to specify the sources from which they will accept multicast content. This has the dual benefit of reducing demands on the network while also improving network security. Any device that has the host address can try to send traffic to the multicast group, but only content from a specified source will be forwarded to the group. This is in contrast to any-source multicast (ASM), which sends all multicast traffic sent to the host address to all subscribed clients.

Protocol-Independent Multicast

With the amount of configuration required to implement multicast streaming, many IT managers are hesitant to implement it beyond a single LAN. Multicast routing beyond the LAN is made possible by Protocol-Independent Multicast (PIM).

PIM allows multicast routing over LANs, WANs, or even, theoretically, the open Internet. Rather than routing information on their own, PIM protocols use the routing information supplied by whatever routing protocol the network is already using, which is why it's protocol independent. PIM is generally divided into two categories: dense mode and sparse mode.

Dense mode sends multicast traffic to every router of the network and then prunes any routers that aren't actually using the stream. By default, dense mode refloods the network every 3 minutes. PIM-DM is easier to implement than sparse mode but scales poorly. It's suitable only for applications where the majority of users will join each multicast group.

Sparse mode sends multicast traffic only to those routers that explicitly request it. This is the fastest and most scalable multicast implementation for WANs. We'll examine it further.

PIM Sparse Mode

PIM sparse mode (PIM-SM) sends multicast feeds only to routers that specifically request the feed using a PIM join message. The multicast source sends its stream to an adjacent router. That router must be configured to send the multicast traffic onward to a specialized multicast router called a *rendezvous point* (RP). There can be only one RP per multicast group, although there can be several multicast groups per RP.

Here's how it works:

1. The destination host sends a join message toward the multicast source or toward the rendezvous point.
2. The join message is forwarded until it reaches a router receiving a copy of the multicast stream—all the way back to the source if necessary.
3. The router sends a copy of the stream back to the host.
4. When the host is ready to leave the multicast stream, it sends a prune message to its adjacent router.
5. The router receiving the prune message checks to see whether it still needs to send the multicast stream to any other hosts. If it doesn't, it sends its own prune message on to the next router.

Packets are sent only to the network segments that need them. If only certain users will access the multicast stream or if many multicast streams will be broadcast at once, sparse mode is the most efficient use of network resources. IPTV, for example, is typically implemented using PIM-SM.

However, PIM-SM also requires the most configuration (see Figure 17-2). All routers on the network must be aware of the multicast RPs. Routers should also be configured with access lists so that only designated users will have access to the multicast group. RPs themselves must be configured with a traffic threshold. Once the threshold is exceeded, join messages will bypass the RP and go directly to the source. This threshold

Figure 17-2 Setting up a multicast network using PIM sparse mode

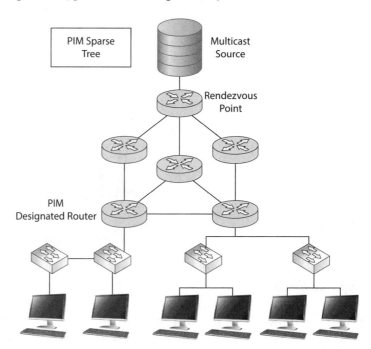

is usually set to zero automatically, which means the RP will direct all join requests to the source by default. It can also be lifted entirely. The decision depends on how much multicast traffic each network segment can bear.

IPv4 Multicast Addressing

The Internet Assigned Numbers Authority (IANA), the same organization responsible for assigning port numbers, has divided the IPv4 multicast IP address range (224.0.0.0/4) into a series of blocks allocated to different types of multicast messaging. See Table 17-1.

Each block in the address assignment table serves a different function:

- The local network control block, or link-local block, is used for network protocols that will remain within the same subnet. Control or announcement messages that must be sent to all the routers in a subnet are addressed to destinations in this range.
- The internetwork control block is used to send multicast control messages beyond the local network segment. For example, in PIM sparse mode, rendezvous point information may be communicated across broadcast domains using an address in this range.
- Ad hoc addresses are assigned by IANA to multicast control protocols that don't fit clearly into the first two blocks. However, much of the ad hoc block was assigned prior to the existence of usage guidelines, so many addresses in this range are simply reserved for commercial uses.
- The SAP/SDP block is reserved for applications that send session announcements.
- The SSM block provides multicast host addresses for SSM applications.
- The GLOP block (see the following note) provides multicast host addresses for commercial content providers.
- Some address ranges remain unassigned for experimental or future uses.

Address Range	CIDR Mask	Description
224.0.0.0–224.0.0.255	224.0.0/24	Local network control block
224.0.1.0–224.0.1.255	224.0.1/24	Internetwork control block
224.0.2.0–224.0.255.255	n/a	Ad hoc block
224.1.0.0–224.1.255.255	n/a	Unassigned
224.2.0.0–224.2.255.255	224.2/16	SAP (session announcement protocol)/SDP (session description protocol) block
224.3.0.0–231.255.255.255	n/a	Unassigned
232.0.0.0–232.255.255.255	232/8	SSM (source-specific multicast) block
233.0.0.0–233.255.255.255	233/8	GLOP block
234.0.0.0–238.255.255.255	n/a	Unassigned
239.0.0.0–239.255.255.255	239/8	Administratively scoped block

Table 17-1 IANA IPv4 Multicast Address Assignments

- The administratively scoped block is reserved for private multicast domains. These addresses will never be assigned by IANA to any specific multicast technology or content provider. You're free to use these as multicast host addresses within the enterprise network, but they can't be used to send multicast content over the Internet.

Any multicast host address used by a private enterprise (other than a commercial content provider, such as an ISP or television network) will come from either the SSM or the administratively scoped block.

NOTE GLOP is not an acronym. It turns out the original authors of this request for comment (RFC) needed to refer to this mechanism by something other than "that address allocation method where you put your autonomous system in the middle two octets." Lacking anything better to call it, one of the authors simply began to refer to this as "GLOP" addressing, and the name stuck.

IPv6 Multicast Addressing

IANA has assigned a very different series of blocks for multicasting in the more extensive IPv6 address range. See Table 17-2.

Streaming Reflectors

A streaming video reflector (sometimes called a *relay*) subscribes to a video stream and retransmits it to another address. This retransmission can be any combination of multicast or unicast inputs and outputs. In the case of forwarding multicast across a VPN, a pair of reflectors can be used. In such a situation, a streaming source outputs a multicast stream. The reflector service subscribes to the stream and de-encapsulates layers 3 and 4 (IP and UDP headers). It then re-encapsulates the data with new TCP and IP headers and forwards the packet. The receiving end receives the unicast stream and performs the reverse process, forwarding a multicast stream.

When implementing multicast reflecting, you need to consider the following:

- **Administration** The reflector is typically configured by the same administrator who is responsible for the streaming service. Beyond the initial configuration at installation, no network configuration is required to change the reflector service. For occasional use, such as company announcements, a reflector at the source site can be configured and left active with the receive-side reflectors for the duration of the event.

- **Bandwidth** Because the reflected streams are individually configured, there is no risk of inadvertently forwarding multicast streams. The receive-side reflector requests the stream based on local administration. While the receive-side reflector service is enabled, the bandwidth is occupied even if no one is subscribed to the multicast stream. There is a small packet overhead increase with the conversion from UDP to TCP.

Address	Description
ff02::1	All nodes on the local network segment
ff02::2	All routers on the local network segment
ff02::5	OSPFv3 (Open Shortest Path First)—All SPF (Shortest Path First) routers
ff02::6	OSPFv3—All DR (Designated Router) routers
ff02::8	IS-IS (Intermediate System–Intermediate System) for IPv6 routers
ff02::9	RIP (Routing Information Protocol) routers
ff02::a	EIGRP (Enhanced Interior Gateway Routing Protocol) routers
ff02::d	PIM routers
ff02::12	VRRP v3 (Virtual Router Redundancy Protocol)
ff02::16	MLDv2 reports
ff02::1:2	All DHCPv6 servers and relay agents on the local network segment
ff02::1:3	All LLMNR (Link-Local Multicast Name Resolution) hosts on the local network segment
ff05::1:3	All DHCPv6 servers on the local network site
ff0x::c	SSDP (Simple Service Discovery Protocol)
ff0x::fb	mDNS (multicast DNS)
ff0x::101	NTP (Network Time Protocol)
ff0x::108	NIS (Network Information Service)
ff0x::181	PTP v2 (Precision Time Protocol) messages (Sync, Announce, etc.) except peer delay measurement
ff02::6b	PTP v2 peer delay measurement messages
ff0x::114	Experimental use only

Table 17-2 IANA IPv6 Multicast Address Assignments

- **Scalability** A single reflector at the source location can reflect to multiple receive-site reflectors, but the bandwidth is unicast, so each receive-side reflector gets a separate stream.
- **Configuration** A separate PIM rendezvous point will need to be configured at the receiving site. Multicast addresses do not need to map between the sites. Each site can have its own addressing scheme. For IPTV applications, a separate channel guide will have to be implemented if the multicast addresses do not map between the sites.

Chapter Review

Streaming is an increasingly important component of AV systems design. Once you've performed the requisite analyses for a streaming system, you need to be able to identify the impact of bandwidth restrictions and network policies and the impact of inherent network latency on the streaming application. You should also calculate the required bandwidth for uncompressed digital AV streams and factor in the appropriate transport and distribution protocols for the client's streaming needs.

Always keep in mind that streaming AV consumes network resources—perhaps more resources than a network manager anticipated. Document everything, from the needs analysis, to network restrictions, to the technical specifications of your proposed solution. If you can make the case that your streaming system meets the customer's needs and you've created a design that respects the requirements of the larger enterprise network, you will be successful.

Review Questions

The following questions are based on the content covered in this chapter and are intended to help reinforce the knowledge you have assimilated. These questions are similar to the questions presented on the CTS-D exam. See Appendix E for more information on how to access the free online sample questions.

1. If your streaming AV system does not support High-Bandwidth Digital Content Protection (HDCP), it's likely the system _____.

    A. Won't be able to play high-definition video content

    B. Won't be able to play standard-definition video content

    C. Won't be able to play live video streams

    D. Won't be able to play protected audio streams

2. In a converged network, where 70 percent of the bandwidth capacity is available and 30 percent of the available capacity is available for streaming, approximately what percentage of the network's rated capacity is available for streaming?

    A. 20 percent

    B. 30 percent

    C. 50 percent

    D. 70 percent

3. What might be considered acceptable latency for streaming desktop video?

    A. 50 microseconds

    B. 30 milliseconds

    C. 200 milliseconds

    D. 1 second

4. Which of the following protocols might you use in a streaming AV system? (Select all that apply.)

    A. User Datagram Protocol

    B. Real-Time Transport Protocol

C. Real-Time Transport Control Protocol

D. MPEG

5. Which of the following statements applies to unicast streaming?

    A. It consumes more bandwidth with the more people who need to view the stream.

    B. It uses group-management protocols to control streaming channels.

    C. It requires special routers called rendezvous points.

    D. It uses special IP addresses assigned by the Internet Assigned Numbers Authority.

6. If your multicast streaming AV system runs on an IPv6-based network, the group-management protocol to use is called _____.

    A. Internet Group Management Protocol

    B. Multicast Listener Discovery

    C. Source-Specific Multicast

    D. Protocol-Independent Multicast

Answers

1. **A.** If your streaming AV system does not support HDCP, it's likely the system won't be able to play high-definition video content.

2. **A.** In a converged network, where 70 percent of the bandwidth capacity is available and 30 percent of the available capacity is available for streaming, approximately 20 percent of the network's rated capacity is available for streaming.

3. **D.** In general, streaming desktop video can tolerate 1 second of latency.

4. **A, B, C.** When designing a streaming AV system, you might use User Datagram, Real-Time Transport, and Real-Time Transport Control Protocols.

5. **A.** Unicast consumes more bandwidth with the more people who need to view the stream. The other three statements apply to multicast.

6. **B.** If your multicast streaming AV system runs on an IPv6-based network, the group-management protocol to use is called Multicast Listener Discovery.

CHAPTER 18

Security for Networked AV Applications

In this chapter, you will learn about
- Documenting security objectives
- Evaluating a client's security posture
- Creating a risk register for an AV/IT network
- Implementing a risk response for common security risks

Consider how and why you protect your home, whether you use an alarm system or simply lock your doors. You secure your home for the same reason your customers protect their IT networks—to protect the valuables inside. As AV systems are frequently part of a client's enterprise data network, the client expects them to maintain a security posture in alignment with their overall information security goals. AV designers must understand their clients' security requirements and take them into account when designing a system.

Every client's security needs are unique and evolve over time. Like the all-important needs assessment that designers conduct before creating an AV system, there is a process they should follow to ensure they discover clients' security needs and design a system that meets them.

> **Duty Check**
>
> This chapter relates directly to the following task on the CTS-D Exam Content Outline:
>
> - Duty A, Task 5: Identify Client Expectations
> - Duty B, Task 7: Coordinate with IT and Network Security Professionals
>
> In addition to skills related to these tasks, this chapter may also relate to other tasks.

Security Objectives

Clients expect the AV system you design to be in alignment with their information security goals. In many cases, those goals may be more than just best practices; they may be required for regulatory compliance. In most jurisdictions there are stringent requirements for data security, data privacy, mandatory reporting of data breaches, and in many cases, for the housing of data within specified geographic regions. Industrial companies and others around the world may seek to comply with ISO-27000, a security framework that is now required for all ISO-9000 series–certified companies. To the extent that an AV system interfaces with IT systems or handles data that falls under various compliance frameworks, design and integration choices that put them out of compliance may impede installation.

Even when regulatory compliance isn't a central concern, most clients have a similar set of security objectives:

- **Confidentiality**   Ensuring that people who shouldn't have access to information are actively prevented from obtaining it.
- **Integrity**   Ensuring that people who are permitted to access information can trust that the information has not been altered. It is correct, authentic, and reliable.
- **Availability**   Ensuring that authorized access to information is unimpeded. This extends to every type of information, including multimedia and video resources.

These objectives are widely referred to as the CIA triad. No security system is perfect, and each organization weighs these objectives differently. Some companies can trade off tighter security measures for easier access to files. Others can tolerate downtime if it means keeping their data secure. Typically, as you communicate with clients, you may find that two of the three aspects of the CIA triad are most important.

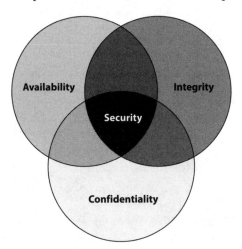

Any technology that is connected to an enterprise network could potentially impact one or more of these security objectives. As an AV designer, you are not likely to specify or configure firewalls, intrusion detection devices, or other network security technologies.

Your job is to be aware of a client's security requirements, communicate that information to the AV team, and, where possible, address and mitigate security risk in your AV design.

Perhaps the most important aspect of discovering security objectives and requirements, regardless of how extensive they may be, is documenting them. During any security needs analysis, designers should document everything they can about security and risk. When planning a project, allocate more time than you think necessary to document security requirements, issues, and strategies. You'll be glad you did.

Identifying Security Requirements

A recurring issue in the AV industry—and an impediment to getting paid—is that security requirements are often discovered during installation or commissioning. There is nothing worse than performing your final systems verification, expecting final sign-off, and finding there's a security requirement you didn't know about, rendering the project unacceptable. Therefore, it's critical to identify security requirements and set expectations up-front, before the client signs off on the initial design (see Figure 18-1). This will give you time to communicate the requirements and, if necessary, submit change orders to ensure equipment is specified that can support those requirements. You may also need to budget cost and time to complete International Organization for Standardization (ISO) surveys and risk mitigation plans.

If possible, move the identification of security requirements into the needs analysis phase of the AV design. This may not always be achievable; a company may not want to expose its security profile or requirements to just anyone who bids. But you can at least begin to understand what compliance frameworks the client operates under and which aspects of the CIA triad are priorities.

At a minimum, designers should identify security requirements and address them in the design prior to sign-off. This is usually the last time in the process when it is reasonably easy to make changes without impacting schedules or equipment costs.

Security requirements and the steps needed to meet them should be appended to the statement of work and listed as system sign-off criteria. This is to protect both the client and the AV designer/integrator.

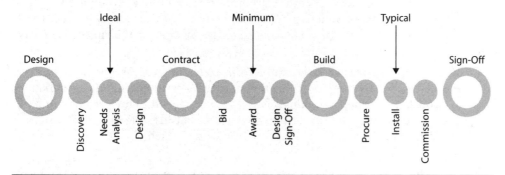

Figure 18-1 Ideally, you should identify security requirements up-front. Typically, AV pros leave it until too late.

Determining a Security Posture

AV systems help facilitate communication, but they can also create vulnerabilities in your client's system. The client's *security posture* may limit what equipment can be in an AV installation. For example, some customers may insist on the use of encrypted wireless microphones because their signals can be intercepted. Other customers may not allow you to use a projector with nonvolatile memory (NVM) for fear that images could be retrieved from the projector if it were stolen.

With respect to networked AV systems, you have to consider whether the AV system provides a way to access the rest of the network or any other information the customer wants to protect. Can you guarantee that a network intruder can't use the digital signage outside the boardroom to access the chief executive officer's (CEO's) schedule or use the videoconferencing system to eavesdrop on confidential conversations and meetings? You need to understand the elements of a security posture that are relevant to your AV design.

A security posture describes what an organization is trying to protect and how vigorously it needs to do so. When it comes to AV systems, the security posture should be set by the client. AV designers don't create security postures; they make their systems compliant with an existing posture.

Security postures will likely be unique to each job because each customer will have different requirements. You might hear security postures described as lax, realistic, or paranoid. Security postures are often influenced by external oversight, such as an industry or regulatory authority that sets security standards, or internal operations, such as a company security policy. The posture may also take into account the customer's willingness to accept possible reputational harm from the revelation that security breaches have occurred.

There are several steps to determining a security posture:

- *Learn the client's mission.* The AV system you're designing should help achieve an organization's mission. Knowing the vision or goals of the organization will help you see the value of the AV system from the client's perspective and prioritize which aspects the client needs to secure.

- *Learn the concept of operations.* The concept of operations describes how the client fulfills its mission. How does the organization function? How will the AV system help the organization function better? Examine its current security processes and procedures. Can any of them be applied to the AV system? Will any of them have to change?

- *Assess the data's importance.* Specifically, this concerns data traveling through the AV system. Where does the information in your AV system fit into the organization's mission and concept of operations? Is the AV system essential to operations, or does it simply carry casual information among departments? Is there an existing system that carries similar information that can be used for reference?

- *Learn the client's risk profile.* This may involve signing a nondisclosure agreement. A risk profile is how tolerant a customer is of risk. How worried is the customer about security breaches? A customer's risk profile may vary for different content.

What type of intrusions does the customer worry about most? For instance, a university may not care much if an outsider hacks into its streaming video server and watches all the recorded lectures. It may be concerned, however, about protecting the content servers that store footage from its medical research facility. The organization's security policies will be a reflection of its risk profile.

- *Identify stakeholders.* This especially includes those with the power to approve and negotiate. The project stakeholders are the best sources of information about security posture. Document the role of each stakeholder so that various members of the project team know whom they can contact to learn different information about security posture. Negotiation is an integral part of the security and risk management process. You will need to identify who among your stakeholders has approval authority or negotiation power.

- *Identify governance structure.* You need to know the sections of your AV system that will probably need to exist within the customer's IT framework. Because of the complexity of IT systems, mature organizations create control objectives that govern how to run the system. They may adhere to a standardized IT governance structure, such as Information Technology Infrastructure Library (ITIL) or Control Objectives for Information and Related Technology (COBIT). Even if the client doesn't use a specific standardized governance structure, its IT policies should cover similar categories, including the following:

    - How do the systems and processes help the organization meet its mission?
    - What are the usage policies, and which best practices does your customer follow?
    - Who are your user groups, and will any of them use the system remotely?
    - What are the audit guidelines and processes?
    - How does the system compare to industry best practices for similar systems?
    - How does your customer measure the success of a system?

- *Identify project constraints.* Constraints include budget, time for completion, and policies that affect the eventual installation. The client's security policies may impose certain constraints on the project. For example, there may be a policy against using Wi-Fi. In that case, you may only select devices without Wi-Fi or with the capability to disable Wi-Fi. Constraints will likely impose financial burdens on the project. A cost-benefit analysis of those burdens may result in the revision of policies or requirements. For instance, the client might agree that, given the cost of running new cable, a few encrypted Wi-Fi tablets as room control panels are OK after all.

TIP You've landed two projects with the same client. One is a videoconferencing system for the international board of directors. The other is a video-streaming system for HR training. To learn the risk profile for the videoconferencing system, research security measures taken on other systems used by the board of directors. For the video-streaming system, research how the client protects other HR educational materials.

Stakeholder Input

Information security always supports business goals, and not every department in an organization has the same business goals. The amount of time, effort, and cost spent on security depends on those goals. There is no one-size-fits-all, "secure enough" system. Designers need to come to an agreement with various parties on what security requirements are important.

Verify that you have input from all relevant stakeholders. Typically, an AV designer/integrator needs to address three areas of security, so it is important to speak to stakeholders with experience in each area.

- **Operational** These are the stakeholders who own data. They determine what needs to be protected and how vigorously. The other two areas support operational security.
- **Network** These are the IT stakeholders who specify and administrate network policies. There are generally two types of IT network stakeholders: those responsible for the ports and protocols, firewalls, and routers in a network and those responsible for access control (see "The Triple-A of Access Control").
- **Physical** These are traditional security stakeholders, responsible for physical access to gear, spaces, and more. They can help you create policies on how to secure gear and cabinets to prevent theft or tampering (unauthorized "adjustments").

Here is an example of an AV-specific operational security concern that may not involve network security: Your client has just finished renovating their office building. They have a beautiful conference room on their lobby level with three glass walls. If the CEO displays the company's three-year financial plan on the conference room videowall, anyone in the lobby could read it. These are the types of security concerns an AV designer must be aware of.

The Triple-A of Access Control

When specifying IT security for AV applications, you need to consider access control, which can be described by three As:

- **Authentication** The person using the system is who they say they are. This is proven with certificates, passwords, and tokens.
- **Authorization** The person using the system is allowed to use it and take specific actions. This is managed through permissions in the system and directories or authorized users.
- **Accounting** Those who manage authentication and authorization also have an accurate record of what happens with the system and over the network in general. This accounting supports a concept called *nonrepudiation,* which includes producing records proving who was using a networked system and what they did while they had access.

Without all three As, nonrepudiation can't work.

Assessing Risk

Put yourself in this position: You are considering for your design an AV device that supports a bunch of standard network protocols, including Telnet, Secure Shell (SSH), HTTP Secure (HTTPS), Hypertext Transfer Protocol (HTTP), and File Transfer Protocol (FTP). It has a default password of "password" and only one user account named "admin," which can't be changed. Here are some potential vulnerabilities:

- HTTP, Telnet, and FTP are clear-text protocols, meaning anyone with access and a simple network analyzer can see exactly what's going across the wire, including usernames and passwords.
- The default password could be found easily on Google.
- The default username can't be changed.
- The client can't set up multiple accounts for nonrepudiation.
- The client can't set up accounts with different user permissions.

Now you're in a meeting with representatives of the client. You bring up the possible vulnerabilities of your proposed AV device and engage the client in the question of "Big deal or not a big deal?" What you're doing is helping to identify risk or possible threats to the AV system.

A threat usually requires a vulnerability and someone or something that might want to exploit that vulnerability. In most cases, if there's a vulnerability, you should assume someone will want to exploit it. Depending on the value of what the client must protect or the ease with which a vulnerability might be exploited, you may determine that a threat is larger or smaller and requires commensurate protective measures.

AV designers should attend risk-response meetings with a list of potential vulnerabilities. This is not necessarily a list of bugs and exploits, but rather a list of services and functions handled by AV devices that may need a security policy. Depending on the security posture and any required procedures within the organization, this may be a general list of functions or a list of specific gear.

Risk Registers

Once you're aware of the threats and risks to an AV system, you should create a *risk register*. A risk register is a methodology for prioritizing the threats that you can mitigate. It usually takes the form of a comprehensive table or spreadsheet and helps you assign value to risks depending on two factors: probability (how likely something bad will happen) and impact (how bad things will be if it does).

Risk registers are used across all forms of business, not just security. There are many methodologies and templates for assigning risk, and the process can be simple or complex, depending on the details and the client's security posture. A risk register might include information such as a description of the risk, the type of risk, the likelihood it will occur, the severity of its effects, a current status of the risk, and any internal stakeholders responsible for managing the risk. In general, if your client already has a risk management process or template, use theirs.

A risk register also includes an entry for countermeasures—basically how the risk might be handled. Typically, there are four ways to handle a risk:

- *Avoid it.* Limit or avert the risky activity.
- *Accept it.* Because the probability or impact is low enough, the client is willing to take the risk.
- *Transfer it.* The client can purchase insurance or maintenance plans or adopt cloud-based technology services.
- *Mitigate it.* Make changes to design, configuration, or operational procedures to lower the probability or impact of a risk to the point where it's acceptable.

Designers should consult with clients about their options and assign a response to each risk in the risk register.

Mitigation Planning

The most common risk response is mitigation, one of the final steps in a security strategy (see Figure 18-2). A mitigation plan includes the action proposed to mitigate a risk, a contingency (what to do if something bad happens even after you've mitigated the risk), the name of someone responsible for the mitigation plan, and a due date.

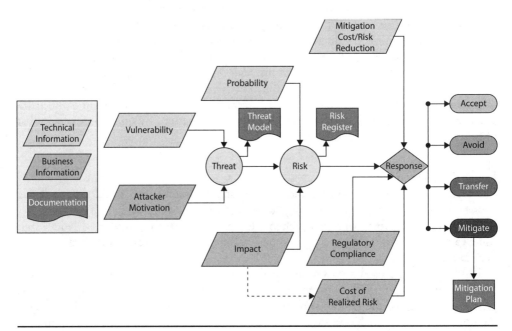

Figure 18-2 The process of planning for risk mitigation in AV/IT system

The responsible party and due date will usually come from the client, not the designer/integrator. For example, consider the situation where you have a legacy projector that can be controlled only via Telnet. The vulnerability is that Telnet usernames and passwords are sent via clear text and can therefore be discovered by network snooping and the system accessed. The potential impact is fairly low because the worst thing an intruder could do is turn off a projector. But the probability it will happen is fairly high because of how easy it is to find out the password and break in.

A mitigation plan for this system may be to create an AV-specific virtual LAN (VLAN) for just the projector and the devices that need to talk to it, such as a control system. Then create an access-control list on the router to that VLAN that blocks all Telnet traffic. That way, only a computer or device that is physically connected to that VLAN can see the network traffic and Telnet into the vulnerable projector. Typically, the job of creating the VLAN would be assigned to the client's network team, with a due date before the installation date.

In general, you may find that segregating AV applications onto separate VLANs is a good mitigation strategy for a variety of situations. In most cases, this is a design-level consideration rather than a configuration task, so address it early.

The following sections examine some other risk-mitigation strategies.

Change Default Passwords

Default passwords should never be left in a system. This is a huge security risk that is easy to fix. Before integration, change all default passwords to a job-specific password. You can use some standard combination of numbers and job name. For example, 768*TyrellCorp could be your job password. This gives you a measure of security and makes it easy for project team members to access the system.

After systems verification and commissioning, when you turn over systems documentation to the client, give them a list of devices with usernames, job passwords, and instructions for changing the passwords. That way, they can implement password security that meets their own company policy.

Use Two-Factor Authentication

When assigning permissions for administrative privileges or access to sensitive systems and applications, some organizations require two-factor authentication. Two-factor authentication combines two different validation methods to address a single authentication request.

These authentication methods require the user to provide one of three basic types of identifier:

- Something you know (such as a password)
- Something you have (such as a digital token, smart device, or smart card)
- Something you are (usually a biometric identifier such as a fingerprint or facial pattern)

Combining two of these identifiers improves the security of the authentication process. Any single one of these factors may be defeated by an intruder attempting to gain unauthorized access to AV systems, such as guessing or phishing for a password or using a lost or stolen smart card. Combining the factors requires potential intruders to approach from two very different vectors, making access a much more difficult task. If your client's access control system supports two-factor authentication, it substantially reduces the system's vulnerability.

Create Multiple User Roles

If a piece of AV equipment is going to be accessed over the network for day-to-day use—as opposed to just configuration—create separate accounts for users, other devices, and administrators with only the permissions required to perform their job functions. It's bad practice to allow anyone who might need to use or schedule an AV system to have configuration rights, too.

Accounts for Every User

The most secure password management strategy is to set an account for every user and then assign individual roles based on their access requirements. The challenge is scalability. What if you have 100 projectors that require two admins and eight users each? And what if the eight users are different based on where each projector is used? And what if somebody quits?

This is where directory servers come in handy. Directory servers are a centralized system used to keep, change, and store passwords. Most clients usually have a directory server for enforcing company password policies on laptops or other computing devices. You can integrate AV systems with an existing directory server for authentication and authorization.

Authentication is the username and password that proves the person requesting access to a system is who they say they are. Authorization takes that identity and matches it to a group membership. By being a member of the group, the person inherits the group's permissions. Say a member group is called *projector admins*. If someone wants to administer a projector, the projector authenticates that person and then checks to see whether the person is a member of the projector admin group. If so, the projector grants access.

The following are among the advantages of using a centralized directory:

- It greatly lowers the amount of administration required. You can administer policies for 100 projectors in one place.
- It is much more likely that password policy will be adhered to because users have to change only one password for every device they access.
- Because you may be able to leverage directory servers that are already in place for IT purposes, you can hand off password policy to the organization.

Keep in mind, however, the ability to use a centralized directory is not supported by all AV products.

Disable Unnecessary Services

AV devices tend to be "Swiss army knives," with many different functions and connection capabilities, which may or may not be used on a given project. Best practice is to disable services such as Telnet, FTP, IGMP, or HTTP when they are not being used for a particular installation. This gives a potential intruder fewer points of entry.

Keeping Systems Updated

The ongoing battle between network security and network intruders continues to escalate. Would-be intruders are constantly seeking weaknesses and flaws in existing security measures and in the infrastructure (software, hardware, device drivers, operating systems) that underlies AV and IT systems. As exploits for these flaws become known to infrastructure suppliers, measures are sometimes taken to mitigate the flaw by issuing updates or patches for the affected software and firmware. Enterprise IT departments have policies covering when such updates should be applied, often being concerned about balancing the timeliness of the fix against the stability of the new patch. Mitigating a minor risk but inadvertently bringing the entire enterprise to a standstill in the process is every IT department's nightmare. In establishing the security posture of an organization, discussions should include establishing policies for the updating of the AV-specific components of the system you design.

Air-Gapping

Where the AV system is used for highly sensitive applications, it is worth considering implementing an air-gapped network as an additional layer of security for the project. Air-gapping is a mitigation strategy that entails providing a completely independent network for the AV system. To achieve the necessary separation, an air-gapped network cannot be connected to any of the surrounding enterprise network infrastructure, including mutual wireless networks. This includes securing physical access to the air-gapped networking components such as hardware and cabling. While this strategy provides a very high level of security, it can be complex and expensive to implement.

Enable Encryption and Auditing

Use encryption where possible. This prevents people from reading passwords and potentially compromising data. Use HTTPS instead of HTTP, SSH instead of Telnet, and so on. Then, disable the nonencrypted version.

Plus, any systems that have an audit function, which tracks access and security failures, should be enabled. This offers a record of who logged in when, any failed login attempts, and more. Such an audit report not only can help catch and punish intruders, it can also help the client figure out how they got into a system to further mitigate risk down the road.

AVIXA RP-C303.01:2018 Recommended Practices for Security in Networked AV Systems

To assist AV professionals with developing the security for their networked AV systems, AVIXA has produced the recommended practice RP-C303.01:2018 *Recommended Practices for Security in Networked AV Systems*. This document provides guidance and current (in 2018) best practices for securing networked AV systems of all sizes. Regardless of their size or experience, organizations that take the steps outlined in the document can form a baseline for establishing a robust AV security management program. Best practices are described for

- Identifying vulnerabilities and potential threats created by the network integration of AV systems
- Assessing risk (identify, analyze, and evaluate)
- Developing a plan to mitigate and respond to identified (and unidentified) risks
- Deploying controls to continuously address and manage security risks in AV systems

There is also a section on applications for AV security in

- Conferencing and collaboration
- Smart buildings and IoT systems
- Streaming media

RP-C303.01:2018 *Recommended Practices for Security in Networked AV Systems* is available from AVIXA's Standards website at www.avixa.org/standards.

Chapter Review

AV designers aren't necessarily in the business of securing their clients' systems. But it is their job to understand and document clients' security requirements, recognize threats and vulnerabilities, help assess risk, and offer plans for mitigating or otherwise addressing risk as part of their designs.

In this chapter, you learned how to identify a client's security objectives, requirements, and posture. You also learned how to use a risk register not only to describe the possible threats to a client's information systems but also to plan for ways of addressing those threats. If you come away with nothing else from this chapter, remember always to document everything about a client's security requirements and how they might be impacted by your AV design. You never want to learn too late that you've created an insecure AV system.

Review Questions

The following questions are based on the content covered in this chapter and are intended to help reinforce the knowledge you have assimilated. These questions are similar to the questions presented on the CTS-D exam. See Appendix E for more information on how to access the free online sample questions.

1. A client's security objectives can often be described with respect to the CIA triad. What does CIA stand for?
    A. Central Intelligence Agency
    B. Confidentiality, integrity, availability
    C. Confidentiality, information, access
    D. Communication, infrastructure, authorization

2. Which of the following is not a step in assessing a client's security posture?
    A. Assessing how important the client's data is
    B. Testing the client's network for vulnerabilities
    C. Identifying stakeholders
    D. Understanding project constraints

3. When you've identified a security risk, you can avoid it, accept it, transfer it, or _____.
    A. Eliminate it
    B. Change it
    C. Mitigate it
    D. Reconfigure it

4. Which is an important consideration for AV designers concerning a client's wireless network?
    A. QoS could be minimized.
    B. Integration of POS equipment into the system.
    C. Data security/encryption issues.
    D. Variable bandwidth dynamics issues.

5. Which of the following is a mitigation strategy for addressing security risks introduced by AV systems?
    A. Building a firewall
    B. Setting all passwords to "password"
    C. Hiring a security expert
    D. Putting AV systems on a VLAN

Answers

1. **B.** A client's security objectives can often be described with respect to confidentiality, integrity, and availability.
2. **B.** When assessing a client's security posture, it's important to assess the importance of the client's data, identify stakeholders, and understand project constraints. Testing the network for vulnerabilities is not a designer's job.
3. **C.** When you've identified a security risk, you can avoid it, accept it, transfer it, or mitigate it.
4. **C.** With respect to a client's wireless network, AV designers need to consider data security and encryption issues.
5. **D.** To mitigate security issues introduced by AV systems, consider putting those systems on a virtual LAN (VLAN).

CHAPTER 19

Conducting Project Implementation Activities

In this chapter, you will learn about
- The audiovisual performance verification process, including for discrete audio and video systems
- Troubleshooting methods
- The verification standard for system closeout
- Closeout documentation
- Customer training
- Obtaining customer sign-off

As the AV industry establishes standards and best practices on and off the job site, owners, consultants, and integrators will need to conform to the standards and follow the guidelines for the proper design, fabrication, installation, and integration of AV systems. In this chapter, you will learn about the system performance verification standard and resources provided by several organizations. You will also learn how to troubleshoot common problems.

For starters, you should familiarize yourself with various AV-related standards. This is an excellent time to verify that the system that was installed is compliant with all relevant standards. You will also need some specific test and measurement tools for calibration and verification of both audio and video.

Your knowledge of the basic principles and theories involved in electronics, audio, video, and networking technologies will be well utilized during the final stretch of AV system integration. Once you have studied the theory and conducted verification of the system on the job site, you will be able to set up and verify new AV systems with confidence.

> **Duty Check**
>
> This chapter relates directly to the following tasks on the CTS-D Exam Content Outline:
>
> - Duty C, Task 5: Finalize Project Documentation
> - Duty D, Task 1: Participate in Project Implementation Communication
> - Duty D, Task 2: Conduct System Performance Verifications
> - Duty D, Task 3: Conduct Project Closeout Activities
>
> In addition to skills related to these tasks, this chapter may also relate to other tasks.

Performance Verification Standard

The ANSI/AVIXA 10:2013 standard, *Audiovisual Systems Performance Verification,* can help determine whether an AV system has met your client's objectives and is performing to design expectations.

Benefits of using this standard include

- Streamlining verification tests and reporting
- Providing a verifiable outcome
- Creating a common language between all parties
- Aligning outcome and performance expectations at an early stage in the project
- Creating reporting that completes the project documentation
- Reducing project risk through early identification of problems, thereby reducing the likelihood of remedial work

This standard should be used in conjunction with ANSI/AVIXA D401.01:202X, *Standard Guide for Audiovisual Systems Design and Coordination Processes,* as well as other relevant performance standards, such as A102.01 *Audio Coverage Uniformity* and A103.01 *Sound System Spectral Balance.* The D401.01 standard (originally released as ANSI/AVIXA 2M-2010 and under revision at the time of publishing) provides a framework and supporting processes for determining elements of an audiovisual system that need to be verified, the timing of that verification within the project delivery cycle, a process for determining verification metrics, and reporting procedures. The D401.01 standard includes approximately 150 reference verification items.

As shown in Figure 19-1, verification of conformance to the D401.01 standard must include the delivery of the Audiovisual Systems Design and Coordination Processes Checklist.

Figure 19-1 Title block of Audiovisual Systems Design and Coordination Processes Checklist

The D401.01 standard serves to verify the following:

- **Documentation of applicability** An outline that indicates which sections of the standard are applicable to the project according to contractual agreements, as well as any services not applicable (indicated as N/A)
- **Consideration** Written verification that the service providers have read and understand this standard and agree to be referred to as the Responsible Party for the applicable sections
- **Completion** Written verification that the service providers have completed the approved services, indicated by the party authorized to sign as Accepted By

The verification can be as simple as checking off boxes when you complete a task. By documenting your work along the way, you are setting yourself up for success when you hand over the system to the client.

The Audiovisual Systems Design and Coordination Processes Checklist includes three blank fields for each item for the project team to address:

- **The Activity Code** This identifies the type of item, such as deliverable (D), coordination (C), task (T), meeting (M), other (O), or not applicable (NA).
- **The Responsible Party** This is the designer, contractor, or integrator who is delegated and contracted to perform the activity. This may be more than one party and should be identified as such.
- **Accepted By** The client, designer, contractor, or integrator who is authorized and contracted to verify that the activity has been performed. This may be more than one party and should be identified as such.

Figure 19-2 provides an example of how a checklist line item could be filled out. The task is "schedule and agreement for meetings." This is an activity code M, C, and D, which stands for "meeting, coordination, and deliverable." The integrator has been

Standard Guide for Audiovisual Systems Design and Coordination Processes Checklist
Project Title _____ Description _____
Location _____ Architect _____
Designer _____ Date _____
Integrator _____ Client _____
Activity Codes: Meeting – M   Coordination – C   Task – T   Deliverable – D   Other – O   NA – Not Applicable
Note: There may be more than one activity code on each line

Date	Activity Code	Responsible Party	Accepted By	
1. Project Planning and Coordination Meetings				
21.7.2022	M, C, D	Integrator	Consultant	1. Schedule and agreement for meetings

Figure 19-2   A sample checklist line item filled out in accordance with the *Standard Guide for Audiovisual Systems Design and Coordination Processes*

contracted to perform the activity, and the consultant is authorized to verify that the activity has been performed.

System Verification Process

The performance verification process involves many phases, starting with pre-integration and continuing all the way through to final acceptance by the client, as shown in Figure 19-3. The actual testing of system performance starts with testing and verifying system components and subsystems in the workshop and continues through to handing over the controllers and final documentation to your client.

The verification process in each of the phases is as follows:

- **Pre-integration verification**   Refers to items that take place prior to systems integration. These items will generally verify existing conditions, such as the presence of device enclosures, or items such as backing/blocking/framing that require coordination among the trades for AV system installation.

- **Systems integration verification**   Refers to items that take place while the audiovisual systems are being integrated or built, including offsite and onsite work. These items will generally verify proper operation or configuration so that the system can function (for example, equipment mounted level, phantom power, termination stress, AV rack thermal gradient performance).

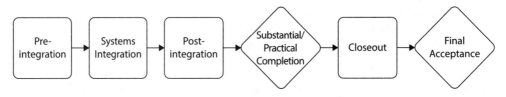

Figure 19-3   Performance verification phases and milestones

- **Post-integration verification**   Refers to items that take place after the audiovisual system integration has been completed. These items will generally verify system performance against verification metrics, as defined in the project documentation (for example, image contrast ratio, audio and video recording, and control system automated functions).

- **Substantial/practical completion verification**   Indicates conditional acceptance of the project has been issued by the owner or owner's representative, acknowledging that the project or a designated portion is substantially or practically complete and ready for use by the owner; however, some requirements and/or deliverables defined in the project documentation may not be complete. This milestone occurs at the end of the post-integration verification phase.

- **Closeout verification**   Refers to items involved with closing out the project. These items will generally be related to documenting the as-built/as-is status of the systems and transfer of system software, among other items, such as control system test reporting, as-built drawings complete, and warranties.

- **Final acceptance verification**   Indicates that acceptance of the project has been issued by the owner or owner's representative, acknowledging that the project is 100 percent complete; all required deliverables, services, verification lists, testing, performance metrics, and sign-offs have been received; and all requirements defined in the project documentation that occur at the completion of the closeout verification phase have been satisfied and completed. No further project activity will take place after this milestone is verified.

Certain items may need verification multiple times during a project. The reference verification items provided in the ANSI/AVIXA 10:2013 *Standard for Audiovisual Systems Performance Verification* define the verification phases at which those items should be tested. Where additional items are added on a project-specific basis, they will also need to be allocated a verification phase.

Regional Regulations

You need to know how codes, regulations, and safety procedures apply to your job site. To do that, you need to identify in the location of your project the *authority having jurisdiction* (AHJ), or the *regional regulatory authorities,* as they are known in some places. These organizations typically monitor compliance with codes and laws. Standards and best practices are typically established by organizations consisting of representation from various sectors of the industry.

Codes, regulations, and laws are mandated methods, practices, and collections of standards that are enforceable by law. You can be legally punished for not following them. Generally, an inspector will be present to verify that the work is being done according to the law. Each jurisdiction may have its own set of regulations.

If you encounter conflicting regulations, follow the most restrictive regulatory code for the region in which you are working. In other words, if you find the interpretations of applicable regulatory codes and standards are in conflict, follow the requirements of the more stringent code or standard.

Standards are documents that provide requirements, specifications, guidelines, or characteristics that can be used consistently to ensure that materials, products, processes, and services are fit for their purpose. They are often prepared by a standards organization or group and published with an established procedure.

Best practice is the best choice of the available methods for accomplishing a specific task in an industry. Best practices are recognized as the optimum way to do a certain task. They can be a generally accepted industry practice or unique to a specific company.

Resources for Regional Codes

Many different codes and standards apply to various regions in the world. Every country has its own resources for standards. If you need to access standards by your region or industry, always remember that web search engines are your friends. You might begin by researching on the following websites:

- **The International Organization for Standardization** ISO (www.iso.org) publishes standards to ensure that products and services are safe, reliable, and of good quality. For businesses, they are strategic tools that reduce costs by minimizing waste and errors and increasing productivity. The ISO has technical committees working on standards for a large number of industries.
- **International Electrotechnical Commission** IEC (www.iec.ch) publishes consensus-based international standards and manages conformity assessment systems for electric and electronic products, systems, and services, collectively known as electrotechnology.
- **Institute of Electrical and Electronics Engineers Standards Association** IEEE SA (standards.ieee.org) is the standards group within the IEEE, a worldwide professional association for electrical and electronic engineering, telecommunications, computer engineering, and allied disciplines.
- **StandardsPortal** StandardsPortal (www.standardsportal.org) is owned and maintained by the American National Standards Institute (ANSI) and provides standards information on the United States, the People's Republic of China, the Republic of India, the Republic of Korea, and the Federative Republic of Brazil.

Verification Tools

You're not necessarily an installer, but during pre-installation preparations, your team should have developed a verification checklist; revisit it to make sure you or someone on the team has the tools for calibration and verification of audio and video signals listed in Table 19-1, including general items such as crimpers, cable ties, and tape. In addition, you will want to use software-based signal analyzers for some of the components in your design. Note that Table 19-1 does not provide an exhaustive list; you may have more tools that are specific to certain work on various projects.

Performance verification checklist	Impedance meter	SPL meter
Computer device with interfaces/adapters for Wi-Fi, fiber and wired LAN, serial data, audio, and video	Copper and fiber-optic cable certification testers	Multimeter (RMS)
Signal analyzer and/or signal analysis software	Network cable testers	Continuity tester
Test signal generators and/or signal generation software	Measurement microphone	Headphones
Spot photometer	Camera alignment charts	Reference AV program material

Table 19-1 Test and Measurement Tools for AV System Verification

Audio System Verification

The audio system can be one of the most challenging aspects of an AV system design. It can be challenging to ensure a system provides an undistorted, high-quality signal with adequate levels to all specified destinations.

An audio system requires final adjustment after all the components have been installed for it to produce the desired or specified volume and quality. It is part of your job to set the audio gain and system equalization (EQ) and to adjust the various digital signal processing (DSP) components for the system to sound as good as it possibly can.

You will need to verify the performance of many components in an audio system and at many points along the signal path. You will want to use the ANSI/AVIXA 10:2013 *Audiovisual Systems Performance Verification* standard. The list of audio verification items found in that standard will make it easier for you to document the status of the AV system and to keep track of what you and your team have already inspected.

Before any performance verification is attempted, the installation team should have already verified the full operational status of the system.

Audio-Testing Tools

You will need to use different test instruments to measure and verify various aspects of the audio signal along its path from a microphone or other audio source to the loudspeakers. Audio-testing tools range from handheld devices to computer applications; some are available for single-function testing, and others have a suite of testing functions.

Piezo Tweeter

A tweeter is a term often used for a loudspeaker, often horn or dome shaped, that produces audio frequencies in the range of 2kHz to 20kHz (which is considered to be the upper limit of human hearing). Special tweeters can deliver high frequencies up to 100kHz.

A piezo tweeter contains a piezo-electric crystal mechanically coupled to a sound diaphragm. When a voltage is applied to the surface of a piezo-electric crystal, it flexes in proportion to the voltage applied, thus converting electrical energy into mechanical movement. Figure 19-4 shows a piezo horn-tweeter connected to an XLR audio connector.

Figure 19-4
Piezo horn-tweeter connected to an XLR connector

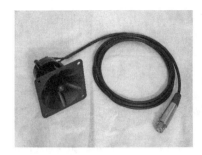

Audio Signal Generators

An audio signal generator is a test device that generates calibrated electronic waveforms in the audio frequency range for the testing or alignment of electronic circuits or systems.

Audio signal generators, such as the one shown in Figure 19-5, generate sine, square, triangle, or other waves at specific frequencies or combinations of frequencies. Many audio signal generators also generate *pink noise* (a quasi-random noise source characterized by an equal amount of energy per octave of frequency—it rolls off at 3dB per octave) and *white noise* (a quasi-random sound that has the same energy level at all frequencies). In addition to generating frequency sweeps, these devices can be used to test polarity.

Sine waves are used as steady references for setting signal levels and to reveal distortion caused by the equipment being evaluated. Common sine wave frequencies used for testing and verification include

- **1kHz** Used for setting levels and setting system gain using the unity gain method
- **400Hz** Used in conjunction with a *piezo tweeter* to listen for clipping, as well as setting system gain using the optimization method

Figure 19-5
A signal generator

Chapter 19: Conducting Project Implementation Activities

513

A wide range of professional-quality audio test signal generator applications are available for smart personal devices and laptop operating systems. Together with a quality audio interface (either built in or an external adapter) and some time invested in calibrating the signal output, you may be able to configure an audio installation without requiring a dedicated, stand-alone signal generator.

A tone generator can generate a stable, constant signal (such as a 1kHz sine wave at 0dBu) that can be measured with a signal analyzer to establish the baseline measurements for setting gain. Whether you are setting unity gain or using the system optimization method, you will need at least a cable tester, signal generator, and signal measurement device.

Sound Pressure Level Meters

Sound pressure level (SPL) is a measurement of all the acoustic energy present in an environment. As discussed in Chapter 6, this is typically expressed in decibels (dB SPL), with reference to 0dB, the threshold of human hearing.

An SPL meter, as shown in Figure 19-6, gives a single-number measurement of the sound pressure at the measurement location. The meter consists of a calibrated microphone and the necessary circuitry to detect and display the sound level in decibels. Its function is simple: it converts the sound pressure levels in the air into corresponding electrical signals. These signals are measured and processed through filters, and the results are displayed in decibels.

SPL Meter Classification   When selecting an SPL meter, you want to use one that can take readings as accurately as possible. The IEC 61672-1:2013 *Electroacoustics – Sound Level Meters* standard defines standards for sound measurement devices (also known

Figure 19-6
Meters for measuring sound pressure level

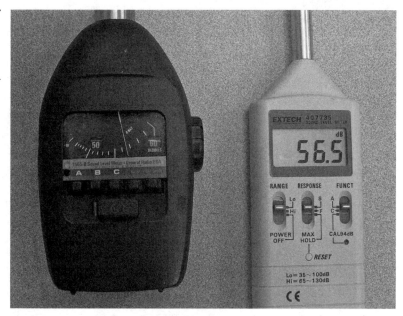

as ANSI/ASA S1.4-2014 in the United States). Meters are classified by the accuracy of their measurements:

- **Type 1/Class 1** Precision-grade instruments for laboratory and field use, with measurement tolerance of ±0.7dB. These are intended for use in environmental applications, building acoustics, and road vehicle noise measurements.
- **Type 2/Class 2** General-purpose-grade instruments for field use, with measurement tolerance of ±1.0dB. These are intended for use in measuring noise in the workplace, basic environmental applications, and motor sport measurements.

Devices that are not classified do not conform to a standard, so they are not reliable for measurement and testing purposes. For many audio purposes, a Class 2 meter is acceptable.

Always reference the project specifications for the SPL meter settings necessary for proper verification. If the verification requirements do not specify settings for the SPL meter, you can use SPL meter weighting guidelines to select one.

SPL Meter Weightings You can apply weighting to the SPL meter measurement to correlate the meter readings with how people perceive loudness. As you saw in Chapter 6, the *equal loudness curve* shown in Figure 19-7 is a measure of sound pressure across the frequency spectrum of human hearing. This curve shows how loud different frequencies must actually be for the human ear to perceive them to be of equal loudness. The curve with dashes represents the threshold of human hearing: the minimum sound pressure required

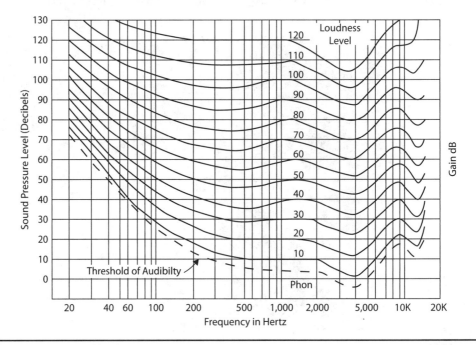

Figure 19-7 Equal loudness curve

for an average human to hear the different frequencies within the audio spectrum. The x-axis of the graph shows the frequency in hertz, and the y-axis shows the SPL in decibels. From this graph you can see that at the threshold of human hearing, the 40Hz tone must be about 50dB SPL louder before the human ear can perceive it equally as loud as a 1kHz tone. A 200Hz tone, however, would have to be only 15dB SPL louder to be perceived as equally as loud as a 1kHz tone. These SPLs use 0dB SPL (20µPa at 1kHz) as their reference.

The *weighting curves* represent standard filter contours designed to make the output of test instruments approximate the response of the human ear at various SPLs. An SPL meter will typically have three weighting settings: A-, C-, and Z- (or zero) weighting. All standards-compliant SPL meters are required to have the capacity to display levels with an A-weighting, while a C-weighting capability is also required in Class 1 meters.

- **A-weighting** This filter gives readings that closely reflect the response of the human ear to noise and its relative insensitivity to lower frequencies at lower listening levels.
- **C-weighting** This filter produces a more uniform response over the entire frequency range, but with –3dB roll-off points at 31.5Hz and 8kHz.
- **Z- (or zero) weighting** This filter produces a flat frequency response (±1.5dB) between 10Hz and 20kHz.

You can see the differences in SPL weighting by examining the weighting curves in Figure 19-8. You can see that discriminates against low-frequency energy. The A-weighting curve is almost the inverse of the equal loudness curve at a low listening level.

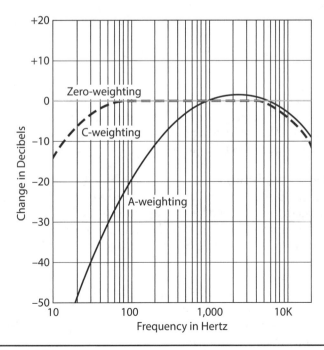

Figure 19-8 SPL weighting curves

The A-weighting curve reflects how the human ear perceives low-frequency energy as being at a lower SPL than it is in actuality.

A-weighting is useful in situations with low listening levels, including most speech applications. As the listening level increases to 85 to 140dB SPL, the human ear's response flattens out (and starts to cause hearing loss at SPLs above 90dB). At that point you may then choose a C-weighted filter, whose low-frequency curve is far less steep.

Because they reflect the response of the human ear at low listening levels, A-weighted measurements are frequently used to quantify ambient background noise. While there are better metrics for quantifying background noise levels and their effects on the listener, an SPL measurement provides a simple one-number rating.

The ANSI/ASA S12.60 standard shows that maximum 1-hour SPLs in learning spaces, including those from building services such as heating, ventilating, and air conditioning (HVAC), should not exceed 35dB SPL A-weighting.

The fact that a space should have a minimum 25dB signal-to-noise (S/N) ratio acoustically provides a goal of speech level of at least 60dB SPL A-weighted. However, a sound reinforcement system for speech may more typically operate in the range of 70 to 75dB SPL.

SPL Response Times In addition to choosing a weighting for your SPL meter, you will need to select a response time. A fast response is used for capturing transient, momentary levels. A slow response will capture consistent noise levels. This is useful for averaging rapid fluctuations in sound pressure levels and more closely mimics the way your ears react.

Multimeter

Since electronic audio signal levels are typically measured in dBu or dBV, both of which are referenced to a known voltage, an alternating current (AC) voltmeter or multimeter can be used to measure or set signal levels. You can also use a multimeter for testing cable continuity, connector pinouts, and polarity; measuring direct current (DC) voltage, current, and resistance; and testing equipment batteries.

When selecting a multimeter, look for a true root mean square (RMS) capability, which will enable you to measure the RMS (effective) value of an AC voltage or current. RMS, rather than peak voltage, is the most accurate means to compare voltage readings in audio applications. The multimeter must also have wide enough measurement bandwidth to accurately measure both the reference and the signal.

Measurement Microphone

Test and measurement microphones are used to accurately capture system performance. A measurement microphone should be omnidirectional and have a flat on-axis frequency response across the entire audio spectrum. Typical measurement microphones used for system alignment have a nominal diaphragm diameter of ½ inch (13 millimeters).

Oscilloscope

An oscilloscope is a sensitive test device that enables measurement and evaluation of electronic signals by displaying the signal as a time-referenced waveform. In audio system verification, an oscilloscope may be used to identify clipped signal waveforms when using the optimization method for setting system gain.

These devices are the basic tools for testing and evaluating the audio signal pathway and listening environment.

Video System Verification

Ensuring appropriate signal levels across all video equipment, adjusting video cameras and displays, and checking audio/video synchronization are all part of the video verification process. As you close out a project, you should test and verify several items using the Video System Performance Verification Items and the Audio/Video System Performance Verification Items from the ANSI/AVIXA 10:2013 *Audiovisual Systems Performance Verification* standard.

Verifying the Video Signal Path

Before any performance verification is attempted, the installation team should have already verified the full operational status of the system. When verifying a video system, you first should check the wiring and cabling of the video equipment. Once all connections have been completed, verify that the wiring methods and cable pathways are correct. This is typically a visual inspection process because you will compare the technical drawings, such as a signal flow drawing, against what you discover in the rack and elsewhere.

As with audio system wiring, during the inspection process, you need to look for the following:

- Correctly terminated connectors. Look for defects such as bent pins or frayed shields.
- Correctly connected cables. They should be attached to terminals as indicated on your system drawings.
- Cabling in the pathways that are properly organized in the walls, raceways, and the ceiling. Also, check for proper bend radius at all junction and pull boxes.

As you work through this process, record any errors or discrepancies for immediate repair. Permanent changes that occurred during installation should be documented and submitted to you for updating.

Signal Extenders

Due to the resistive and reactive properties of all cables, long cable runs can cause significant signal attenuation. The high-frequency components in video signals limit the effective length of DVI, HDMI, and DisplayPort cable runs to a theoretical maximum of 5 meters (16 feet), which is often not sufficient for a professional AV system. One of the solutions to this problem is the use of an extender technology, which not only reaches farther than a direct video connection but is often cheaper than a long run of quality digital video cable.

Figure 19-9
A pair of HDMI baluns

Passive Extenders

Passive extenders (see Figure 19-9) generally use *balun* technology to interface between the digital video cable and one or two twisted-pair network cables such as Cat 6+ to extend the video signal over distances up to about 50 meters (165 feet). Short for *bal*anced-to-*un*balanced, a balun is a transformer used to connect between a balanced circuit and an unbalanced circuit. In an extender application, one balun is required at the point of connection between the video signal and the twisted-pair network cable, and a second balun is required at the far end of the twisted-pair extension cable to match the video signal to the receiving device.

Some extension systems use a single twisted-pair network cable, while others use two cables. As baluns are less than 100 percent efficient and there is signal attenuation in the extension cables, the specified distance limitations of extension systems must be observed. It is important that the extender devices at each end of the extension system are of completely compatible make and model, as there is no standard for pinouts between manufacturers and different product ranges.

Active Extenders

Active extenders contain processing circuitry to boost and regenerate the video signals before transmission down the extension line and to regenerate them again at the receiving end. This enables active extenders to operate over greater distances than passive extenders. These extenders may use one or two twisted-pair network cables (Cat 6+) or one or two optical fiber cables as their extension medium. HDMI and DisplayPort signals may easily be extended for distances up to about 100 meters (330 feet) with wired extensions and up to 30 kilometers (19 miles) with fiber-optic extensions.

Active extenders require power to drive the processing electronics in both the transmitting and receiving devices. Some wire-connected systems send power down the cable(s) from the transmitter to the receiver, while others, including fiber-based systems, require a power supply at both ends. As with passive extenders, it is important that the transmitters and receivers are of completely compatible make and model, as there is no standard for pinouts and signal levels between manufacturers and different product ranges. HDBaseT is a widely used active extender technology, with a full ecosystem of HDMI extension and distribution devices available.

It is also a common practice in event, large venue, and broadcast applications to use standard video conversion devices to convert HDMI or DisplayPort to XX-SDI, run the XX-SDI over coax or fiber to the destination, and then convert the XX-SDI back to HDMI or DisplayPort for display. This is a useful solution for locally originating content, but replay of copy-protected material can be problematic, as XX-SDI does not support HDCP.

Network Extenders

As discussed in Chapter 16, a range of AV over IP technologies can be used to extend video signal delivery to any point on a TCP/IP network.

NOTE Active extenders reclock the signal and output a duplicate of the original signal. Passive extenders change the physical medium but do not reclock the signal.

Verifying Video Sources

After installation, the AV team will need to fine-tune camera adjustments and complete display calibration.

The location, intensity, and shadow quality of light sources in the picture area play a major part in the quality of the images picked up by cameras. When low-contrast lighting is used, images often look washed out and flat, while appropriately designed lighting will create images with contrast, dimension, and good exposure. The design of the lighting for video camera image capture is an integral part of any AV system design, and checking alignment, light levels, coverage, color temperature, color rendering, and contrast ratios should form part of the verification of the installation. The system design documents should include details of all lighting alignment parameters. It is quite common for a lighting designer, gaffer, director of photography, or lighting technician to be brought in for final alignments and level setting before system handover on projects with many luminaires and complex requirements.

Camera Adjustments

Video cameras are common in meeting rooms, classrooms, offices, and training centers. Clients tend to use cameras for videoconferences and to record events in their facilities. Similar to other AV devices, cameras should be set up for the environment in which they operate. Often, cameras are connected, powered on, and checked that they are operating, and that is the extent of the setup. The designer or integrator should make some standard adjustments to a camera's image to ensure that the images produced are of adequate quality.

Figure 19-10
A sample focus chart

Focus

The focus adjustment allows the camera to deliver a sharp, clearly visible image of the subject. To set the focus:

1. Have a subject stand in the general location where the camera will normally be capturing images.
2. Zoom the lens in tightly to the subject and adjust the main focus until the subject image appears sharp and crisp. A focus chart (see Figure 19-10) can make the process easier; the contrasting black and white sections allow for very accurate focusing.
3. Zoom back out to frame the shot for the best coverage.

Back-focus

Lenses must stay in focus as they are zoomed from wide-angle shots to narrow fields of view. Adjusting the back-focus on the lens will keep zoomed-out images focused. To adjust back-focus:

1. Locate the back-focus setting on the camera lens. If the camera has this setting, it will likely be a manual, hardware-dependent adjustment near the focus setting.
2. At the target area, zoom into the previously used focus chart, and focus as before.
3. Zoom all the way out and adjust the back-focus until the shot is again in focus.

Repeat this process until the lens stays in focus throughout its zoom range. If possible, lock the setting in place.

Pan, Tilt, Zoom, and Focus Presets

In system designs that include preset scenes or setups for cameras with remote robotic pan/tilt/zoom (PTZ) capabilities, each preset should be set up as specified in the design and the preset labeled with the name, such as "wide shot," "presenter closeup," "white board," or "full stage," as indicated in the design documents. Fixed-position cameras with remote zoom and focus preset capabilities should be set up in a similar way.

Some more advanced robotic camera systems may include capabilities for remote dolly/pedestal position, camera elevation, and live moves. These systems will generally be set up in collaboration with the end user during system training prior to sign-off.

Iris Settings (Exposure)

The iris in the camera's lens controls how much light passes through the lens to reach the camera's imaging device(s). Some adjustment to the iris will usually be required to correctly set the exposure of the image. If light levels change in the area covered by the camera, the aperture (opening) of the iris will require adjustment to maintain the correct exposure of the image. This adjustment is generally found on the body of the camera, as shown in Figure 19-11.

The automatic iris adjustment found on many AV cameras adjusts the aperture of the iris based on the output of the camera's image pickup device(s). Auto-iris responses may be based on the light falling on the center of the pickup (center-weighted) or on the average of the light across the entire pickup area. Some cameras' auto-iris systems can be switched between several exposure modes. Most lenses will also have a manually operated iris mode, which allows an operator, either local or remote, to set the lens aperture for the desired exposure.

There is a strong interaction in the exposure of an image between the level of light on the subject of the image and the level of light on the background behind the subject (often a wall):

- If the subject is substantially brighter than the background, it is possible for the auto iris to expose the subject correctly, while the background will be less prominent.

- If the subject is much darker than the background, the auto-iris systems may expose for the level of the background and leave the subject underexposed. However, if the auto-iris system is center-weighted, it may instead expose the central subject correctly and overexpose the background. Neither of these results are likely to be satisfactory pictures.

- If the subject and the background are of similar brightness, the auto-iris system will be able to get a good exposure of the subject, but the image may lack sufficient contrast to emphasize the subject and allow it to stand out clearly from the background.

Figure 19-11 Typical controls on a professional AV camera. (Courtesy of Panasonic Corporation.)

If the system design includes adequate lighting control, you will be able to expose a camera with an auto iris using the following simple procedure.

With the background lighting at its minimum level, set the subject lighting (including fill lighting and backlighting) at close to maximum intensity, and allow the auto iris to expose the camera. Now, adjust the level of the background lighting to make the background visible, but not dominant, in the image. If there is no simple dimming control, it is possible to adjust the intensity of the background lighting by switching some of it off or on or by inserting neutral density filters/meshes or diffusion material in the luminaires illuminating the background.

The balance between subject and background lighting for a good exposure will vary substantially depending on the color and brightness of the subject material. If the subject is a person, the reflectivity of their skin will greatly affect the balance. You will most likely have to adjust the lighting to account for significant changes between darker and lighter skin tones of subjects. If the lighting control system has preset level memory, it is advisable to set up a number of lighting states to accommodate the range of skin tones that may be present in the subject area.

Backlight Adjustment

Some cameras have a *backlight adjustment* mode, which surprisingly has nothing to do with the lighting focused on the back and edges of a subject to give separation from the background. The backlight adjustment is a compensation control for when a subject is seen against a bright background, such as a daylit window, which usually tricks the auto-iris system into underexposing the less-lit subject.

When activated, the backlight adjustment attempts to compensate for the over-bright background by increasing the aperture of the camera above the auto-iris setting to bring the subject up to a better exposure. This inevitably overexposes the background in the process of solving the subject's exposure. Backlight adjustment is, at best, a temporary measure to be used under duress. The solution to the problem is to improve the balance of light in the image by either reducing the light in the background (such as closing the curtain or blind over the window) or increasing the light level on the subject. Light-reducing window tinting can be used as a permanent solution to allow a subject to be seen against a brightly daylit exterior.

Automatic Gain Control

In situations where insufficient light is available to properly expose an image, it is possible to lift the brightness of the image by increasing the output of the camera's imaging device(s) via the video gain control functions. As with any amplification process, increasing gain also increases the noise in the signal, which results in random sparkles in the video output. Engaging the auto gain control (AGC), if available, will usually result in images that are of sufficient brightness, but will often be quite noisy. The best solution is to make sure that the subject is in an adequately lit area, which may be as simple as moving the subject into an existing pool of light or swinging a couple of lights around to cover the subject area.

Shutter Adjustments

The camera's shutter system controls the amount of light that the imaging system must process. In some cameras this is achieved by varying the size or speed of an actual rotating mechanical shutter that sits in the optical train before the imaging device(s). In other cameras the shutter control is a setting that varies the time between successive readings of the data from the imaging device(s). If a lot of light is entering the camera from a very bright scene, an automatic shutter system will reduce the amount of time the shutter is open, which may cause any fast movements in the scene to appear stuttered or jerky. The best solution to jerkiness caused by the shutter is to use some means other than the shutter to reduce the exposure of the imaging device(s). These solutions include reducing image gain, inserting a neutral density filter, or possibly closing down the iris.

White Balancing

Many cameras have an automatic white balance function. This circuit looks for bright objects in the image, assumes that they are white, and adjusts the color levels in the camera's output to produce a white signal for those bright objects. It is therefore critical that all automatic white balancing is performed using a matte white reference object, such as a piece of white cardstock or paper, illuminated by the same lighting as the subject. If the automatic white balance feature is off, manual white balancing is required. This is often performed using a grayscale chart to fine-tune the color balance across the camera's full dynamic range.

To automatically white-balance a video camera:

1. Place a large piece of white paper at the object location in the lighting that will be used for the subject.
2. Zoom into the white paper until it fills as much of the image as possible.
3. Press the white balance button. This will set the color reference levels for the current lighting conditions.

An automatic white balance is the best attempt the camera's electronics can make under limited conditions, so it is unwise to expect that two cameras capturing images of the same object will completely match, even if they are of the same model and have the same lens. Where color matching between multiple cameras is important, expert human intervention is usually required.

Some cameras have a number of preset color balance settings that may be selected to accommodate known color temperature conditions.

Framing the Image

After the camera has been correctly set up to capture the image accurately, an AV professional needs to frame the picture according to the designer's specifications. The method for doing so depends on the type of camera system used. If the camera is on a pan-tilt-zoom mount or robotic tracking and elevation system, the technician needs to make sure that any specified presets point and focus the camera to the proper points in the room. If the camera is on a fixed mounting, however, the technician will need to pan, tilt, zoom, and focus the camera to the specified framing.

Display Setup

Displays should be set up and adjusted to produce the best image based on the environment in which they are to be viewed. The viewing environment can have a major effect on the quality of the displayed image. Viewing a movie in the theater, watching a presentation in a boardroom, viewing the IMAG on a live event, and viewing digital signage in bright daylight are very different viewing experiences.

Display setup requires general knowledge of signal generators and the purposes of common test patterns. A knowledgeable technician can make the necessary adjustments based on the viewing environment. Technicians are working with multiple types of displays. All these display types require different procedures for correct setup. The job is only partially complete if the technician connects the displays, turns them on, and walks away.

Identify Display Parameters

The first step is to identify the parameters of the display device, which could be a projector, an LED screen, or a flat-panel display. Next, determine the input signal types that will be used on the display. The signal will probably be HDMI, SDI, or DisplayPort, although DVI and RGBHV may still occasionally be used in existing systems.

Determine the aspect ratio of the displayed image. The most common ratio is 16:9, although other ratios may be used in cinematic; videowall; and multipanel, multimodule, or multiprojector systems. This information for each display device may be in the owner's manual, or it can be calculated by dividing the width by the height of the displayed image. As a guide, a ratio of width to height of 1.77:1 is equivalent to 16:9.

It is possible to display an image that was designed for a legacy 4:3 display on a 16:9 display, but depending on the display system settings, the image will be either horizontally stretched to fit or have vertical black bars on either side of the image. Similarly, images intended for widescreen (2.39:1) cinematic display will usually display with horizontal black bars at the top and bottom. It is usually preferable to set up the display to show the entire image unstretched with black bars than with the image either cropped or distorted by stretching. If these settings are not defined in the system design documents, you should consult with the client or end user.

Image Geometry

Once you've identified the display's viewing parameters, make sure that the source signal fills the screen correctly. You can use two patterns for this step: the *crosshatch* and *geometry* patterns. The crosshatch pattern is used to determine the maximum viewable screen size, ensuring that the image fits the screen.

Using a crosshatch pattern (Figure 19-12), adjust the horizontal and vertical controls until the outer lines of the pattern are exactly at the edge of the screen or display. You can use electronic controls on a direct view display. If you are adjusting a projector, you can physically position the projector or use lens shift. The crosshatch pattern verifies that the image fills the screen and is centered. You can also use it to check linearity.

A geometry pattern (Figure 19-13) is used to indicate the correct aspect ratio. The image contains a large circle or several circles. Using the geometry pattern that matches

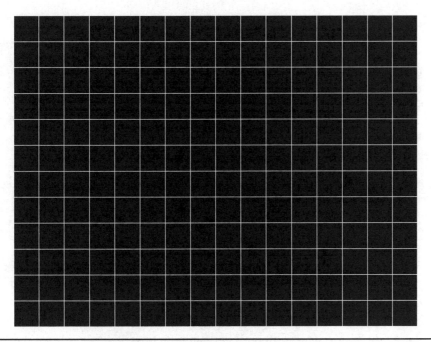

Figure 19-12 Crosshatch pattern

the aspect ratio of the video source (not the display), examine the display. If the circles appear to be an oval, then either the display is set to the wrong aspect ratio or the display may be stretching or compressing the pattern. There are two common versions of this pattern: 4:3 and 16:9. The one depicted in Figure 19-13 is the 16:9 version.

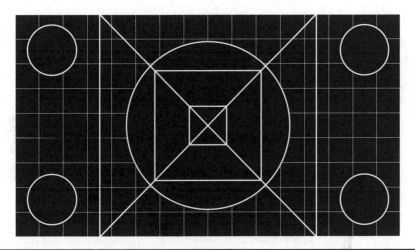

Figure 19-13 16:9 geometry pattern

Set Contrast

The test pattern shown in Figure 19-14 is used to set contrast. The white rectangle in the center is super white (step 255), or the brightest item on the screen. The two very black rectangles beside it are super black (step 0), and the black background that surrounds the white and blacker rectangles is normal video black (step 16). The goal is to strike a balance between all the shades in the image. Make adjustments until all the bars are clearly distinct and visible. Ensure the white bars look white and the black bars look black. If the display's grayscale transfer function (gamma) is set correctly, the bars should appear in even, gradual steps.

If the black sections of the image are too dark, increase the brightness control. If the image sections are too bright, increase the contrast. It may take several adjustments to balance the image and ensure that all variations of white and gray can be clearly seen.

The procedure for adjusting brightness and contrast is as follows:

1. Display a grayscale test pattern.
2. Adjust the contrast level down and then increase the level while watching the white rectangle in the center of the screen. Adjust up until the white does not get any whiter.
3. Adjust brightness until you cannot see any difference between black and super black.
4. Repeat the adjustments until both settings are achieved.

Once complete, the gray bars should gradually increase/decrease in value in a linear fashion. Repeat these steps until all bars appear in even, gradual steps.

Set Chroma Level

Color bars are test patterns that provide a standard image for color alignment of displays. They can be used to set a display's chroma level (color saturation) and, where present, hue shift. Hue shift control is generally only found on systems that display images in the North American NTSC format, which is prone to color phase errors.

Adjust the display for accurate monochrome brightness and contrast before color adjustments.

Figure 19-14
Test pattern to set contrast

A commonly used test pattern is the HD Society of Motion Picture and Television Engineers (SMPTE) version. Alternatively, you may come across the Association of Radio Industries and Businesses (ARIB) version. You can use these interchangeably, and the process for using them is almost identical.

Check Appendix D for the link to the AVIXA video library, which includes a video explaining the process of setting the contrast, brightness, and chroma.

Image System Contrast

Image quality can be assessed using criteria such as contrast, luminance, color rendition, resolution, video motion rendition, image uniformity, and even how glossy a screen is. However, contrast remains the fundamental metric to determine image quality. Taking the viewing environment into consideration, the difference between system black and the brightest possible image is the system *contrast ratio*.

Some time ago AVIXA developed an ANSI standard that set out how to measure contrast ratios for projected images and what contrast ratios are suitable for different viewing requirements. That standard has now been replaced by ANSI/AVIXA V201.01:2021 *Image System Contrast Ratio* (ISCR), which extends beyond projection to cover system contrast for viewing all images, including both projection and direct-viewed displays. The complete standard is available from the Standards section of the AVIXA website at www.avixa.org, but here is a brief summary of the four viewing requirement categories and their required minimum contrast ratios:

- *Passive viewing* is where the content does not require assimilation and retention of detail, but the general intent is to be understood (e.g., noncritical or informal viewing of video and data). This requires a minimum contrast ratio of 7:1.

- *Basic decision-making* (BDM) requires that a viewer can make decisions from the displayed image but that comprehending the informational content is not dependent upon being able to resolve every element detail (e.g., information displays, presentations containing detailed images, classrooms, boardrooms, multipurpose rooms, product illustrations). This requires a minimum contrast ratio of 15:1.

- *Analytical decision-making* (ADM) is where the viewer is fully analytically engaged with making decisions based on the details of the content right down to the pixel level (e.g., medical imaging, architectural/engineering drawings, fine arts, forensic evidence, photographic image inspection). This requires a minimum contrast ratio of 50:1.

- *Full-motion video* is where the viewer is able to discern key elements present in the full-motion video, including detail provided by the cinematographer or videographer necessary to support the story line and intent (e.g., home theater, business screening room, live event production, broadcast post-production). This requires a minimum contrast ratio of 80:1.

The standard includes a simple measurement procedure to verify that the system conforms to the desired contrast ratio of the viewing task. First, you must identify the five

Figure 19-15
Contrast ratio should be measured from five locations within the viewing area.

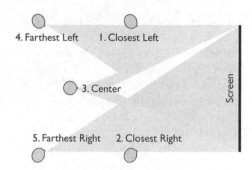

measurement locations, as shown in Figure 19-15. The five locations are to be recorded on a viewing area plan.

- **Viewing location 1**   Viewing location closest to the image and farthest to the left in the plan view.
- **Viewing location 2**   Viewing location closest to the screen and farthest to the right in the plan view.
- **Viewing location 3**   Viewing location at the central point of viewing locations 1, 2, 4, and 5. In the case where this central viewing location is obstructed (such as by a conference table), the measurement location will be the first available viewing location on the screen center line behind the obstruction.
- **Viewing location 4**   Viewing location farthest from the screen and farthest to the left in the plan view.
- **Viewing location 5**   Viewing location farthest from the screen and farthest to the right in the plan view.

After you have identified the viewing locations, measure the contrast ratio of the system using a 16-zone black-and-white checkerboard (intraframe) pattern, as shown in Figure 19-16. You will also need a luminance meter or spot photometer with up-to-date calibration.

The procedure for verifying the image contrast ratio is as follows:

1. Display a 16-zone black-and-white checkerboard test pattern on the projection screen, as illustrated in Figure 19-16, under conditions that represent the actual viewing environment.

Figure 19-16
Checkerboard pattern used to measure contrast ratio

2. From the first measurement position identified on the viewing area plan (viewing location 1), measure and record the luminance values at the center of each of the eight white rectangles.

3. From the same measurement position, measure and record the luminance values at the center of each of the eight black rectangles.

4. Calculate the average of the eight white measurements and the average of the eight black measurements.

5. Divide the resulting average white value by the average black value to obtain the contrast ratio at that measurement position.

6. Repeat the contrast measurement procedure at each of the five measurement positions identified on the viewing area plan.

7. Record the resulting contrast ratios for each of the measurement positions on the viewing area plan.

Contrast ratio = average maximum luminance ÷ average minimum luminance

Next determine the sequential contrast ratio for each measurement position:

1. Measure and record the luminance value at the center of a white-on-black window test pattern with 18 percent of the screen area being 100 percent white (see Figure 19-17).

2. Measure and record the luminance value at the center of a full-screen black pattern.

3. Divide the resulting white value by the black value to obtain the sequential contrast ratio for the viewing location.

If the contrast ratio meets or exceeds the minimum laid out in the ISCR standard at all five viewing locations, then the system conforms to the standard. If the contrast ratio at any one of the measured locations falls below the identified viewing category by more than 10 percent, the system fails to conform to the ISCR standard.

Check Appendix D for the link to the AVIXA video library, which includes a video demonstrating how to measure the contrast ratio of an image.

Figure 19-17
Black/white window test pattern from the ISCR standard

Audio/Video Sync

In many AV systems, the audio and video signals may take significantly different paths to get from the source to the point at which the user will experience them, such as in a seat within a venue, at a remote location, or distribution of recorded program material. The audio and video signals may also undergo very different numbers of processing steps, each with their own inherent latency.

In systems where associated video and audio signals may be transported or processed separately and are then subsequently combined for transmission or display, one of those signals could be delayed more than the other, creating synchronization errors sometimes known as *lip-sync errors*. These errors are typically corrected by applying delay to the least delayed signal so that it aligns with the most delayed signal. For a good user experience, the signals should arrive in synchronization at the point at which the user will experience them, regardless of whether that point is local or remote.

Sync Standards

As part of the metric selection process, the following documentation should be referred to in order of precedence to define the metric required for testing:

- ITU-R BT.1359-1
- EBU Recommendation R37
- ATSC IS-19

In the absence of defined requirements in the project documentation, the recommended maximum interval by which the audio should lead the video is 40 milliseconds and by which audio should lag behind the video is 60 milliseconds throughout an entire signal chain. This equates to approximately two or three frames on video systems.

For television applications, the recommended maximum synchronization error is audio leading by 15 milliseconds or lagging by 45 milliseconds.

A "blip-and-flash" test is a common method for measuring the synchronization of audio and video signals. A test signal consisting of one full frame of white video, accompanied by an audio tone of the same time length, is required. An alternative signal may be an image changing from full-frame white to full-frame black at regular intervals, with each change accompanied by a brief click or tone burst. In the absence of an appropriate signal generator, this can be produced by a computer running a slideshow. The signal at the point of reception can then be analyzed using a dual-channel digital storage oscilloscope to record the timings at which the audio and video signals are received.

If the sync alignment is not being measured, then a subjective reception test using the same test signal may be used, in addition to replaying video material that clearly depicts a person, such as a news presenter, speaking directly to camera. The subjective effect of any delay can then be evaluated for a pass/fail assessment.

Tests should be undertaken at the inputs of recording devices, at the inputs to transmission devices such as videoconferencing codecs and broadcast circuits, and locally within a presentation space at both near-screen and farthest-viewer positions.

Verifying Audio/Video Sync

After the AV system has been installed and all recording devices, presentation equipment, and link equipment/circuits are operational, the procedure for measuring the time alignment of the audio and video signals is as follows:

1. Set up a blip-and-flash source.
2. Measure delay (synchronization error) between audio and video at nearest and farthest viewer positions within the room.
3. Measure delay at inputs to all recording devices.
4. Measure delay at inputs to all transmission devices, such as codecs, network streaming interfaces, and broadcast circuits.
5. Note if the measured delay is within the specifications as stated in the project documentation or within +15 to –45 milliseconds (audio/video) in the absence of other information.

Alternative subjective tests may be undertaken using source video with no synchronization error, where a person is seen speaking. Validate subjective synchronization delay at locations listed earlier.

Correcting Audio/Video Sync Errors

Downstream video processing can cause delay. Some displays have built-in video processors, which, when used in conjunction with audio processors, scalers, and other processors, will cause delay in the signal.

Unfortunately, there is no clear way to predict how much, if any, delay will occur in an AV system. The latency introduced depends on the digital signal processors within the system, and the specific amount varies by manufacturer and device type. In some cases, you may be able to depend on the HDMI lip-sync feature, which will automatically compensate for delays in the video signal. In these cases, EDID can be used as a tool for correcting the sync issues. HDMI uses EDID to communicate delay information to upstream devices. During negotiation it will measure the delay introduced and send that information to the sink for automatic correction. This may be sufficient for systems with a single display, but with systems that involve multiple displays, when EDID detects different devices with different delay values, it is unlikely to be able to derive a single delay correction that will work for all sink devices.

In these cases, the integrator can manually introduce delay into an audio distribution system. Delay is a function typically found within digital signal processors (DSPs), switchers, or as a stand-alone, dedicated device.

Delay allows you to extract the audio at the beginning of the signal chain and send it right to the infrastructure. If you are responsible for selecting DSPs for a system, you should consider including a DSP that has the capacity to compensate for delay. Then, if you do encounter this problem, you can compensate for it by using a *blip-and-flash* test reel and adjusting the delay until the signals are back in sync.

Conducting System Closeout

After verifying that the system is working properly, you must demonstrate to the client or the client's representative that it meets the performance specifications of the AV design. To obtain customer sign-off, you will need to complete several tasks as part of the system closeout process.

Handing over documentation to your client is an essential part of system closeout. If troubleshooting or requested changes remain undocumented or incomplete, the project remains open and could require additional staff hours to resolve. This could possibly delay payments and project conclusion. On some projects, your work may involve training the client's staff on how to operate the systems.

Closeout Documentation

The importance of documentation at every stage of an AV project can scarcely be overstated. From the first meetings through installation and verification, documentation of every change is crucial for several reasons, including future maintenance of the system. Closeout documentation includes all the information gathered during the project, as well as drawings of record, operational documentation, and possibly a completed punch/snag/problem list.

Verification Standard for System Closeout

Adhering to the ANSI/AVIXA 10:2013 *Audiovisual Systems Performance Verification* standard can help you smoothly close out your project. This standard requires you to meet with project stakeholders to determine what verification tests will be performed on the system and what metrics will be used to assess system performance. It includes guidance as to when in the installation process each step should be performed. The standard requires the installation team to turn in verification testing reports for each phase of construction.

When you use this standard to manage system verification, you are also forming a shared understanding with your client, installation team, and other stakeholders of what constitutes a complete, fully functional system. Your performance and outcome expectations are explicitly stated and aligned early in the project. This reduces risk by identifying potential problems or disagreements early, thereby reducing the need for remedial work. The standard also helps both you and your client verify the project outcome by mandating documentation of each stage of construction and testing as it is completed. Both you and the client know when the system is ready for use.

The System and Record Documentation items from the ANSI/AVIXA 10:2013 standard are listed here in Figure 19-18, followed by a description of each list item.

9.6 System and Record Documentation Reference Verification Items

Item Number	Item Name	Pre-integration	Systems Integration	Post-integration	Closeout
DOC-100	Final Inventory of AV Equipment	X	X	X	
DOC-101	Approval of Samples	X			
DOC-102	Delivered Product Against Samples		X		
DOC-103	Wireless Frequency Licensing		X		
DOC-104	Consultant's Testing			X	
DOC-105	General Contractor's Testing			X	
DOC-106	Integrator's Testing			X	
DOC-107	Manufacturer's Testing			X	
DOC-108	Owner's Testing			X	
DOC-109	Third-Party Testing			X	
DOC-110	Substantial/Practical Completion			X	
DOC-111	As-Built Drawings Complete				X
DOC-112	Audio System Test Reporting				X
DOC-113	Control System Test Reporting				X
DOC-114	Final Commissioning Report and System Turnover				X
DOC-115	Required Closeout Documentation				X
DOC-116	Software Licensing				X
DOC-117	User Manuals				X
DOC-118	Video System Test Reporting				X
DOC-119	Warranties				X
DOC-120	Final Acceptance				X

Figure 19-18 System and Record Documentation verification items from the ANSI/AVIXA 10:2013 *Standard for Audiovisual Systems Performance Verification*

You may find some of these verification processes applicable to an installation you are undertaking, or you may find some processes that can form the basis of your own verification procedures.

- **DOC-100** Verify that all equipment has been delivered as defined in the project documentation.
- **DOC-101** Verify that samples of all equipment to be used as defined in the project documentation have been submitted for approval.
- **DOC-102** Where samples of products have been required for approval, verify that the products that are delivered are the same and of the same quality.
- **DOC-103** Verify that the correct and valid wireless frequency licensing permits have been obtained for legal operation of the system.
- **DOC-104** Verify that any consultant's testing requirements defined in the project documentation have been performed and approved.
- **DOC-105** Verify that any builder's/building contractor's/general contractor's testing requirements defined in the project documentation have been performed and approved.

- **DOC-106** Verify that any integrator's testing requirements have been performed and approved as defined in the project documentation.
- **DOC-107** Verify that any manufacturer's testing requirements defined in the project documentation have been performed and approved.
- **DOC-108** Verify that any owner's testing requirements defined in the project documentation have been performed and approved.
- **DOC-109** Verify that any third-party testing requirements have been performed and approved as defined in the project documentation.
- **DOC-110** Verify that a conditional acceptance of the project has been issued by the owner or owner's representative, acknowledging that the project or a designated portion is substantially/practically complete and ready for use by the owner; however, some requirements and/or deliverables defined in the project documentation may not be complete.
- **DOC-111** Verify that a complete set of accurate as-built drawings indicating all AV devices, AV device locations, mounting details, system wiring and cabling interconnects, and all other details has been provided as defined in the project documentation.
- **DOC-112** Verify that the audio system test report has been completed and issued as defined in the project documentation.
- **DOC-113** Verify that the control system test report has been completed and issued as defined in the project documentation.
- **DOC-114** Verify that the final commissioning report has been completed, issued to the proper entity, and accepted as defined in the project documentation.
- **DOC-115** Verify that a complete set of as-built system documentation has been provided as defined in the project documentation.
- **DOC-116** Verify that software usage and ownership rights have been assigned as defined in the project documentation.
- **DOC-117** Verify that manufacturer's user manuals are delivered to the owner in a format defined in the project documentation (e.g., binders, PDFs), or dispose of the manuals in a responsible manner (recycling) if the owner specifies that they do not wish to receive the manuals.
- **DOC-118** Verify that the video system test report has been completed and issued as defined in the project documentation.
- **DOC-119** Verify that all warranties are activated and that all warranty details have been passed to the owner as defined in the project documentation.
- **DOC-120** Verify that a final acceptance of the project has been issued by the owner or owner's representative, acknowledging that the project is 100 percent complete; that all required deliverables, services, project-specific verification lists, testing, verification, and sign-offs have been received; and that all requirements defined in the project documentation have been satisfied and completed.

A critical factor in ensuring that documentation deliverables are up to date is determining who is responsible for each document. You must understand both your specific role and the tasks involved in your installation and commissioning teams' closeout documentation. Knowing who is responsible for which items will enable you to successfully gather all of the relevant documentation for closeout.

Drawings of Record

After an installation has been completed and verified, the installation team should hand over comprehensive documentation, including the *drawings of record*, to the project manager.

Drawings and documents of record reflect changes made onsite to the original AV design and represent the actual installation as it exists at the point of handover. The original plans often cannot take into account product changes that may require changes on the job site. By the installation team carefully keeping track of changes to wiring, changes in model numbers of devices, audio and video signal modifications, and other system design elements, these drawings of record become valuable references during service calls and routine maintenance in the future.

The installation team should document every single change, even if it was just changing a connection point. The changes should be noted on the system plans. This step is vital not only in communicating system changes to all parties currently involved in the project but also to those who will work on changes to the system in the future.

Operational Documentation

Operational documentation should be included with a contract at the end of a project. Even if user documentation is not specified in a contract, training or job reference manuals should be prepared for the customer.

System documentation may include the following:

- System-specific operating instructions and manufacturer equipment manuals
- System design and block drawings
- Control system configuration, scripts, settings, and MAC and IP addresses
- DSP software and configuration files
- Network device inventory information
- Drawings of record
- Spreadsheet listing all equipment provided, including model numbers, firmware versions, serial numbers, and physical locations
- Description of recommended service needs and schedules of maintenance
- Warranty information (both manufacturer and AV system as a whole)
- Support telephone numbers/messaging addresses/contact URLs

Other deliverables should include remote-control devices, adapters, cables, spare parts, and media (optical media, flash, or cloud storage) with software programs and user documentation.

> **Best Practice for Equipment Manuals and Software Documentation**
>
> To ensure ease of system maintenance, organize system documentation as follows:
>
> - Arrange equipment manuals in alphabetical order and group by function. For example, all display device manuals would be placed in one group.
> - Leave a copy of the equipment operation documents with the equipment racks, the project manager, or the client's support personnel or end user.
> - Create backups on archive-quality media of all computer files associated with the equipment, especially if the files contain records of the system settings.

Punch/Snag/Problem List

As you may need to answer questions about the AV system and address issues as they are discovered, a punch/snag/problem list is useful. It contains the corrections or changes that need to be made to conform an installation to the scope of the project.

Some of the items such a list may contain include

- Minor changes such as correction of illegible labels on equipment inputs
- Physical changes such as moving a wall-mounted connector to a higher location
- Functional changes such as adjusting a monitor
- Programming changes such as renaming input buttons on the control UI
- Technical changes such as rewiring a rack that was not properly dressed
- Who is responsible for resolving each item

Even if your client does not require this document, the use of such a list has become standard practice in the construction industry. In fact, it could be the document that protects you and your company from any future misunderstandings. When all the work on the punch list cannot be completed immediately, make plans for follow-up and arrange to meet with the project manager to review the changes the next day.

Troubleshooting

There are going to be times when your AV system will not work correctly. When this happens, you may need to help troubleshoot the systems to find where the issues are coming from. Troubleshooting is a process for investigating, determining, and settling problems. The most important process in troubleshooting problems is to be systematic and logical.

The following are some basic troubleshooting best practices:

- Assume nothing. Anything that *can* go wrong *will* go wrong.
- Change only one thing at a time.
- Test after each change.

- Use signal generators to provide a known reference to measure against.
- Use signal analyzers to obtain quantifiable information.
- Document your procedures and findings.

A maintenance log is essential for fixed AV systems. It should include every item examined and repaired and the service performed. Maintenance logs are easy to update, especially right after service has been performed. They may include the date and time the system was checked, who checked it, and the status of the system. They may also include problems that were discovered and what measures were taken to fix them. You should make it standard practice to record corrective measures that were taken to restore system performance or recommended actions and whether they were acted upon. From time to time, equipment may need to be updated. For example, in a boardroom, a streaming device may replace a Blu-ray Disc player. The date this took place should be documented, as well as how the signal was rerouted.

Regularly check the system by running test signals to ensure they are up to specifications. Despite rigorous preventive care, from time to time a piece of equipment will need repair. This, too, should be entered on a maintenance log. By documenting exactly what needs repair and how much it might cost, your client will know when a device needs to be repaired or replaced and perhaps save money in the long run. Inspect connectors and wiring. Also make sure the area is clean.

Customer Training

After you have completed all the procedures, work with the client to schedule a training session for the typical users of the system. Several levels of training may be required for different groups of users. Each training session should be catered to the needs and interests of the audience. Different training groups might include the following:

- Technical staff who will be responsible for operations in the future and require in-depth training on all systems.
- Help desk or support staff who will be responsible for new-user or inexperienced-user hand-holding.
- Power users to nontechnical users who take full advantage of a system's functionality. They may hold frequent videoconferences, teach every day, or conduct regular multimedia presentations.
- Casual users who make limited use of the system's functionality.

You may also want to conduct in-depth "train the trainer" sessions with individuals from within the owner's organization so that training can continue into the future.

First study the system. If you are training people on the system's operations, you should be completely familiar with them yourself. Lack of familiarity or uncertainty on the part of the trainer can result in similar uncertainty in the trainees. More operational mistakes are likely if the trainer is uncertain or ineffective.

The client briefing should be informal. It is preferable to demonstrate operations by working with the newly installed system in front of the end users. Sometimes this will be done with the designer as part of the commissioning process. You, the project manager, or a sales team member may also brief the client.

Not all users require the complete and detailed operating procedures. Knowing the appropriate end users to receive extended training is most important.

It is also necessary to provide some level of documentation to customers describing how to operate a system. Detailed operating instructions for every piece of the AV system are only one portion of the total package.

End-user training should at least include the following:

- How to turn the system on and off
- How to configure and initiate a multisite communications session
- How to switch between the various program sources
- How to start, pause, rewind, and stop the various program sources, including cloud streams
- How to turn on and mute microphones
- How to adjust the volume levels of audio sources
- How to use the control system interface
- How to adjust internal HVAC and lighting presets and settings
- How to access technical support for system failures

When creating this documentation, also keep in mind that presenters using the system may only have time to glance at instructions. They will only need simplified, bullet-point instructions.

You may at times need to connect your customer with the manufacturer for additional training or support during closeout. The technology manager in charge often requires more extensive training than you may be able to provide. Remember your steps for escalating questions or concerns to address this type of briefing.

During training, some end users may be disappointed that the system does not include functions or features that they desire, or they may request the trainer to add functions or features to the system that lie outside the system's scope or design specifications. Such requests for changes to the system should be directed to the project manager for consideration through the formal change request process.

Best Practice for User Training

Create an outline of the topics you are going to cover. An outline will keep you on track and make sure you do not overlook important points. It will also engage the trainees in taking notes on points they need to remember.

It is best to conduct training as a separate scheduled event for the appropriate end users so that they are able to focus their attention on learning how to use the AV system.

Client Sign-Off

To obtain client sign-off, you should review the overall project. Before beginning this review process, you need to have completed the following tasks:

- You have completed a step-by-step approach to testing all the parts of the AV system.
- You have identified any problems or issues associated with the system and setup of equipment and have addressed them prior to turning over the system to the customer.
- You have provided a means for your company to demonstrate to the customer that the system meets both design and performance specifications and standards.

Either at or following the training, ask the consultant or client to sign closeout documentation as specified. Having everything at the training provides an opportunity for sign-off with all parties present.

Once signed off, the project now enters the support and maintenance phase of its life cycle. This phase is covered by the CTS-I (Installation) certification.

Chapter Review

When an AV project has been competed—when your design has become a reality—a designer's job is not over. Because your most important task is ensuring the client gets exactly what they expected, it's critical to oversee a comprehensive process of systems verification. This includes everything from determining whether the systems were properly installed to whether they are performing as expected. An AV designer's knowledge of basic AV and familiarity with certain testing and verification equipment will come in handy.

And once the system has been determined to be performing properly, it is important to deliver all documentation to the client, train users on the system as necessary, and obtain final signoff. Then, and only then, is your AV design complete.

Review Questions

The following questions are based on the content covered in this chapter and are intended to help reinforce the knowledge you have assimilated. These questions are similar to the questions presented on the CTS-D exam. See Appendix E for more information on how to access the free online sample questions.

1. What are the benefits of using the ANSI/AVIXA 10:2013 *Audiovisual Systems Performance Verification* standard? (Select all that apply.)
    A. Reducing project risk
    B. Aligning outcome and performance expectations
    C. Providing a verifiable outcome
    D. Creating reporting that completes the project documentation

2. Which of the following are features of an audio signal generator? (Select all that apply.)
   A. Generates sine wave and other signals used for setting system levels
   B. Usually has both 1kHz and 400Hz signal outputs
   C. With a properly calibrated signal can determine signal bit rate
   D. Can be helpful in measuring distortion

3. According to the ANSI/AVIXA standard V201.01:2021 *Image System Contrast Ratio,* what is the minimum contrast ratio for basic decision-making tasks?
   A. 7:1
   B. 15:1
   C. 50:1
   D. 80:1

4. Which of the following are associated with sound pressure levels? (Select all that apply.)
   A. Is typically measured in decibels (dB SPL)
   B. Is the ultimate method for quantifying background noise levels
   C. Eliminates distortion in speech reinforcement
   D. Can most accurately be measured with a Class 1 precision SPL meter

5. What type of video is best used to verify audio/video synchronization?
   A. Landscapes with orchestra music
   B. News anchors speaking
   C. Team sports
   D. A blip-and-flash test reel

6. Contrast ratio should be measured from _____ locations within the viewing area.
   A. All
   B. Two
   C. Four
   D. Five

7. After a project has been completed and verified, the AV team should hand over comprehensive documentation, including _____.
   A. Drawings of record
   B. Copies of relevant standards
   C. Test patterns
   D. A business card

8. To "train the trainer" is to _____.
   A. Learn the AV system so you can train the client
   B. Train individuals at the client location who can then train other users
   C. Train the client on aspects of the user interface
   D. Host classes at the designer's office

9. Comprehensive training of end users and support staff is essential to customer satisfaction and also _____.
   A. Is not expensive or time-consuming because the manufacturer pays for it
   B. Reduces service calls as well as improper use and equipment damage
   C. Reduces AV contractor costs because it is a tax-deductible expense
   D. Enables the AV contractor to continue to keep paid personnel at the site

10. Where would the person in charge of verifying system performance locate performance criteria?
    A. Call the installers of the system.
    B. Typically, there isn't any documentation; using personal experience is best.
    C. Project specification documentation.
    D. Owner's manuals.

Answers

1. **A, B, C, D.** Using the ANSI/AVIXA 10:2013 *Audiovisual Systems Performance Verification* standard will help reduce project risk, align outcome and performance expectations, provide a verifiable outcome, and create reporting that completes project documentation.

2. **A, B, D.** An audio signal generator will generate a sine wave and other signals used for setting system levels. It usually has both 1kHz and 400kHz signal outputs and can be helpful in measuring distortion.

3. **B.** 15:1 is the minimum contrast ratio for projected images used in basic decision-making tasks.

4. **A, D.** Sound pressure levels are typically measured in decibels and can most accurately be measured with a Class 1 precision SPL meter.

5. **D.** A blip-and-flash test will verify video synchronization.

6. **D.** Contrast ratio should be measured from five locations within the viewing area.

7. **A.** After a project has been completed and verified, the AV team should hand over comprehensive documentation, including drawings of record.

8. **B.** To "train the trainer" is to train individuals at the client location who can then train other users.

9. **B.** Comprehensive training of end users and support staff is essential to customer satisfaction and also reduces service calls, as well as improper use and equipment damage.

10. **C.** The person in charge of verifying system performance should be able to locate performance criteria in the project specification documentation.

PART V

Appendixes and Glossary

- **Appendix A** Math Formulas Used in AV Design
- **Appendix B** AVIXA Standards
- **Appendix C** AVIXA AV Standards Clearinghouse
- **Appendix D** Video References
- **Appendix E** About the Online Content
- **Glossary**

APPENDIX A

Math Formulas Used in AV Design

How long has it been since you solved a word problem or used a math formula? Many CTS-D exam candidates have not done math in a formal setting in years. You may be familiar with the skills and tools in this appendix, but then again, you may need a refresher.

Using the Proper Order of Operations

Most AV math formulas use only the four common operators: add, subtract, multiply, and divide. However, some formulas require a solid foundation in the order of operations. The order of operations helps you correctly solve formulas by prioritizing which part of the formula to solve first. It is a way to rank the order in which you work your way through a formula. This section will review and test your knowledge of applying the order of operations.

This is the order of operations:

1. Any numbers within a pair of parentheses or brackets
2. Any exponents, indices, or orders
3. Any multiplication or division
4. Any addition or subtraction

If there are multiple operations with the same priority, then proceed from left to right: parentheses, exponents, multiplication, division, addition, subtraction. The order of operations can be remembered by using one of these acronyms: PEMDAS, BEMDAS, BIDMAS, or BODMAS.

Steps to Solving Word Problems

All math formulas summarize relationships between concepts. Word problems are designed to test how well an individual can apply that relationship to a new situation.

By following a simple strategy, you can turn a complicated word problem into a few straightforward steps. This section provides a structured approach to solving problems. Within this structure, you will find many strategies for solving different types of problems. This strategy is based on *How to Solve It: A New Aspect of Mathematical Method* by G. Pólya (Princeton University Press, 2015).

Step 1: Understand the Problem

As typical within the AV industry (and in general), the first step is to understand the problem you're trying to solve. Here are the tasks to complete for this step:

1. Read the entire math problem.
2. Identify your goal or unknown. What information are you trying to determine?
3. Identify what you have been given. What data, numbers, or other information in the problem can help you determine the answer?
4. Predict the answer if you can. What range of values would make sense as an answer?

Example: Calculate the current in a circuit where the voltage is 2 volts and the resistance is 8 ohms.

First, identify your goal. What are you trying to solve for? When you see the word *calculate,* generally the word that follows is your goal. Other words that identify the goal include *determine, find,* and *solve for.* Your goal in this problem is to calculate current.

An easy way to identify your given information is to find the numbers in the problem. In this example, the numbers are 2 and 8. Look for context clues or units to identify what those numbers represent. "*The voltage is*" identifies 2 as the voltage. The *ohms* after the 8 identifies 8 as the resistance.

Sometimes it is unclear what each number represents. In that case, drawing a diagram can help you make sense of what the problem is trying to say. You may want to make a chart of your given and unknown information for quick reference. For more complex problems, tables of given information can be extremely helpful.

Givens and Goal	Values
Current	?
Voltage	2V
Resistance	8Ω

Step 2: Create a Plan

The second step in this process is to translate the words in the problem into numbers you can enter into a formula. Begin with the following:

1. Assign appropriate values to the goal and given information.
2. Determine a formula that describes the relationships between your variables.

If a single formula has all your variables in it, move on to the next step.

In some cases, you may be unable to identify a formula that will determine your goal based on your given information. If you're stuck, consider the following strategies:

- Use an intermediate formula to solve for the information you are missing.
- Use an outside reference, such as a chart or graph, to find information not listed in the problem.
- Use a strategy that has worked to solve similar problems in the past.
- Diagram the scenario described in the word problem. Use the diagram to keep track of the relationships between values. For instance, as listeners move farther away from a sound source, you know to expect a decrease in sound pressure.

Example: Calculate the current in a circuit where the voltage is 2 volts and the resistance is 8 ohms.

Again, start by assigning variables to your given and unknown information. Note that some items are represented by different variables in different contexts. If you are having trouble determining which variables to use, consider drawing a diagram and labeling it with your given information, like this:

Givens and Goal	Values	Variables
Current	?	I
Voltage	2V	V
Resistance	8Ω	R

Once you have assigned variables, think about the relationship between the information. There should be a formula that describes the relationship. For complex problems, you may need to use several formulas.

You may know formulas from memory. If you don't, look them up (there are several useful formulas at the end of this appendix). It is helpful to think about a time when you solved for the value previously or a similar problem you may have solved in the past. Try to think of a problem that used the same givens and unknowns.

For example, you might not remember how to solve for current using voltage and resistance. But if you remember how to solve for voltage using current and resistance, you can solve this problem. Voltage is equal to current multiplied by resistance.

Step 3: Execute Your Plan

The third step is to put your plan in action, as follows:

1. Write the formulas.
2. Substitute the given information for the variables.
3. Perform the calculation.
4. Assign units to your final answer.

Example: Calculate the current in a circuit where the voltage is 2 volts and the resistance is 8 ohms.

Once you have determined the appropriate formula, write it down and then replace the placeholders with the numbers for this problem. People who skip this step are prone to making mistakes.

Formula: $V = I \times R$
Substitution: $2 = I \times 8$

To solve this equation, you need to get the *I* by itself. The 8 is currently being multiplied. To move it to the other side of the equation, perform the opposite mathematical function. In this case, that function is division.

$2 \div 8 = (I \times 8) \div 8$
$2 \div 8 = I$
$0.25 = I$

You need to assign units to your answer before it is final. Because *I* represents current, you would assign 0.25 the unit for current, which is amps (A).

I = 0.25A = 250mA

Step 4: Check Your Answer

Your final step is to make sure the numbers you've calculated still make sense when translated back into words. Compare your answer to the scenario described in the problem. Is the result reasonable? Is it within the range you originally predicted?

For example, suppose you are calculating the voltage present in a boardroom loudspeaker circuit. A result of 95V is probably a reasonable answer. A result of 50,000V indicates that you made a mistake in your calculations.

If you have an incorrect answer and use it to solve other parts of a process, it will result in cascading problems.

Example: Calculate the current in a circuit where the voltage is 2 volts and the resistance is 8 ohms.

In this example, is "less than 1A" a reasonable answer? Understanding the problem is essential here. An AA battery has a voltage of 1.5V. When an AA battery is attached to a circuit, there is not much current. So, the small number of 0.25 amps is a reasonable answer.

Rounding

Many of the results listed have been rounded to the nearest tenth. When solving multistep problems, you may be tempted to round at each step. The earlier or more often you round in a multistep problem, the less accurate your result will be. Round only your final result.

AV Math Formulas

This section presents some math formulas that may be useful for AV design professionals.

 NOTE On test day, CTS-D candidates have access to some relevant math formulas on the computer screen while they test. The formula sheet does not cover every possible formula that you may encounter on the exam, but it does contain some common (and complicated) formulas. Formulas not on the exam formula sheet will need to be memorized for use during the exam. Find it at www.avixa.org/ctsd.

Estimated Projector Throw

The formula for estimating projector throw distance is:

$$Distance = Screen\ Width \times Throw\ Ratio$$

where

- *Distance* is the distance from the front of the lens to the closest point on the screen
- *Screen Width* is the width of the projected image
- *Throw Ratio* is the ratio of throw distance to image width

Refer to the owner's manual of your projector and lens combination to find an accurate formula for your specific projector.

Projector Lumens Output

The formula for estimating required projector brightness is:

$$Lumens = [(L_A \times C \times A) \div S_G] \div D_R$$

where

- *Lumens* = The required brightness of the projector in lumens.
- L_A = The ambient light level.
- *A* = The screen area.
- *C* = The required contrast ratio as defined in ANSI/AVIXA V201.01:2021 *Image System Contrast Ratio*.
- S_G = The gain of the screen. Assume a screen gain of 1 unless otherwise noted.
- D_R = The projector derating value. Assume a derating value of 0.75 unless otherwise noted.

The units for ambient light (L_A) and screen area (A) measurements must match. If the ambient light measurement is in lux (lm/sqm), the screen area must be in square meters. If the ambient light measurement is in footcandles (lm/sqft), the screen dimensions must be in square feet.

ANSI Brightness of a Projector

The formula for calculating the ANSI brightness of a projector is:

$$ANSI\ lumens = [(Z_1 + Z_2 ... Z_9) \div 9] \times (S_H \times S_W)$$

where

- Z_1 to Z_9 = The incident light measured on the nine zones of an ANSI test screen
- S_H = Height of the image
- S_W = Width of the image

When calculating projector brightness, the units used for incident light (Z_n) and area measurements (S_H and S_W) must match. If the incident light measurement is in lux (lm/sqm), the image dimensions must be in meters. If the incident light measurement is in footcandles (lm/sqft), the image dimensions must be in feet.

Luminance from Illuminance

Incident light measurements can be converted to a reasonable approximation of the luminance of an illuminated matte surface if the light reflectance value (LRV) of that surface is known.

The conversion formula is:

$$L = E \times LRV \times \frac{\pi}{100}$$

where

- L = luminance in cd/m²
- E = illuminance at the surface, in lux
- LRV = the LRV of the illuminated surface (0 to 100)
- π = pi ~3.14

This formula only works for matte surfaces. If a surface has any specularity (shininess or sheen), this relationship is not valid.

Decibel Formula for Distance

The formula for decibel changes in sound pressure level over distance is:

$$dB = 20 \times log(D_1 \div D_2)$$

where

- dB is the change in decibels
- D_1 is the original or reference distance
- D_2 is the new or measured distance

The result of this calculation will be either positive or negative. If it is positive, the result is an increase, or gain. If it is negative, the result is a decrease, or loss.

Decibel Formula for Voltage

The formula for determining decibel changes for voltage is:

$$dB = 20 \times log(V_1 \div V_R)$$

where

- dB is the change in decibels
- V_1 is the new or measured voltage
- V_R is the original or reference voltage

The result of this calculation will be either positive or negative. If it is positive, the result is an increase, or *gain*. If it is negative, the result is a decrease, or *loss*.

Decibel Formula for Power

The formula for calculating decibel changes for power is:

$$dB = 10 \times log(P_1 \div P_R)$$

where

- dB is the change in decibels
- P_1 is the new or measured power measurement
- P_R is the original or reference power measurement

The result of this calculation will be either positive or negative. If it is positive, the result is an increase, or *gain*. If it is negative, the result is a decrease, or *loss*.

Current Formula (Ohm's Law)

The formula for calculating current using Ohm's law is:

$$I = V \div R$$

where

- I is the current
- V is the voltage
- R is the resistance

Power Formula

The formula to solve for power is:

$$P = I \times V$$

where

- P is the power
- I is the current
- V is the voltage

Series Circuit Impedance Formula

The formula for calculating the total impedance of a series loudspeaker circuit is:

$$Z_T = Z_1 + Z_2 + Z_3 \ldots + Z_N$$

where

- Z_T is the total impedance of the loudspeaker circuit
- $Z_1 \ldots Z_N$ is the impedance of each loudspeaker

Parallel Circuit Impedance Formula: Loudspeakers with the Same Impedance

The formula to find the circuit impedance for loudspeakers wired in parallel with the same impedance is:

$$Z_T = Z_1 \div N$$

where

- Z_T is the total impedance of the loudspeaker system
- Z_1 is the impedance of each loudspeaker
- N is the number of loudspeakers in the circuit

Parallel Circuit Impedance Formula: Loudspeakers with Different Impedances

The formula to find the circuit impedance for loudspeakers wired in parallel with differing impedance is:

$$Z_T = \frac{1}{\frac{1}{Z_1} + \frac{1}{Z_2} + \frac{1}{Z_3} \cdots + \frac{1}{Z_N}}$$

where

- Z_T is the total impedance of the loudspeaker circuit
- Z_x is the impedance of each individual loudspeaker

Series/Parallel Circuit Impedance Formulas

Two formulas are used to calculate the expected total impedance of a series/parallel circuit.

First, the series circuit impedance formula is used to calculate the impedance of each branch, as follows:

$$Z_T = Z_1 + Z_2 + Z_3 \ldots + Z_N$$

where

- Z_T is the total impedance of the branch
- Z_x is the impedance of each loudspeaker

Then, the parallel circuit impedance formula is used to calculate the total impedance of the series/parallel circuit, as follows:

$$Z_T = \frac{1}{\frac{1}{Z_1} + \frac{1}{Z_2} + \frac{1}{Z_3} \cdots + \frac{1}{Z_N}}$$

where

- Z_T is the total impedance of the loudspeaker circuit
- Z_x is the total impedance of each branch

Needed Acoustic Gain

The formula for calculating needed acoustic gain—how loud speakers need to be for listeners to hear the intended audio—is:

$$NAG = 20 \times \log(D_0 \div EAD)$$

where

- D_0 is the distance from the source to the listener

- *EAD* is the equivalent acoustic distance, or the farthest distance one can go from the source without needing sound amplification or reinforcement to maintain intelligibility

Potential Acoustic Gain

The formula for calculating potential acoustic gain (gain before feedback) is:

$$PAG = 20 \times \log[(D_0 \times D_1) \div (D_2 \times D_S)]$$

where

- D_0 is the distance between the talker and the farthest listener
- D_1 is the distance between the microphone and the closest loudspeaker to it
- D_2 is the distance between the farthest listener and the loudspeaker closest to them
- D_S is the distance between the sound source (talker) and the microphone

Audio System Stability (PAG/NAG)

The formula for checking audio system stability by determining that needed acoustic gain (NAG) is less than potential acoustic gain (PAG) is:

$$20 \times \log(D_0 \div EAD) < 20 \times \log[(D_0 \times D_1) \div (D_2 \times D_S)] - 10 \times \log(NOM) - FSM$$

where

- *NOM* is the number of open microphones
- *FSM* is the feedback stability margin
- *EAD* is the equivalent acoustic distance
- D_0 is the distance between the talker and the farthest listener
- D_1 is the distance between the microphone and the closest loudspeaker to it
- D_2 is the distance between the farthest listener and the loudspeaker closest to them
- D_S is the distance between the sound source (talker) and the microphone

Conduit Capacity

The formula for calculating conduit capacity is:

$$id = \sqrt{\frac{od_1^2 + od_2^2 + od_3^2 \ldots}{fp}}$$

where

- od_x is the outer diameter of the cables

- *id* is the minimum inner diameter of the conduit
- *fp* is the permissible fill percentage (expressed as a decimal fraction) of the conduit based on the number of cables

Permissible fill percentage is determined by the authority having jurisdiction (AHJ). For example, in the United States the National Electrical Code (NEC) requirements are as follows:

- One cable: 53 percent (0.53)
- Two cables: 31 percent (0.31)
- Three or more cables: 40 percent (0.40)

Jam Ratio

The formula for calculating jam ratio is:

$$Jam = id \div \left(\frac{od_1 + od_2 + od_3}{3} \right)$$

where

- od_x is the outer diameter of the cables
- *id* is the inner diameter of the conduit

Jam ratio is applicable only to conduits with exactly three cables.

Heat Load Formula (Btu)

The formula for calculating heat load in Btu is:

$$\text{Total Btu} = W_E \times 3.4$$

where

- W_E is the total watts of all equipment used in the room
- 3.4 is the conversion factor, where 1 watt of power generates 3.4 Btu of heat per hour

This formula does not account for the heat load generated by amplifiers.

Heat Load Formula (kJ)

The formula for calculating heat load in kilojoules is:

$$\text{Total kJ} = W_E \times 3.6$$

where

- W_E is the total watts of all equipment used in the room

- 3.6 is the conversion factor, where 1 watt of power generates 3.6 kilojoules of heat per hour

This formula does not account for the heat load generated by amplifiers.

Power Amplifier Heat Load (kJ)

The formula for calculating the heat load of a power amplifier is:

$$\text{Total } kJ = W \times 3.6 \times (1 - E_D)$$

where

- W is the wattage of the amplifier
- E_D is the efficiency of the device

Power Amplifier Heat Load (Btu)

The formula for calculating the heat load of a power amplifier is:

$$\text{Total } Btu = W \times 3.4 \times (1 - E_D)$$

where

- W is the wattage of the amplifier
- E_D is the efficiency of the device

Required Amplifier Power (Constant Voltage Loudspeaker Systems)

The formula for calculating required amplifier power for constant voltage loudspeaker systems is:

$$W_T = W \times N \times 1.5$$

where

- W_T is required power in watts
- W is nominal wattage tap used at the individual loudspeaker
- N is total number of loudspeakers
- 1.5 is the factor for an allowance of 50 percent for power headroom

Wattage at the Loudspeaker

The formula for the electrical power (EPR) required at a loudspeaker is:

$$EPR = 10^{\left(\frac{[L_p + H - L_s + (20 \times \log(D_2 \div D_r))]}{10}\right)} \times W_{ref}$$

where

- L_p is the SPL required at distance D_2
- H is the headroom required
- L_s is the loudspeaker sensitivity reference, usually 1 watt at 1 meter
- D_2 is the distance from the loudspeaker to the farthest listener
- D_r is the distance reference value, usually 1 meter
- W_{ref} is the power reference value; assume a W_{ref} of 1 watt, unless otherwise noted

Simplified Room Mode Calculation

A simple formula to approximate the critical (*Schroeder*) frequency where room modes begin to dominate the low-frequency performance of the room is:

$$F_c = 3 \times \frac{C_{air}}{RSD}$$

where

- F_c is critical frequency
- C_{air} is the velocity of sound in air (343m/s or 1,125 ft/s)
- RSD is the room's smallest dimension (in meters or feet to match C_{air})

Loudspeaker Coverage Pattern (Ceiling Mounted)

The formula for calculating the diameter (twice the radius) of the circle that represents the coverage area of a loudspeaker is:

$$D = 2 \times (H - h) \times \tan(C\angle \div 2)$$

where

- D is the diameter of the coverage area
- H is the ceiling height
- h is the height of the listeners' ears
- $C\angle$ is the loudspeaker's angle of coverage in degrees

Loudspeaker Spacing (Ceiling Mounted)

The formula for calculating the space between ceiling-mounted speakers depends on how much overlap of each speaker's coverage pattern is desired. Here are formulas for three typical coverage patterns, based on overlap:

- Edge-to-edge (no overlap): $D = 2 \times r$
- Minimal overlap: $D = r \times \sqrt{2}$
- Center-to-center (maximum overlap): $D = r$

In all cases

- D is the distance between loudspeakers
- r is the radius of the loudspeakers' coverage circles

Digital Video Bandwidth

The formula for calculating digital signal bandwidth for full bit depth (4:4:4), RGB, or YUV video is:

Bandwidth (bps) = Horizontal Pixels × Vertical Pixels × Bit Depth × Frame Rate × 3

The formula for calculating the digital signal bandwidth for 4:2:2 sampled digital video is:

Bandwidth (bps) = Horizontal Pixels × Vertical Pixels × Bit Depth × Frame Rate × 2

The formula for calculating the digital signal bandwidth for 4:1:1 or 4:2:0 sampled digital video is:

Bandwidth (bps) = Horizontal Pixels × Vertical Pixels × Bit Depth × Frame Rate × 1.5

The formula for calculating the digital signal bandwidth for uncompressed RGB video with an alpha channel (4:4:4:4) is:

Bandwidth (bps) = Horizontal Pixels × Vertical Pixels × Bit Depth × Frame Rate × 4

APPENDIX B

AVIXA Standards

The American National Standards Institute (ANSI) is the official U.S. representative to the International Organization for Standardization (ISO). AVIXA's Certified Technology Specialist (CTS) certification exam, for which you are studying, is ANSI-accredited under the ISO and the ISO/IEC 17024:2012 *Conformity Assessment—General Requirements for Bodies Operating Certification Schemes of Persons* standard.

In addition, AVIXA is an ANSI-accredited Standards Developer (ASD), developing voluntary standards for the commercial AV industry. Accreditation by ANSI signifies that the processes used by standards development organizations (SDOs) to develop ANSI standards meet the ANSI's requirements for openness, balance, consensus, right to appeal, and due process. A diverse group of subject matter experts work cooperatively to develop voluntary ANSI/AVIXA standards.

AVIXA develops performance, documentation, management, and verification standards for all aspects of AV system design and integration. AVIXA standards are nontechnical standards that are both vendor and product neutral. They take into account technology, human factors, architecture, and other variables in determining the best way to design, implement, and manage the performance of all types of AV systems. You can keep up with news about current standards, planned revisions, and those in development by visiting AVIXA's website (www.avixa.org/standards).

Don't be surprised if you see references to some of these standards on the CTS-D exam. It is important for CTS-D professionals to be able to understand and apply relevant standards. As a CTS-D–certified AV professional, you should consider standards when designing AV projects.

In this appendix, we offer a brief synopsis of the existing ANSI/AVIXA standards and those currently under development.

Published Standards, Recommended Practices, and Technical Reports

The following standards were available from AVIXA at the time of publishing.

559

A102.01:2017 Audio Coverage Uniformity in Listener Areas (ACU) This standard defines measurement requirements and parameters for characterizing a sound system's coverage in listener areas. It provides performance classifications to describe the uniformity of coverage of a sound system's early arriving sound with the goal of achieving consistent sound pressure levels throughout defined listener areas. At the time of publication, this standard was under revision.

S601.01:2021 Audiovisual Systems Energy Management This standard defines and prescribes processes and requirements for ongoing energy consumption management of the audiovisual (AV) system. It identifies requirements for the control and continuous monitoring of electrical power for AV systems, whereby energy is conserved whenever possible and components operate at the lowest energy-consuming state possible without compromising the system's performance for user needs. AV systems that conform to this standard will meet the defined requirements for automation, monitoring, reporting, and documentation.

10:2013 Audiovisual Systems Performance Verification This standard provides a framework and supporting processes for determining elements of an audiovisual system that need to be verified, the timing of that verification within the project delivery cycle, a process for determining verification criteria/metrics, and reporting procedures. At the time of publication, this standard was under revision.

J-STD 710 – 2015 (CTA/AVIXA/CEDIA) Audio, Video, and Control Architectural Drawing Symbols This standard defines architectural floor plan and reflected ceiling plan symbols for audio, video, and control systems, with associated technologies such as environmental control and communication networks. It also includes descriptions and guidelines for the use of these symbols. At the time of publication, this standard was under revision.

F501.01:2015 Cable Labeling for Audiovisual Systems This standard defines requirements for audiovisual system cable labeling for a variety of venues. It provides requirements to easily identify all power and signal paths in a completed audiovisual system to aid in operation, support, maintenance, and troubleshooting. At the time of publication, this standard was under revision.

V202.01:2016 Display Image Size for 2D Content in Audiovisual Systems This standard determines required display image size and relative viewing positions based on user need. It can be used to design a new space or to assess or modify an existing space, from either drawings or the space itself. It applies to permanently installed and temporary systems. The standard does not apply to the performance or efficiency of any component. A free online calculator is available at www.avixa.org/discascalc (requires one-time registration).

V201.01:2021 Image System Contrast Ratio This standard defines contrast ratios based on user viewing requirements. It is designed to facilitate informed decision-making for any display, projector, and screen selection relative to location and purpose. It applies to permanently installed systems and live events, front and rear projection, and direct-view displays.

RP-38-15:2018 (IES/AVIXA), Lighting Performance for Small-to-Medium-Sized Videoconferencing Rooms This standard provides lighting parameters and performance criteria for small-to-medium-sized single-axis videoconferencing spaces (with 3 to 25 primary seating locations), defined as one set of video displays and cameras oriented toward a group of seated participants. It provides guidance to professionals involved in the design, construction, assessment, and support of videoconferencing environments by establishing performance criteria for the design and testing of room lighting and finishes that will provide appropriate picture quality.

F502.01:2018 Rack Building for Audiovisual Systems This standard defines requirements for building AV equipment racks, which are defined as assembly of rack(s), mounting of AV equipment and accessories, cable management, and finishing.

F502.02:2020 Rack Design for Audiovisual Systems This standard defines minimum requirements for the audiovisual rack planning and design process, including required process inputs and outputs. Key performance criteria validate the impact to internal and external integration with the facility requirements.

RP-C303.01:2018 Recommended Practices for Security in Networked Audiovisual Systems This recommended practice provides guidance and current best practices for securing networked AV systems, including how to recognize risks and develop a risk mitigation management plan to address those risks.

2M-2010 Standard Guide for Audiovisual Systems Design and Coordination Processes This standard provides a framework for the methods, procedures, tasks, and deliverables typically recommended or applied by industry professionals in the design and implementation of audiovisual communication systems. The framework enables clients and other design and construction team members to assess whether responsible parties are providing expected services. At the time of publication, this standard was under revision. The revision will define minimum documentation requirements for AV systems design.

TR-111.01:2019 Unified Automation for Buildings This technical report provides a detailed overview of the building automation environment and identifies the need for a unified set of standards to integrate multiple building systems, including but not limited to traditional building automation systems (BASs), into cohesive and functional systems and/or subsystems for increased benefits.

Standards in Development

The following standards were in development at the time of publishing and may now be completed and available.

A103.01 Sound System Spectral Balance This standard defines a measurement and verification process for sound system reproduction of spectral balance, also known as uniform frequency response, accomplished by documenting the frequency response from the sound system across a specified bandwidth within a low- to high-frequency range within the listening area.

A104.01 Sound System Dynamic Range   This standard provides a procedure to measure and classify the dynamic range, or signal-to-noise ratio, of early arriving sound from a sound system across a listener area.

UX701.01 User Experience Design for Audiovisual Systems   This standard defines processes that optimize user experience for AV-equipped spaces. Processes include user engagement, design, testing, deployment, and continuous refinement.

NOTE   AVIXA standards are constantly being reviewed and updated to keep them relevant and timely. To ensure that you are working with the current versions of standards, you should always check the standards section of the AVIXA website (www.avixa.org/standards) for updates and revisions.

Scan For AVIXA Standards

APPENDIX C

AVIXA AV Standards Clearinghouse

AVIXA is not the only provider of audiovisual (AV) standards. Other organizations around the world also develop AV standards. The process of researching these standards to determine which ones apply to your project can become a challenge.

In response to the need for an easier way to find AV-related standards, AVIXA offers the AV Standards Clearinghouse, an extensive, searchable spreadsheet of standards that are applicable to audiovisual technologies. This free resource was developed to help you spend less time researching and more time working toward delivering a high-quality, standards-compliant end product.

AVIXA's AV Standards Clearinghouse is a powerful tool that enables you to identify applicable standards by title, publication date, publisher, functional category (audio, video, control, other), interest category (integrator, tech manager, consultant, and so on), and regional applicability. The clearinghouse is updated periodically to include the latest standards.

The AV Standards Clearinghouse is not just a useful tool on the job. If you are focused on mastering certain areas of the CTS-D Exam Content Outline, the clearinghouse is a great resource for potential further study materials. Examples include standards for the following:

- Measuring jitter in digital systems
- Cable mapping documentation symbology
- Measuring projector performance
- Electrical safety precautions

You can find standards relating to almost any task on the CTS-D Exam Content Outline in the AV Standards Clearinghouse.

You can download the clearinghouse spreadsheet and the accompanying guide document from the AVIXA website. Use the following link, provided as both a URL and a QR code:

www.avixa.org/standards/av-standards-clearinghouse

APPENDIX D

Video References

AVIXA has produced several short videos that can help you prepare for the CTS-D exam, practice installation skills, and learn best practices. Most of these videos are less than ten minutes long. Use the following link, provided as both a URL and a QR code, for the current list of videos:

www.avixa.org/cts-supplemental-training-videos

APPENDIX E

About the Online Content

This book comes complete with TotalTester Online customizable practice exam software with 55 sample exam questions from AVIXA.

System Requirements

The current and previous major versions of the following desktop browsers are recommended and supported: Chrome, Microsoft Edge, Firefox, and Safari. These browsers update frequently, and sometimes an update may cause compatibility issues with the TotalTester Online or other content hosted on the Training Hub. If you run into a problem using one of these browsers, please try using another until the problem is resolved.

Your Total Seminars Training Hub Account

To get access to the online content, you will need to create an account on the Total Seminars Training Hub. Registration is free, and you will be able to track all your online content using your account. You may also opt in if you wish to receive marketing information from McGraw Hill or Total Seminars, but this is not required for you to gain access to the online content.

Privacy Notice

McGraw Hill values your privacy. Please be sure to read the Privacy Notice available during registration to see how the information you have provided will be used. You may view our Corporate Customer Privacy Policy by visiting the McGraw Hill Privacy Center. Visit the **mheducation.com** site and click **Privacy** at the bottom of the page.

Single User License Terms and Conditions

Online access to the digital content included with this book is governed by the McGraw Hill License Agreement outlined next. By using this digital content, you agree to the terms of that license.

Access To register and activate your Total Seminars Training Hub account, simply follow these easy steps.

1. Go to this URL: **hub.totalsem.com/mheclaim**
2. To register and create a new Training Hub account, enter your e-mail address, name, and password on the **Register** tab. No further personal information (such as credit card number) is required to create an account.

   If you already have a Total Seminars Training Hub account, enter your e-mail address and password on the **Log in** tab.
3. Enter your Product Key: **crnz-mt2r-js3g**
4. Click to accept the user license terms.
5. For new users, click the **Register and Claim** button to create your account. For existing users, click the **Log in and Claim** button.

   You will be taken to the Training Hub and have access to the content for this book.

Duration of License Access to your online content through the Total Seminars Training Hub will expire one year from the date the publisher declares the book out of print.

Your purchase of this McGraw Hill product, including its access code, through a retail store is subject to the refund policy of that store.

The Content is a copyrighted work of McGraw Hill, and McGraw Hill reserves all rights in and to the Content. The Work is © 2023 by McGraw Hill.

Restrictions on Transfer The user is receiving only a limited right to use the Content for the user's own internal and personal use, dependent on purchase and continued ownership of this book. The user may not reproduce, forward, modify, create derivative works based upon, transmit, distribute, disseminate, sell, publish, or sublicense the Content or in any way commingle the Content with other third-party content without McGraw Hill's consent.

Limited Warranty The McGraw Hill Content is provided on an "as is" basis. Neither McGraw Hill nor its licensors make any guarantees or warranties of any kind, either express or implied, including, but not limited to, implied warranties of merchantability or fitness for a particular purpose or use as to any McGraw Hill Content or the information therein or any warranties as to the accuracy, completeness, correctness, or results to be obtained from, accessing or using the McGraw Hill Content, or any material referenced in such Content or any information entered into licensee's product by users or other persons and/or any material available on or that can be accessed through the licensee's product (including via any hyperlink or otherwise) or as to non-infringement of third-party rights. Any warranties of any kind, whether express or implied, are disclaimed. Any material or data obtained through use of the McGraw Hill Content is at your own discretion and risk and user understands that it will be solely responsible for any resulting damage to its computer system or loss of data.

Neither McGraw Hill nor its licensors shall be liable to any subscriber or to any user or anyone else for any inaccuracy, delay, interruption in service, error or omission, regardless of cause, or for any damage resulting therefrom.

In no event will McGraw Hill or its licensors be liable for any indirect, special or consequential damages, including but not limited to, lost time, lost money, lost profits or good will, whether in contract, tort, strict liability or otherwise, and whether or not such damages are foreseen or unforeseen with respect to any use of the McGraw Hill Content.

TotalTester Online

TotalTester Online provides you with sample questions from the CTS-D exam. The sample questions can be used in Practice Mode or Exam Mode. Practice Mode provides an assistance window with hints, references to the book, explanations of the correct answers, and the option to check your answer as you take the test. Exam Mode provides a timed review of the questions. The number of questions, the types of questions, and the time allowed are intended to be an accurate representation of the exam environment. The option to customize your quiz allows you to create custom exams from selected duties (domains) or chapters, and you can further customize the number of questions and time allowed. Note that not all chapters have questions associated with them.

To take a test, follow the instructions provided in the previous section to register and activate your Total Seminars Training Hub account. When you register, you will be taken to the Total Seminars Training Hub. From the Training Hub Home page, select your certification from the list of "Your Topics" on the Home page, and then click the TotalTester link to launch the TotalTester. Once you've launched your TotalTester, you can select the option to customize your quiz and begin testing yourself in Practice Mode or Exam Mode. All exams provide an overall grade and a grade broken down by domain.

Technical Support

For questions regarding the TotalTester or operation of the Training Hub, visit **www.totalsem.com** or e-mail **support@totalsem.com**.

For questions regarding book content, visit **www.mheducation.com/customerservice**.

GLOSSARY

1/3 octave equalizer A graphic equalizer that provides 30 or 31 slider adjustments corresponding to specific fixed frequencies with fixed bandwidths, with the frequencies centered at every one-third of an octave. The adjustment points shape the overall frequency response of the system.

1/4-inch phone connector A connector typically used to transport unbalanced line-level audio signals from musical instruments and between processing devices. Also known as a *6.35mm phone connector*.

1/8-inch phone connector A connector typically used to transport unbalanced line-level audio signals from portable devices and computers. Also known as a *3.5mm phone connector*.

3.5mm phone connector A connector typically used to transport unbalanced line-level audio signals from portable devices and computers. Also known as a 1/8-inch phone connector.

4K An ultra-high-resolution video format with a minimum resolution of 3840×2160 pixels in a 16:9 (1.77:1) aspect ratio, approximately four times that of full HD (1920×1080 pixels). 4K formats include 3840×2160 pixels (UHDTV-1) and 4096×2160 pixels (digital cinema, DCI).

4K ecosystem The video cameras, recorders, editors, processors, servers, distribution networks, and display technologies used for the production, distribution, and display of 4K ultra-high-resolution video.

6.35mm phone connector A connector typically used to transport unbalanced line-level audio signals from musical instruments and between processing devices. Also known as a *1/4-inch phone connector*.

8K An ultra-high-resolution video format with a minimum resolution of 7680×4320 pixels, which is 16 times the resolution of full HD (1920×1080 pixels). The related consumer television format is known as 7680p, 8K Ultra HD, and UHDTV-2.

8K ecosystem The video cameras, recorders, editors, processors, servers, distribution networks, and display technologies used for the production, distribution, and display of 8K ultra-high-resolution video.

8P8C connector An eight-pin (eight position, eight conductor) modular connector typically used for the termination of multipair cables. Often used in Ethernet data networks. It is commonly incorrectly referred to as an RJ-45 connector, which is actually an 8P2C telephone connector.

acceptable viewing area   The viewing range for a screen, suggested as a 60-degree arc off the far vertical edge of the screen being viewed.

access   The ability to access and use a system or resource. This ability may allow a user to handle information and control system components and functions.

access point   A network device that allows other devices to connect to a data network.

acoustic echo canceller   An echo cancelling device used in conferencing systems that attempts to remove environmental echoes that are created at the far site from sound reflected off hard surfaces and returned to conferencing microphones.

acoustics   The properties or qualities of a room or building that determine how sound is transmitted and reflected. The study of the properties of sound.

AES   Audio Engineering Society.

allied trade   A business that collaborates with AV professionals to complete a project.

alternating current (AC)   An electric current that reverses its direction periodically.

ambient light   The sum of all lighting in an area.

ambient noise   The sum of all sounds in an area.

amperage   The amount of electric current flowing in a circuit. Current is measured in amperes (amps), abbreviated as A. The preferred term is "current."

amplifier   A device for increasing the strength of signals.

amplitude   The height of the waveform of a signal.

analog (analogue)   A continuously variable signal.

analog-to-digital/digital-to-analog (AD/DA or A to D/D to A) converters   Devices that convert signals from analog to digital or from digital to analog.

angularly reflective screen   A screen that reflects light at the same angle at which it arrived.

ANSI   American National Standards Institute.

anthropometrics   Results of the study of the measurements and proportions of the human body.

aperture   The opening in an optical train that controls the amount of light passing through.

arc-fault circuit interrupter (AFCI)   Also known as an arc-fault detection device (AFDD), a type of circuit breaker that triggers on detecting an electrical-arc fault in its load circuit.

arrayed loudspeaker system   A loudspeaker arrangement that delivers sound from a single point in space.

artifact An element introduced into a signal during processing. Not always a good thing.

aspect ratio The ratio of image width to image height.

attack time The time taken for an action to complete its effect once triggered.

attenuate To reduce the amplitude of a signal.

Audio Coverage Uniformity measurement locations (ACUMLs) The test points within a venue that have been determined to carry out the measurements for the AVIXA Audio Coverage Uniformity test.

Audio Coverage Uniformity Plan (ACUP) A stand-alone document that identifies the Audio Coverage Uniformity measurement locations for a particular venue, using the AVIXA indication symbol.

audio processor A device used to manipulate audio signals.

Audio Return Channel (ARC) Introduced to the HDMI standard with version 1.4. ARC allows a display to send audio data upstream to a receiver or surround-sound controller, eliminating the need for a separate audio connection.

audio signal An electrical representation of sound.

audio transduction The process of converting acoustical energy into electrical energy or electrical energy back into acoustical energy.

Audio Video Bridging (AVB) A standards-based audiovisual Data Link-layer (layer 2) protocol defined under IEEE 802.1-AVB. It runs across a standard Ethernet network but requires AVB-enabled switches and network components that handle QoS prioritization of data. AVB has been renamed Time-Sensitive Networking to reflect the standard's applicability to communication among different types of devices, such as network sensors.

audiovisual infrastructure The physical building components that make up the pathways, supports, and architectural elements required for audiovisual technical equipment installations.

audiovisual rack A housing unit for electronic equipment. The inside of a typical AV industry rack is 19in. (482.6mm) wide. Many of the technical specifications for a rack, including size and equipment height, are determined by standards that have been established by numerous standards-setting organizations. The outside width of the rack varies from approximately 530mm to 630mm (21in. to 25in.).

authentication The ability or process of verifying the identity of an entity, such as a user, process, or device, for providing access to resources such as AV systems.

authority having jurisdiction (AHJ) An organization, office, or individual responsible for enforcing the requirements of a code or standard or for approving equipment, materials, installation, or procedures. In some places the authority having jurisdiction may be known as the regional regulatory authority.

authorization The process used to determine whether a user is granted access to a specific resource by evaluating relevant access control information.

automatic gain control (AGC) A circuit or process that maintains a constant output gain in response to input variables such as signal strength or ambient noise level.

Automatic Private IP Addressing (APIPA) A Windows function that assigns locally routable addresses from the reserved network 169.254.0.0/16 to devices that do not have or cannot obtain an IP address. This allows devices to communicate with other devices on the same LAN. APIPA operates at the Network layer (layer 3) of the OSI Model and the Internet layer of the TCP/IP protocol stack.

balanced circuit A two-conductor circuit in which both conductors and all circuits connected to them have the same impedance with respect to ground.

balun A contraction of *bal*anced-to-*un*balanced. A device used to connect a balanced circuit to an unbalanced circuit. For example, a transformer used to connect a 300-ohm television antenna cable (balanced) to a 75-ohm antenna cable (unbalanced).

band A grouping or range of frequencies.

bandwidth 1. A range of frequencies. 2. A measure of the amount of data or signal that can pass through a system during a given time interval.

bandwidth limiting The process of limiting the bandwidth of a signal, usually to allow the signal to be transmitted over a lower-bandwidth path.

bandwidth (networking) The available or consumed data communication resources of a communication path, measured in bits per second (bps). It is also called data throughput or *bit rate*.

baseband A signal that has not been modulated onto a higher-frequency carrier.

bass trap An acoustic energy absorber designed to dampen low-frequency sound energy.

benchmarking The process of examining methods, techniques, and principles to establish a standard to which comparisons can be made.

bend radius The radial measure of a curve in a cable, conductor, waveguide, or interconnect that defines the physical limit beyond which further bending has a measurable effect on the signal being transported.

bi-directional polar pattern The shape of the region where some microphones will be most sensitive to sound from the front and rear, while rejecting sound from the top, bottom, and sides.

bill of materials (BOM) A complete equipment list of components that must be procured in order to build the system as specified. The BOM also lists the costs associated with each aspect of designing and implementing the system.

bit A contraction of the words *binary digit*. The smallest unit of binary digital information. May have the value 1 or 0.

bit depth The number of bits used to specify a parameter.

bit error rate (BER) The number of error bits present in a signal stream per unit of time.

bit rate The measurement of the quantity of data transmitted over a digital signal stream. It is measured in bits per second (bps).

block diagram A diagram of a system or device in which the principal parts are represented by suitably annotated geometrical figures to show both the functions of the parts and their functional relationships.

Bluetooth A wireless technology for low-cost, short-range radio links between devices. It operates in the 2.402GHz to 2.480GHz industrial, scientific, and medical (ISM) frequency band.

BNC connector A type of connector featuring a two-pin bayonet-type lock. Available in 75Ω and 50Ω impedances. The most common professional coaxial cable connector because of its reliability and ruggedness. It is used to terminate cables transporting signals such as SDI, RF, video, and time code. Originally named Bayonet Neill-Concelman after its inventors Paul Neill and Carl Concelman.

bonding Joining conductive material by a low-impedance connection, thus ensuring that they are at the same electrical potential.

boundary microphone A microphone design where the diaphragm is placed close to a sonic "boundary" such as a wall, ceiling, or other flat surface. This arrangement prevents the acoustic reflections from the surface from mixing with the direct waveform and causing phase distortions. It is used in conference and telepresence systems. Also known as a *pressure zone microphone (PZM)*.

branch circuit The circuit conductors between the final overcurrent device protecting the circuit and the load connection.

breaker box Another name for an electrical load distribution panel. *See* panelboard.

bring-your-own-device (BYOD) An approach that allows users to access resources such as a network with personal devices.

broadcast domain A set of devices that can send Data Link-layer (layer 2) frames to each other directly, without passing through a Network-layer (layer 3) device. Broadcast traffic sent by one device in a broadcast domain is received by all devices in the domain.

buffer amplifier An electronic device that provides isolation and some load independence between circuit components.

building information modeling (BIM) A data repository for building design, construction, and maintenance data shared by multiple disciplines on a single project.

bus A wiring system that links multiple devices.

busbar An electrically conductive path that serves as a common connection for two or more circuits.

buzz A noise generated by the higher-order harmonics of the hum (50Hz or 60Hz) generated by the electrical mains.

byte A data word containing 8 bits, also known as an *octet*. The symbol for byte is B.

cable An assembly of more than one signal carrier (wire or optical fiber).

cable tray A structure to provide rigid continuous support for cables.

campus area network (CAN) A type of network linking multiple LANs in a limited geographical area such as a university campus or a cluster of buildings.

candela The unit of luminous intensity. A candela is the luminous intensity emitted by a reference point light source in a given direction. The symbol for candela is cd. By sheer "coincidence," the luminous intensity of a common wax candle is approximately 1 candela.

cannon connector An alternative name for the XLR family of secure, low-voltage connectors used in professional audiovisual systems. The three-pin version is the standard audio signal cable for the production and AV industries. The four-pin version is widely used for communication headsets, and the five-pin version is the standard connector for the DMX512 digital lighting protocol.

capacitance The ability of a material to store an electrical charge. Capacitance is measured in farads (F). The symbol for capacitance is C.

capacitive reactance The opposition a capacitive device offers to alternating current flow. It is measured in ohms (Ω). The symbol for capacitive reactance is X_C. Capacitive reactance is inversely proportional to the frequency of the current in a circuit.

capacitor A passive electrical component used to store electric charge. Constructed from electrodes of conductive material separated by a dielectric.

captive screw connector Also known as a Euroblock or Phoenix connector. A termination method where a stripped wire is inserted into the connector and secured by a set screw that pushes down a gate to form the electrical bond and clamp the wire in place.

cardioid polar pattern A heart-shaped region where some microphones will be most sensitive to sound predominately from the front of the microphone diaphragm and reject sound coming from the sides and rear.

carrier Modulated frequency that carries a communication signal.

Category 5 (Cat 5) The designation for 100Ω unshielded twisted-pair (UTP) cables and associated connecting hardware whose characteristics are specified for data transmission up to 100Mbps (part of the EIA/TIA 568A standard).

Category 5e (Cat 5e)   An enhanced version of the Cat 5 UTP cable standard that adds specifications to reduce far-end crosstalk for data transmission up to 1Gbps (part of the EIA/TIA 568A standard).

Category 6 (Cat 6)   A UTP cable standard for Gigabit Ethernet and other interconnections that is backward-compatible with Cat 5, Cat 5e, and Cat 3 cable (part of the EIA/TIA 568A standard). Cat 6 features more stringent specifications for crosstalk and system noise.

Category 6A (Cat 6A)   A twisted-pair cable standard for 10 Gigabit Ethernet and other interconnections that is backward-compatible with Cat 5, Cat 5e, and Cat 6 cable. Constructed using a mechanism to physically separate the twisted-pairs, Cat 6A features more stringent specifications for alien and near-end crosstalk.

Category 7 (Cat 7)   A twisted-pair cable standard for 10 Gigabit Ethernet and other interconnections that is backward-compatible with Cat 5e and Cat 6A cable. To further reduce crosstalk, in Cat 7 cables, each twisted-pair is shielded and the entire cable is shielded.

Category 8 (Cat 8)   A twisted-pair cable standard for 40 Gigabit Ethernet and other interconnections that is backward-compatible with Cat 5e, Cat 6, and Cat 6A cable (part of the EIA/TIA 568A standard). Cat 8 features a shield around each twisted-pair and a further shield enclosing all twisted-pairs. It is only specified at 40Gbps for runs up to 30m (100ft).

cathode ray tube (CRT)   A high-vacuum glass tube containing an electron gun to produce the images seen on the phosphor-coated face of the tube. This video display technology was used in early-generation video monitors, television receivers, radar displays, and oscilloscopes.

CATV   Community antenna television. A system where broadcast television signals are received by a single central antenna and distributed to multiple end users.

center tap   A connection point located halfway along the track or winding of an electronic device such as an inductor or resistor.

central cluster   A single-source configuration of loudspeakers. In a central cluster, the sound is perceived as coming from one point in the room. The central cluster is normally located directly above (on the proscenium) and slightly in front of the primary microphone location.

central processing unit (CPU)   The portion of a computer system that reads and executes commands.

charged-coupled device (CCD)   A light-sensitive semiconductor device, commonly used in video and digital cameras, that converts optical images into electronic signals.

chassis   Also called a *cabinet* or *frame,* an enclosure that houses electronic equipment and is frequently electrically conductive (metal). The conductive enclosure acts as a Faraday cage/shield and is connected to the safety-grounding conductor of the AC power supply to provide protection against electric shock.

chassis ground   A 0V (zero volt) connection point of any electrically conductive chassis or enclosure surrounding an electronic device. This connection point may be extended to the earth/ground.

chroma   The saturation, or intensity, of a specific color. It is one of the three attributes that define a color; the other two are hue and value or luminance.

chrominance   The color component of a composite or S-video signal.

Classless Inter-Domain Routing (CIDR) notation   A compact representation of an IP address and its subnet mask using a slash and a decimal number to indicate how many leading 1 bits are in the network mask, with the remaining bits being the network host identifiers (e.g., 192.168.220.16/24). The subnet mask of 255.255.255.0 (24 mask bits) is represented as /24 in CIDR notation. CIDR replaces the class designations (A, B, and C) for IP address ranges.

cliff effect   The sudden loss of digital signal reception. When a digital signal is attenuated to the point where signal for a digital 1 is indistinguishable from the signal for a digital 0.

clipping   The distortion of a signal when a device's peak amplitude is exceeded.

clock adjustment   The process used to align the timing of digital signals between devices.

CMOS   Complementary metal oxide semiconductor is a semiconductor fabrication process that uses symmetrical pairs of complementary p-type and n-type field effect transistors (FETs) for logic functions. CMOS devices have high noise immunity and draw low current in a static state. This technology is extensively used in the production of digital electronic devices, including camera pickup chips.

coaxial (coax) cable   A cable consisting of a center conductor surrounded by insulating material, concentric outer conductor, and optional protective covering, all of circular cross section.

CobraNet   CobraNet is a proprietary digital audio Data Link-layer (layer 2) protocol designed by Cirrus Logic. It uses standard Fast Ethernet cabling, switches, and other components. CobraNet signals are nonroutable.

codec   A contraction of the term *coder/decoder*. An electronic device or software process that encodes or decodes a data stream for transmission and reception over a communications medium.

collision domain   A set of devices on a carrier-sense multiple access (CSMA) local area network whose packets may collide with one another if they transmit data at the same time. Only found in nonswitched Ethernet networks.

color difference signal   A signal that conveys color information such as hue and saturation in a composite format. Two such signals are needed. These color difference signals are R-Y and B-Y. Sometimes referred to as Pr and Pb or Cr and Cb.

color rendering index (CRI)   The effect a light source has on the perceived color of objects indexed against an incandescent source (CRI 100) of the same correlated color temperature.

color space   A color space is the range of spectral colors that can be interpreted or displayed by a device. The range is usually identified by being mapped against an area on the CIE chromaticity diagram. A range of well-known color spaces is used in imaging and display.

color temperature   The quantification of the color of "white" light in reference to the light emitted by a standardized object at a specified temperature on the Kelvin scale. Measured in kelvins (K). Low color temperature light (~2,000K) has a warm (reddish) look, while light with a high color temperature (>4,000K) has a colder (bluish) appearance.

combiner   A device or process that combines signals of different frequencies together in a single medium. Used to combine multiple RF television signals into one cable for use in broadband cable television distribution.

common mode   **1.** Voltage fed in phase to both inputs of a differential amplifier. **2.** The signal voltage that appears equally and in phase from each current carrying conductor to ground.

common-mode rejection ratio (CMRR)   The ratio of the differential voltage gain to the common-mode voltage gain; expressed in decibels.

compander   An audio processing device that combines compression and expansion.

component video   Color video in which the brightness (luminance), color hue, and saturation (chrominance) are handled independently. The red, green, and blue signals—or, more commonly, the Y, R-Y, and B-Y signals—are encoded onto three wires.

composite video signal   A single video signal that carries the complete color picture information and all synchronization signals.

compression   **1.** An increase in density and pressure in a compressible medium such as air. **2.** A process that reduces data size.

compression ratio   **1.** How much the volume on an audio compressor reduces depending on how far above the threshold the signal is. **2.** The ratio in size between the original signal and its compressed form.

compressor (audio)   A device that controls the overall amplitude of a signal by reducing the part of the signal that exceeds an adjustable threshold level set by the user. When the signal exceeds the threshold level, the overall amplitude is reduced by a ratio, also usually adjustable by the user.

compressor threshold   Sets the point at which the automatic volume reduction kicks in. When the input goes above the threshold, an audio compressor automatically reduces the volume to keep the signal from getting too loud.

condenser microphone   Also called a *capacitor microphone,* a microphone that transduces sound into an electric current using capacitive principles.

conductor   In electronics, a material that easily conducts an electric current because some electrons in the material are easy to move.

conduit   A circular tube that houses cable.

cone   A lightweight, semi-rigid, conical diaphragm structure attached to the voice coil of a loudspeaker.

constant voltage distribution   25V, 70V, 100V; a method of distributing signals to loudspeakers over a large area with lower losses than typical direct-coupled connections. Also known as a high-impedance distribution system.

Consumer Electronic Control (CEC)   A single-wire, bi-directional serial bus that uses AV link protocols to perform remote-control functions. It is an optional feature of the HDMI specification that allows for system-level automation when all devices in an AV system support it.

contact closure   A simple signaling system based on whether a contact or switch is open or closed. The conventional protocol is to interpret a closed contact (on) as a binary 1 and an open contact (off) as a binary 0.

contactor   An electrically controlled switch used for switching a power circuit, similar to a relay, but with higher switched-current ratings. May be single-pole or multipole.

content delivery network (CDN)   A distributed network of caching servers that can provide hosted unicast distribution of media for an organization. They are most often utilized by organizations whose content is in high demand.

contrast   The difference in luminance between dark and light elements of an image.

contrast ratio   Describes the dynamic video range of a display device as a numeric relationship between the brightest color (typically white) and the darkest color (typically black) that the system is capable of producing. Two methods are used to specify contrast ratio; the full on/full off method describes the dynamic contrast ratio, and the ANSI method measures the static contrast ratio.

control system   A system that controls subsystems such as audio, video, winches, drapes, mechanical devices, lighting, and atmospheric effects.

correlated color temperature (CCT)   The color appearance of a light source as compared with a standardized object heated to a temperature measured on the Kelvin scale. CCT is measured kelvins (K).

coverage pattern   The pattern of sound energy that a loudspeaker emits. The coverage pattern is dependent on the frequency of the sounds and the dimensions of the loudspeaker.

critical distance (d_c)   The point where the sound pressure level of the direct and reverberant sound fields is equal.

critical path schedule   A scheduling methodology that reveals the interdependence of activities and assesses resource and time requirements and trade-offs. It also determines the project's completion date and provides the capability to evaluate activity performance.

crossover   An electronic device that separates the frequency bands of an audio signal so that each driver in a multidriver loudspeaker system is sent only those frequencies that it will transduce accurately.

crosspoint   A matrix-based switching device with multiple inputs and outputs wherein any input may be connected to any, many, or all outputs.

crosstalk   Any phenomenon by which a signal transmitted on one circuit or channel of a transmission system creates an undesired effect in another circuit or channel.

current   The quantity of electrical charge flowing in a circuit, measured in amperes (A). The symbol for current is *I*.

curvature of field   A blurry appearance around the edge of an otherwise in-focus object (or the reverse) when the angle incidence of light passing through the edges of a lens is different from the angle of incidence at the center of the surface. Curvature of field is due to the shape of the lens.

cybersecurity   A process or capability intended to protect information and communications systems against unauthorized use or modification, as well as against damage and exploitation. This includes policies, strategies, and standards regarding the security and operations of these systems.

Dante   A proprietary digital audio Network-layer (layer 3) protocol created by Audinate. Dante sends audio and video information as Internet Protocol (IP) packets. It is fully routable over IP networks using standard Ethernet switches, routers, and other components. Dante controller software manages data prioritization and signal routes.

dB SPL   A measure of sound pressure level with reference to 0dB SPL (a sound pressure of 20 micropascal [μPa]).

decay time   The time taken for an effect to diminish.

decibel (dB)   A base-10 logarithmic representation of the ratio between two numbers. This ratio is used for quantifying differences in voltage, distance, sound pressure, and power.

decryption   The process of translating encrypted communications or data into its original form.

DECT   Digital enhanced cordless telecommunications (DECT). A wireless communications protocol widely used for cordless telephone systems and some other portable devices.

delay   A signal-processing device or circuit used to delay the delivery of a signal.

demodulator A device or process that extracts the information from a modulated signal.

depth of field The area in front of a lens that is in focus. The distance from the closest focused item to the focused item farthest away.

detail drawing A detail drawing that shows small details at a larger scale than the main drawing.

dielectric constant (relative permittivity) The ratio of the permittivity of a material compared to the permittivity of a vacuum. A measure of the dielectric strength of a material.

dielectric strength The ability of a material to act as an insulator: how much potential difference it can withstand before breaking down and becoming a conductor.

differential mode A signaling mode where two wires carry the same signal but one is in reverse polarity.

differentiated service (DiffServ) A network quality-of-service (QoS) strategy wherein data from specific applications or protocols is assigned a class of service. Flows assigned a high priority are given preferential treatment at the router, but delivery is not guaranteed.

diffusion The scattering or random reflection or transmission of a wave from a surface.

digital media player Device or software application that replays recorded or streamed audio and video content.

digital signage Digital video displays used for signage. Sometimes referred to as dynamic signage to differentiate it from conventional static signage.

digital signage media player A specialized media player used to store and forward or play back digital signage content.

digital signage template Customizable layout and content templates used for the content of digital signage. Used to simplify the design and maintenance of standardized content.

digital signal processor (DSP) A digital device designed to process signal streams such as audio, video, and RF data.

digital-to-analog converter A device that converts a digital signal into an analog form.

direct coupled A loudspeaker signal distribution system in which the amplifier is connected directly to the loudspeaker(s).

direct current (DC) Electric current that does not reverse direction, unlike alternating current (AC). DC may vary in amplitude but not direction.

direct sound Also known as *near-field*. Sound that is received directly from the source and not colored by the acoustics of its surroundings.

direct view display A display such as an LCD, LED, OLED, vacuum fluorescent, or plasma screen where the viewer is looking directly at the image source, not at a projection screen.

directivity The coverage pattern of loudspeakers or microphones.

dispersion The separation of light into different frequencies/wavelengths, such as when a white light beam passes through a triangular prism. The different wavelengths of light refract at different angles, dispersing the light into its individual wavelengths.

Display Data Channel (DDC) A data channel used between video sources and video displays to carry information about display capabilities and video formats. The channel used for exchanging EDID and HDCP data.

DisplayID A standard developed by VESA outlining how video display data is structured to describe the display's performance and capabilities when communicating with other devices. By structuring data in a flexible, modular way, DisplayID enables devices to identify new display resolutions, refresh rates, audio standards, and other formats as they become available. For example, the standard can support a single image segmented across tiled displays using multiple video processors.

DisplayPort A VESA-developed, high-speed digital data transport protocol used to connect a video source to display devices. It can also carry audio, USB, and other data.

distributed sound system A sound system using multiple distributed loudspeakers to provide sound coverage across an area at a constant sound pressure level.

distribution amplifier (DA) An active device used to split one input signal into multiple isolated output signals at a constant amplitude.

diversity receiver An RF receiver that uses multiple antennas to receive a single RF transmission. The receiver calculates phase differences between the received signals to dynamically shift between antennas to avoid multipath signal cancellation.

DLP Digital Light Processing. A projection technology based on the MEMS digital micromirror device (DMD) chip family. It uses a matrix of thousands of movable microscopic mirrors on a chip to display images on a screen.

Domain Name System (DNS) A hierarchical, distributed database that maps names to data such as IP addresses, name server addresses, and mail exchange addresses.

dome A type of loudspeaker driver construction. Fabric, thin metallic, or woven materials are used to create a dome-shaped diaphragm. The voice coil is attached to the edge of the dome-shaped diaphragm.

dotted-decimal/dotted-quad notation A format commonly used for expressing 32-bit IPv4 addresses using four decimal numbers in the range 0 to 255, separated by decimal points (dots) (e.g., 192.168.12.254).

driver **1.** In audio, an individual loudspeaker unit. **2.** In electronics, a piece of software or firmware that takes a high-level command set and implements the commands in a specific format for the actual hardware or software present in the system.

dual channel In test equipment, refers to a test device with two independent input channels.

DVD Digital Video Disc or Digital Versatile Disc. An optical storage medium for digital data or video. Widely used for video content distribution from the late 1990s to approximately 2015.

DVI Digital Visual Interface. A digital interface to connect an uncompressed video source to a display device. DVI has largely been replaced by HDMI, DisplayPort, and other formats.

Dynamic Host Configuration Protocol (DHCP) An IP address management process that automates the assignment of IP addresses and networking parameters to devices on an IP network.

dynamic microphone A microphone with a diaphragm attached to a moveable coil of wire located in a magnetic field. Sound pressure waves cause the diaphragm to move the coil in the magnetic field, inducing a small electric current in the coil.

dynamic range The difference between the highest and lowest levels of a signal. Usually expressed in decibels (dB).

early reflected sound The sound waves that arrive at the listener's ear closely following (<30ms) the direct sound wave. These are the sound waves that are reflected off surfaces between the source and the listener.

earthing conductor A conductor used to connect equipment or the grounded circuit of a wiring system to a grounding/earthing electrode (planet Earth).

echo A reflected or duplicated version of a signal that arrives at the listener with sufficient delay and separation from the original signal to allow the delayed signal to be perceived distinctly and later in time from the original signal.

echo cancellation A means of eliminating echo from a signal path.

eggcrate A light baffle with rectangular cells that direct, block, and diffuse light.

electret microphone A type of condenser microphone using a prepolarized material, called an *electret,* which is applied to the microphone's diaphragm or backplate.

electrical service The conductors and equipment for delivering energy from the electricity supply system to the wiring system of the site served.

electromagnetic interference (EMI) Interference signals produced by electromagnetic fields.

elevation drawing   A side view of an object or surface taken in the vertical plane from outside the object or surface.

emissive technology   Any display technology that emits light to create an image.

encoded   A signal that has been converted into another format.

encryption   The process of transforming communications or data into a form that cannot easily be understood by recipients who have not been granted access to it.

energy management plan (EMP)   A document that details a systematic approach to implementing the most effective power consumption methods and procedures to achieve and maintain optimum energy usage.

equalizer   Electronic equipment that modifies the frequency characteristics of a signal.

equipment grounding   The connection to ground (planet Earth) from the conductive, user-accessible parts of equipment.

equipment grounding conductor (EGC)   North American term for the safety earth, the conductive path installed to connect normally noncurrent-carrying conductive parts of equipment to the grounding/earthing electrode (planet Earth).

equipment rack   An equipment housing unit that protects and organizes electronic equipment.

equivalent acoustic distance (EAD)   The farthest distance one can go from the source without the need for sound amplification or reinforcement to maintain good speech intelligibility. The distance is dependent on the level of the presenter and the noise level in the room.

ergonomics   Also known as human factors or human factors engineering. This is the scientific study of the way people interact with a system. It focuses on effectiveness, efficiency, reducing errors, increasing productivity, improving safety, reducing fatigue and stress, increasing comfort, increasing user acceptance, increasing job satisfaction, and improving quality of life.

Ethernet   A set of network cabling, signaling, and network access protocol standards. Before the introduction of switch-based networks, Ethernet was based on carrier-sense multiple access technology with collision detection (CSMA/CD).

EtherSound   A proprietary digital audio Data Link-layer (layer 2) protocol designed by Digigram. It uses standard 100Mbps or 1Gbps Ethernet cabling, switches, and other components. It requires a separate network with dedicated bandwidth. EtherSound signals are nonroutable.

expander   An audio processor that increases the dynamic range of an audio signal. It comes in two types: a downward expander and as part of a compander.

exploit   An attack technique intended to take advantage of a vulnerability to breach the security of a network or gain information.

extended display identification data (EDID)   A data structure within a video sink device that is used to describe the sink's characteristics to a video source. The characteristics described include the sink's native resolution, color space information, and audio type (mono or stereo).

external configuration   Refers to the ability of one device to configure other devices and subsystems.

Faraday cage   A conductive solid or mesh enclosure used to block electromagnetic fields. Also known as a Faraday shield. Named after its inventor Michael Faraday.

far field   The sound field distant enough from the sound source that the SPL decreases by 6dB for each doubling of the distance from the source.

farthest viewer   The viewer positioned at the farthest distance from the screen as defined by the viewing area.

feedback   **1.** In audio, unwanted noise caused by the loop of an audio system's output back to its input. **2.** In a control system, data supplied to give an indication of system performance or status.

feedback stability margin (FSM)   The extra margin added (usually 6dB) into the needed acoustic gain (NAG) formula that represents extra gain that a sound system may have available before the onset of feedback.

fiber-optic   A technology that uses total internal reflection of light along a transparent fiber to transmit information.

field   In interlaced video, one-half of a video frame containing every other line of information. Each interlaced video frame contains two fields.

filter   A device or process that blocks or passes certain elements of a signal.

firewall   Any technology, hardware, or software that regulates the traffic permitted to enter or exit a network. Firewalls control access across network boundaries.

firewall traversal   Any firewall or application-layer gateway (ALG) is expected to provide mechanisms that allows traffic to traverse through the firewall/ALG to reach an intended destination. Firewall traversal is provided in multiple ways, including NAT traversal, IPSec tunnels, IP ACLs (access control lists), or port-based ACLs.

firmware   A type of software that is stored in nonvolatile memory in a piece of hardware.

fixture   A luminaire that is mounted or fixed in place.

flex life   The number of times a cable can be bent before it breaks. A wire with more strands or twists per meter will have a greater flex life than one with a lower number of strands or fewer twists per meter.

focal length   The distance between the center of a lens and the point where the image comes into focus. The shorter the focal length, the wider the angle of the image will be.

foot-candle The U.S. customary unit of illuminance. The incident light measured when 1 lumen of light is spread over an area of 1 square foot. Its symbol is fc. 1 foot-candle = 10.76 lux.

footlambert The footlambert is the U.S. customary unit for luminance. It is equal to $1/\pi$ candela per square foot. Its symbol is fl. 1 footlambert = 3.43 candela per square meter.

frame rate The number of frames per second sent from a source.

frequency The number of complete cycles in a specified period of time. It is measured in hertz (Hz). 1Hz = 1 cycle per second.

frequency domain A signal viewed as frequency versus amplitude is in the frequency domain. This allows you to view the amount of energy present at different frequencies.

frequency response The amplitude response versus frequency for a given device.

Fresnel lens A flat lens in which the curvature of a normal lens surface has been collapsed in such a way that concentric circles are impressed on the lens surface. A Fresnel lens is often used for the condenser lens in overhead projectors, in rear-screen projection systems, and in Fresnel spotlights.

front-screen projection An image projection system where the image is projected from a source on the viewer's side of the screen.

full-duplex communication A form of bi-directional data transmission in which messages may simultaneously travel in both directions.

full HD A high-definition video mode with a resolution of 1920×1080 pixels.

fundamental frequency Known as *pure tone,* the lowest frequency in a harmonic series.

gain The change in the amplitude of a signal.

gain control A gain control is an electronic adjustment through which the amplitude of a signal can be increased or decreased.

gain-sharing automatic mixers A gain-sharing automatic mixer is an audio mixer that automatically increases the gain for microphone channels that are in use and attenuates the microphone channels that are not in use.

gate An audio processor that allows signals to pass only above a certain setting or threshold.

gated automatic mixer An audio mixer that turns microphone channels either "on" or "off" automatically, based on the amplitude of their signal.

gateway A router device that connects a local network to an external network. All outgoing network traffic must travel through it. A gateway will pass incoming traffic to the routers below. The routers below look to the gateway to resolve DNS addresses not found on the local network.

gauge The thickness or diameter of a wire or plate.

genlock To lock the synchronization signals of multiple devices to a single synchronization source.

graphic equalizer An equalizer with an interface that has a graph comparing amplitude on the vertical axis with frequency on the horizontal axis.

graphical user interface (GUI) Often pronounced "gooey," provides a graphical interface for a user to view a system's elements and functions.

graphics adapter Commonly referred to as a *video card*, a device that generates and outputs video signals from a computer.

graphics processing unit (GPU) A specialized circuit designed for processing display functions. The processor is optimized to render and manipulate images in a video frame buffer.

grayscale The luminance (brightness and darkness) of a color. It is sometimes called *value*. It is one of the three attributes of color; the other two are hue and chroma.

grayscale test pattern A test pattern that displays a range of known gray values between black and white.

ground 1. The earth. 2. In the context of an electrical circuit, the earth or some conductive body that extends the ground (earth) connection. 3. In the context of electronics, the 0V (zero volt) circuit reference point. This electronic circuit reference point may or may not have a connection to the earth.

ground fault An unintentional electrical connection between any ungrounded conductor of the electrical system and any noncurrent-carrying metal object.

ground loop An electrically conductive loop that has two or more ground reference connections. The loop can be detrimental when the reference connections are at different potentials, which causes current flow within the loop.

ground plane A continuous conductive area. The fundamental property of a ground plane is that every point on its surface is at the same potential (low impedance) at all frequencies of concern.

ground potential A point in a circuit that is at the same potential as the earth/ground.

ground reference The 0V (zero volt) reference point for a circuit.

grounded conductor A system or circuit conductor that is connected to the earth/ground.

ground-fault circuit interrupter (GFCI) A safety device that de-energizes a circuit within an established period of time when a current to ground exceeds a specified level. Triggered by an imbalance between the supply and return currents. Similar in function to an earth leakage circuit breaker (ELCB), a residual current device (RCD), or a core balance relay (CBR).

ground-fault current path   An electrically conductive path from the point of a ground fault on a wiring system through normally noncurrent-carrying conductors, equipment, or earth to the electrical supply source.

grounding   Connecting to ground or to a conductive body that extends the ground connection. The connected connection is referred to as *grounded*.

grounding conductor   A conductor used to connect equipment or the grounding circuit of a wiring system to a grounding/earth electrode or electrodes.

grounding electrode   A conducting object through which a direct connection to earth (planet Earth) is established.

grounding electrode conductor   The conductor used to connect the system grounded conductor or the equipment to a grounding electrode or to a point on the grounding electrode system.

Group Management Protocol (GMP)   A protocol that allows a host to inform its neighboring routers of its desire to start or stop receiving multicast transmissions.

group of pictures (GoP)   A set of successive frames that are required to display a complete series in a digital video signal. It includes the visible picture, timing/sync information, and compression frames.

half-duplex communication   A form of bi-directional data transmission in which messages can only travel in one direction at any time.

harmonic distortion   Signal distortion that arises when harmonics of an input signal are generated during processing and appear in the output together with the processed input signal.

harmonics   Higher-frequency waves that are a multiple of the fundamental frequency.

HD-15 connector   An HD-15 connector, sometimes called a VGA connector, is a video connector that is typically associated with the output of an analog computer graphics card. It has three rows of five pins, which carry analog red, green, blue, and sync signals along with display data channel information.

HDBaseT   A connectivity standard for the transmission of high-definition video, audio, DC power, Ethernet, and serial signaling, including USB and other protocols, over standard twisted-pair data cables such as Cat 5e and above.

HDCP   High-bandwidth digital content protection.

HDCP key   A number that a program uses to verify authenticity and encode/decode content. HDCP processes use multiple types of keys. These keys are strongly protected by Digital Content Protection, LLC.

HDCP receiver/sink   A device that can receive and decode HDCP signals. A television is an example of a receiver.

HDCP repeater   A device that can receive HDCP signals and transmit them to another device, such as a switcher or distribution amplifier.

HDCP transmitter/source   A device that can send HDCP-encoded signals and content. A Blu-ray Disc player is an example of an HDCP transmitter.

HDMI Ethernet Audio Control (HEAC)   In HDMI, the combining of HDMI Ethernet Channel (HEC) and Audio Return Channel (ARC) into one port or cable. *See* HDMI Ethernet Channel (HEC) and Audio Return Channel (ARC).

HDMI Ethernet Channel (HEC)   Consolidates video, audio, and data streams into a single HDMI cable. A dedicated data channel enables high-speed, bi-directional networking to support future IP solutions and allow multiple devices to share an Internet connection.

HDTV   High-definition television. Generally includes image resolutions above 1280×720.

headend   The equipment located at the start of a cable television distribution system where the signals are processed and combined prior to distribution.

headroom   The difference between the average performance level of a system and the maximum level it can produce. Usually measured in dB.

heat load   Heat load is the heat released by a device during operation. It is measured in joules (or British thermal units).

heat sink   A device that absorbs and dissipates the heat produced by an object or process.

hemispheric polar pattern   The dome shape of the region where some microphones will be most sensitive to sound. This is the pattern of boundary/pressure zone microphones.

hertz (Hz)   The unit of frequency. 1Hz = 1 cycle per second.

hextet   A group of 16 bits, usually written as four lowercase hexadecimal digits (e.g., 0fe8). The 128-bit addresses used in IPv6 are written as eight hextets separated by colons (e.g., 2006:0fe8:85a3:0000:0002:8a2e:0a77:c082).

High-Bandwidth Digital Content Protection (HDCP)   A form of encryption developed by Intel to control digital audio and video content as it travels across Digital Video Interface (DVI), DisplayPort (DP), or High-Definition Multimedia Interface (HDMI) connections. It prevents transmission or interception of unencrypted HD content. Does not work across Serial Digital Interface (SDI) streams.

High Definition Multimedia Interface (HDMI)   A point-to-point connection protocol between video devices for digital video and audio. HDMI signals include audio, control, Ethernet, and digital asset rights management information.

high dynamic range (HDR)   Digital images having a bit depth of at least 10 bits (1,024 levels) per channel. Standard dynamic range images have a bit depth of 8 bits (256 levels) per channel.

high-pass filter   A circuit that allows signals above a specified frequency to pass unaltered while simultaneously attenuating frequencies below the specified limit.

hiss   Broadband higher-frequency noise generated by random electron movement in the amplification stages of a system. In audio systems it is typically associated with poor system gain structure.

horn   A flared-shape loudspeaker that reproduces mid to high frequencies.

hot plug   **1.** A low-level signal sent by an EDID source that indicates whether a sink or display is connected. **2.** A system that can detect and respond to devices being connected or disconnected during normal operations (e.g., USB). **3.** Plugging or unplugging equipment while it is still powered up.

hue   The attribute of a color that represents its position in a defined color space or on the visible spectrum. Hue is usually described with a color name such as "red," "blue," "yellow," or "purple." It is one of the three attributes that define color; the other two are luminance and chroma.

hum   Undesirable (usually 50 to 60Hz-plus harmonics) noise emanating from an audio device or evidenced by a rolling hum bar on a display.

IEEE   The Institute of Electrical and Electronics Engineers.

illuminance   The amount of light falling on a surface, measured in lux (lx) or foot-candle (fc). 1 foot-candle = 10.76 lux.

IGBT   The insulated-gate bipolar transistor (IGBT) is a high-power, three-terminal semiconductor device similar in structure to a thyristor. It is primarily used for high-speed switching applications such as inverters for variable-speed motor control, switch mode power supplies, and reverse phase-control dimming.

image constraint token (ICT)   A digital flag signal incorporated into some high-definition digital video streams. If present, the video stream can be decoded at full resolution only on HDCP-enabled devices.

image resolution   The total number of pixels in the image. It is normally expressed as the number of horizontal pixels multiplied by the number of vertical pixels.

imager   A light-sensitive electronic device behind a camera's lens, usually made up of thousands of sensors that convert the light input into an electrical output.

impedance   The total opposition that a circuit presents to an alternating current. It includes resistance (R), inductive reactance (X_L), and capacitive reactance (X_C). Impedance is measured in ohms (Ω). Its symbol is Z. Formula: $Z = \sqrt{R^2 + (X_L - X_C)^2}$

impedance matching Having an impedance value on an input that an output is expecting. It does not necessarily mean having comparable impedances on an input and an output.

impedance meter Device used to measure the impedance of an electrical circuit.

incident (security) An event that threatens or adversely affects AV and information systems or the data that resides on them.

inductance The magnetic property of a circuit that opposes any change in current, represented by the symbol L and measured in henrys (H).

induction The influence exerted on a conductor by a changing magnetic field.

inductive reactance (X_L) Opposition to the current flow offered by the inductance of a circuit. Inductive reactance is measured in ohms (Ω). Its symbol is X_L. The inductive reactance in a circuit is directly proportional to the frequency of the current.

InfoFrames Structured packets of data that carry information regarding aspects of audio and video transmission, as defined by the EIA/CEA-861 standard. Using this structure, a frame-by-frame summary is sent to the display, permitting selection of appropriate display modes automatically. InfoFrames typically include auxiliary video information, generic vendor-specific source product description, MPEG, and audio information.

infrared (IR) The range of nonvisible light frequencies below the red end of the visible spectrum. IR signal transmission requires an uninterrupted line of sight between transmitter and receiver.

input A connection point that receives information from another piece of equipment.

insider threat A vector of risk to an organization's AV or information systems that comes from within the organization, such as employees, contractors, or other associates. These parties generally have increased access to systems and a greater knowledge of their defenses.

insulation A material of high dielectric strength used to isolate the flow of electric current between conductors.

Integrated Services Digital Network (ISDN) A communications standard for transmitting data over digital telephone lines.

intellectual property (IP) A type of property that includes the creations of the human intellect. It includes ideas; written material such as books, articles, poems, and essays; designs including plans, specifications, diagrams, sketches, and program code; images such as photographs, graphic arts, and other artworks; music and songs; and movie and video contents. Intellectual property rights are usually asserted through the use of patents, trademarks, copyrights, and trade secrets. All IP used in the design, execution, and operation of an AV installation must be correctly licensed from each of the owners of that IP.

intelligibility A sound system's ability to produce an accurate reproduction of sound, allowing listeners to identify words and sentence structure.

Glossary

Internet Corporation for Assigned Names and Numbers (ICANN) An organization chartered to oversee several Internet-related tasks. ICANN manages Domain Name System (DNS) policy, including the top-level domain space for the Internet.

Internet Group Management Protocol (IGMP) The group management protocol for IPv4. IGMPv1 allowed individual clients to subscribe to a multicast channel. IGMPv2 and IGMPv3 added the ability to unsubscribe from a multicast channel.

interlaced scanning The scanning process that alternately displays the odd and even lines of a video frame to construct a full frame of video signal.

internal configuration Refers to the local setup and customization of management or control of a device.

Internet of Things (IoT) Refers to a situation where network connectivity and computing capability have been extended to include objects, sensors, and other devices, allowing these devices to exchange data.

Internet Protocol (IP) A TCP/IP protocol defined in the IETF standard RFC 791. IP defines rules for addressing, packaging, fragmenting, and routing data sent across an IP network. IP falls under the Internet layer of the TCP/IP protocol stack and the Network layer (layer 3) of the OSI Model.

Internet Protocol Television (IPTV) A system that delivers television services over an IP network such as a LAN or the Internet.

inverse square law The law of physics stating that a physical quantity or strength is inversely proportional to the square of the distance from the source of that physical quantity.

I/O port Input and/or output port. A connection port on a device for handling input and/or output signals.

IoT *See* Internet of Things (IoT).

IP *See* Internet Protocol (IP) *and* intellectual property (IP).

IR *See* infrared (IR).

isolated ground (IG) An equipment grounding method permitted by the North American NEC for reducing electrical noise (electromagnetic interference) on the grounding circuit. The isolation between IG receptacles and circuits and the normal equipment grounding is maintained up to the point of the service entrance (or a separately derived system) where the grounded (neutral) conductor, equipment grounding, and isolated equipment grounding conductor are bonded together and to earth/ground (planet Earth).

isolated grounding circuit A circuit that allows an equipment enclosure to be isolated from the raceway containing circuits. The equipment on the circuit is grounded via an insulated earthing/grounding conductor.

isolated receptacle   A power receptacle or mains power outlet in which the grounding terminal is purposely insulated from the receptacle-mounting means. In North America isolated receptacles are identified by a triangle engraved on the face. The receptacle (and so the equipment plugged into the receptacle) is grounded via an insulated earthing/grounding conductor.

jacket   Outside covering used to protect cable wires and their shielding.

junction box   A structure for enclosing the junction of electrical wires and cables. In North America a junction box can be used as a termination point with a custom connector plate or interface plate. A junction box can also be installed and used as a pull box for longer cable runs.

keystone error   The trapezoidal distortion of an image due to the projection device being at an angle to the plane of the screen.

lamp   The light source in some luminaires and projectors.

latency   Response time of a system. The delay between an input being received by a system and the corresponding output being generated. Measured in seconds.

lavalier   A small microphone usually worn either around the neck or attached to apparel. Derived from the name of an item of jewelry worn as a pendant.

law of conservation of energy   States that energy cannot be created or destroyed. Energy can be transformed from one form to another and transferred from one body to another, but the total amount of energy remains constant.

lenticular screen   A screen surface characterized by embossing designed to transmit or reflect maximum light over wide horizontal, but narrow vertical, angles.

lighting fixture   A lighting instrument or luminaire. Sometimes called a fixture.

Li-Fi   A wireless communication technology that carries data using modulated light, often using high-frequency modulation of the luminaires illuminating a space.

limiter   **1.** A signal processor that limits the amplitude of a signal to a preset level. **2.** An audio signal processor that functions like a compressor except that signals exceeding the threshold level are reduced at ratios of 10:1 or greater.

limiter ratio   Defines how much an audio limiter will compress signals that exceed its threshold. The limiter compresses only the portion of the signal that exceeds its threshold.

limiter threshold   Defines which portions of a signal the limiter will affect. All signals at levels below or equal to the threshold will pass through the limiter unchanged. All signals above the threshold will be compressed.

line conditioner   *See* power conditioner.

line driver   An amplifier used to compensate for signal attenuation created by cable impedance.

line level The specified strength of an audio signal used to transmit analog audio between the elements of an audio system. This is generally considered to be approximately 1V at 1kHz into a 600Ω impedance.

liquid crystal display (LCD) A video display technology that uses light transmission through polarizing liquid crystals to display an image.

liquid crystal on silicon (LCoS) A reflective liquid crystal imaging technology. A liquid crystal layer is applied to a reflective complementary metal oxide semiconductor (CMOS) mirror substrate. The embedded CMOS circuitry controls the reflectivity of the liquid crystal pixels.

listed Equipment, materials, or services included in a list published by a recognized testing laboratory. The term is usually applied to products or processes tested by the U.S. Underwriters Laboratories (UL).

load center An electrical industry term used to identify a lighting and appliance electrical distribution board in residential and light-commercial applications.

local area network (LAN) A computer network connecting devices within a confined geographical area, such as a building.

local monitor A local device used to monitor the output signal from a system.

logarithm The exponent of 10 that equals the value of a number.

logic network diagram A project management tool that aids in sequencing and ultimately scheduling a project's activities and milestones. It represents a project's critical path as well as the scope for the project.

lossless compression A process that compresses data without losing any information.

lossy compression A form of compression that produces an approximation of the original data by eliminating noncritical or redundant information.

loudspeaker A transducer that converts electrical energy into acoustical energy.

loudspeaker circuit A group of wired loudspeakers.

low-pass filter A circuit that allows signals below a specified frequency to pass unaltered while simultaneously attenuating frequencies above the one specified.

low smoke zero halogen (LSZH) cable A type of cable with insulation and sheathing that produces only low levels of smoke and no halogen products on combustion. Suitable for use in ventilation plenum spaces.

low voltage An ambiguous term. It may mean less than 70V AC to an AV technician, while an electrician may use the same term to describe circuits less than 600V or 1kV AC. The meaning of the term may also be determined by the authority having jurisdiction (AHJ).

lumen The unit of luminous flux. A measure of the total quantity of visible light emitted from light source per unit of time. Its symbol is lm.

luminaire A lighting instrument. Consists of a light source, optical system, housing, and mounting mechanism.

luminance (Y) Also called *luma,* the brightness component of a combined video signal that includes synchronization, color, and brightness information. Its symbol is Y.

lux The international unit of illuminance. The incident light measured when 1 lumen of light is spread over an area of 1 square meter. Its symbol is lx. 10.76 lux = 1 foot-candle.

MAC address The unique 48-bit hardware address of the media access controller on a network device. Usually written as six groups of two hexadecimal numbers, separated by a hyphen or colon (e.g., 01-23-45-67-89-ab or 01:23:45:67:89:ab).

mains buzz A mixture of higher-order harmonics of the 50Hz or 60Hz noise (hum) originating from the AC mains power system and audible in a sound system.

malware Software that threatens the operation of an AV or information system by performing an unauthorized function.

matrix decoder A video decoder that extracts red, green, and blue signals from either composite or Y, R-Y, and B-Y signals.

matrix switcher An electronic switching device with multiple inputs and outputs. The matrix allows any input to be connected to any one or more of the outputs.

matte-white screen A screen that uniformly disperses light, both horizontally and vertically, creating a wide viewing cone and wide viewing angle.

maximum transmission unit (MTU) The size in bytes of the largest frame that can pass over a Data Link-layer (layer 2) connection. Header information must be included within the MTU.

MCB Miniature circuit breaker. Small-format circuit breakers used to protect branch loads in electrical distribution panels.

mechanical switcher A switch that mechanically opens and closes to connect circuit elements. It functions like a wall switch, meaning there is a mechanical connection or disconnection.

MEMS *See* microelectromechanical systems (MEMS).

MEMS microphone A microphone built using MEMS technology. Generally, MEMS implementations of either condenser or piezo-electric microphones.

mesh topology A network where each node is connected via bridges, switches, or routers to at least one other node.

metropolitan area network (MAN) A communications network that covers a single geographic area, such as a suburb or city.

mic level   The very low-level signal from a microphone. Typically only a few millivolts.

microphone sensitivity   A specification that indicates the electrical output of a microphone when it is subjected to a known sound pressure level. Usually measured in dBV/Pa.

microelectromechanical systems (MEMS)   Mechanical devices built directly onto silicon chips using the same fabrication processes as microprocessors and memory systems. Best known as the digital micromirror devices (DMD) used for light switching in DLP projectors.

middleware   Software that provides services to applications that aren't available from the operating system. In a streaming system, for example, middleware may perform transcoding, compression, or remote access authentication.

midrange   A loudspeaker that reproduces midrange frequencies, typically 300Hz to 8kHz.

milestone   A significant or key event in a project, usually the completion of a major deliverable or the occurrence of an important event. It can often be associated with payment milestones and client approvals.

millwork   Carpentry work produced in a mill or factory. Usually refers to finished woodwork such as doors, molding, trim, flooring, cupboards, and wall paneling.

mitigation   The process of applying measures that diminish the likelihood of an undesirable occurrence or reduce the impact if it occurs.

mixer   A device for blending multiple signal sources.

mix-minus system   An audio mix with some channels omitted. Also known as a clean feed mix. Used in speech reinforcement, interruptible foldback (IFB), and conferencing systems to allow multiple participants to be heard without echoes or feedback. Each participant is provided with a submix which includes all signals except the closest microphone.

Mobile High-Definition Link (MHL)   A standard audio/video interface for connecting mobile electronics to high-definition televisions and audio receivers. SuperMHL is capable of 36Gbps with HDR and WCG video up to 8K at 120fps.

modular connector   A latching electrical connector with four, six, or eight pins. Common modular connectors are RJ-11(6P6C) and RJ-45 (8P8C).

modulator   A device that varies one or more properties of a carrier signal (frequency, amplitude, phase) with information from another signal.

MOSFET   Metal oxide semiconductor field-effect transistor. A semiconductor device with an insulated gate capable of significantly changing its load current with a small change in gate voltage. The power MOSFET is used for controlling and switching large currents. Also known as an insulated-gate field-effect transistor (IGFET).

Multicast Listener Discovery (MLD) The IPv6 group management protocol. Multicast is natively supported by IPv6; any IPv6 router will support MLD. MLDv1 performs roughly the same functions as IGMPv2, and MLDv2 supports roughly the same functions as IGMPv3.

multicast streaming A one-to-many transmission, meaning one server sends out a single stream that can then be accessed by multiple clients. In IPv4, the IP address range (224.0.0.0 to 239.255.255.255) is reserved for multicast transmissions. In IPv6, multicast addresses have the prefix ff00::/8.

multimeter A test instrument with multiple ranges for measuring current, voltage, and resistance. Many instruments also include a simple continuity test capability.

multiplexing The sharing of a single communications channel for multiple signals.

multipoint Also called *continuous presence,* videoconferencing that links many sites to a common gateway service, allowing all sites to see, hear, and interact at the same time. Multipoint requires a bridge or bridging service.

Multi-Protocol Label Switching (MPLS) A networking protocol that allows any combination of Data Link-layer (layer 2) protocols to be transported over any type of Network layer (layer 3). MPLS routes data by examining each packet's MPLS label without examining packet contents. Implementing MPLS improves interoperability and routing speed.

Murphy's law Also known as Sod's law. *If anything can go wrong, it will go wrong.* A reminder that no assumptions should ever be made about the cause of a fault in a system. If you saw the identical fault last week, it will almost certainly have an entirely different cause this week.

native resolution The number of rows of horizontal and vertical pixels that create the picture. The native resolution describes the actual resolution of the imaging device and not the resolution of the delivery signal.

near-field The sound field close to the sound source that has not been colored by room reflections. This is also known as *direct sound.*

needed acoustic gain (NAG) The gain the sound system requires to achieve an equivalent acoustic level at the farthest listener equal to what the nearest listener would hear without sound reinforcement.

needs analysis A needs analysis identifies the activities that the end users need to perform, then develops the functional descriptions of the systems that support those needs.

network address translation (NAT) Any method of altering IP address information in IP packet headers as the packet traverses a routing device. NAT is typically implemented as part of a firewall strategy. The most common form of NAT is port address translation (PAT).

network bridge A network device that connects between two networks. It may translate between different protocols on the bridged networks.

network interface card (NIC) An interface that allows a device to be connected to a network.

network segment A network segment is any single section of a network that is physically separated from the rest of the network by a networking device such as a switch or router. A segment may contain one or more hosts.

network switch A networking device that connects multiple network segments by storing the data packets sent from the transmitting segment and forwarding them to the receiving segment. Hosts on the segments are identified by their MAC addresses.

network topology The physical arrangement of the elements connected to a network.

neutral conductor The conductor in an electrical supply system that is connected to the earth/ground for current return. This is not part of the protective earth/grounding system. *See also* grounded conductor (for U.S. usage).

nibble/nybble A group of 4 bits. Half a byte or half an octet. Usually written as a single hexadecimal digit (values 0 to f).

nine-pin connector The DB 9 is the most common type of connector used in RS-232 control systems.

nit The nonstandard, and now-deprecated, name for the candela per square meter (cd/m^2), the unit of luminance. Formerly used to describe screen or surface brightness.

noise Any signal present in a system other than the desired signal.

noise criterion (NC) *See* noise rating (NR).

noise rating (NR) A rating system developed to establish satisfactory conditions for speech intelligibility and general living environments. Measurements are taken at eight center octave frequencies from 63Hz to 8kHz and plotted against a standardized curve. Known as noise criterion (NC) in the United States.

noise-masking system Also known as a sound-masking or *speech privacy system*. A sound system that deliberately introduces background noise into a space to raise the threshold of hearing and thus increase privacy between occupants in a shared space.

noisy ground An electrical connection to a ground point that produces or injects spurious voltages into the computer system through the connection to ground (IEEE Std. 142-1991).

nominal impedance An estimate of the typical impedance of a device or system.

nonvolatile memory Computer memory that retains its data when not powered up. Includes flash memory; read-only memory (ROM); erasable programmable ROM (EPROM); electrically erasable programmable ROM (EEPROM); optical disks; and magnetic storage such as floppy disks, hard disks, and magnetic tape.

notch filter A filter that notches out, or eliminates, a specific band of frequencies.

number of open microphones (NOM) A measure that takes into account the increased possibility of feedback by adding more live microphones in a space. Each time the number of open microphones is doubled, you lose 3dB of gain before feedback.

Nyquist-Shannon sampling theorem States that an analog signal can be reconstructed if it is encoded using a sampling rate that is greater than twice the highest frequency sampled. For example, since the range of human hearing extends to 20kHz, the minimum sampling rate for digital audio should be greater than 40kHz. The higher the sampling rate above this minimum, the more accurate the digital sample.

octave A band, or group, of frequencies in which the highest frequency in the band is double the lowest frequency. For example, 200Hz to 400Hz is an octave, 6kHz to 12kHz is an octave, and so on.

octet A group of eight binary digits (bits), often called a byte.

Ohm's law A law that defines the relationship between current, voltage, and resistance in an electrical circuit. The current is proportional to the applied voltage and inversely proportional to the resistance, giving the formula $I = V \div R$, where I is the current (in amps), V is the voltage (in volts), and R is the resistance (in ohms).

omnidirectional Receiving signals from or transmitting in all directions. Used to describe the sensitivity or radiation pattern for devices that operate equally in nearly all directions.

on-axis Along the axis of a symmetrical pattern. In projection, the center line of a screen, perpendicular to the viewing area for a displayed image. In audio, along the central axis of a microphone's pick-up pattern or along the central axis of a loudspeaker's dispersion pattern.

open port A subaddress in Transmission Control Protocol or User Datagram Protocol that is configured to accept packets of data in network traffic.

organic light-emitting diode (OLED) A semiconductor light-emitting diode constructed from organic compounds. Displays built from OLEDs generally use separate layers for emitting the red, green, and blue components of an image.

oscilloscope A device that allows the viewing and measurement of electronic signals on a visual display.

OSI Model Open Systems Interconnection Model. This is a reference model developed by ISO in 1984 as a conceptual framework of standards for communication in the data network across different equipment and applications by different vendors. Under this model network communication protocols are divided into seven categories, or layers.

overcurrent Any current in excess of the rated current for equipment or a conductor. It may result from overload, a short circuit, or a ground fault.

overcurrent protection device   A safety device designed to disconnect a circuit if the current exceeds a predetermined value. Examples are circuit breakers and fuses.

packet filtering   A network data filtering process that uses rules to determine whether a data packet will be allowed to pass through a router or gateway. The filtering rules are based on the contents of the protocol header of each packet.

panelboard   The North American name for an electrical load-distribution board.

parallel circuit   A circuit in which the voltage is the same across each load, but the current divides and takes all the available paths and returns to the source.

parametric equalizer   An equalizer that allows discrete selection of a center frequency and adjustment of the width of the frequency range that will be affected. This can allow for precise manipulation with minimal impact of adjacent frequencies.

password   A technique used to authenticate the identity of a user by requiring the input of a predetermined set of characters, such as letters, numbers, or other symbols.

peak   The highest level of signal amplitude, determined by the height of the signal's waveform.

peaking   An adjustment method that allows compensation for high-frequency loss in cables.

peaking control   Electronic adjustments within a video component that can be used to compensate for system losses.

penetration testing   A method of assessing the security posture of an organization's security systems by searching for vulnerabilities that an attacker could exploit.

permissible area   The maximum amount of space that cables should occupy inside an electrical conduit.

personal area network (PAN)   A limited-range, usually wireless, network that serves a single person or small workgroup.

phantom power   A DC power supply delivered as an "invisible" overlay on the signal wires of a system. In audio phantom power systems, a voltage is overlaid on microphone signal lines to enable the remote powering of devices such as condenser microphones and active direct input boxes. In power over Ethernet (PoE) systems, the voltage is overlaid across both wires in a twisted-pair.

phase   A particular value of time for any periodic function. For a point on a sine wave, it is a measure of that point's distance from the most recent positive-going zero crossing of the waveform. It is measured in degrees; 0 to 360 degrees is a complete cycle.

phase-control dimming   A method of electronic dimming that progressively removes the leading edge of each half-cycle of the AC power to a lamp.

phishing The fraudulent practice of sending messages purporting to be from a known or reputable sender in order to induce individuals to reveal confidential information.

phono connector The international name for the type of coaxial connector known as RCA in North America.

phosphor A substance that produces light when stimulated by radiation. Phosphors are used to produce some colors of visible light in fluorescent lamps, LED sources, some laser light sources, and CRT and plasma displays.

physical security The practice of protecting AV and information systems from material threats such as actions and events that could damage them or compromise their data.

ping A computer network software utility that measures the round-trip time of test messages between hosts on a computer network. Implemented on IP networks using the Internet Control Message Protocol (ICMP). The name derives from the use of sonic "pings" in sonar for underwater ranging.

pink noise A signal with a broad spectrum of random frequencies that has equal energy in each octave band.

pink noise generator (PNG) A device to generate pink noise. An audio PNG is commonly used in conjunction with an audio spectrum analyzer to evaluate and align a sound system in an environment.

pixel A contraction of the words *picture* and *element*. The smallest element used to build an image.

plan view A drawing of a space from the "top view" taken directly from above. Examples include a floor plan and site plan.

plane of screen Identification of image position on a plan or drawing relative to other plotted locations. It is a notional line, whether in plan view or elevation, that aligns with the front surface of the screen (that is, image position) used as a datum to define viewers' relative positions.

plasma display panel (PDP) A direct-view display technology consisting of an array of pixels, which are composed of three subpixels, corresponding to the colors red, green, and blue. Gas in the plasma state is used to react with phosphors in each subpixel to produce colored light (red, green, or blue) from a phosphor in each subpixel.

playback system A system designed specifically to replay recorded material.

plenum space The plenum space is also called *environmental air space*. It is an area connected to air ducts that forms part of the air distribution system.

PoE *See* Power over Ethernet.

point source A sound system that has an apparent central location for the loudspeakers. This type of sound system is typically used in a performance venue or a large house of worship.

point-to-point   Conferencing where sites are directly linked.

polar pattern   Also known as pickup pattern or transmission pattern. The shape of the pattern covered by the pickup or transmission device.

polar plot   A polar plot is a graphical representation of the relationship between a device's directionality and its input or output.

port   **1.** An input and/or output socket on an electronic device. **2.** In a TCP/IP network, a 16-bit number included in the TCP or UDP Transport-layer (layer 4) header. The port number typically indicates the Application-layer (layer 7) protocol that generated a data packet. A port may also be called by its associated service (e.g., port 80 may be called HTTP, or port 23 may be called Telnet). **3.** To relocate an application or function to a new platform.

port address translation (PAT)   A method of network address translation (NAT) whereby devices with private, unregistered IP addresses can access the Internet through a device with a registered IP address. Unregistered clients send datagrams to a NAT server with a globally routable address (typically a firewall, a gateway, or a router). The NAT server forwards the data to its destination and relays responses to the original client.

post tension type construction   A type of structure that uses metal cables embedded within a concrete slab to support a structure. The cables act as a suspension support system that allows for wider spacing of support structures within a building.

potential acoustic gain (PAG)   The potential gain that can be delivered by the sound system without ringing and before feedback occurs. It is based upon the number of open microphones and the distances between sources (like a presenter) and microphones, microphones and loudspeakers, and listeners and loudspeakers.

power   The rate at which work is done. Measured in watts (W); 1 watt = 1 joule/second. The symbol for power is P.

power amplifier   Amplifies an audio signal to a level sufficient to drive loudspeakers.

power conditioner   Also known as a line conditioner or power line conditioner. A device that conditions the quality of power being fed to equipment by regulating the voltage and eliminating line noise.

power distribution unit (PDU)   An electrical device that distributes mains power to multiple electrical devices. A PDU may contain switches, overcurrent protection, voltage and/or current monitoring, remote circuit-controllers, and power receptacles. *See also* power strip.

Power over Ethernet (PoE)   A DC power supply delivered as an "invisible" overlay on the data pairs of an Ethernet network system. The voltage is overlaid in common mode across both wires in a twisted-pair to eliminate any effect on the data. PoE is used to power a wide range of Ethernet-connected devices.

power sequencing The act of powering on and off equipment that requires a progressive startup or shutdown sequence for safe or convenient operations. Sequencing may help prevent tripping circuit breakers by limiting surge or inrush currents when devices are powered up.

power sourcing equipment Power sourcing equipment (PSE) are devices that provide power into a PoE system.

power strip A block of electrical outlets attached to a power lead, designed to enable multiple devices to be powered from a single electrical outlet. Power strips may incorporate power indicators, individual switches, overload protection, and surge protection.

powered device A powered device (PD) is a device powered by PoE.

preamplifier An amplifier that boosts a low-level electronic signal before it is sent to other processing equipment.

preset **1.** A recallable state of settings and/or levels. **2.** In lighting, a recallable state of lighting levels for one or more zones.

pressure zone microphone (PZM) An alternative name for a *boundary microphone*.

primary optic Also known as the primary lens. The major lens in an optical system. In a projector, the primary lens usually controls the focus of the image on a screen.

program report A document that describes the client's specific needs, the system purpose and functionality, and the designer's best estimate of probable cost in a nontechnical format for review and approval by the owner. Also known as the AV narrative, the discovery phase report, the return brief, or the concept design report.

progressive scanning Scanning that traces the image's scan lines sequentially.

Protocol Independent Multicast (PIM) Allows multicast routing over LANs, WANs, or the open Internet. Rather than routing information on their own, PIM protocols use the routing information supplied by whatever routing protocol the network is already using.

pulling tension The maximum amount of tension that can be applied to a cable or conductor before it is damaged.

pure tone *See* fundamental frequency.

Q factor The quality factor of an audio filter is the ratio of the height of the center frequency of the filter to the bandwidth of the filter at the 3dB point.

quality of service (QoS) Any method of managing network data traffic to preserve system functionality and provide the required bandwidth for critical applications. Typically, QoS involves some combination of bandwidth allocation and data prioritization.

quiet ground A point on a ground system that does not inject spurious voltages into the electronic system. There are no standards to measure how quiet a quiet ground is.

raceway An enclosed channel of metal or nonmetallic materials designed for carrying wires, cables, or busbars.

rack *See* equipment rack.

rack elevation diagram A rack elevation diagram is a pictorial representation of the front of a rack and the location of each piece of equipment within that rack, typically labeling the number of RUs used for each piece of equipment.

rack unit (RU) A unit of measure of the vertical space in a rack. One RU equals 1.75 inches (44.5mm).

radio frequency (RF) The portion of the electromagnetic spectrum that is suitable for radio communications. Generally, this is considered to be from 10kHz up to 300MHz. This range extends to 300GHz if the microwave portion of the spectrum is included.

radio frequency interference (RFI) Radiated electromagnetic energy that interferes with or disturbs an electrical circuit.

rarefaction A decrease of density and pressure in a compressible medium such as air.

ratio A mathematical expression that represents the relationship between the quantities of numbers of the same kind. A ratio is typically written as X:Y or X/Y.

RCA connector The North American name for the *phono connector*, a coaxial connector most often used with line-level audio signals and consumer composite video signals.

RCBO Residual current breaker with overcurrent protection. An electrical protection device that combines the functionality of a miniature circuit breaker (MCB) and a residual current detector/residual current circuit breaker (RCD/RCCB).

RCD *See* residual current device.

reactance Opposition to current flow in a circuit resulting from the reaction of the capacitance and inductance in the circuit. Measured in ohms (Ω). The symbol for reactance is X.

rear-screen projection A system in which the image is projected toward the audience through a translucent screen material for viewing from the opposite side.

reference level In the context of decibel measurements, the reference level is the established starting point represented by 0dB. The reference level varies according to unit and application.

reference point The point of zero potential used as the voltage reference for a circuit.

reflected ceiling plan (RCP) A plan used to illustrate elements in the ceiling with respect to the floor. It should be interpreted as though the floor is a mirrored surface, reflecting the features within the ceiling.

reflecting server A caching server in a content delivery network (CDN). Ingests a unicast stream and broadcasts a multicast stream.

reflection Electromagnetic energy (light, radio waves, etc.) or sound energy that has been redirected by a surface.

reflective technology A display device that reflects light to create an image.

refraction The bending or changing of the direction of a light ray when passing between transparent mediums such as air, water, glass, or a vacuum. The *refractive index* of a material is a measure of the speed that light travels through the medium in comparison to its speed through a vacuum.

refresh rate The number of times per second a display device will update the display of a received image.

release time The release time of an audio compressor determines how quickly the volume increases when an audio signal returns below the threshold.

relocatable power tap The North American term for a cord-connected product rated 250V AC or less and 20A or less with multiple receptacles. This tap is intended only for indoor use and plugged directly into a branch circuit. It is not intended to be connected to another relocatable power tap. *See* power strip.

reserve DHCP A hybrid approach to IP address allocation. Using reserve DHCP, a block of addresses is reserved for devices requiring a static IP address. The remaining IP addresses in the subnet pool are assigned dynamically using DHCP.

residual current device (RCD) A safety device that de-energizes a circuit (or a portion of that circuit) within an established period of time when a current to ground exceeds a specified level. Also known as a residual current circuit breaker (RCCB). Similar in function to an earth leakage circuit breaker (ELCB), a ground-fault circuit interrupter (GFCI), or a core balance relay (CBR).

resistance The property of a material to impede the flow of electrical current. Measured in ohms (Ω). The symbol for resistance is R.

resistor A passive electrical device that produces opposition to current flow. Current passes through a resistor in direct proportion to voltage, independent of frequency. The relationship between voltage across and current through a resistor is defined in Ohm's law.

resolution **1.** The amount of detail in an image. **2.** The number of picture elements (pixels) in a display.

reverberant sound Sound waves that bounce off multiple surfaces before reaching the listener but arrive at the listener's ears later than early reflected sound.

reverberation Numerous persistent reflections of sound energy.

reverse phase-control dimming   A method of electronic dimming that progressively removes the trailing edge of each half-cycle of the AC power.

RF   *See* radio frequency.

RF control   A method of control employing RF wireless signaling. RF control systems vary in complexity from simple one-way on/off signals to high-bandwidth, multichannel, bi-directional systems with complex user interfaces and rich feedback. Wireless control may use many formats, including Wi-Fi, DECT, ISM, Bluetooth, Zigbee, UWB, or LTE frequencies and signaling protocols.

RF distribution system   A closed-circuit television distribution system with each of the composite video and audio program signals modulated onto a radio frequency carrier signal for distribution via RF cables. Receiving devices must have a demodulator capable of extracting the separate program channels.

RGBHV signal   A high-bandwidth video signal with separate conductors for the red signal, green signal, blue signal, horizontal sync, and vertical sync.

RGBS signal   A four-component video signal composed of a red signal, a green signal, a blue signal, and a composite sync signal.

rigid metal conduit   A rigid metal conduit, called just *rigid* in North America, is the heaviest electrical conduit and offers the best physical and EMI protection.

rigid nonmetallic tubing   Rigid nonmetallic tubing is very stiff with a thick wall but lightweight. It is similar to plumbing tubing. Because it is not flexible, it is available in preformed pieces at various angles.

ring   A network topology that connects terminals, computers, or nodes in a continuous loop.

risk   The potential for an unwanted or adverse impact on organizational operations resulting from an occurrence involving AV and information systems, given the potential impact of a specific threat and the likelihood of that threat taking place.

risk analysis   The process of collecting information regarding the risks of an AV system and assigning values to each of those risks.

risk mitigation plan   The set of steps an organization takes to evaluate and prioritize its risks and implement countermeasures intended to reduce the likelihood of a risk occurring or its impact.

risk register   A document created to record the risks an organization faces to its AV and information systems. This information includes a detailed description of each risk, including its probability, severity and impact, and steps to mitigate it.

room criteria (RC) rating   A single number rating that quantifies the tonal characteristics of background noise, including the characteristics of ventilation noise, such as rumble or hiss.

room mode An acoustic wave-interference phenomenon that occurs between the parallel surfaces of an enclosure where the dimension between those parallel surfaces equals one-half wavelength (and the harmonics thereof) of a frequency. The wave is thus reflected on itself out of phase, creating location-specific areas of maximum and minimum pressure.

router A network device that forwards data packets between computer networks. It operates on the OSI Network layer (layer 3).

RS-232 A point-to-point serial data protocol. The interface between data terminal equipment and data circuit-terminating equipment employing serial binary data interchange. It supports a half-duplex mode of operation with one driver and one receiver. At a cable length of 15m (50ft), RS-232 supports a data rate of up to 19.2kbps. At its maximum cable length of 900m (3,000ft), it supports a data rate of 2.4kbps.

RS-422 A multidrop serial data protocol. Provides the electrical characteristics of balanced-voltage digital interface circuits. It is a balanced signal with one driver and up to 10 receivers with multidrop capability. The maximum cable length for RS-422 is 1,220m (4,000ft) with a data rate of 10Mbps.

RS-485 A transmission line serial data protocol. Supports a differential mode of operation with 32 tri-state drivers and 32 tri-state receivers and multidrop capability. The maximum cable length for RS-485 is 1.2km (4,000ft) with a data rate of 10Mbps.

RsGsBs A video transmission system using red, green, and blue signals with composite sync added to each color channel. This requires three cables to carry the entire signal. It is often referred to as *RGB sync on all three*.

RT_{60} The time taken for the energy in an initially steady reverberant sound field to decay by 60dB after the source of the sound ceases.

RU **1.** *See* rack unit (RU). **2.** A CTS renewal unit. The completion of renewal unit–accredited training is required by CTS, CTS-I, and CTS-D certification holders to maintain the currency of their certification.

sabin An index of sound absorption efficiency. Possible values range from 1.00 (absorbs everything) to 0.00 (absorbs nothing).

sampling rate The number of samples taken per unit of time when converting a continuous analog signal to a digital signal.

scale The representation of a number by another number differing by a fixed ratio.

scale drawing A drawing that shows objects in accurate proportion, with all dimensions enlarged or reduced by a fixed ratio.

scaler A processor that changes the resolution of an image without changing its apparent content. Scaling may be required when the image resolution does not match the resolution of the display device.

scan rate The rate at which a raster-scan image is displayed. Horizontal scan rate is the rate at which a single horizontal line is displayed. Vertical scan rate (refresh rate) is the rate at which an entire screen image is displayed.

scattering When a wave strikes a textured surface, the incoming waves are reflected in multiple directions because the surface is uneven.

scene A recallable preset or state of lighting levels for one or more zones.

scope statement A written agreement between the client, the project sponsor, the key stakeholders, and the project team that defines the boundaries of the project.

SCR The silicon-controlled rectifier (SCR) is a semiconductor from the thyristor family of four-layer power-switching devices. As it is a rectifier, it must be used in an inverse-pair to control both the positive and negative half-cycles of mains power. It was the first type of solid-state device to be used for power control and phase-control dimming.

screen gain The ability of a projection screen to concentrate the light reflected from it. Gain is compared to the reflection of a matte-white screen, which reflects light uniformly in all directions (a gain of 1).

SDTV Standard-definition television. It has a 4:3 aspect ratio and a resolution of 576i for PAL and SECAM and 480i for NTSC.

section drawing A section drawing is a view of the interior of a building in the vertical plane. Section drawings show a bisected wall, which allows you to view what is behind it.

security policy A set of rules that establish the acceptable use of an organization's AV and IT systems based on the risk that each system faces and the measures taken to address those risks.

sensitivity specification The measure of a device's capacity to convert one form of energy into another form of energy. Usually stated as a ratio of input units to output units, such as mV/Pa for microphones.

serial digital interface (SDI) A set of serial data standards to transport digital video data over BNC-terminated 75Ω coaxial cable or over optical fiber. Variants include HD-SDI for HD video, 6G-SDI (6Gbps) for 4kp30, 12G-SDI (12Gbps) for 4kp60, and 24G-SDI (24Gbps) for 8kp30 video.

series circuit A circuit arrangement where all the current supplied by the source will flow through the entire circuit. While all the current flows through all the circuit elements, the voltage is divided between the loads and the wires that connect them.

series/parallel loudspeaker circuit An arrangement of loudspeakers where some are wired in parallel and some are wired in series. Typically, groups of loudspeakers are wired in parallel, then those groups are connected to the amplifier in series.

server A computer system that shares resources and services with other connected devices.

service-level agreement (SLA)   A document used to record an agreement between a service provider and a customer. It describes the services to be provided, documents service-level targets, and specifies the roles and responsibilities of the service provider(s) and the customer(s).

session border controller (SBC)   A device used in VoIP networks to control the signaling and media streams involved in setting up, conducting, and tearing down calls. The SBC enforces security, QoS, and admission control over VoIP sessions.

shear force   The force exerted on an object in the direction of the object's cross section. In the case of a wall-mounting bolt, the shear force across the bolt caused by the gravitational load of the object it is supporting.

shield or screen   A grounded conductive partition placed between two regions of space to control the propagation of electric and magnetic fields between them. The screen or shield acts as a Faraday cage. It can be the chassis (metallic box) that houses an electronic device or the conductive enclosure (aluminum foil, conductive polymer, or copper braid) that surrounds a screened/shielded wire or cable.

shielded twisted-pair cable   A cable that contains one or more twisted-pairs of conductors inside an overall shield. Known variously as STP, or FTP when the shield is constructed of foil.

shotgun microphone   A long, cylindrical, highly sensitive microphone with a very narrow pickup pattern.

sightline   The line of sight between a viewer and an object that needs to be seen.

signal flow   The traceable path of signals through a system.

signal generator   A test instrument that produces calibrated electronic signals for the testing or alignment of electronic circuits or systems.

signal ground   **1.** A 0V (zero volt) point of no potential that serves as the circuit reference. **2.** A low-impedance path for the current to return to the source.

signal-to-noise (S/N) ratio   The ratio, usually measured in decibels, between an information signal and the accompanying noninformation noise. The higher this ratio, the cleaner the signal.

Simple Network Management Protocol (SNMP)   A set of Internet Engineering Task Force (IETF) standards for network management.

single-phase power   Alternating-current electrical power supplied by two current-carrying conductors. This type of supply is used for residential and many light-commercial applications.

single-point ground (SPG)   In the context of IEEE Standard 1100, refers to implementation of an isolated equipment grounding configuration for the purposes of minimizing problems caused by circulating current in ground loops.

Glossary

socket   In a TCP/IP network, the combined port number, Transport-layer (layer 4) protocol identifier, and IP addresses of communicating end systems. A socket uniquely identifies a session of a given transport protocol.

software patch   A piece of computer code used to update an AV or IT program with the intention of fixing or improving it. The update may address known security vulnerabilities to limit their exploitation by intruders.

sound pressure level (SPL)   The effective pressure level of a sound, usually stated in relation to a reference pressure such as 20µPa (the threshold of human hearing). In the context of AVIXA Standard A102.01:2017, expressed in unweighted decibels.

sound reinforcement system   The combination of microphones, audio mixers, signal processors, power amplifiers, and loudspeakers that is used to electronically amplify and distribute sound.

source-specific multicast (SSM)   In data streaming SSM allows clients to specify the sources from which they will accept content. This has the dual benefits of reducing demands on the network while improving network security.

specification   A written, precise description of the design criteria for a device or a piece of work. Specifications define the level of qualitative and/or quantitative parameters to be met and the criteria for their acceptance. All specifications must be formulated in terms that are specific, measurable, verifiable, and unambiguous.

specular reflection   The mirror-like reflection of electromagnetic radiation, in which most of the radiation is reflected in a single direction.

speech privacy system   A sound system that adds background noise to an environment to raise the hearing threshold and thus make it more difficult to hear low-level sounds such as traffic noise, machinery, and distant human speech. Used to assist with speech privacy in open environments.

speech-reinforcement system   An audio system that reinforces or amplifies a presenter's voice.

splitter   An electronic device that splits a signal to route it to different devices.

spot photometer   Also known as a spot meter. A type of photometer used to measure illuminance over a narrow angle.

standing wave   The phenomenon that occurs between parallel reflecting surfaces (e.g., walls in a room or a loudspeaker cabinet) where the distance between those parallel surfaces equals one-half wavelength of a wave (and the harmonics thereof). The wave is thus reflected on itself out of phase, creating an interference pattern with location-specific areas of maximum and minimum amplitude.

star topology   A network topology where all network devices are connected to a central network device, usually a switch or bridge.

static IP address A permanently assigned IP address.

stereophonic An audio reproduction system where multiple outputs are designed to create the illusion of sound perspective.

streaming video/streaming audio Sequence of moving images or sounds sent in a continuous stream over the Internet and displayed by the viewer as they arrive. With streaming video or audio, a web user does not need to wait to download a large file before seeing the video or hearing the sound.

subnet A logical group of hosts within a local area network (LAN). A LAN may consist of a single subnet, or it may be divided into several subnets. Additional subnets may be created by modifying the subnet mask on the network devices and hosts.

subnet mask A number that identifies which part of an IP address corresponds to the network address and which corresponds to a host address on that network. In CIDR notation the mask is represented by a slash (/), followed by the number of bits as a decimal number (e.g., /24). In dotted-decimal notation the mask is represented by a dotted-decimal representation of the bit pattern for network and host address bits, where bits equal to 1 indicate that the corresponding bits in the IP address identify the network address, and bits equal to 0 identify the host address (e.g., 255.255.255.0, which is equivalent to /24). IP addresses with the same network identifier bits are on the same subnet.

subwoofer A loudspeaker that reproduces lower frequencies, typically 20Hz to 200Hz.

supercardioid polar pattern The exaggerated heart shape of the area where a highly directional microphone is most sensitive to sound.

surface-mount microphone Also called a *boundary microphone* or *pressure zone microphone*. A microphone designed to be mounted directly against a hard boundary or surface, such as a conference table, a stage floor, or a wall, to pick up sound.

surround-sound system An audio system that uses multiple channels to produce an acoustic experience where the sound appears to surround listeners.

switcher A device used to select one of several available signals.

switch-mode power supply A type of DC power supply that uses a switching regulator to control the output voltage.

sync Synchronization. The timing information used to coordinate signals and events.

system In the AV industry, a compilation of multiple individual AV components and subsystems interconnected to achieve a communication goal.

system black The lowest level of luminance a video system is capable of producing for its operating conditions.

system ground   The point at which the safety earth/ground for an electrical system is connected to the earth, usually through a highly conductive earthing/grounding spike driven into the ground (planet Earth).

tap   A connection to a transformer winding that allows you to select a different voltage from the transformer.

task lighting   Lighting directed to a specific surface or area that provides illumination for visual tasks.

tensile strength   The maximum force that a material can withstand before deforming or stretching.

threat   An individual, group, circumstance, or event that may exploit vulnerabilities in AV or information systems to adversely affect them.

threat/vulnerability model   A listing of AV services and devices that may contain vulnerabilities and therefore could be threatened with exploitation. The assessment should also account for threat agents who may attempt to compromise these systems.

three-phase power   Alternating-current electrical power supplied by three current-carrying conductors, each carrying an AC voltage with a phase offset of 120 degrees from one another. A fourth conductor, a neutral, is used as the return conductor. This type of supply is used for commercial and industrial applications.

threshold   The point at which a function or effect becomes active.

throw distance   The distance between a light source, such as a projector or luminaire, and a focusing surface, such as a stage or a screen.

Thunderbolt   Interface technology that transfers audio, video, power, and data over one cable in two directions. Thunderbolt versions 1 and 2 use the same connector as Mini DisplayPort (MDP), while Thunderbolt versions 3 and 4 use USB Type-C.

time code   A sequence of numeric codes generated at fixed intervals to provide a time synchronization signal. The Society of Motion Picture and Television Engineers (SMPTE) time code used throughout the AV and production industries uses an eight-digit data scheme, representing the hour, minute, second, and frame number for each frame of a video sequence. The SMPTE time code is encoded in a wide variety of formats, including being embedded in audio, video, and data streams.

time domain   A view of a signal as amplitude versus time. The display on a time-based oscilloscope shows the input signal in the time domain.

Time-Sensitive Networking (TSN)   The IEEE working group overseeing the Audio Video Bridging standard has been renamed Time-Sensitive Networking to reflect the standard's applicability to communication among different types of devices, such as network sensors. Also known as *Audio Video Bridging (AVB)*.

transduction The process by which one form of energy is changed into another.

transformer A passive electrical device that electromagnetically transfers energy between two AC circuits. Commonly constructed of at least two electrically isolated induction coils sharing a common core.

transient disturbance A momentary variation in a signal, such as a surge, spike, sag, dropout, or spurious noise.

transition-minimized differential signaling (TMDS) A technology for transmitting high-speed serial data. The signaling method used in the HDMI and DVI interfaces.

transmission The passage of a wave through a medium. Examples include the transmission of soundwaves through air and the transmission of electromagnetic waves through space.

Transmission Control Protocol (TCP) A connection-oriented, reliable Transport-layer (layer 4) protocol. TCP transport uses two-way communication to provide guaranteed delivery of information to a remote host. It is connection-oriented, meaning it creates and verifies a connection with the remote host before sending it any data. It is reliable because it tracks each packet and ensures that it arrives intact. TCP is the most common transport protocol for sending data across the Internet.

transmission loss The attenuation that occurs as a signal moves through a medium. Usually expressed in decibels.

transmissive technology Any display device that creates images by controlling the passage of light.

triac A bi-directional semiconductor device from the thyristor power-switching family that can switch both half-cycles of an AC current. The equivalent of a pair of SCRs in inverse parallel. The switching device used in many commercial phase-control dimmers.

tweeter A loudspeaker that is specifically designed to reproduce frequencies above 3kHz.

twisted-pair A pair of wires that are twisted around each other to facilitate common-mode noise rejection.

two-factor authentication An authentication method that combines two different validation methods to address a single authentication request.

two-way/three-way loudspeaker A loudspeaker enclosure containing two or three separate loudspeakers, each designed to optimally reproduce a portion of the audio spectrum. The multispeaker enclosure is intended to cover the entire spectrum from a single cabinet. Each loudspeaker may be fed from a separate amplifier (bi-amplification or tri-amplification), or the entire enclosure may be fed from a single amplifier with an internal crossover filter network used to send the optimal frequency band to each of the loudspeakers.

Glossary

UHD ecosystem The video cameras, recorders, servers, media players, displays, distribution, processing, and networking technologies used for recording, editing, producing, delivering, and displaying ultra-high-definition video.

ultra-high-definition (UHD or ultra HD) A term used to describe video formats with a minimum resolution of 3840×2160 pixels in a 16×9 or wider aspect ratio.

ultra-wideband (UWB) UWB is a low-power communications protocol that transmits an extremely wide bandwidth (500+MHz) signal using time modulation to encode the data. Its signals are highly immune to interference from other RF systems, yet produce very little interference with them. UWB can detect the distance between the transmitter and receiver, enabling proximity detection and triggering.

unbalanced circuit A circuit in which one conductor carries the signal and another conductor carries the return. The return conductor is often the cable shield or drain wire and is a low-impedance connection connected to the signal ground. As the impedances of the two conductors are quite different, they are unbalanced with respect to one another.

unicast streaming A one-to-one connection between the streaming server sending out the data and client devices listening to the stream. Each client has a direct relationship with the server. Since the server is sending out a separate stream for each client, unicast streaming of media to three clients at 100kbps actually uses 300kbps of bandwidth. IP unicast streams may use either UDP or TCP transport, although TCP transport will inevitably require some buffering.

unity gain Derived from the number 1; refers to no change in gain.

Universal Serial Bus (USB) A standard for connecting, communicating, and supplying power between electronic devices. Version 3.2 of USB is capable of communicating at 20Gbps and can utilize a USB Type-C (USB-C) connector, which supports DisplayPort, HDMI, VGA, power, USB 2.0, and USB3.2. USB4 includes handling the Thunderbolt protocol. USB4 version 1 communicates at up to 40Gbps, while version 2 communicates at up to 80Gbps in symmetric mode, and at up to 120/40Gbps in asymmetric mode.

unshielded twisted-pair (UTP) cable Typically used for data transfer, UTP cable contains multiple two-conductor pairs twisted at regular intervals, employing no external shielding.

UX User experience. How a user interacts with and experiences a system or an object.

vectorscope A specialized oscilloscope-type display used in video systems to display and measure chrominance accuracy and levels. A vectorscope mode may be included in video waveform monitoring systems.

vertical scan rate The number of complete fields a device draws in a second. This may also be called the *frame rate*, *vertical sync rate*, or *refresh rate*. The vertical scan rate is measured in hertz (Hz).

videowall   A video display composed of a matrix of smaller video displays linked to display a contiguous image.

viewing angle   The angle at which a viewer is located in reference to the center axis of a display.

viewing area plan   A plan-view drawing of the viewing environment that identifies five viewing locations as defined in the requirements section of the ANSI/AVIXA V201.01 Standard *Image System Contrast Ratio*.

viewing cone   The volume of space containing the audience viewing a display. The term *cone* is used because there are width, height, and depth to the viewing space, which emanate from the center of the display.

virtual local area network (VLAN)   A network created when network devices on separate LAN segments are linked to form a logical group.

virtual private network (VPN)   A virtual point-to-point private connection established across a network via an encrypted tunneling protocol. VPNs are used for secure remote access, monitoring, troubleshooting, and control.

visual acuity   The eye's ability to discern fine details. There are several different kinds of acuity, including resolution acuity, which is the ability to detect that there are two stimuli, rather than one, in a visual field, and recognition acuity, which is the ability to identify correctly a visual target, such as differentiating between a *G* and a *C*.

visual field   The volume of space that can be seen when a person's head and eyes are absolutely still. It is specified as an angle, usually in degrees. The visual field of a single eye is termed *monocular vision*, and the combined visual field where the perceived image from both eyes overlap is called *binocular vision*.

Voice over Internet Protocol (VoIP)   A suite of technologies and protocols that allow the transmission of telephone calls and multimedia over Internet Protocol (IP) networks.

volatile memory   Computer memory that loses its data when no longer powered. This type of memory includes the fast-access, dynamic random-access memory (DRAM) associated with most processor CPUs and GPUs.

volt   The basic international unit of potential difference or electromotive force. It is represented by the symbol V.

voltage   The electrical potential difference across a circuit.

vulnerability   A weakness in a system, such as an AV or IT service, that may be exploited to adversely affect an organization.

vulnerability testing   An assessment process that involves tools that connect to networked AV or IT systems to determine how they are configured and what vulnerabilities may exist on them.

watt The international unit of power. It is represented by the symbol W.

waveform monitor An oscilloscope-type monitor used to display the waveforms of signals. A video waveform monitor is a specialized monitor used to display and analyze a video signal's sync, luminance, and chroma. Some waveform monitors include a *vectorscope* mode.

wavelength The distance between the corresponding points on two consecutive cycles of a wave. Measured in meters.

wayfinding The use of audiovisual guides or signage to assist with navigation to a destination.

webcasting The broadcasting of digital media such as audio or video over the World Wide Web, which audience members can stream live or access on demand.

white noise A signal with a broad spectrum of random frequencies at the same energy level.

wide area network (WAN) A data communications network that links local area networks (LANs) that are distributed over large geographic areas, such as cities, states, countries, regions, and continents.

wire A single conductive element intended to carry a current.

wireless access point A network device that allows other devices to access a wireless network.

wireless local area network (WLAN) A network that shares information by radio frequency (RF) wireless transmission.

woofer A loudspeaker that produces low frequencies, typically 20Hz to 200Hz.

work breakdown structure (WBS) A deliverable-oriented grouping of project elements that will ultimately organize and define the total scope of the project. Each descending level represents an increasingly detailed definition of a project component.

XLR connector A secure, low-voltage connector used in professional audiovisual systems. The three-pin version is the standard audio signal cable for the production and AV industries. The four-pin version is widely used for communication headsets, and the five-pin version is the standard connector for the DMX512 digital lighting protocol. Also known as a *cannon connector*.

zone **1.** A defined area within a system. **2.** In lighting, a zone is a grouping of luminaires (lighting fixtures) that are focused on the same area. **3.** In digital signage, a zone is an area where specific content may be placed. **4.** In audio, a zone is an area where the same program is delivered.

INDEX

A

A-weighting in sound pressure levels, 141, 514–515
AACS (Advanced Access Content System), 469
abbreviations in architectural drawings, 50–51
absorption of sound, 306–307
    absorption coefficient, 308
    in air, 308
    noise reduction coefficient, 307–308
    porous absorbers, 308–309
    resonant absorbers, 309
    sabin, 307
    sound absorption average, 308
AC. *See* alternating current (AC)
acceptance strategy for risk handling, 498
access control in network security, 496
access control lists (ACLs), 455
access issues with lecterns, 91
accounting in access control, 496
ACLs (access control lists), 455
acoustic echo, 383
acoustic mass law in sound transmission, 309–311
acoustic S/N ratio, 175
acoustics, 295
    acoustic engineering property, 296
    duty check, 296
    interaction property, 296, 299–300
    questions, 318–319
    review, 318
    sound absorption, 306–309
    sound production, 296–297
    sound propagation, 297–299
    sound reception, 313–317
    sound reflections, 300–306
    sound transmission, 309–313
active analog loudspeaker wiring, 148
active crossover filters, 386
active video signal extenders, 518–519
acuity
    text size, 99
    visual, 97, 99–100
ADA (Americans with Disabilities Act)
    CTS-D exam, 24
    provisions, 230–232
addition in mathematical formulas, 14–15
addresses
    IP. *See* IP addresses
    MAC, 413, 426, 443

ADM. *See* analytical decision-making (ADM)
administration considerations in streaming reflectors, 486
Advanced Access Content System (AACS), 469
Advanced Video Coding (AVC), 478–479
AES67 interoperability, 458
AES67 standard, 458–461
affordance factor in human-centered interface design, 398
AGC (automatic gain control)
    cameras, 522
    description, 377
    videoconference lighting, 224
AHJ (authority having jurisdiction)
    codes prescribed by, 233, 291, 509
    electrical infrastructure, 264–265
air, absorption of sound in, 308
air-gapped networks, 501
aligned seating arrangements, 86–87
allied trades
    communicating with. *See* communicating with allied trades
    description, 188
alternating current (AC)
    vs. DC, 256–257
    dimmers, 218–219
    ground loops, 286
    impedance, 147, 258–261
    inductance, 260–261
    measuring, 516
    Ohm's law, 147
    overview, 257
    parallel circuits, 262
    power-generation stations, 265
    resistance, 258–261
    series circuits, 261–262
    space layout planning, 60
    three-phase, 274
    three-to-two-pin adapter dangers, 271
ambient level control, 377
ambient light
    projector screens, 125, 127
    rear projection displays, 111
    space lighting, 207
ambient light rejection in front projection displays, 109
American National Standards Institute (ANSI)
    AV drawing symbols, 51–53
    AVIXA accreditation, 4–5, 11
    regional regulations, 510

Americans with Disabilities Act (ADA)
    CTS-D exam, 24
    provisions, 230–232
amplifiers
    heat loads, 247
    impedance, 147
    loudspeaker systems, 155–158
amplitude of sound, 298–299
AMX SVSI technology, 460
analog audio
    vs. digital, 368–369
    signal monitoring, 372–373
analog loudspeaker wiring, 148
analog voltage control points, 399
analytical decision-making (ADM)
    DISCAS calculations, 100
    image size, 97–98
    image system contrast, 527
    viewing angles, 101–102
    viewing distance, 103
angles, viewing, 101–102
ANSI (American National Standards Institute)
    AV drawing symbols, 51–53
    AVIXA accreditation, 4–5, 11
    regional regulations, 510
ANSI/ASA standards for sound pressure levels, 514, 516
ANSI/AVIXA standards
    Audio Coverage Uniformity, 506
    Audiovisual Systems Energy Management, 252
    Audiovisual Systems Performance Verification, 38, 196, 405, 511, 517, 532–533
    Display Image Size for 2D Content in Audiovisual Systems (DISCAS), 84, 97–98
    Image System Contrast Ratio (ISCR), 97, 121, 207, 527
    Recommended Practice for Lighting Performance for Small to Medium Sized Videoconferencing Rooms, 211, 221
    Sound System Spectral Balance, 506
    Standard for Audiovisual Systems Performance Verification, 509
    Standard Guide for Audiovisual Systems Design and Coordination Processes, 33, 196, 506–508
ANSI brightness for projectors, 128–129
ANSI/IES TM-30-20 (TM-30) standard, 205
ANSI-J-STD-710 standard, 51–53
anthropometrics, 85
any-source multicast (ASM), 483
APIPA (Automatic Private IP Addressing), 445–446
Application layer
    functions, 452
    OSI Model, 427
application process for CTS-D exam, 20–24
applications
    fiber-optic cable, 419–420
    videowalls, 116
    VLANs, 434
approval in concept design/program report sections, 75–76
architects, collaboration with, 189

architectural drawings, 55–58
    abbreviations, 50–51
    symbols, 46, 48–50
ARCs (audio return channels), 343
ARIB (Association of Radio Industries and Businesses)
    test pattern, 527
array microphones, 162
artificial intelligence–based voice assistant platforms, 395
ask questions step in concept design/program report development, 66–67
ASM (any-source multicast), 483
aspect ratio
    calculating, 105–106
    description, 105
    display setup, 524–525
    EDID, 352
    media formats, 104
    screen diagonal, 106
assessment, risk, 497–498
assigning IP addresses, 443–448
Association of Radio Industries and Businesses (ARIB)
    test pattern, 527
ASTM International, mounting considerations, 238
ATSC IS-19 sync standard, 530
attack time
    compressor settings, 378–379
    description, 377
    gate settings, 381
    limiter settings, 380
attenuation in equalization filters, 384
audience
    loudspeaker system coverage, 144
    streaming design, 467–468, 470
audience area, 60–61
audio. *See also* sound
    acoustics. *See* acoustics
    CTS-D exam, 4
audio bandwidth for digital signals, 324
audio CD sampling rates, 324
audio compression, 330
audio coverage in presenter area, 60
Audio Coverage Uniformity standard, 506
audio design, 367–368
    analog vs. digital, 368–369
    DSP architectures, 369–371
    duty check, 368
    equalization, 383–389
    questions, 389–391
    review, 389
    signal monitoring, 371–373
    system setup, 373–383
Audio Engineering Society (AES67) interoperability, 458
audio principles, 135
    background noise, 140
    decibels, 136–139
    duty check, 136
    loudspeaker systems. *See* loudspeaker systems
    microphones. *See* microphones
    perceived sound pressure level, 139–141

Index

quality, 174–180
questions, 181–183
review, 180–181
transformers, 153–154
audio return channels (ARCs), 343
audio signal generators, 512–513
audio system quality, 174
    acoustic gain, 176–180
    intelligibility, 175–176
    loudness, 174–175
    stability, 176
audio system verification, 511
    audio signal generators, 512–513
    sound pressure level meters, 513–516
    testing tools, 511–517
Audio Video Bridging (AVB) suite, 457–458
audio/video sync, 530–532
Audiovisual Systems Design and Coordination Processes Checklist, 506–507
Audiovisual Systems Energy Management standard, 252
Audiovisual Systems Performance Verification standard, 38, 196, 405, 511, 517, 532–533
auditing, 501
authentication, 499–500
    access control, 496
    HDCP. *See* high-bandwidth digital content protection (HDCP)
    user accounts, 500
authority having jurisdiction (AHJ)
    codes prescribed by, 233, 291, 509
    electrical infrastructure, 264–265
authorization, 496
auto-iris systems for videoconference lighting, 223–224
automatic exposure controls
    videoconference lighting, 223–224
    wall and table finishes, 224–225
automatic gain control (AGC)
    cameras, 522
    description, 377
    videoconference lighting, 224
automatic mixers for microphones, 168–169
Automatic Private IP Addressing (APIPA), 445–446
automation for energy management, 252
AV circuits, 276
    branch circuit loads, 277–278
    documentation reviews, 277
    number calculations, 278–280
AV Design Levels 1–3 Online course, 11
AV design package, 54
    architectural and infrastructure drawings, 55–58
    front-end documentation, 54–55
    system drawings, 58–59
AV drawing symbols, 51–53
AV-enabled spaces, 60
    audience and presenter areas, 60–61
    control and projection areas, 62–63
AV managers, needs analysis for, 65–66
AV Math for Design Online course, 11
AV Math Online course, 11, 17

AV system drawings, 58–59
availability in network security, 492
AVB (Audio Video Bridging) suite, 457–458
AVC (Advanced Video Coding), 478–479
AVIXA
    background, 3–4
    standards. *See* ANSI/AVIXA standards
AVIXA RP-C303.01 standard, 502
AVIXA Testing page, 22
avoidance strategy for risk handling, 498
AVoIP media server–based control systems, 394
axial modes in sound reflection, 303

B

B (bel) unit, 136
B-Frames (bi-directional Frames) in interframe video compression, 331
back-focus in cameras, 520
Back Frames in interframe video compression, 331
background noise
    HVAC, 249
    measuring, 140
    sound reception, 316–317
backlight
    camera adjustments, 522
    LCD displays, 107
    videoconferences, 223
balanced and unbalanced circuits, 286–288
balloon plots for loudspeaker systems, 142–143
balun technology for passive signal extenders, 518
bandpass filters
    description, 386
    series circuits, 262
bandwidth
    audio, 324
    cable categories, 415–416
    codecs, 452
    digital signals, 324–327
    Ethernet networks, 432
    multicast streaming, 482
    networks, 452–454
    streaming design, 472–474, 476
    streaming reflectors, 486
    topologies, 475
    video, 325–327
    VPNs, 436
basic decision-making (BDM)
    image system contrast, 527
    viewing angles, 101
    viewing distance, 102–103
    viewing requirements, 96–98
bass traps, 302
BD+ content protection, 469
BDM. *See* basic decision-making (BDM)
beamforming array microphones, 162
bel (B) unit, 136
benchmarking evaluation in concept design/program report development, 68–71

BER (bit error rate) of signals, 340
best effort classes, 453
best practices
    codes as, 230
    CTS-D exam, 12
    equipment manuals and software documentation, 536
    network security, 502
    power cable management, 281
    regional regulations, 509–510
    troubleshooting, 536–537
    user experience design, 401
bezels for videowalls, 117
bi-directional Frames (B-Frames) in interframe video compression, 331
bid response forms, 54
bidirectional microphone pickup patterns, 163
BIM (building information modeling), 56
binocular vision, 80
bit depth
    audio, 324
    signal monitoring, 373
    video, 325–327
bit error rate (BER) of signals, 340
black level
    rear projection displays, 112
    system, 120
blackout drapes, 206–207
blended network devices, 414
blip-and-flash tests, 530–532
block diagrams for racks, 245–246
blocking for wall mounting, 234–235
Blu-ray discs audio sampling rates, 324
BlueRiver Field-Programmable Gate Array (FPGA), 460–461
BODMAS acronym, 14
bolts for mounting, 237
boost shelves, 386
boundary microphones, 161
braided cable shields, 285
branch circuits
    loads, 277–278
    loudspeaker, 151–152
bridges in networks, 415
briefing phase in AV design projects, 35–36
brightness
    lighting specifications, 200
    OFE projectors, 128–129
    projectors, 124–128
bring your own device (BYOD) systems, 356
broadcast addresses, 442
broadcast messages, 433
broadcast network control points, 400
broadcast video DiffServ classes, 453
budget recommendations in concept design/program report sections, 74–75
buffering in unicast streaming, 480
building by committees, 66
building codes, 232–233
building information modeling (BIM), 56
building materials in sound transmission, 312

building professionals, collaboration with, 189
built-in control systems, 394
buyers, needs analysis for, 66
buzz, electrical, 282
BYOD (bring your own device) systems, 356
bytes
    IPv4 addresses, 438
    IPv6 addresses, 442

C

C-weighting in sound pressure levels, 514–515
cable
    cliff effect, 339
    fiber-optic, 417–421
    shielding, 285
    twisted-pair, 415–417
cable support systems
    cable ducts, 289
    cable trays, 288
    conduit, 289
    hook suspension systems, 289–290
CAD drawings
    AV design, 56
    scaled, 41
calculator use in CTS-D exam, 13–14
cameras
    adjustments, 519–523
    color balance, 224
    color rendering index, 203–204
    color temperature, 203, 211
    glare, 221–222
    right color of light, 211
    videoconference lighting, 221–224
    wall mounting, 234
campus area networks (CANs), 434
candela units, 118, 120
capacitance
    condenser microphones, 159
    description, 258–259
    electric-field coupling, 284
capacitive reactance
    description, 258
    impedance, 147
capacitors
    condenser microphones, 159
    description, 258–259
    electric-field coupling, 284
    parallel circuits, 262–263
    series circuits, 261–262
capacity of conduits, 290–292
cardioid pickup patterns for microphones, 163
carpentry work in detail drawings, 57–58
carts, 91
casework in detail drawings, 57
categories of cable, 415–416
CBR (constant bit rate) encoding, 330
CBRs (core balance relays), 265, 270
CEA (Consumer Electronics Association)
    AV drawing symbols, 51

CEC (Consumer Electronics Control) data, 343
CEDIA (Custom Electronic Design and Installation Association) AV drawing symbols, 51
ceiling plans, 42, 44
ceilings
    loudspeaker system coverage, 144
    reflectance values, 225
    sound reception, 315–316
CEN (European Committee for Standardization), 232
central processing units (CPUs)
    configurations, 402–403
    control requirements, 395–396
centralized CPU configurations, 402
Certified Technology Specialist (CTS), 3–5
Certified Technology Specialist-Design (CTS-D)
    AVIXA administration, 3–4
    certification benefits, 4–6
    exam eligibility, 7
    review, 7
    tasks, 6–7
Certified Technology Specialist-Installation (CTS-I), 5
chairs
    AV design projects, 89–90
    reflectance values, 225
charts
    EAD, 177–178
    eye charts, 99–100
    flip, 91
    focus, 520
    Gantt, 190–193
    Macbeth, 203
chemical fire protection systems, 251
chroma levels in display setup, 526–527
chromaticity diagrams, 201–202
CIA security triad, 492
CIDR (Classless Inter-Domain Routing) notation, 439
CIE (International Commission on Illumination)
    color rendering index, 204
CinemaScope display format, 338
circuit breakers, 265–268, 270
circuit theory, 256
    alternating current, 257
    direct current, 256–257
    impedance and resistance, 258–261
    parallel circuits, 262–263
    series circuits, 261–262
circuits, balanced and unbalanced, 286–288
circulation space, 90
cladding for fiber-optic cable, 418
Class 1 sound pressure levels, 514
Class 2 sound pressure levels, 514
Classless Inter-Domain Routing (CIDR) notation, 439
client mission in security posture, 494
client risk profile in security posture, 494–495
client-server CPU configurations, 402
clients
    needs analysis, 64–65
    networks, 412–413
    sign-off, 539

cliff effect, 339–340
closeout
    client sign-off, 539
    customer training, 537–538
    documentation, 532–536
    drawings of record, 535
    troubleshooting, 536–537
    verification, 509
closest viewers factor in viewing distance, 102
clothing considerations for microphones, 172–173
cloud-based controls, 394
CMR (common mode rejection), 287
CMRR (common-mode rejection ratio), 288
coatings for fiber-optic cable, 418
COBIT (Control Objectives for Information and Related Technology), 495
Code of Ethics and Conduct, 5
codecs, 328–329, 451–452
codes and regulations, 230–231
    electric and building codes, 232–233
    electrical infrastructure, 264–265
    equal access, 231–232
    regional, 509–510
collaboration AV experience type, 32–33
colons (:) in IPv6 addresses, 442
color
    light, 211–212
    YUV processing system, 326
color balance for videoconferences, 224
color bars for display setup, 526–527
color checker charts, 203
color discrimination limits, 81
color LCD displays, 107
Color Quality Scale (CQS), 204
color rendering index (CRI), 203–205
color schemes, 58
color temperature
    cameras, 211
    lighting specifications, 201–203
column system symbols in architectural drawings, 46, 48
comb filters for sound reception, 314–315
combination cable shields, 285
comfort concerns, 60–61
comfortable head movement, 83
committees, building by, 66
common-carrier topologies in WANs, 435
common mode rejection (CMR), 287
common-mode rejection ratio (CMRR), 288
communicating design intent, 31
    AV design package, 54–59
    AV design project phases, 33–38
    AV-enabled spaces, 60–63
    AV experience components, 33
    AV experience types, 32–33
    concept design/program report development, 66–72
    concept design/program report sections, 72–76
    construction drawings. *See* construction drawings
    duty check, 32
    needs analysis, 63–66

communicating design intent (cont.)
  programming, 63
  questions, 76–78
  review, 76
communicating with allied trades, 187
  consultation hierarchy, 195–196
  duty check, 187–188
  industry standards, 193–194
  integrators, 196
  project tracking, 189–193
  questions, 197–198
  review, 196
  stakeholders, 188–189
compander settings in DSPs, 380
compression of digital signals, 328–331
compressor settings in DSPs, 377–379
compressors, HVAC noise from, 249
computer-based calculator use in CTS-D exam, 13–14
computer-driven control systems, 394
computer graphics video formats, 325
concept design phase in AV design projects, 35–36
concept design/program report development, 66
  creation step, 71–72
  documentation review step, 67–68
  program meetings, 71
  questions step, 66–67
  site environment/benchmarking evaluation, 68–71
concept design/program report sections, 72
  budget recommendations, 74–75
  distribution and approval, 75–76
  executive summaries, 73
  infrastructure considerations, 74
  space planning, 73
  system descriptions, 73
concept of operations in security posture, 494
condenser microphones, 158–159
conductors
  dynamic microphones, 158
  grounding, 268–269
  three-phase power distribution, 273–274
conduits
  cable, 289
  capacity, 290–292
  jam ratio, 291–292
conference tables, microphone placement on, 169–170
conferences
  color balance, 224
  glare, 222–223
  latency, 475
  light balance, 223–224
  lighting, 221–225
confidentiality in network security, 492
configuration
  CPUs, 402–403
  streaming reflectors, 487
conformance, ISCR, 121–122
connection points in audience and presenter areas, 60
connectionless protocols, 448–449
connectors
  fiber-optic cable, 420–421
  USB, 345–347

constant bit rate (CBR) encoding, 330
constant current electronic dimmers, 220
constant voltage amplifiers, 157
constant voltage electronic dimmers, 219
constraints
  human-centered interface design, 398
  security posture, 495
construction drawings, 34, 38–39
  architectural drawing abbreviations, 50–51
  architectural drawing symbols, 46, 48–50
  AV drawing symbols, 51–53
  scaled drawings, 39–41
  types, 42–47
construction phase in AV design projects, 37
construction techniques in sound transmission, 312
consultation hierarchy, 195–196
consumer audio decibel levels, 139
Consumer Electronics Association (CEA)
  AV drawing symbols, 51
Consumer Electronics Control (CEC) data, 343
contact closure control points, 399
contactors, 218
containers in digital media formats, 329
content component in AV experience, 33
content compression and encoding, 328–329
content developers, collaboration with, 189
content sources for streaming design, 468–471
contract representatives, needs analysis for, 66
contrast
  display setup, 526
  image systems, 527–529
  rear projection displays, 112
contrast ratio
  display environment, 120–121
  standard, 207
control area, 62–63
control functions scripts, 403–404
Control Objectives for Information and Related Technology (COBIT), 495
Control of Noise at Work Regulations, 230
control panels, 397
control property for sound, 296
control requirements, 393
  central processing unit configurations, 402–403
  central processing units, 395–396
  control points, 399–400, 405
  design for user experience, 401
  duty check, 393
  interfaces, 396–399
  needs analysis, 401–402
  programming, 403–404
  questions, 405–407
  review, 405
  system performance verification, 405
  system types, 394–395
converged network bandwidth, 472
converting dimensions, 40–41
cooling
  control and projection areas, 62
  racks, 247–249

Index

copyrighted content in streaming design, 469–470
core balance relays (CBRs), 265, 270
cores in fiber-optic cable, 418
correlated color temperature, 202
costs
    budget recommendations, 74–75
    control needs analysis, 402
    rear projection displays, 112
    wireless connections, 424
counters, 90
coverage in loudspeaker systems, 142–146
CQS (Color Quality Scale), 204
Crestron DM NVX technology, 460
CRI (color rendering index), 203–205
critical frequency in sound reflection, 303–304
crosshatch patterns for display setup, 524–525
crossover
    filters, 386–387
    parallel circuits, 262–263
    series circuits, 262
crosstalk
    cable categories, 415–416
    induced currents, 261
    racks, 244
CTS-D. *See* Certified Technology Specialist-Design (CTS-D)
CTS-D Candidate Handbook, 6–7, 11, 20–21, 24
CTS-D Certified Technology Specialist-Design Exam Guide, 9–10, 12
CTS-D exam
    application process, 20–24
    content outline, 10
    contents, 21–22
    dismissal or removal from, 23
    electrical calculations, 16–19
    eligibility, 7
    guidelines, 22–23
    hazardous weather and local emergencies, 23
    identification requirements, 21
    mathematical strategies, 13–15
    practice questions, 25–26
    preparation strategies, 11–12
    restricted items, 21
    retesting, 24
    review, 27
    scope, 9–10
    scoring, 24
    special accommodations, 24
    testing center arrival time, 20
CTS-I (Certified Technology Specialist-Installation), 5
current
    branch circuit loads, 277–278
    electric-field coupling, 284
    induced, 261
    magnetic-field coupling, 283
    number of circuits calculations, 278–280
    Ohm's law, 16–19
    power distribution systems, 274
custom-built control applications, 395
custom carpentry work in detail drawings, 57–58
Custom Electronic Design and Installation Association (CEDIA) AV drawing symbols, 51
customer training, 537–538

D

Dante protocol, 458–459
data
    control and projection areas, 62
    presenter area, 61
    security posture, 494
Data Link layer in OSI Model, 426
data tables in EDID, 353–354
data throughput
    bandwidth, 452
    digital audio signals, 324
data transmission in OSI Model, 427–428
dates on architectural drawings, 49–50
dB. *See* decibels (dB)
DC (direct current)
    circuit theory, 256–257
    dimmers, 219–220
    impedance, 258
DCI 4K (Digital Cinema Initiatives 4K), 338
DDC (Display Data Channel), 349–350
DDNS (dynamic DNS), 447
decibels (dB)
    change calculations, 137
    overview, 136–137
    reference levels, 137–139
    sound, 299
decibels per octave in equalization filters, 384
decoding, 328
default passwords, 499
delay
    audio/video sync, 530–532
    DSP settings, 382–383
    reflections, 314–315
deliverables in logic network diagrams, 192–193
delivery in rear projection screens, 113
dense mode in multicast streaming, 483
design intent, communicating. *See* communicating design intent
design phase in AV design projects, 36–37
design specifications, 34
desktop video, latency in, 475
destinations in TCP transport, 449
detail drawings
    carpentry work, 57–58
    description, 46
    detail flags, 49
    example, 48
details in visual acuity, 97
device authentication in HDCP, 358
device automation in control needs analysis, 401
device limits in HDCP, 361
devices, network, 414–415
DHCP (Dynamic Host Configuration Protocol), 444–445
dielectric strength in electric-field coupling, 284
differentiated service (DiffServ) class, 453

differentiated service (DiffServ) QoS, 472
diffuse reflections in sound, 300–302
diffusers
    HVAC noise, 249–250
    sound, 301–302
diffusion screen material for rear projection displays, 112–113
digital audio
    vs. analog, 368–369
    compression, 330
    signal monitoring, 372–373
Digital Cinema Initiatives 4K (DCI 4K), 338
digital light processing (DLP) projectors, 160
digital micromirror devices (DMDs), 160
digital-only tokens (DOTs) in HDCP, 361
digital photos in CTS-D exam, 21
Digital Rights Management (DRM)
    copyrighted content, 469
    description, 357
    HDCP. *See* high-bandwidth digital content protection (HDCP)
digital signal processors (DSPs)
    architectures, 369–370
    audio systems, 511
    audio/video sync, 531–532
    common settings, 377
    compressor settings, 377–379
    delay, 382–383
    echo cancelers, 383
    expander settings, 380
    gate settings, 381
    limiter settings, 379–380
digital signals, 323–324
    audio bandwidth, 324
    content compression and encoding, 328–331
    digital audio compression, 330
    digital media formats, 329
    digital video compression, 330–331
    duty check, 323
    questions, 332–333
    review, 332
    video bandwidth, 325–327
digital video design, 335
    basics, 336
    cliff effect, 339–340
    DRM, 357–362
    duty check, 336
    EDID, 348–356
    high-definition and ultra high-definition video, 337–339
    questions, 363–365
    review, 362–363
    video signal types, 340–348
Digital Visual Interface (DVI), 342
dimension conversions, 40–41
dimmers
    phase distortion, 274
    types, 218–220

direct current (DC)
    circuit theory, 256–257
    dimmers, 219–220
    impedance, 258
direct glare in videoconferences, 222
direct light measurement, 119
direct sound, 300
direct sound reflections, 300
direct-to-reflected sound ratio, 175
direct-view displays
    color temperature, 203
    types, 107–108
direction in space lighting, 212–213
directional microphones, 173
directly connected loudspeakers, power amplifiers for, 157
directory servers, 500
Disability Discrimination Act, 230
disability glare in luminance, 130
disabling unnecessary services, 501
discharge lamps, 208
discoverability expectations for devices, 398
dismissal from exam, 23
displacement amplitude of sound, 299
Display Data Channel (DDC), 349–350
display device selection, 103–104
    aspect ratio, 105–106
    video resolution, 104
display environment
    contrast ratio, 120–121
    Image System Contrast Ratio (ISCR) standard, 121–122
    light measuring, 118–120
    system black level, 120
Display Image Size for 2D Content in Audiovisual Systems (DISCAS), 84, 97–98
display types, 106–107
    direct-view, 107–108
    front projection, 108–111
    rear projection, 111–114
    videowalls, 115–118
DisplayID, 348
DisplayPort
    description, 344–345
    signal extension, 518–519
    Thunderbolt technology, 347
displays
    EDID, 356
    native resolution, 336
    setup, 524–529
    visual principles. *See* visual principles
distances
    control needs analysis, 402
    decibels, 137
    fiber-optic cable limitations, 419
    viewing, 101–103
distributed audio systems, power amplifiers for, 157–158
distributed loudspeaker systems
    description, 142
    layout options, 145–146

distributed processing CPU configurations, 403
distribution and approval in concept design/program
    report sections, 75–76
division in mathematical formulas, 14–15
DLP (digital light processing) projectors, 160
DM NVX technology, 460
DMDs (digital micromirror devices), 160
DNS (Domain Name System), 446–447
    dynamic, 447
    internal, 447–448
documentation
    energy management, 252
    front-end, 54–55
    luminaires, 216–217
    system closeout, 532–536
documentation reviews
    AV circuits, 277
    concept design/program report development, 67–68
documents in design phase, 36
Domain Name System (DNS), 446–447
    dynamic, 447
    internal, 447–448
DOTs (digital-only tokens) in HDCP, 361
dotted-decimal notation in IP addresses, 438
downward expander settings in DSPs, 380
drapes, 206–207
drawings
    construction. *See* construction drawings
    construction phase, 37
    design phase, 36
drawings of record in system closeout, 535
dressing lights for videoconferences, 223
DRM (Digital Rights Management)
    copyrighted content, 469
    description, 357
    HDCP. *See* high-bandwidth digital content protection (HDCP)
dry fire protection systems, 251
DSPs. *See* digital signal processors (DSPs)
dual-bus architecture, 346
ducts
    cable, 289
    HVAC noise from, 249
duties
    exam contents, 9–10
    JTA, 6–7
Duty A, 12
Duty B, 12
Duty C, 11
duty check
    acoustics, 296
    audio design, 368
    audio principles, 136
    communicating design intent, 32
    communicating with allied trades, 187–188
    control requirements, 393
    digital signals, 323
    digital video design, 336
    electrical infrastructure, 256

ergonomics, 79
lighting specifications, 200
networks, 410
project implementation activities, 506
security for networked AV applications, 491
streaming design, 466
structural and mechanical considerations, 230
visual principles, 95–96
duty cycles in power amplifiers, 247
DVD audio sampling rates, 324
DVI (Digital Visual Interface), 342
dynamic DNS (DDNS), 447
dynamic filters in DSPs, 381
Dynamic Host Configuration Protocol (DHCP), 444–445
dynamic IP addresses, 443–444
dynamic microphones, 158
dynamic ports, 450
dynamic range
    bit depth, 326
    signal monitoring, 373

E

E-EDID (Enhanced EDID), 348
EAD (equivalent acoustic distance), 177–179
ear-worn microphones, 172
earth faults, 269–271
earth leakage circuit breakers (ELCBs), 265, 270
earth-leakage protection devices, 270–271
EASE (Enhanced Acoustic Simulator for Engineers), 142–143
ease of movement in audience area, 61
EBU Recommendation R37 sync standard, 530
echo cancelers in DSPs, 383
echoes in sound reflection, 302
edge-to-center coverage in loudspeaker systems, 146
edge-to-edge coverage in loudspeaker systems, 145–146
EDID. *See* Extended Display Identification Data (EDID)
efficacy in energy consumption, 205–206
efficiency
    luminaires, 209
    power amplifiers, 247
efforts in construction phase, 37
EGC (equipment-grounding conductor) connections, 266–269, 272–273
EIA (Electronics Industries Association) cable categories, 415–416
eighth-order equalization filters, 384–385
ELCBs (earth leakage circuit breakers), 265, 270
electret microphones, 159
electric codes, 232–233
electric fields
    coupling, 284
    shielding from, 286
electrical calculations in Ohm's law, 16–19
electrical distribution systems, 265–266
    basics, 266
    ground faults, 269–271

electrical distribution systems (*cont.*)
  grounds, 271–272
  neutral/return conductors, 273–274
  power distribution systems, 272–276
  protective connections, 266–269
  three-to-two-pin adapters, 271
electrical infrastructure, 255
  AV circuits, 276–280
  balanced and unbalanced circuits, 286–288
  cable support systems, 288–290
  circuit theory, 256–263
  codes and regulations, 264–265
  conduit capacity, 290–292
  distribution systems. *See* electrical distribution systems
  duty check, 256
  ground loops, 283–286
  interference prevention and noise defense, 282–286
  isolated grounds, 282
  power, 263–265
  power strips and leads, 280–281
  questions, 292–294
  review, 292
electrical power required (EPR) in loudspeaker systems, 155
electrical professionals, collaboration with, 189
electrical signal levels in microphones, 167
electromagnetic interference (EMI)
  conduits, 289
  fiber-optic cable, 417
  shielded twisted-pair network cables, 416–417
electromagnetic shielding, 285–286
electronic cameras, 203
electronic delay in DSPs, 382
electronic dimmers, 218–220
electronic echo, 383
electronic S/N ratio, 175
electronic whiteboards, 91
Electronics Industries Association (EIA) cable categories, 415–416
element height, 99–101
elevation drawings, 42, 45
elevation flags in architectural drawing symbols, 49
ELV (extra low voltage), 257
emergency issues and CTS-D exam, 23
EMI (electromagnetic interference)
  conduits, 289
  fiber-optic cable, 417
  shielded twisted-pair network cables, 416–417
emulators, EDID, 355
enabling encryption and auditing, 501
encoding, 328
encryption
  enabling, 501
  HDCP, 357–359
  VPNs, 436
end points in streaming design, 468, 470
end users, needs analysis for, 63–65
energy consumption in lighting specifications, 205–206
energy management, 252
Enhanced Acoustic Simulator for Engineers (EASE), 142–143

Enhanced EDID (E-EDID), 348
enterprise networks, 436
ephemeral ports, 450
EPR (electrical power required) in loudspeaker systems, 155
equal access, design for, 231–232
equal loudness curves, 139–140, 514
equalization, 383–384
  audio systems, 511
  crossover filters, 386–387
  feedback-suppression filters, 387
  graphic equalizers, 388–389
  noise-suppression filters, 388
  parametric, 385
  pass filters, 386
equipment
  audience and presenter areas, 60
  control and projection areas, 62
  manuals, 536
  rear projection displays, 111
equipment-grounding conductor (EGC) connections, 266–269, 272–273
equivalent acoustic distance (EAD), 177–179
ergonomics, 79–80
  duty check, 79
  eye height, 85–86
  floor layouts, 87–88
  furniture, 88–91
  human dimensions and visual field, 80–84
  questions, 92–93
  racks, 242–243
  review, 92
  seating layouts, 86–87
  sightlines, 84–85
Essential Video Coding (EVC) codec, 479
estimates in budget recommendations, 75
Ethernet networks, 428
  AV over, 457–458
  LAN isolation, 432–433
  LANs, 428
  speed, 431–432
  standard, 431
  topologies, 428–431
  types, 432
  VLANs, 433–434
ethics, 5
European Broadcasting Union (Eurovision), 204–205
European Committee for Standardization (CEN), 232
evacuation rack cooling, 248
evaluate phase in design-thinking principles, 401
EVC (Essential Video Coding) codec, 479
executive summaries in concept design/program report sections, 73
exit signs, 225–226
expanded beam connectors in fiber-optic cable, 421
expander settings in DSPs, 380
exploration AV experience type, 32–33
exposure controls in videoconference lighting, 223–224
exposure settings for cameras, 521–522

Index

629

Extended Display Identification Data (EDID), 348–349
    audio/video sync, 531
    data tables, 353–354
    displays, 356
    handshakes, 355
    issues, 354–356
    operation, 349–350
    packets, 349
    resolution, 355
    strategies, 352
    switching sources, 356
    tables, 350–351
    tools, 354–355
extended star LAN topologies, 430–431
extra drop for front projection displays, 110
extra low voltage (ELV), 257
extractors in EDID, 355
Extron NAV video, 460
eye charts, 99–100
eyes
    height, 85–86
    visual field, 80–84

F

f-stops for projector lenses, 126
facility drawings, 58–59
facility managers, needs analysis for, 65
fans, HVAC noise from, 249
far-end crosstalk (FEXT), 415
farthest viewers factor in viewing distance, 102
fasteners for mounting, 238
faults in electrical distribution systems, 269–271
feedback in human-centered interface design, 398
feedback stability margin (FSM), 179–180
feedback-suppression filters, 387
FEXT (far-end crosstalk), 415
fiber-optic cable
    applications, 419–420
    connectors, 420–421
    distance limitations, 419
    overview, 417–419
    types and data ranges, 419
Field-Programmable Gate Arrays (FPGAs), 460–461
field theory factors in electrical interference, 283
fill lights for videoconferences, 223
filters
    crossover, 386–387
    DSPs, 381
    equalization, 384–385
    feedback-suppression, 387
    firewalls, 456–457
    LCD displays, 107
    noise-suppression, 388
    parallel circuits, 262–263
    parametric equalizers, 385
    pass, 386
    series circuits, 262
    sound pressure levels, 141
    sound reception, 314–315

final acceptance verification, 509
fire protection, 250–251
firestop materials, 251
firewalls, 455–457
first-order equalization filters, 384
fixed-architecture processors in DSPs, 370
fixed-multifunction (hybrid) processors in DSPs, 370–371
fixtures, 206
flat floors, 87–88
flexible-architecture processors in DSPs, 369–370
flip charts, 91
floor mounting, 233–234
floor space for rear projection displays, 112, 114
floor standing racks, 239
floors
    layouts, 87–88
    reflectance values, 225
    sound reception, 315
fluorescent lamps, 208
flutter echoes, 302
focal length in rear projection displays, 114
focal ratio in projector lenses, 126
focus settings for cameras, 520
foil cable shields, 285
foot-candles units, 119, 200
foot-lamberts units, 120, 200
formats
    bit depth, 326
    dates, 49–50
    digital media, 329–330
    digital video, 103, 340–342
    high-definition, 337–338
    network transmissions, 400
    video bandwidth, 325
    video resolution, 104
formulas in CTS-D exam, 13
4:1:1 video subsampling, 327
4:2:0 video subsampling, 327
4:2:2 video subsampling, 327
4:4:4 video subsampling, 327
fourth-order equalization filters, 384
FPGAs (Field-Programmable Gate Arrays), 460–461
fragmenting, IP, 437
frame rates for video, 325, 336
frames for rear projection screens, 114
framing images for cameras, 523
frequencies
    acoustic mass law, 309–311
    audio signal generators, 512–513
    equalization, 383–389
    inductive reactance, 260
    microphone response, 165
    Nyquist-Shannon sampling theorem, 324
    parallel circuits, 262–263
    polar plots, 163–164
    series circuits, 261–262
    sound absorption, 306–309
    sound reception, 314–316
    sound reflection, 303–305
    sound transmission class, 311–312

Fresnel lenses in rear projection displays, 113
front-end documentation in AV design package, 54–55
front projection displays, 108
    mapped projection, 110–111
    screens, 109
    specifying, 109–110
FSM (feedback stability margin), 179–180
full HD display format, 337
full-motion video
    description, 97
    image system contrast, 527
fundamental light measurement units, 118–119
furniture, 88–89
    lecterns, 90–91
    tables and chairs, 89–90

G

gain
    AGC, 224, 377, 522
    audio settings, 374–376
    equalization. *See* equalization
    expander settings, 380
    front projection displays, 109
    microphone pre-amp, 167–168
    PAG and NAG, 176–180
    projector screens, 125–126
    verifying, 512
gain before feedback in audio system quality, 176
gain sharing in microphones, 169
galvanic isolation, 261
Gantt, Henry, 191
Gantt charts, 191–192
gate settings in DSPs, 381
gated sharing for microphones, 168–169
gateways, network, 414
general conditions documentation, 54–55
geometry patterns for display setup, 524–525
GFCIs (ground-fault circuit interrupters), 265, 270
glare
    luminance, 130
    videoconferences, 222–223
glass cores in fiber-optic cable, 418
glass treatments for space lighting, 207
global IP addresses, 440–441
golden room ratios in sound reflection, 305
gooseneck microphones, 161
governance structure in security, 495
government-issued IDs for CTS-D exam, 21
graphic equalizers, 385, 388–389
grayscale test patterns for display setup, 526
grid lines in architectural drawing symbols, 48
ground-fault circuit interrupters (GFCIs), 265, 270
ground-lifting adapter dangers, 271
ground loops, 282–286
grounds in electrical distribution systems
    connections, 266–269
    faults, 269–271
    isolated, 282
    types, 271–272

group-management protocol requests in multicast streaming, 483
group standards, 194

H

H.222/H.262 protocol, 478
H.264 protocol, 328, 478–479
H.265 protocol, 479
H.266 protocol, 328
Haas effect
    delay for, 382
    sound reception, 314
habitual-learning control systems, 395
handheld microphones
    description, 161
    for presenters, 172
handheld remote control interfaces, 398
handshakes
    EDID, 348–349, 355
    HDCP, 357
    TCP transport, 448–449
hardware, mounting, 237–238
harmonic distortion, 176
hazardous weather and CTS-D exam, 23
HDBaseT video signal types, 344
HDCP. *See* high-bandwidth digital content protection (HDCP)
HDMI ethernet and audio return channels (HEACs), 343
HDTV display aspect ratio, 105
head rotation, 83–84
headmic microphones, 161
headroom in loudspeaker systems, 155–156
Health and Safety Executive, 230
hearing
    decibels, 136–139
    perceived sound pressure level, 139–141
heat loads for racks, 245–247
heating in control and projection areas, 62
heating, ventilation, and air conditioning (HVAC) professionals, collaboration with, 189
heating, ventilation, and air conditioning (HVAC) systems
    issues, 249–250
    rack heat loads, 246
height
    aspect ratio, 105–106
    element, 99–101
    listener ear levels, 144
    media formats, 104
    video formats, 325
hexadecimal groups in IPv6 addresses, 442–443
hierarchy, consultation, 195–196
high-bandwidth digital content protection (HDCP), 357
    device authentication and key exchange, 358
    device limits, 361
    interfaces, 357–358
    locality checks, 359

operation, 358
repeaters, 360
session key exchange, 359
streaming design, 469–470
switchers, 359
troubleshooting, 361–362
high-definition (HD) video
   aspect ratio, 105
   digital video design, 337–339
High-Definition Multimedia Interface (HDMI)
   description, 341
   display aspect ratio, 105
   overview, 342–344
   signal extenders, 518–519
   signal format, 337–338
high dynamic range (HDR)
   bit depth, 326
High Efficiency Video Coding (HEVC), 479
high-fidelity audio, latency in, 475
high frame rate (HFR) cliff effect, 339
high-pass filters, 386
high Q filters, 385
high-quality streaming video, 478–480
high-throughput data class in DiffServ, 454
hiss settings, 376
hole spacing in racks, 242
home addresses in IP addresses, 442
hook suspension cable systems, 289–290
horizontal grid lines in architectural drawing symbols, 48
horizontal visual field, 81
host identifier bits in IP addresses, 438
host layer protocols, 451–452
hot plug detection (HPD), 349, 355
hot plugs, 349
hot spotting in projector screens, 125
hotspots for wireless connections, 423
hours of operation for videowalls, 117
hum, electrical, 282
human-centered interface design, 398–399
human dimensions
   ergonomics, 80
   head rotation, 83–84
   horizontal visual field, 81
   vertical visual field, 82
human-factors engineering. *See* ergonomics
human sightlines, 85
HVAC (heating, ventilation, and air conditioning) professionals, collaboration with, 189
HVAC (heating, ventilation, and air conditioning) systems
   issues, 249–250
   rack heat loads, 246
hybrid fiber-optic cable, 419
hybrid (fixed-multifunction) processors in DSPs, 370–371
hypercardioid pickup patterns, 163
hypoxic air fire protection systems, 251

I

I-Frames (intraframes) in video compression, 331
IANA (Internet Assigned Numbers Authority)
   IP addresses, 439–440
   ports, 450
IBC (International Building Code), 233
ICTs (image constraint tokens) in HDCP, 361
identification requirements for CTS-D exam, 21
IEC (International Electrotechnical Commission)
   AVIXA accreditation, 5
   regional regulations, 510
IEEE (Institute of Electrical and Electronics Engineers Standards Association) regional regulations, 510
IES (Illuminating Engineering Society), 205
IGBTs (insulated-gate bipolar transistors), 218–219
IGMP (Independent Group Management Protocol), 483
IIC (impact insulation class) for sound transmission, 312–313
illuminance light measurements, 119
illuminated exit signs, 225–226
Illuminating Engineering Society (IES), 205
image constraint tokens (ICTs) in HDCP, 361
image geometry for display setup, 524–525
Image System Contrast Ratio (ISCR), 207
   conformance, 121–122
   contrast, 120, 527
   description, 97
images
   bandwidth factors in, 472–473
   contrast, 527–529
   height, 101
   sizing, 97–98
   specifications, 96
impact insulation class (IIC) for sound transmission, 312–313
impedance
   description, 258
   loudspeaker systems, 147–148
   measuring, 152–153
   Ohm's law, 16–19
   parallel circuits, 149–151
   power amplifiers, 157
   series circuits, 149
   series/parallel circuits, 152
   transformers, 153–154
impellers, HVAC noise from, 249
imperial measurements in scaled drawings, 40–41
incandescent lamps, 208
incident meters for light, 118, 200
Independent Group Management Protocol (IGMP), 483
indirect glare in videoconferences, 222
induced currents, 261
inductance, 260–261
inductive reactance
   AC circuits, 260
   description, 258
   loudspeakers, 147

inductors
    description, 260
    parallel circuits, 262–263
    series circuits, 261–262
industry standards, 193–194
Information Technology Infrastructure Library (ITIL), 495
infrared (IR) control points, 400
infrastructure considerations in concept design/program report sections, 74
infrastructure drawings, 55–58
ingested content in streaming design, 469
input equalization, 383
insertion loss in transformers, 153
installation drawings in construction phase, 37
Institute of Electrical and Electronics Engineers Standards Association (IEEE) regional regulations, 510
instruction AV experience type, 32–33
instrument microphones, 161
insulated-gate bipolar transistors (IGBTs), 218–219
integrated carpentry work in detail drawings, 57–58
integration intervals in sound reception, 314
integration process in sound reception, 313–316
integrity in network security, 492
intelligibility in audio systems, 175–176
intensity of sound, 298–299
interaction property of sound, 296, 299–300
    absorption, 306–309
    reflections, 300–306
    transmission, 309–313
interactive smart boards, 91
interfaces, control, 396–399
interference and interference prevention
    electric-field coupling, 284
    electrical infrastructure, 282–286
    electromagnetic shielding, 285–286
    electronic dimmers, 218
    fiber-optic cable, 417
    magnetic-field coupling, 283–284
    shielded twisted-pair network cables, 416–417
interframe video compression, 330–331
interior designers, collaboration with, 189
internal DNS, 447–448
International Building Code (IBC), 233
International Commission on Illumination (CIE) color rendering index, 204
International Electrotechnical Commission (IEC)
    AVIXA accreditation, 5
    regional regulations, 510
International Organization for Standardization (ISO)
    AVIXA accreditation, 5
    background noise, 316
    BIM, 56
    mounting considerations, 238
    network security surveys, 493
    regional regulations, 510
    security, 492–493
    twisted-pair cable, 417
International System of Units (SI) in scaled drawings, 40–41
International Telecommunication Union (ITU)
    standards, 338

Internet Assigned Numbers Authority (IANA)
    IP addresses, 439–440
    ports, 450
Internet Protocol (IP)
    control points, 400
    functions, 437
Internet Protocol television (IPTV), 465
intraframe video compression, 330
intraframes (I-Frames) in video compression, 331
invitations to bid, 54
IP (Internet Protocol)
    control points, 400
    functions, 437
IP addresses
    addressing, 437–438
    APIPA, 445–446
    assigning, 443–448
    broadcast, 442
    DHCP, 444–445
    DNS, 446–447
    IPv4, 438–439
    IPv6, 442–443
    loopback, 442
    multicast streaming, 480–481, 485–486
    NAT, 441
    static and dynamic, 443–444
    types, 439–440
IP masquerading, 456
IP network-based control systems, 394
IP networks, AV over, 458–461
IPTV (Internet Protocol television), 465
IR (infrared) control points, 400
iris settings for cameras, 521–522
irregular material in videoconference lighting, 225
ISCR. *See* Image System Contrast Ratio (ISCR)
ISO. *See* International Organization for Standardization (ISO)
ISO-27000 security framework, 492
ISO/IEC 17024:2012 personnel standard, 5
isolated grounds, 282
IT professionals, collaboration with, 189
ITIL (Information Technology Infrastructure Library), 495
ITU (International Telecommunication Union)
    standards, 338
ITU-R BT.1359-1 sync standard, 530

J

jam ratio for conduits, 291–292
jamming wireless connections, 424
jitter in DiffServ classes, 453
job task analysis (JTA)
    categories, 6–7
    CTS-D exam, 9
    exam preparation strategies, 11

K

key exchange in HDCP, 358–359
key lights for videoconferences, 223

Index

L

lamps, 206
LANs (local area networks)
    data travel in, 428
    isolation, 432–433
    topologies, 430–431
latency
    codecs, 452
    content compression, 329
    DiffServ classes, 453–454
    DSPs, 369, 382
    network, 457, 460–461
    SDI, 341
    streaming design, 474–476
lavalier microphones
    description, 161
    for presenters, 172
Law of Conservation of Energy in sound interaction, 299
laws, regional, 509–510
layers in OSI Model, 425–427
layouts
    floor, 87–88
    seating, 86–87
LC (Lucent) fiber-optic cable connectors, 421
LCDs (liquid-crystal displays), 107–108
LCEVC (Low Complexity Enhancement Video Coding), 479–480
leases in DHCP, 444
least privilege philosophy for networks, 454
lecterns, 90–91
LEDs. *See* light-emitting diodes (LEDs)
lenses
    projectors, 126
    rear projection displays, 113–114
lenticular lenses in rear projection displays, 113
Li-Fi data, 400
licenses
    codecs, 451
    copyrighted content, 469–470
    DRM, 357
    HDCP, 358
    streaming design, 469–470
life safety protection, 250–251
light
    color, 211–212
    measuring, 118–120
light balance for videoconferences, 223–224
light-emitting diodes (LEDs)
    CRI, 203–204
    direct-view displays, 107–108
    efficacy, 205
    light sources, 208
    videowalls, 117
light paths for projectors, 123
light reflectance value (LRV), 201
light source life for projectors, 125–126
light transmission in fiber-optic cable, 417–421
lighting control
    on/off vs. dimmable, 218–220
    scenes, 220–221

Lighting Control Protocols, 220
lighting in control and projection areas, 62
lighting specifications, 199
    brightness, 200
    color rendering index, 203–205
    color temperature, 201–203
    duty check, 200
    energy consumption, 205–206
    illuminated exit signs, 225–226
    lighting control, 218–221
    questions, 226–228
    review, 226
    space lighting. *See* space lighting
    videoconferences, 221–225
limiter settings in DSPs, 379–380
line levels in decibels, 138–139
links, network, 415–417
lip-sync errors, 530
liquid-crystal displays (LCDs), 107–108
listener ear levels in loudspeaker systems, 143–144
live production professionals, collaboration with, 189
load limit mounting considerations, 236–237
local area networks (LANs)
    data travel in, 428
    isolation, 432–433
    topologies, 430–431
local emergencies and CTS-D exam, 23
local IP addresses, 440–441
locality checks in HDCP, 359
logic network diagrams in project tracking, 192–193
logical topologies for Ethernet networks, 429
loopback IP addresses, 442
loops, ground, 283–286
lossless compression, 328
lossy compression, 328
loudness
    audio systems, 174–175
    sound, 298–299
loudspeaker systems
    coverage, 142–144
    distributed, 142
    distributed layout options, 145–146
    edge-to-center coverage, 146
    edge-to-edge coverage, 145–146
    headroom requirements, 156
    impedance, 147
    partial overlap coverage, 146
    point-source, 142
    power amplifiers, 155–158
    sensitivity, 156–157
    taps, 154
    wiring, 148–152
Low Complexity Enhancement Video Coding (LCEVC), 479–480
low latency DiffServ classes, 453–454
low loss DiffServ classes, 453
low-pass filters, 386
low-priority data DiffServ classes, 454
low Q filters, 386
low smoke zero halogen (LSZH) cable options, 419
low voltage (LV), 257

LRV (light reflectance value), 201
LSZH (low smoke zero halogen) cable options, 419
Lucent (LC) fiber-optic cable connectors, 421
lumen units
    light measurements, 119
    OFE projectors, 128–129
luminaires
    choosing, 209–210
    description, 206
    documenting, 216–217
    specifications, 209–210
luminance
    measuring, 200
    specifications, 201
    task-light levels, 129–130
    YUV processing system, 326
lux units, 119, 200
LV (low voltage), 257

M

MAC (Media Access Control) addresses
    IPv6 addresses, 443
    NICs, 413
    OSI Model, 426
Macbeth charts, 203
magnetic-field coupling, 283–284
magnetic fields, shielding from, 285–286
magnetic lines of flux, 283
mains voltage and ground faults, 269
maintenance logs, 537
managed switches in networks, 414
MANs (metropolitan area networks), 434
mapped projection in front projection displays, 110–111
mapping in human-centered interface design, 398
masking systems in front projection displays, 110
masks in IPv4 addresses, 438–439
master unit control requirements, 395–396
match lines in architectural drawing symbols, 48
matching levels in microphones, 167
materials
    audience and presenter areas, 60
    rear projection screens, 112–114
mathematical formulas
    CTS-D exam, 13
    order of operations, 14–15
mathematical strategies in CTS-D exam, 13–15
matte materials in videoconference lighting, 225
matte white screens in front projection displays, 109
Measurement and Classification of Spectral Balance of Sound Systems in Listener Areas, 384
measurement microphones, 516
measuring
    alternating current, 516
    background noise, 140
    impedance, 152–153
    light, 118–120
    luminance, 200
Mechanical Transfer Registered Jack (MT-RJ) fiber-optic cable connectors, 421
mechanical vibration, HVAC noise in, 250

Media Access Control (MAC) addresses
    IPv6 addresses, 443
    NICs, 413
    OSI Model, 426
meetings
    AV design projects, 34–35
    concept design/program report development, 71
MEMS (microelectromechanical system) microphones, 160
mesh-earth systems, 282
mesh topologies
    LANs, 430–431
    WANs, 435
meter weighting in sound pressure levels, 141
meters
    audio signal levels, 516
    light measurements, 118, 200
    scaled drawings, 40
    sound pressure levels, 513–516
metric measurements in scaled drawings, 40–41
metropolitan area networks (MANs), 434
mic levels
    description, 165–166
    dynamic microphones, 158
microelectromechanical system (MEMS) microphones, 160
microLEDs, 108
microphones
    audio settings, 375
    clothing considerations, 172–173
    condenser, 158–159
    decibel levels, 138
    dynamic, 158
    electret, 159
    frequency response, 165
    measurement, 516
    MEMS, 160
    mixers, 168–169
    physical design and placement, 161–162
    placement, 169–171
    polar plots, 163–164
    polar response, 162–163
    pre-amp gain, 167–168
    presenter reinforcement, 171–173
    sensitivity, 166–167
    signal levels, 165–166
    sound reflection, 304–305
milestones in project tracking, 193
millimeters in scaled drawings, 39–40
millwork in detail drawings, 57
Mini DisplayPort, 347
mirrors in rear projection displays, 114
mitigate strategy for risk handling, 498
mitigation planning for network security, 498–501
mixers
    audio settings, 374
    microphones, 168–169
MLD (Multicast Listener Discovery), 483
mobile racks, 241
mobility issues for lecterns, 91
modes
    fiber-optic cable, 418
    sound reflection, 302–305

Index

money matters in contracts, 55
monitoring
    control and projection areas, 62
    energy management, 252
monocular vision, 80
mounting considerations, 233
    hardware, 237–238
    load limits, 236–237
    options, 233–236
MP3 audio encoding format, 330
MPEG-2 protocol, 478
MPEG-4 protocol, 478–479
MPEG-5 protocol, 479
MPEG-H protocol, 479
MPEG-I protocol, 479
MPEG Transport Stream (MPEG-TS), 477
MST (multistream transport) in Thunderbolt, 347
MT-RJ (Mechanical Transfer Registered Jack) fiber-optic cable connectors, 421
multi-tap speaker transformers, 154
Multicast Listener Discovery (MLD), 483
multicast streaming, 480–481
    implementing, 483–485
    IP addresses, 485–486
    vs. unicast, 482
multichannel power amplifiers, 247
multidrop network control points, 400
multimedia conferencing DiffServ classes, 453
multimedia streaming DiffServ classes, 454
multimeters for audio signal levels, 516
multimode mode transmission in fiber-optic cable, 418
multiple device integration, control needs analysis for, 401
multiple locations, control needs analysis for, 402
multiple user roles, 500
multiplication in mathematical formulas, 14–15
multipurpose chairs, 89
multistream transport (MST) in Thunderbolt, 347
Music, Physics, and Engineering (Olson), 297

N

NAC (Network Access Control), 454–455
NAG (needed acoustic gain), 176–180
NAPT (network and port translation), 456
NAT (network address translation), 441
National Building Code of India, 230
National Construction Code, 231
National Electrical Code (NEC), 232
National Fire Protection Association (NFPA), 232
National Institute of Standards and Technology (NIST) Color Quality Scale, 204
native video resolution
    description, 104
    digital video design, 336
natural light in space lighting, 207
NC (noise criterion), 316–317
NDI (Network Device Interface) platform, 461
near-end crosstalk (NEXT), 416
NEC (National Electrical Code), 232
needed acoustic gain (NAG), 176–180

needs analysis, 34, 63
    AV managers and technology managers, 65–66
    buyers, purchasing agents, and contract representatives, 66
    clients, 64–65
    control requirements, 401–402
    end users, 63–65
    facility managers, 65
    owners, 65
    streaming design, 466–470
Network Access Control (NAC), 454–455
network address translation (NAT), 441
network and port translation (NAPT), 456
network-based QoS (NQoS), 474
Network Device Interface (NDI) platform, 461
network diagrams in project tracking, 192–193
network environment in streaming design, 471–473
network identifier bits in IP addresses, 437
network interface cards (NICs), 413
Network layer
    OSI Model, 426
    protocols, 437–443
network stakeholders in network security, 496
networked loudspeaker wiring, 148
networks, 409
    address assignments, 443–448
    AV over, 457–461
    bandwidth, 452–454
    clients and servers, 412–413
    components overview, 410–412
    control points, 400
    description, 410
    devices, 414–415
    duty check, 410
    Ethernet, 428–434
    fiber-optic cable, 417–421
    host layers, 451–452
    links, 415–417
    NICs, 413
    OSI Model, 425–428
    policies and restrictions, 475–476
    protocols, 437–443
    questions, 462–464
    review, 461
    security. *See* security for networked AV applications
    transport protocols, 448–450
    WANs, 434–436
    wireless connections, 422–425
neutral/return conductors in three-phase power distribution, 273–274
NewTek NDI, 461
NEXT (near-end crosstalk), 416
NFPA (National Fire Protection Association), 232
NICs (network interface cards), 413
NIST (National Institute of Standards and Technology) Color Quality Scale, 204
nit light units, 120
noise
    audio signal generators, 512
    audio systems, 174–175

noise (cont.)
　　background, 140
　　balanced and unbalanced circuits, 286–288
　　cable categories, 415–416
　　control and projection areas, 62–63
　　defense against, 282–286
　　dimmers, 218–219
　　expander settings, 380
　　ground loops, 282
　　HVAC, 249–250
　　rear projection displays, 112
　　sound reception, 316–317
noise criterion (NC), 316–317
noise rating (NR), 316–317
noise reduction coefficient (NRC), 307–308
noise-suppression filters, 388
NOM (number of open microphones) in audio system quality, 179–180
nonvolatile memory (NVM) in projectors, 494
North America, power distribution systems in, 275–276
notch filters, 262–263, 385, 387
notches in frequency response, 314–315
NQoS (network-based QoS), 474
NR (noise rating), 316–317
NRC (noise reduction coefficient), 307–308
number of open microphones (NOM) in audio system quality, 179–180
NVM (nonvolatile memory) in projectors, 494
Nyquist-Shannon sampling theorem, 324

O

octaves in equalization filters, 384
octets in IP addresses, 438
OFE (owner-furnished equipment) projector brightness, 128–129
offsets in projector positioning, 122
Ohm's law, 16–19
Ohm's Law and the Power Formula video, 17
OLED displays, 108
Olson, Harry F., 297
omnidirectional microphones for presenters, 173
omnidirectional pickup patterns for microphones, 163
on/off vs. dimmable lighting, 218–220
one-to-many NAT, 456
1:1 transformers, 153
1080i display format, 337
1080p display format, 337
online courses, 11
open frame racks, 239–240
Open Systems Interconnection (OSI) Model
　　data transmission, 427–428
　　host layers, 451–452
　　layers overview, 425–427
　　network layer, 437–443
　　transport protocols, 448–450
operational documentation in system closeout, 535–536
operational stakeholders in network security, 496
operations order in mathematical formulas, 14–15
optical-pattern rear-projection screens, 112
opticalCON fiber-optic cable connectors, 421
optimization in audio settings, 374

optotypes, 99
order of operations in mathematical formulas, 14–15
origins in TCP transport, 449
oscilloscopes, 516–517
OSI Model. *See* Open Systems Interconnection (OSI) Model
outlets for AV circuits, 278–280
output equalization, 383
overhead lighting, 212
overhead mounting, 235–236
owner-furnished equipment (OFE) projector brightness, 128–129
owners in needs analysis, 65

P

P-Frames (predictive frames) in interframe video compression, 331
Pa (pascal) units for pressure, 137–138
packaging, IP, 437
packet filtering rules in firewalls, 456
packets, EDID, 349
PAG (potential acoustic gain), 176–180
pan/tilt/zoom (PTZ) capabilities in cameras, 520
parallel circuits
　　circuit theory, 262–263
　　loudspeaker wiring, 149–151
parametric equalizers, 385
parentheses () in mathematical formulas, 14–15
partial overlap coverage in loudspeaker systems, 146
partially mesh LAN topologies, 430–431
particle displacement in sound, 299
pascal (Pa) units for pressure, 137–138
pass filters, 386
passive crossover filters, 386
passive loudspeaker wiring, 148
passive video signal extenders, 518
passive viewing
　　description, 96
　　image system contrast, 527
passwords
　　access control, 454, 496
　　default, 499
　　risk from, 497
　　user accounts, 500
PAT (port address translation), 456–457
PCI Express, 347
peak program meters (PPMs), 371–372
PEMDAS acronym, 14
performance verification standard, 506–508
permissible area in conduits, 290–291
permissions for networks, 455
phantom power
　　condenser microphones, 159
　　defining, 160
phase control electronic dimmers, 218–219
phases in AV design projects, 33–35
　　briefing, 35–36
　　construction, 37
　　design, 36–37
　　summary, 38
　　verification, 37–38

Index

637

photometric distribution plots, 210
physical design for microphones, 161–162
physical Ethernet topologies, 429
Physical layer in OSI Model, 426
physical stakeholders in network security, 496
pickup patterns for microphones, 162–163
piezo tweeters, 511–512
PIM (Protocol-Independent Multicast), 483
PIM sparse mode (PIM-SM) for multicast streaming, 483–485
pink noise in audio signal generators, 512
pipes
    HVAC noise from, 250
    mounting considerations, 239
placement of microphones, 161–171
plan/specify phase in design-thinking principles, 401
plan view drawings, 42–43
Planckian locus, 201–202
plastic cores in fiber-optic cable, 418
plenums for fiber-optic cable, 420
point-source loudspeaker systems, 142
point-to-point networks, 400
polar-pattern directivity in loudspeaker systems, 142
polar plots
    microphones, 163–164
    presenters, 173
polar response of microphones, 162–163
polarized light in LCDs, 107
policies
    codecs, 329, 451
    firewalls, 455
    network, 475–476
    security, 494–495
    user accounts, 500
policy-based QoS rules, 474
polished material in videoconference lighting, 224
porous absorbers, 308–309
port address translation (PAT), 456–457
portable racks, 241
portraiture, 211–212
ports
    firewalls, 456
    transport protocols, 449–450
positioning projectors, 122–124
post-integration verification, 509
posture, security, 494–495
potential acoustic gain (PAG), 176–180
power
    branch circuit load consumption, 277–278
    codes and regulations, 264–265
    condenser microphones, 159
    control and projection areas, 62
    decibels, 137–138
    electrical. *See* electrical infrastructure
    Ohm's law, 16–19
    presenter area, 61
    sound, 298–299
    terms, 264
power amplifiers
    heat loads, 247
    loudspeaker systems, 155–158

power cables, 281
power metal oxide silicon field-effect transistors (power MOSFETs) in electronic dimmers, 218–219
power strips and leads, 280–281
powered analog loudspeaker wiring, 148
PPMs (peak program meters), 371–372
practical completion verification, 509
practice questions for CTS-D exam, 25–26
pre-amp gain for microphones, 167–168
pre-construction meetings, 34
pre-integration verification, 508
preamplifiers
    audio settings, 374–376
    condenser microphones, 159
precedence effect in sound reception, 314
predictive frames (P-Frames) in interframe video compression, 331
preparation strategies for CTS-D exam, 11–12
Presentation layer
    codecs, 451–452
    OSI Model, 427
presenter area, 60–61
presenter microphone reinforcement, 171–173
presets for lighting, 220–221
pressure
    in sound intensity, 298–299
    SPL. *See* sound pressure levels (SPLs)
pressure zone microphones (PZMs), 161
pressurization in rack cooling, 248
private IP addresses, 440–441
private WANs, 436
probable costs in budget recommendations, 74–75
problem lists, 536
processors
    configurations, 402–403
    control requirements, 395–396
produce phase in design-thinking principles, 401
production property of sound, 296–297
program meetings in concept design/program report development, 71
program phase in AV design projects, 35–36
programmable lighting control panels, 221
programming
    AV design, 63
    device control, 403–404
project constraints in security posture, 495
project implementation activities, 505
    audio system verification, 511–517
    duty check, 506
    performance verification standard, 506–508
    questions, 539–542
    regional regulations, 509–511
    review, 539
    system closeout, 532–536
    system verification process, 508–509
    video system verification, 517–532
project tracking, 189–190
    Gantt charts, 191–192
    logic network diagrams, 192–193
    work breakdown structures, 190–191

projection area, 62–63
projection mapping in front projection displays, 110
projection throw in projectors, 123–124
projector blending in videowalls, 118
projectors
  ambient light levels, 127
  brightness predictions, 124–128
  color temperature, 203
  lenses, 126
  light paths, 123
  light source life, 125–126
  OFE, 128–129
  positioning, 122–124
  projection throw, 123–124
  screen gain, 125
  security, 494
propagation property of sound, 296–299
property classification in mounting considerations, 238
protection, earth-leakage, 270–271
protective-earth connections in electrical distribution systems, 266–269, 272–273
Protocol-Independent Multicast (PIM), 483
protocols
  streaming design, 477–479
  transport, 448–450
PTZ (pan/tilt/zoom) capabilities in cameras, 520
Public Health and Disability Act, 230
public IP addresses, 441
public WANs, 436
pull-out mounting considerations, 237
pulse-width modulation (PWM) for electronic dimmers, 219
pumps, HVAC noise from, 249
punch lists, 536
purchasing agents, needs analysis for, 66
PWM (pulse-width modulation) for electronic dimmers, 219
PZMs (pressure zone microphones), 161

Q

Q-SYS platform, 461
quality in audio principles, 174–180
quality of service (QoS)
  bandwidth, 452–454
  streaming design, 473–476
questions
  concept design/program report development, 66–67
  streaming design, 467, 470–471
quotes in budget recommendations, 75

R

Rack Building for Audiovisual Systems, 239
Rack Design for Audiovisual Systems, 239, 244
rack-mounted power distribution strips, 281
rack units (RUs), 242
racks, 239–241
  block diagrams, 245–246
  cooling, 247–249
  ergonomics, 242–243
  heat loads, 245–247

signal separation, 244–245
sizes, 241–242
weight distribution, 242–244
radio frequency (RF) control points, 400
radio frequency interference (RFI)
  electronic dimmers, 218
  fiber-optic cable, 417
  shielded twisted-pair network cables, 416–417
range of wireless connections, 424
ratio
  compressor settings, 378–379
  limiter settings, 379–380
  signal-to-noise, 174–175, 375–376
RAVENNA (Realtime Audio Video Enhanced Next generation Network Architecture) protocol, 459
RC (room criteria) background noise measurement, 316–317
RCBO (residual current breaker with over-current protection), 270
RCDs (residual current devices), 265, 270
reactance
  capacitive, 259
  description, 258
  inductive, 260
  loudspeakers, 147
readers, EDID, 355
real-time interactive DiffServ classes, 453
Real-Time Streaming Protocol (RTSP), 477
Real-Time Transport Control Protocol (RTCP), 477
Real-Time Transport Protocol (RTP), 477
Realtime Audio Video Enhanced Next generation Network Architecture (RAVENNA) protocol, 459
rear projection displays, 111–112
  design considerations, 114
  screens, 112–114
receivers of electrical interference, 282–283
receptacles for AV circuits, 278–280
reception property of sound, 296
recognition acuity, 99
Recommended Practice for Lighting Performance for Small to Medium Sized Videoconferencing Rooms, 211, 221
Recommended Practice for Powering and Grounding Electronic Equipment, 264
Recommended Practices for Security in Networked AV Systems, 502
reference decibel levels, 137–139
reflectance values in videoconference lighting, 225
reflected ceiling plans, 42, 44, 217
reflected light measurements, 120
reflected meters, 118, 200
reflection of sound, 300
  diffuse, 300–302
  reverberation, 305–306
  room modes, 302–306
  small rooms, 305
  specular, 300
reflections
  rear projection displays, 114
  videoconferences, 222
refresh rates in digital video design, 336
regional regulations, 509–511

Index

639

registers, risk, 497–498
regulations. *See* codes and regulations
relays in multicast reflecting, 486–487
release time
    compressor settings, 378–379
    description, 377
    gate settings, 381
    limiter settings, 380
removal from exam, 23
repeaters in HDCP, 360
reporting
    design intent. *See* concept design/program report development
    energy management, 252
requests for qualifications, 54
reserve DHCP, 445
reserved IP addresses, 440
residual current breaker with over-current protection (RCBO), 270
residual current devices (RCDs), 265, 270
resistance
    description, 258
    Ohm's law, 16–19
resistors
    parallel circuits, 262–263
    series circuits, 261–262
resolution
    codecs, 451–452
    digital video design, 336–339
    EDID, 352, 355
    native, 104, 336
    video, 104, 325
resolution acuity, 99
resonant absorbers, 309
Resource Reservation Protocol (RSVP), 472–474
resources for regional codes, 510
response times in sound pressure level meters, 516
restricted items in testing center, 21
restrictions in streaming design, 475–476
retesting CTS-D exam, 24
return air plenums, HVAC noise from, 250
return conductors in single-phase power distribution, 266
reverberation, 305–306
reverberation time (RT), 306
reverse phase control electronic dimmers, 218–219
reverse wrap in front projection displays, 110
reviewing documentation, 67–68
RFI (radio frequency interference)
    fiber-optic cable, 417
    shielded twisted-pair network cables, 416–417
RGB video signals in YUV subsampling, 326
right amount of light, 210–211
right color of light, 211–212
right direction in space lighting, 212–213
riser-rated fiber-optic cable, 419–420
risk
    assessment, 497–498
    mitigation plans, 493
risk registers, 497–498
RMS (root mean square) values
    AC measurements, 257
    multimeter measurements, 516

room criteria (RC) background noise measurement, 316–317
room modes in sound reflection, 302–305
room's smallest dimension (RSD) in sound reflection, 303–304
root mean square (RMS) values
    AC measurements, 257
    multimeter measurements, 516
routers
    networks, 414
    WANs, 434
routing
    IP, 437
    microphones, 168–169
RSD (room's smallest dimension) in sound reflection, 303–304
RSVP (Resource Reservation Protocol), 472–474
RT (reverberation time), 306
RTCP (Real-Time Transport Control Protocol), 477
RTP (Real-Time Transport Protocol), 477
RTSP (Real-Time Streaming Protocol), 477
rulers for scaled drawings, 39
rules for firewalls, 456

S

S/N (signal-to-noise ratio)
    audio systems, 174–175
    gain, 375–376
    speech, 175
SAA (sound absorption average), 308
sabins for sound absorption, 307
SACD (Super Audio CD) audio sampling rates, 324
SAE International mounting considerations, 238
Safe Working Load (SWL) mounting considerations, 236
safety factor in mounting, 236
safety grounds in electrical distribution systems, 266–269, 272
safety protection, 250–251
sampling
    audio signals, 324
    color, 344
    MP3 audio, 330
    video signals, 326–327, 338
SC (subscriber) fiber-optic cable connectors, 421
scalability
    streaming reflectors, 487
    wireless connections, 423
scaled drawings, 39
    CAD systems, 41
    dimension conversions, 40–41
    rulers, 39
scattered sound reflections, 300–302
scene lighting, 220–221
schedules in Gantt charts, 191–192
schematic design phase in AV design projects, 36
Schroeder frequency, 303–304
scientific calculators for CTS-D exam, 13–14
scope of work in concept design/program report development, 72
scoring in CTS-D exam, 24
screen diagonal, 106

screen gain
  front projection displays, 109
  projector brightness, 125–126
screens
  front projection displays, 109
  rear projection displays, 112–114
SCRs (silicon-controlled rectifiers) in electronic dimmers, 218
SD (standard-definition) video aspect ratio, 105
SDI (Serial Digital Interface), 341
SDP (Session Description Protocol), 477
SDR (standard dynamic range) bit depth, 326
SDVoE (Software Defined Video over Ethernet) alliance, 460–461
seated sightlines, 85–86
seating layouts, 86–87
second-order equalization filters, 384
section cut flags in architectural drawing symbols, 49
section drawings, 46–47
security for networked AV applications
  access control, 454–455, 496
  duty check, 491
  firewalls, 455–457
  mitigation planning, 498–501
  objectives, 492–493
  questions, 503–504
  recommended practices, 502
  requirements, 493
  review, 502
  risk assessment, 497–498
  security posture, 494–495
  stakeholder input, 496
  technologies, 454–457
security in needs analysis, 402
security professionals, collaboration with, 189
security technologies
  firewalls, 455–457
  NAC, 454–455
segmentation in transport protocols, 448
semigloss materials in videoconference lighting, 225
sensation AV experience type, 32–33
sensitivity
  loudspeakers, 156–157
  microphones, 138, 163, 166–167
separation in racks, 244–245
Serial Digital Interface (SDI), 341
series circuits
  circuit theory, 260–262
  loudspeaker wiring, 148–149
series/parallel loudspeaker wiring, 151–152
servers, network, 412–413
service-level agreements (SLAs) in streaming design, 466
services, disabling, 501
Session Description Protocol (SDP), 477
session key exchange in HDCP, 359
Session layer
  description, 451
  OSI Model, 426
720p display format, 337
shades, 206–207
shadows in space lighting, 211–212
shearing as mounting consideration, 236

shelving-pass filters, 386
shielded twisted-pair cable, 287, 416–417
shielding, electromagnetic, 285–286
shock-mounted racks, 241
shop drawings in construction phase, 37
shotgun microphones, 161
shutter adjustments in cameras, 523
SI (International System of Units) in scaled drawings, 40–41
sides of rear projection screens, 113
sightlines
  factors, 84–85
  floor layouts, 87–88
  presenter area, 60–61
  seated, 85–86
sign-off, client, 539
signal delay in DSPs, 382
signal extenders for video systems, 517–518
signal generators, 512–513
signal grounds, 271
signal levels for microphones, 165–166
signal monitoring
  analog vs. digital, 372–373
  audio design, 371–373
signal paths in video systems, 517
signal separation in racks, 244–245
signal-to-noise ratio (S/N)
  audio systems, 174–175
  gain, 375–376
  speech, 175
signaling service DiffServ classes, 453
signifiers in human-centered interface design, 398
silicon-controlled rectifiers (SCRs) in electronic dimmers, 218
simple wired panel control interfaces, 397
simplicity of operation in control needs analysis, 401
sine wave electronic dimmers, 219
single-ended circuits, 287
single-mode transmission in fiber-optic cable, 418
single-phase power distribution
  description, 266
  North America, 275–276
  vs. three-phase, 272, 276
single points of failure in Ethernet networks, 430
sinks, EDID, 349–350, 352
site environment evaluation in concept design/program report development, 68–71
size
  direct-view displays, 108
  front projection displays, 110
  images, 97–98
  media formats, 104
  racks, 241–242
  rear projection screens, 113
  text, 99
  videowalls, 115
SLAs (service-level agreements) in streaming design, 466
slide-out racks, 240–241
sloped floors, 87–88
small rooms, sound reflection in, 305
smart boards, 91
smart device control interfaces, 397–398

Index

641

SMEs (subject-matter experts)
    collaboration with, 189
    job task analysis participation, 9
SMPTE (Society of Motion Picture and Television Engineers)
    SMPTE ST 2110, 459
    test patterns, 527
    video standard, 341
snag lists, 536
Snellen eye chart, 99–100
Society of Motion Picture and Television Engineers (SMPTE)
    SMPTE ST 2110, 459
    test patterns, 527
    video standard, 341
software
    documentation, 536
    EDID, 354–355
Software Defined Video over Ethernet (SDVoE) alliance, 460–461
solid-state lights, 208
sound. *See also* audio
    absorption. *See* absorption of sound
    acoustics. *See* acoustics
    audience area, 61
sound absorption average (SAA), 308
sound isolation in control and projection areas, 62
sound pressure levels (SPLs)
    background noise, 140
    decibels, 136–138
    loudness, 174–175
    meter weighting, 141
    microphones, 166–167
    perceived, 139–141
sound pressure levels (SPLs) meters, 513
    classifications, 513–514
    response times, 516
    weightings, 514–515
sound reception, 313
    background noise, 316–317
    integration process, 313–316
Sound System Spectral Balance, 506
sound transmission class (STC), 311–312
source-specific multicast (SSM), 483
sources
    EDID, 349–350, 352
    electrical interference, 282–283
    space lighting, 207–209
    video systems, 519–523
    videowall inputs, 117
space component in AV experience, 33
space lighting
    design, 210–213
    luminaires, 209–210, 216–217
    right amount of light, 210–211
    right color of light, 211–212
    right direction, 212–213
    shades and blackout drapes, 206–207
    sources, 207–209
    task, 206
    zoning plans, 213–216
space planning in concept design/program report sections, 73
sparse mode in multicast streaming, 483–485
SPD (spectral-power distribution) curves, 204–205
special accommodations for CTS-D exam, 24
specification documents in design phase, 36
specifications in system diagrams, 59
spectral-power distribution (SPD) curves, 204–205
spectrum analysis, 141
spectrum management in wireless connections, 424–425
specular sound reflections, 300
speech loudness level, 175
speed
    Ethernet networks, 431–432
    wireless connections, 422, 424
SPLs. *See* sound pressure levels (SPLs)
spot meters, 118, 200
SSM (source-specific multicast), 483
ST (straight-tip) fiber-optic cable connectors, 420–421
stability in audio systems, 176
stackable chairs, 89
staggered seating arrangements, 86–87
stakeholders
    communicating with, 188–189
    network security, 496
    security posture, 495
standard-definition (SD) video aspect ratio, 105
standard DiffServ class, 454
standard dynamic range (SDR) bit depth, 326
Standard for Audiovisual Systems Performance Verification, 509
Standard Guide for Audiovisual Systems Design and Coordination Processes, 33, 196, 506
standards
    industry, 193–194
    regional regulations, 510
StandardsPortal, 510
standing people, 90
standing waves in sound reflection, 302
star configuration for power distribution systems, 273–274
star topologies
    LANs, 430
    WANs, 435
static IP addresses, 443–444
statistical reverberant fields in sound reflection, 306
STC (sound transmission class), 311–312
step-down transformers, 153–154
step-up transformers, 153
storage in streaming design, 471
straight-tip (ST) fiber-optic cable connectors, 420–421
streaming design, 465
    audience considerations, 467–468
    bandwidth, 472–474, 476
    content sources, 468–471
    copyrighted content, 469–470
    duty check, 466
    end points, 468

streaming design (cont.)
   high-quality streaming video, 478–480
   latency, 474–476
   multicast, 480–485
   multicast addressing, 486–488
   needs analysis, 466–471
   network environment, 471–473
   network policies and restrictions, 475–476
   protocols, 477
   quality of service, 473–476
   questions, 467, 470–471, 488–489
   review, 487–488
   RTP, 477
   storage, 471
   tasks, 466–467
   topologies, 471
   unicast, 480–482
streaming desktop video latency, 475
streaming reflectors in multicast, 486–487
structural and mechanical considerations, 229
   codes and regulations, 230–233
   duty check, 230
   energy management, 252
   fire and life safety protection, 250–251
   HVAC, 249–250
   mounting, 233–239
   questions, 253–254
   racks, 239–249
   review, 252
structural engineering professionals, collaboration with, 189
subdomains, 447
subject-matter experts (SMEs)
   collaboration with, 189
   job task analysis participation, 9
submittal drawings in construction phase, 37
subnet masks
   IPv4 addresses, 438–439
   IPv6 addresses, 443
subnets in local IP addresses, 441
subsampling for video, 326–327
subscriber (SC) fiber-optic cable connectors, 421
substantial completion verification, 509
subtraction in mathematical formulas, 14–15
Super Audio CD (SACD) audio sampling rates, 324
supercardioid pickup pattern for microphones, 163
supply conductors in single-phase power distribution, 266
supply grounds in electrical distribution systems, 272
surface mount microphones, 161
switchers in HDCP, 359
switches
   lighting, 218
   networks, 414
switching sources in EDID, 356
swiveling chairs, 89
SWL (Safe Working Load) mounting considerations, 236
symbols
   architectural drawings, 46, 48–50
   AV drawings, 51–53
   block diagrams, 245

Ohm's law, 16
   protected equipment, 268–269
   reflected ceiling plans, 44
symptoms of EDID issues, 354
sync, audio/video, 530–532
System and Record Documentation items, 532–534
system black level
   display environment, 120
   rear projection displays, 112
system closeout
   client sign-off, 539
   customer training, 537–538
   documentation, 532–536
   drawings of record, 535
   troubleshooting, 536–537
system descriptions in concept design/program report sections, 73
system diagrams, 59
system documents in design phase, 36
system drawings, 58–59
system performance verification, 405
system ports in transport protocols, 450
system setup for audio design, 373
   DSP settings, 376–383
   gain, 374–376
system verification process, 508–509
systems, updating, 501
systems integration verification, 508

T

tables
   AV design projects, 89–90
   EDID, 350–351, 353–354
   microphone placement, 169–170
   reflectance values, 225
   videoconferencing, 224–225
tangential modes in sound reflection, 303
taps, transformer, 154
task-light levels, 129–131
task lighting
   control and projection areas, 62
   planning, 206
tasks in streaming design, 466–467, 470
TCP transport, 448–449
tech component in AV experience, 33
technology managers, needs analysis for, 65–66
Telecommunications Industry Association (TIA)
   cable categories, 415–416
telephone, audio sampling rates for, 324
telephony DiffServ classes, 453
Television Lighting Consistency Index (TLCI), 204–205
Telnet vulnerabilities, 499
temperature factor in sound waves, 297
tensile strength mounting consideration, 237
test patterns for display setup, 526
testing center
   arrival time, 20
   restricted items, 21

Index

testing tools for audio system verification, 511–517
text size, 99
THD (total harmonic distortion), 176
thin network servers, 413
third-order equalization filters, 384
three-phase power distribution
    conductors, 272–273
    description, 266
    neutral/return, 273–274
    vs. single-phase, 276
three-to-two-pin adapter dangers, 271
3:1 rule for microphone placement, 170–171
3D modeling drawings, 56
thresholds
    compressor settings, 378–379
    description, 377
    expander settings, 380
    gate settings, 381
    limiter settings, 379–380
throughput
    bandwidth, 452
    DiffServ classes, 454
    digital audio signals, 324
    wireless connections, 422–423
throw distance of projectors, 123–124
Thunderbolt technology, 346–348
TI-30XS MultiView calculator in CTS-D exam, 13–14
TIA (Telecommunications Industry Association) cable categories, 415–416
tiered floors, 87–88
TLCI (Television Lighting Consistency Index), 204–205
TMDS (Transition-Minimized Differential Signaling), 341–343
tone generators, 513
topologies
    bandwidth, 475
    Ethernet networks, 428–431
    streaming design, 471, 475
    WANs, 435–436
total harmonic distortion (THD), 176
total video bandwidth, 327
touch-screen device control interfaces, 397
tracking projects, 189–193
traditional control system processors, 394
traffic shaping, 472, 474
training, 537–538
transfer strategy for risk handling, 498
transformer-balanced circuits, 288
transformers
    induced currents, 261
    taps, 154
    types, 153–154
transient adaptation to light levels, 130
Transition-Minimized Differential Signaling (TMDS), 341–343
transmission in OSI Model, 427–428
transmission of sound, 309
    acoustic mass law, 309–311
    impact insulation class, 312–313
    sound transmission class, 311–312

Transport layer in OSI Model, 426
transport protocols, 448
    ports, 449–450
    TCP, 448–449
    UDP, 449
trays, cable, 288
tree topologies for LANs, 430
triacs in electronic dimmers, 218
troubleshooting
    EDID issues, 354–356
    HDCP, 361–362
    system closeout, 536–537
tweeters for audio system tests, 511–512
twisted-pair cables, 287, 415–417
two-factor authentication, 499–500
two post racks, 240
Type 1 sound pressure level meters, 514
Type 2 sound pressure level meters, 514

U

U.S. customary measurements in scaled drawings, 40–41
UDP transport, 449
Ultra HD display formats, 337
ultra high-definition video, 337–339
unbalanced and balanced circuits, 286–288
understanding expectations for devices, 398
understanding phase in design-thinking principles, 401
unicast streaming
    description, 480
    vs. multicast, 482
unidirectional pickup patterns for microphones, 163
unintentional signal delay in DSPs, 382
unity gain
    audio settings, 374, 513
    front projection displays, 109
unmanaged network switches, 414
unnecessary services, disabling, 501
updating systems, 501
USB connectors, 345–347
user accounts, 500
user mobility with rear projection displays, 112
user ports in transport protocols, 450
user training best practices, 537–538

V

variable bit rate (VBR) encoding, 330
veiling reflections in videoconferences, 222
velocity of sound waves, 297–298
verification
    audio system, 511–517
    audio/video sync, 531
    AV design project phase, 37–38
    performance standard, 506–508
    system, 508–509
    system closeout, 532–535
    system performance, 405
    tools, 510–511
    video sources, 519–523
    video systems, 517–532

verification documents, 34
Versatile Video Coding (VVC), 479
versatility in presenter area, 61
vertical lenticular lenses for rear projection displays, 113
vertical visual field, 82
VESA (Video Electronics Standards Association)
    DisplayPort, 344
vibration, HVAC noise from, 250
victims of electrical interference, 282–283
video
    bandwidth, 325–327
    bit depth, 326
    digital compression, 330–331
    full-motion, 97
    high-quality streaming video, 478–480
    latency, 475
    resolution, 104
    YUV subsampling, 326–327
Video Electronics Standards Association (VESA)
    DisplayPort, 344
Video over IP Networks, 348
video program stream total bandwidth requirements, 327
video signal paths, 517
video signal types, 340–341
    DisplayPort, 344–345
    DVI, 342
    HDBaseT, 344
    HDMI, 342–344
    Serial Digital Interface, 341
    Thunderbolt, 346–348
    Transition-Minimized Differential Signaling, 341–343
    USB connectors, 345–347
    Video over IP Networks, 348
video systems
    audio/video sync, 530–532
    display setup, 524–529
    sources, 519–523
    verification, 517–532
    visual principles. *See* visual principles
videoconferences
    color balance, 224
    glare, 222–223
    latency, 475
    light balance, 223–224
    lighting, 221–225
videowalls, 115–116
    applications, 116
    design, 116–118
viewing angles, 101–102
viewing distance, 101–103
viewing locations in image system contrast, 528
viewing positions, 121
viewing requirements, 96–97
    element height, 99–101
    image sizing, 97–98
    text size, 99
    viewing angles, 101–102
    viewing distance, 101–103
    visual acuity and Snellen eye chart, 99
virtual calculator for CTS-D exam, 13–14
virtual local area networks (VLANs), 433–434, 499

virtual private networks (VPNs), 436
visual acuity
    description, 97
    Snellen eye chart, 99–100
visual field, 80
    head rotation, 83–84
    horizontal, 81
    vertical, 82
visual perception, 200
visual principles, 95
    display device selection, 103–106
    display environment, 118–121
    display types, 106–114
    duty check, 95–96
    image specifications, 96
    OFE projector brightness, 128–129
    projector brightness, 124–128
    projector positioning, 122–124
    questions, 131–133
    review, 131
    task-light levels, 129–131
    viewing requirements, 96–103
visuals in audience area, 61
VLANs (virtual local area networks), 433–434, 499
voice
    control and projection areas, 62
    presenter area, 61
voltage
    AC, 257
    decibel reference levels, 138–139
    Ohm's law, 16–19
    single-phase power distribution, 275–276
voltage ramp control points, 399
volume unit (VU) meters, 371–372
VPNs (virtual private networks), 436
VVC (Versatile Video Coding), 479

W

wall-mounted racks, 239
wall mounting considerations, 234–235
walls
    reflectance values, 225
    sound reception, 315
    videoconferencing, 224–225
WANs (wide area networks), 434
    private and public, 436
    topologies, 435–436
    VPNs, 436
WAPs (wireless access points), 415
wash lights, 223
watts
    decibel reference levels, 138–139
    loudspeaker systems, 155
    sound, 298–299
wavelength division multiplexing (WDM), 418
wavelengths
    loudspeaker coverage, 143
    room modes, 302–305
    scattered sound reflections, 300–301
    sound reception, 314–315
    sound waves, 297–298

WBSs (work breakdown structures), 190–191
WCG (wide color gamut), 339
WDM (wavelength division multiplexing), 418
weather and CTS-D exam, 23
weight distribution in racks, 242–244
weighting sound pressure levels, 141, 514–515
well-known ports, 450
wet fire protection systems, 251
white balancing for cameras, 523
white noise in audio signal generators, 512
whiteboards
    AV design projects, 91
    reflectance values, 225
wide area networks (WANs), 434
    private and public, 436
    topologies, 435–436
    VPNs, 436
wide color gamut (WCG), 339
wide Q filters, 385
widescreen display format, 338
widowmakers, 271
width
    aspect ratio, 105–106
    media formats, 104
    video formats, 325
window reflectance values, 225
window treatments for space lighting, 207
wired control interfaces, 396–397
wireless access points (WAPs), 415
wireless control interfaces, 396–397

wireless network connections, 422
    spectrum management, 424–425
    Wi-Fi advantages, 422–423
    Wi-Fi disadvantages, 424
wireless optical control points, 400
wiring loudspeakers, 148–152
Wiring Rules standard, 232
work breakdown structures (WBSs), 190–191
Working Load Limit (WLL) mounting considerations, 236
workshop drawings in construction phase, 37
workstations, presenter, 61
Wye configuration for power distribution systems, 273–274

X

x-over in series circuits, 262
XX-SDI standard, 519

Y

Y configuration for power distribution systems, 273–274
YUV video subsampling, 326–327

Z

Z-weighting in sound pressure levels, 514
zero reference for decibels, 137
zones for scenes, 220–221
zoning plans in space lighting, 213–216